ESSAYS IN
TIME SERIES AND
ALLIED PROCESSES

Professor E. J. Hannan FAA FASSA

ESSAYS IN TIME SERIES AND ALLIED PROCESSES

Papers in honour of

E. J. HANNAN

Edited by

J. GANI AND M. B. PRIESTLEY

JOURNAL OF APPLIED PROBABILITY
SPECIAL VOLUME 23A (1986)

Published and distributed by the
APPLIED PROBABILITY TRUST
The University
Sheffield S3 7RH
England

ISBN: 0 902016 02 4
ISSN: 0021-9002

Printed in Northern Ireland by
THE UNIVERSITIES PRESS (BELFAST) LTD

Preface

Edward James Hannan, Professor of Statistics at the Australian National University (ANU), Canberra, will be 65 years old on 29 January 1986. To commemorate this happy occasion, Professor Maurice Priestley, a fellow research worker in time series, and I, a friend of many years' standing, invited a number of colleagues, friends, and former students to join us in honouring Ted as a statistician, teacher and research worker. We hope that the papers in this book will reflect the influence he has exerted in time series and allied topics during his distinguished career.

I first met Ted Hannan in 1953 when I joined the Department of Statistics at the ANU as a postgraduate student. A variety of common interests, partly statistical and partly literary, was the basis of a deep friendship which has continued to develop over the past 30 years. Ted was born in Melbourne, Australia, in 1921, and began work at the age of 16 in 1937 as a clerk in the Commonwealth Bank of Australia. During the Second World War, he joined the Australian Army in 1941, fought in the New Guinea campaign and rose to the rank of lieutenant. On demobilization in 1945, he took advantage of the educational opportunities offered to returned Australian servicemen, and studied for a B.Com. Degree which he was awarded by the University of Melbourne in 1949. Among the many courses he attended was one in statistics, delivered by Professor M. H. Belz; this aroused his interest in mathematics and statistics, an interest which grew in strength as time went on.

After graduation, he took up the position of Junior Research Officer in the Economics Department of the Commonwealth Bank in Sydney, and in 1953 was sent to Canberra by the Bank to study for a year in the Economics Department of the ANU. By then, he had acquired considerable expertise in mathematics and statistics through private reading and study; Professor P. A. P. Moran, who had been appointed to the Chair of Statistics at the ANU in 1952, recognized his great mathematical gifts and arranged the award of a Research Fellowship to him in the Department of Statistics. There, he directed Ted's interests towards problems in time series analysis; this was to become his dominant field of research over the next three decades. Ted has frequently mentioned his great good fortune in meeting Professor Moran at this important juncture in his life.

Ted's early work was concerned with tests for correlation in time series, and regression for time series, with applications in meteorology, hydrology, geophysics and economics. By 1959, he had achieved recognition as a statistician of note, and was appointed to the Chair of Statistics in the Canberra University College, later to become the ANU School of General Studies. There, he built up a lively teaching department, while yet finding time to continue his own research and to direct students such as R. D. Terrell and

D. F. Nicholls. By 1970, he had become a world authority on time series, with significant research in such areas as serial correlation, spectral theory, time series regression and estimation of coherence; this research had led to a flow of papers, as well as two books, *Time Series Analysis* (1960) and *Multiple Time Series* (1970), and a monograph, *Group Representations and Applied Probability* (1965).

In 1971, he was appointed to the second Chair of Statistics in Professor Moran's Department at the ANU, and from then on directed most of his energies to research and the training of postgraduate students, among them P. J. Thomson, P. M. Robinson, W. Dunsmuir, J. D. Henstridge, M. A. Cameron, K. Tanaka, B. G. Quinn, V. Solo, Chen Zhao-Guo, An Hong-Zhi and L. Kavalieris.

The originality and distinction of his contributions were recognized by election to the Australian Academy of Science in 1970; the Academy later awarded him its Lyle Medal for mathematical research in 1979. He was also elected to the Academy of the Social Sciences in Australia in 1980, to membership of the International Statistical Institute and to Fellowship of the Econometric Society.

Ted Hannan's research has resolved many difficult problems in time series analysis, but has also occasionally ranged beyond it; he has investigated limit theorems, sampling, Fourier methods, data smoothing, and estimation theory, among other topics. Nor have his interests been entirely academic; in 1964, he joined Norma McArthur, Sir Edward Collingwood (for the London Mathematical Society) and me as one of the four trustees of the newly founded Applied Probability Trust. He has been an editor of the *Journal of Applied Probability* and *Advances in Applied Probability*, and also of the *Journal of Time Series Analysis*, since their inception, and has thus exerted his considerable influence on the development of time series through his editorial role as well as his personal research.

No account of Ted Hannan's activities is complete without reference to his basic kindness and good will, and to the importance he has always attached to the happiness of his wife Irene and his family of two sons and two daughters. The breadth of his intellectual interests is exhilarating: among these is his great love of literature. For many years in Canberra, we shared copies of the *Guardian Weekly* and the *Times Literary Supplement*, exchanged books on every conceivable topic and spent many happy hours each week discussing social and literary issues over morning coffee.

I hope that this book, edited by Maurice Priestley and myself, will reflect the affection and esteem in which this eminent international statistician with deep personal roots in Australia is held by his colleagues, students and friends. We all wish him joy on his sixty-fifth birthday, and many happy returns.

Statistics Program, Department of Mathematics   J. GANI
University of California, Santa Barbara   July 1985

Contents

Part 6 — Allied Stochastic Processes

Part 7 — Algorithms and Computations

Publications of E. J. Hannan

1955

[1] A test for singularities in Sydney rainfall. *Austral. J. Phys.* **8,** 289–297.
[2] Exact tests for serial correlation. *Biometrika* **42,** 133–142.
[3] An exact test for correlation between time series. *Biometrika* **42,** 316–326.

1956

[4] The asymptotic powers of certain tests based on multiple correlations. *J. R. Statist. Soc* B **18,** 227–233.
[5] Exact tests for serial correlation in vector processes. *Proc. Camb. Phil. Soc.* **52,** 482–487.
[6] (with G. S. Watson) Serial correlation in regression analysis. *Biometrika* **43,** 436–448.

1957

[7] Testing for serial correlation in least squares regression. *Biometrika* **44,** 57–66.
[8] The variance of the mean of a stationary process. *J. R. Statist. Soc.* B **19,** 282–285.
[9] Some recent advances in statistics. *Economic Record* **33,** 337–352.

1958

[10] The estimation of the spectral density after trend removal. *J. R. Statist. Soc.* B **20,** 323–333.
[11] The asymptotic powers of certain tests of goodness of fit for time series. *J. R. Statist. Soc.* B **20,** 143–151.

1960

[12] *Time Series Analysis.* Methuen, London. (Published in Russian with an appendix by Yu. A. Rozanov, 1964; also published in Japanese.)
[13] The estimation of seasonal variation. *Austral. J. Statist.* **2,** 1–15.

1961

[14] A central limit theorem for systems of regressions. *Proc. Camb. Phil. Soc.* **57,** 583–588.
[15] Testing for a jump in the spectral function. *J. R. Statist. Soc.* B **23,** 394–404.
[16] The general theory of canonical correlation and its relation to functional analysis. *J. Austral. Math. Soc.* **11,** 229–242.

1962

[17] Systematic sampling. *Biometrika* **49,** 281–283.
[18] Rainfall singularities. *J. Appl. Math.* **1,** 426–429.

1963

[19] Regression for time series. In *Time Series Analysis,* ed. M. Rosenblatt, Wiley, New York, 17–37.

[20] The estimation of seasonal variation in economic time series. *J. Amer. Statist. Assoc.* **18**, 31–44.
[21] Regression for time series with errors of measurement. *Biometrika* **50**, 293–302.
[22] (with B. V. Hamon) Estimating relations between time series. *J. Geophy. Res.* **68**, 6033–6041.

1964

[23] The statistical analysis of hydrological time series. In *Proc. Symp. Water Resources, Use and Management,* Melbourne University Press, 233–243.
[24] The estimation of a changing seasonal pattern. *J. Amer. Statist. Assoc.* **59**, 1063–1077.

1965

[25] *Group Representations and Applied Probability.* Methuen, London. (Published in Russian with an introduction by A. M. Yaglom, 1970.)
[26] The estimation of relations involving distributed lags. *Econometrica* **33**, 206–224.

1966

[27] Spectral analysis for geophysical data. *Geophys. J. R. Astronom. Soc.* **11**, 225–236.

1967

[28] The concept of a filter. *Proc. Camb. Phil. Soc.* **63**, 221–227.
[29] The measurement of a wandering signal amid noise. *J. Appl. Prob.* **4**, 90–102.
[30] Canonical correlation and multiple equation systems in economics. *Econometrica* **35**, 123–138.
[31] The estimation of a lagged regression relation. *Biometrika* **54**, 315–324.

1968

[32] (with R. D. Terrell) Testing for serial correlation after least squares regression. *Econometrica* **36**, 133–150.
[33] (with G. W. Groves) Time series regression of sea level on weather. *Rev. Geophys.* **6**, 129–174.
[34] Least squares efficiency for vector time series. *J. R. Statist. Soc.* B **30**, 490–499.

1969

[35] Fourier methods and random processes. *Bull. Internat. Statist. Inst.* **42**, 475–496.
[36] The identification of vector, mixed autoregressive – moving average, systems. *Biometrika* **56**, 223–225.
[37] A note on an exact test for trend and serial correlation. *Econometrica* **37**, 485–489.
[38] The estimation of mixed moving average autoregressive systems. *Biometrika* **56**, 579–593.
[39] Fourier methods and linear models. *Austral. J. Sci.* **32**, 171–175.

1970

[40] *Multiple Time Series.* Wiley, New York. (Published in Russian, 1974.)
[41] (with R. D. Terrell and N. E. Tuckwell) The seasonal adjustment of economic time series. *Internat. Econom. Rev.* **11**, 24–52.
[42] Data smoothing. In *Data Representation,* ed. R. S. Anderssen and M. R. Osborne, University of Queensland Press, 34–42.

1971

[43] The identification problem for multiple equation systems with moving average errors. *Econometrica* **39**, 751–765.

[44] (with P. J. Thomson) Spectral inference over narrow bands. *J. Appl. Prob.* **8**, 157–169.

[45] Non-linear time series regression. *J. Appl. Prob.* **8**, 767–780.

[46] (with P. J. Thomson) The estimation of coherence and group delay. *Biometrika* **58**, 469–482.

1972

[47] (with R. D. Terrell) Time series regression with linear constraints. *Internat. Econom. Rev.* **13**, 189–200.

[48] (with D. F. Nicholls) The estimation of mixed regression, autoregression, moving average and distributed lag models. *Econometrica* **40**, 529–547.

[49] (with R. C. Boston) The estimation of a non-linear system. In *Proc. Conf. Optimisation,* ed. R. S. Anderssen, E. S. Jennings and D. M. Ryan, University of Queensland Press, 69–85.

[50] Spectra changing over narrow bands. In *Statistical Models and Turbulence,* ed. M. Rosenblatt, Springer-Verlag, New York, 400–469.

[51] (with C. C. Heyde) On limit theorems for quadratic functions of discrete time series. *Ann. Math. Statist.* **43**, 2058–2066.

1973

[52] (with R. D. Terrell) Multiple equation systems with stationary errors. *Econometrica* **41**, 299–320.

[53] (with P. M. Robinson) Lagged regression with unknown lags. *J. R. Statist. Soc.* B **35**, 252–267.

[54] The asymptotic theory of linear time series models. *J. Appl. Prob.* **10**, 130–145.

[55] (with P. J. Thomson) Estimating group delay. *Biometrika* **60**, 241–253.

[56] The estimation of frequency. *J. Appl. Prob.* **10**, 510–519.

[57] (with P. A. P. Moran) The effects of serial correlation on a randomised rainmaking experiment. *Austral. J. Statist.* **15**, 256–261.

[58] Central limit theorems for time series regression. *Z. Wahrscheinlichkeitsth.* **26**, 157–170.

[59] Multivariate time series analysis. *J. Multivariate Anal.* **3**, 395–407.

1974

[60] (with P. J. Thomson) Estimating echo times. *Technometrics* **16**, 77–84.

[61] (with B. V. Hamon) Spectral estimation of time delay for dispersive and non-dispersive systems. *Appl. Statist.* **23**, 134–142.

[62] The uniform convergence of autocovariances. *Ann. Statist.* **2**, 803–806.

[63] Time series analysis. *IEEE Trans. Automatic Control* **AC-19**, 706–715.

1975

[64] Linear regression in continuous time. *J. Austral Math. Soc.* **19**A, 146–159.

[65] The estimation of ARMA models. *Ann. Statist.* **3**, 975–981.

[66] Measuring the velocity of a signal. In *Perspectives in Probability and Statistics, Papers in Honour of M. S. Bartlett,* ed. J. Gani, Applied Probability Trust, Sheffield, 227–238.

[67] New methods of estimating dispersion from stacks of surface waves. *Bull. Seismol. Soc. Amer.* **65**, 1519–1529.

1976

[68] (with W. Dunsmuir) Vector linear time series models. *Adv. Appl. Prob.* **8**, 339–364.
[69] The convergence of some recursions. *Ann. Statist.* **4**, 1258–1270.
[70] The identification and parametrization of ARMAX and state space forms. *Econometrica* **44**, 713–723.
[71] ARMAX and state space systems and recursive calculations. In *Frontiers of Quantitative Economics,* ed. M. Intriligator, North-Holland, Amsterdam, 321–336.

1977

[72] (with M. Kanter) Autoregressive processes with infinite variance. *J. Appl. Prob.* **14**, 411–415.
[73] (with D. F. Nicholls) The estimation of the prediction error variance. *J. Amer. Statist. Assoc.* **72**, 834–840.

1978

[74] (with M. A. Cameron) Measuring the properties of plane waves. *Math. Geol.* **10**, 1–22.
[75] A note on a central limit theorem. *Econometrica* **46**, 451–453.
[76] (with M. Deistler and W. Dunsmuir) Vector linear time series models: corrections and extensions. *Adv. Appl. Prob.* **10**, 360–372.
[77] Multivariate ARMA. In *Stability and Inflation,* ed. A. R. Bergstrom, A. J. L. Catt, M. H. Preston and D. J. Silverstone, Wiley, New York, 201–212.
[78] (with K. Tanaka) ARMAX models and recursive calculations. In *Systems Dynamics and Control in Quantitative Economics,* ed. H. Myoken, Bunshindo, Tokyo, 173–198.
[79] Rates of convergence for time series regression. *Adv. Appl. Prob.* **10**, 740–743.

1979

[80] Statistical theory of linear systems. In *Developments in Statistics* **2**, ed. P. Krishnaiah, Academic Press, New York, 83–121.
[81] (with M. Cameron) Transient signals. *Biometrika* **66**, 243–258.
[82] (with B. G. Quinn) The determination of the order of an autoregression. *J. R. Statist. Soc.* B **41**, 190–195.
[83] The central limit theorem for time series regression. *Stoch. Proc. Appl.* **9**, 281–289.
[84] A note on autoregressive – moving average identification. *Biometrika* **66**, 672–674.
[85] 'Saisonale Variation' and 'Zeitreihenanalyse'. In *Handworterbuch der Mathematischen Wirtschafwissenschaften* **2**, Gabler, Wiesbaden, 157–159; 299–306.

1980

[86] Time series. *Math Chronicle* **9**, 101–119.
[87] Recursive estimation based on ARMA models. *Ann. Statist.* **8**, 762–777.
[88] The estimation of the order of an ARMA process. *Ann. Statist.* **8**, 1071–1081.
[89] (with W. Dunsmuir and M. Deistler) Estimation of vector ARMAX models. *J. Multivariate Anal.* **10**, 275–295.
[90] (with Chen Zhao-Guo) The distribution of periodogram ordinates. *J. Time Series Anal.* **1**, 73–82.

1981

[91] Estimating the dimension of a linear system. *J. Multivariate Anal.* **11**, 459–473.

[92] (with M. Deistler) Some properties of the parametrization of ARMA systems with unknown order. *J. Multivariate Anal.* **11,** 474–497.

[93] System identification. In *Stochastic Systems: The Mathematics of Filtering and Identification and Applications,* ed. M. Hazewinkel and J. C. Willems, Reidel, Dordrecht, 221–226.

[94] (with P. J. Thomson) Delay estimation and the estimation of coherence and phase. *IEEE Trans. Acoust. Speech Signal Processing* **ASSP-29,** 485–490.

1982

[95] Fitting multivariate ARMA models. In *Statistics and Probability: Essays in Honour of C. R. Rao,* ed. G. Kallianpur, P. R. Krishnaiah and J. K. Ghosh, North-Holland, Amsterdam, 307–316.

[96] A note on bilinear time series models. *Stoch. Proc. Appl.* **12,** 221–224.

[97] (with J. Rissanen) Recursive estimation of mixed autoregressive – moving average order. *Biometrika* **69,** 81–94.

[98] Testing for autocorrelation and Akaike's criterion. In *Essays in Statistical Science, Papers in Honour of P. A. P. Moran,* ed. J. Gani and E. J. Hannan, Applied Probability Trust, Sheffield, 403–412.

[99] (with An Hong-Zhi and Chen Zhao-Guo) Autocorrelation, autoregression, and autoregressive approximation. *Ann. Statist.* **10,** 926–936.

[100] (with V. Wertz and M. Gevers) The determination of optimum structures for the state space representation of multivariate stochastic processes. *IEEE Trans. Automatic Control* **AC-27,** 1200–1211.

1983

[101] (with An Hong-Zhi and Chen Zhao-Guo) A note on ARMA estimation. *J. Time Series Anal.* **4,** 9–17.

[102] (with L. Kavalieris) Linear estimation of ARMA processes. *Automatica* **19,** 447–448.

[103] (with L. Kavalieris) The convergence of autocorrelations and autoregressions. *Austral. J. Statist.* **25,** 287–297.

[104] Signal estimation. In *Handbook of Statistics 3,* ed. D. R. Brillinger and P. R. Krishnaiah, North-Holland, Amsterdam, 111–123.

[105] (with An Hong-Zhi and Chen Zhao-Guo) The maximum of the periodogram. *J. Multivariate Anal.* **13,** 383–400.

1984

[106] (with L. Kavalieris) Multivariate linear time series models. *Adv. Appl. Prob.* **16,** 492–561.

[107] (with L. Kavalieris) A method for autoregressive – moving average estimation. *Biometrika* **71,** 273–280.

PART 1

STRUCTURE AND GENERAL METHODS FOR TIME SERIES

Non-Singularity and Asymptotic Independence

PETER BLOOMFIELD

Abstract

A stationary stochastic process must satisfy various requirements to make it a realistic model for a phenomenon in the real world. Some of these requirements are quantitative, such as agreement of distribution or moments. Other, more qualitative requirements deal with the general behavior of the process. Two such requirements are non-singularity and asymptotic independence. Each will be discussed from a variety of points of view, and given precise definition in a succession of progressively stronger forms.

DETERMINISM AND NON-DETERMINISM; PREDICTABILITY; MINIMALITY; CANONICAL AND ESSENTIAL CORRELATION; COMPLETE AND ABSOLUTE REGULARITY

1. Introduction and summary

When a phenomenon in the real world is modeled by a stationary stochastic process, we place various requirements on the process to make it as realistic a model as possible. Some of these requirements are quantitative. For instance, we want the distributions, or at least the moments, of various quantities to match. These requirements are usually met by estimating various parameters in the model from observations on the phenomenon. Other, more qualitative, requirements deal with the general behavior of the process, and have to be met by restricting the class of processes that will be considered as candidate models.

The qualitative properties of real-world phenomena that we wish to be reflected in the model are typically expressed in vague terms, which may be made precise in a variety of ways. In this paper we shall examine the concepts of non-singularity and asymptotic independence. We shall discuss conditions that make the concepts precise, first from the point of view of finite-dimensional distributions, and later introducing infinite-dimensional sets of random variables. A succession of progressively stronger conditions will be given, and characterized as far as possible in terms of the spectrum or autocovariances.

2. Non-singularity

Our first vague concept is that of non-singularity. It is usually safe to assume that a phenomenon does not have redundancy of the form of one variable being almost surely equal to a linear combination of other variables, or of the form that some linear combination of the variables vanishes almost surely.

A finite-dimensional random variable is said to be singular if its distribution is singular with respect to Lebesgue measure. A Gaussian distribution is singular iff it is supported on some hyperplane of reduced dimension. In general, we could say that a distribution is linearly singular if it is singular in this fashion. However, in this paper we shall be concerned only with such linear versions of properties, and we shall usually omit the qualification 'linear'. Thus a distribution is (linearly) singular iff some non-trivial linear combination of the variables vanishes almost surely. Usually, real finite-dimensional phenomena do not satisfy such constraints, and consequently their models are required to be non-singular.

In a similar way, we usually expect real time series not to contain redundancy, and so we may require their models also to be non-singular, by specifying that linear combinations must not vanish almost surely. However, we obtain different definitions of non-singularity by choosing different linear combinations.

In finite-dimensional models, non-singularity also implies that no linear combination of some of the variables can be predicted exactly by a linear combination of the remainder. In the time series case, an analogue of this property is that no linear combination of future values can be perfectly predicted from the past, or even that linear combinations of future values cannot approach perfect predictability. The latter provides a strong form of non-singularity, but one that cannot be expressed in terms of the non-vanishing of linear combinations.

3. Asymptotic independence

A second vague concept that we feel should apply to most real world phenomena is that of asymptotic independence, or in other words that the remote past should have vanishingly little effect on the remote future. This idea may be translated almost immediately into a mathematical condition that can be imposed on a stochastic process model. See Hannan (1970), pp. 200–204, for a discussion of such *mixing conditions* and the closely related concept of *ergodicity*.

Within our linear (or if you wish, Gaussian) framework, we measure dependence by correlation, and hence we are led to consider restricting our models in such a way that the correlation between variables in the remote future and the remote past is small. Again, we find different definitions of asymptotic independence depending on precisely which future and past variables we consider, and on how we choose to measure correlation between sets of random variables.

4. Notation

We shall denote the stochastic process model by $\{x(n): n = 0, \pm 1, \pm 2, \cdots\}$, and we shall assume that it has zero mean and is second-order stationary:

$$\mathscr{E}\{x(n)\} = 0$$

and

$$\mathscr{E}\{x(n)x(m)\} = \gamma(m - n), \quad m, n = 0, \pm 1, \cdots.$$

The autocovariances $\gamma\{(u): u = 0, \pm 1, \cdots\}$ are related to the spectrum F by

$$\gamma(u) = \int_{-\pi}^{\pi} \exp(-iu\lambda)dF(\lambda),$$

where F is a non-decreasing function. Many properties of a stationary time series are most easily described in terms of the spectral density function $f = F'$. The linear combinations that we consider will be combinations of finitely many x's, and their limits in mean square. These are elements of the Hilbert space $\mathscr{H}$ spanned by $\{x(n)\}$ with inner product $(a, b) = \mathscr{E}(ab^*)$.

5. Finite-dimensional properties

The most extreme form of singularity in a time series would be for a finite linear combination to vanish, or equivalently for $x(n)$ to be a linear combination of finitely many past values $x(n - m)$. Now

$$\text{Var} \sum a_m x(n - m) = \int |a(\lambda)|^2 dF(\lambda),$$

where

$$a(\lambda) = \sum a_m \exp(-im\lambda).$$

Since the sum is finite and may without loss of generality be assumed to range over non-negative values of m, $a(\lambda)$ is essentially a polynomial in $\exp(i\lambda)$, and hence can vanish at only finitely many λ's in $(-\pi, \pi]$. Thus (assuming that $a(\lambda)$ is not identically 0) $\text{Var} \sum a_m x(n - m)$ can vanish if and only if F has finite support. Hence our weakest form of non-singularity is defined as follows.

Definition 1. $\{x(n)\}$ is non-singular if no finite linear combination vanishes almost surely.

We have the following characterization.

Proposition 1. $\{x(n)\}$ is non-singular if and only if its spectral measure F has infinite support.

We may also discuss asymptotic independence in finite-dimensional terms. The weakest form would be asymptotic pairwise independence:

$$\operatorname{Cov}\{x(n), x(m)\} = \gamma_{n-m} \to 0 \quad \text{as} \quad |n-m| \to \infty.$$

A sufficient condition for this is that F be absolutely continuous, by the Riemann–Lebesgue lemma, while a necessary condition is that F be continuous (Doob (1953), p. 494). That continuity alone is not sufficient is shown by the fact that the Lebesgue function of the Cantor set has Fourier coefficients that do not converge to 0 (Zygmund (1959), p. 196).

Another approach to asymptotic independence based only on finite-dimensional distributions would be to consider finite sets of variables in the future and the past. Let V and W be two finite sets of integers. The covariance matrix of the sets of variables $\{x(n): n \in V\}$ and $\{x(n+u): n \in W\}$ is of the form

$$\begin{bmatrix} \Sigma_{VW} & \Sigma_{VW}(u) \\ \Sigma_{WV}(u) & \Sigma_{WW} \end{bmatrix}$$

where the diagonal blocks do not depend on u. The entries in the off-diagonal blocks converge to 0 iff $\gamma_u \to 0$ as $u \to \infty$, and in this case all joint measures of correlation such as the largest canonical correlation converge to 0 as well. Thus considering finite sets of random variables in the future and in the past leads to the same ideas about asymptotic independence as looking at just $x(n)$ and $x(n+u)$. Thus we can state the following definition.

Definition 2. $\{x(n)\}$ is asymptotically independent if $\gamma_u \to 0$ as $u \to \infty$.

The corresponding conditions on the spectrum can be stated as follows.

Proposition 2. (a) $\{x(n)\}$ is asymptotically independent if F is absolutely continuous.

(b) If $\{x(n)\}$ is asymptotically independent then F is continuous.

To describe the properties of finite sets, we have the following theorem.

Theorem 1. If V and W are finite sets of integers, then the largest canonical correlation between $\{x(n): n \in V\}$ and $\{x(n+u): n \in W\}$ converges to 0 as $u \to \infty$ iff $\{x(n)\}$ is asymptotically independent.

Proof. Necessity follows from the fact that if $V = W = \{0\}$, then the largest (and only) canonical correlation is $|\gamma_u|/\gamma_0$. Sufficiency follows from the following identity, satisfied by the canonical correlations ρ_i:

$$1-\rho_{\max}^2 \geq \prod (1-\rho_i^2) = \frac{\det \begin{bmatrix} \Sigma_{VV} & \Sigma_{VW}(u) \\ \Sigma_{WV}(u) & \Sigma_{WW} \end{bmatrix}}{\det \Sigma_{VV} \times \det \Sigma_{WW}} \to 1 \quad \text{as} \quad u \to \infty$$

since the determinant is a continuous function of its entries.

6. The infinite past

In the previous section we have discussed a very clear form of singularity, in which some linear combination of finitely many elements of the series vanishes almost surely. This would imply for instance that the entire future of the series could be predicted exactly from a finite number of past values. For a time series to provide a reasonable model for a real situation, we would naturally want to exclude such possibilities. Are we then assured of having a reasonable model? Unfortunately the answer to this question is no. There are many more subtle ways in which types of singularity can appear.

The first such way is that even though there may be no finite linear combination of past values $\{x(n-m): m \geq 0\}$ that exactly predicts $x(n+1)$, there may be combinations that come arbitrarily close. This implies that $x(n+1)$ lies in the closed linear span of $\{x(n-m): m \geq 0\}$, a subspace of $\mathcal{H}$ that we shall denote $\mathcal{P}_n$. Thus a stronger form of non-singularity is contained in the following definition (Hannan (1970), pp. 136–137).

Definition 3. $\{x(n)\}$ is non-deterministic if $x(n+1) \notin \mathcal{P}_n$, and purely deterministic if to the contrary $x(n+1) \in \mathcal{P}_n$.

The following spectral characterization of non-determinism is due to Kolomogorov (1941).

Proposition 3. $\{x(n)\}$ is non-deterministic iff

$$\int \log f(\lambda)d\lambda > -\infty.$$

Notes. (i) $F' = f$ exists for almost all λ, and since $\log(f) < f$, the integral cannot diverge to $+\infty$. If $f=0$ on a set of positive measure, we define the integral to be $-\infty$.

(ii) If F is not absolutely continuous, its singular component does not contribute to the integral, and hence has no impact on non-determinism.

We may also use the notion of the infinite past to obtain a stronger version of asymptotic independence. Suppose that we use the multiple correlation to measure the dependence between $x(n+u)$ and the past, $\mathcal{P}_n$. We define this by analogy with the finite-dimensional version as

$$R^2(u) = \mathscr{E}\{\hat{x}_n(n+u)^2\}/\mathscr{E}\{x(n+u)^2\},$$
$$= [\text{corr}\{x(n+u), \hat{x}_n(n+u)\}]^2$$

where $\hat{x}_n(n+u)$ denotes the projection of $x(n+u)$ onto $\mathscr{P}_n$. This was also called the predictability of $x(n+u)$ by Jewell and Bloomfield (1983). We would usually want our models to have the property that

$$R^2(u) \to 0 \quad \text{as} \quad u \to \infty.$$

This is in fact equivalent to the property of being purely non-deterministic (see for instance Hannan (1970), p. 137).

Definition 4. $\{x(n)\}$ is purely non-deterministic if

$$\bigcap \{\mathscr{P}_n : n < 0\} = \{0\}.$$

This equivalence is contained in the following theorem.

Theorem 2. $\{x(n)\}$ is purely non-deterministic iff $R^2(u) \to 0$ as $u \to \infty$.

Proof. By the Wold decomposition theorem (Hannan (1970), p. 137),

$$x(n+u) = \sum_{j=0}^{\infty} a_j \varepsilon(n+u-j) + v(n+u)$$

where $\varepsilon(m) \in \mathscr{P}_m$, $\varepsilon(m) \perp \varepsilon_{m-j}$ for $j > 0$, and $v(m) \in \mathscr{P}_{m-j}$. It is easily seen that

$$x_n(n+u) = \sum_{j=u}^{\infty} a_j \varepsilon(n+u-j) + v(n+u).$$

Thus $R^2(u) \to 0$ as $u \to \infty$ iff $v(n+u) = 0$. This is clearly implied by $\bigcap \{\mathscr{P}_n\} = \{0\}$. Also if $v(n+u) = 0$ then by stationarity, $v(m) = 0$ for all m, and hence $\{\mathscr{P}_n\} = \{0\}$. This completes the proof.

Thus for our present purposes we shall regard pure non-determinism as a type of asymptotic independence: specifically, of a single future variable and the entire remote past. It is characterized in the following proposition (Kolmogorov (1941)).

Proposition 4. $\{x(n)\}$ is purely non-deterministic iff it is non-deterministic and its spectrum has no singular component. That is, F must be absolutely continuous, and must satisfy

$$\int \log f(\lambda) d\lambda > -\infty.$$

We could also study the dependence on the past of a finite set of future variables $\{x(n+u): u \in W\}$, instead of the single variable $x(n+u)$. However, as in the finite-dimensional case, no new aspects of the problem are revealed.

7. Infinite past and infinite future

Some new kinds of singularity arise when we consider elements of the entire future, rather than finite linear combinations. Historically, the first such problem was discussed by Kolmogorov (1941), in interpolation problems. However, we begin with a slightly weaker notion due to Sarason (see Bloomfield et al. (1983)).

It is easily seen that for a purely non-deterministic process $\{x(n)\}$, no finite linear combination of future values can be in $\mathcal{P}_n$. However, this does not preclude the possibility that a limit of such combinations lies in $\mathcal{P}_n$. Let $\mathcal{F}_{n+1}$ be the subspace of $\mathcal{H}$ spanned by $\{x(n+m): m > 0\}$, that is, the space of such limits. We want to avoid the possibility that there may be a variable in $\mathcal{P}_n$ that is also in $\mathcal{F}_{n+1}$.

Definition 5 (Sarason). $x(n)$ is completely non-deterministic if $\mathcal{P}_n \cap \mathcal{F}_{n+1} = \{0\}$.

The spectral characterization of complete non-determinism is discussed by Bloomfield et al. (1983). No simple characterization is known, but the property can be shown to be related to other mathematical problems, such as determining the exposed points of the unit ball in H^1 (deLeeuw and Rudin (1958)). From the present point of view, the most interesting result is as follows.

Proposition 5. $\{x(n)\}$ is completely non-deterministic iff it is purely non-deterministic and in addition $x(n) \notin \mathcal{P}_{n-1} + \mathcal{F}_{n+1}$.

Thus complete non-determinism, which is a form of non-singularity, implies pure non-determinism, which is a form of asymptotic independence. The additional requirement is that $x(n)$ cannot be written exactly as the sum of elements of the past and the future (both omitting $x(n)$, of course). This is one way of insisting that we cannot interpolate $x(n)$ exactly from the past and future. A more satisfactory version of this property is minimality, introduced by Kolmogorov (1941).

Definition 6. $\{x(n)\}$ is minimal if $x(n) \notin \mathcal{P}_{n-1} \vee \mathcal{F}_{n+1}$.

The subspace involved in this definition is the closed linear span of $\mathcal{P}_{n-1}$ and $\mathcal{P}_{n+1}$, that is, the closure of $\mathcal{P}_{n-1} + \mathcal{F}_{n+1}$. Thus trivially pure non-determinism and minimality together imply complete non-determinism. Minimality alone is not enough only because it does not require the spectrum to be absolutely continuous. Minimality and absolute continuity together imply pure non-determinism, so this seems to be an interesting combination of properties. The difference between minimality and complete non-determinism lies in the fact that $x(n)$ may be a limit of sums of the form $\pi + \phi$ with $\pi \in \mathcal{P}_{n-1}$ and $\phi \in \mathcal{F}_{n+1}$, without being exactly equal to any such sum. Unlike complete non-determinism, minimality has a simple characterization (Kolmogorov (1941)).

Proposition 6. $\{x(n)\}$ is minimal iff

$$\int f(\lambda)^{-1} d\lambda < \infty.$$

8. Angles and canonical correlations

It might appear that with complete non-determinism (or the slightly stronger condition, minimality and absolutely continuity), we have a satisfactorily strong form of non-singularity. However, for some purposes this is not so. Suppose that we wish to discuss the predictability of 'aspects of the future' (by which we mean linear combinations of the future values of $x(n)$, or in other words elements of $\mathcal{F}_{n+1}$). For any $\phi \in \mathcal{F}_{n+1}$ we can define the multiple correlation of ϕ with $\mathcal{P}_n$ by

$$R^2(\phi) = \mathcal{E}(\hat{\phi}_n^2)/\mathcal{E}(\phi^2),$$

where as before $\hat{\phi}_n$ denotes the projection of ϕ onto $\mathcal{P}_n$. If $\{x(n)\}$ is completely non-deterministic then there is no $\phi \in \mathcal{F}_{n+1}$ with $R^2(\phi) = 1$, and conversely. However, there may be ϕ's with $R^2(\phi)$ arbitrarily close to 1.

The multiple correlation is closely connected with angles. We can define the angle θ between ϕ and $\pi \in \mathcal{H}$ by the equation

$$(\phi, \pi) = \|\phi\| \|\pi\| \cos \theta$$

(see also Gelfand and Yaglom (1959)). The angle between $\phi \in \mathcal{F}_{n+1}$ and $\mathcal{P}_n$ is

$$\inf \{\text{angle between } \phi \text{ and } \pi, \pi \in \mathcal{P}_n\}$$

and the angle between $\mathcal{F}_{n+1}$ and $\mathcal{P}_n$ is

$$\inf \{\text{angle between } \phi \text{ and } \mathcal{P}_n, \phi \in \mathcal{F}_{n+1}\}.$$

Thus $\sup \{R^2(\phi): \phi \in \mathcal{F}_{n+1}\} < 1$ iff the angle between $\mathcal{P}_n$ and $\mathcal{F}_{n+1}$ is positive.

Definition 7. $\{x(n)\}$ has the positive angle property if

$$\sup \{R^2(\phi): \phi \in \mathcal{F}_{n+1}\} < 1.$$

Note that the cosine of the angle between two subspaces is an infinite-dimensional analogue of the largest canonical correlation between two sets of variables (see for instance Hannan (1961)). For example, the cosine of the angle between $\mathscr{P}_n$ and $\mathscr{F}_{n+1}$ is

$$\sup\{(\pi,\phi):\pi\in\mathscr{P}_n,\ \phi\in\mathscr{F}_{n+1}\ \|\pi\|=\|\phi\|=1\}$$

that is, the supremum of the correlation between an element of the past and an element of the future.

The spectra of processes with the positive angle property have been characterized by Helson and Szego (1960), and an equivalent condition known as A_2 was given by Hunt et al. (1973).

Proposition 7. (a) $\{x(n)\}$ has the positive angle property iff its spectrum is absolutely continuous and the spectral density f may be written $f=\exp(u+\tilde{v})$, where $u\in L^\infty$ and $\tilde{v}$ is the conjugate harmonic function of some $v\in L^\infty$ with $\|v\|_\infty<\pi/2$.

(b) An equivalent condition on f is that there is a finite constant C such that

$$A_2:\left(|I|^{-1}\int_I f(\lambda)d\lambda\right)\left(|I|^{-1}\int_I f(\lambda)^{-1}d\lambda\right)\leqq C$$

for every subarc I $(-\pi,\pi]$. (A subarc is a set I of the form $(\lambda_1,\lambda_2]$ for $\lambda_1<\lambda_2\in(-\pi,\pi]$, or of the form $(-\pi,\lambda_2]\cup(\lambda_1,\pi]$ for $\lambda_2\leqq\lambda_1\in(-\pi,\pi]$.)

It is clear probabilistically that to have the positive angle property, $\{x(n)\}$ must be completely non-deterministic. It is also clear from A_2 that f^{-1} must be integrable, or in other words that $\{x(n)\}$ must be minimal. In fact it follows from the representation $f=\exp(u+\tilde{v})$ that f^{-p} is integrable for some $p>1$ (Zygmund (1959), p. 254). Thus we can easily construct examples of spectra that are minimal but do not have the positive angle property. On the other hand, Bloomfield et al. (1983) show that the spectrum $f(\lambda)=\cos\lambda/2$, $-\pi<\lambda\leqq\pi$, is completely non-deterministic, but it is clearly not minimal. Thus these three concepts are distinct.

There is also a form of asymptotic independence that depends on the notion of angle. Rosenblatt (1956) defined a strong mixing condition, the linear version of which is that $\mathscr{P}_n$ and $\mathscr{F}_{n+u}$ should become orthogonal as $u\to\infty$.

Definition 8. $\{x(n)\}$ is completely regular if

$$\sup\{\text{corr}(\phi,\pi):\phi\in\mathscr{F}_{n+u},\ \pi\in\mathscr{P}_n\}\to 0\quad\text{as}\quad u\to\infty.$$

The spectra of such processes were characterized by Helson and Sarason (1967) and Sarason (1972).

Proposition 8. $\{x(n)\}$ is completely regular iff its spectrum is absolutely continuous and the spectral density may be written

$$f=|P|^2\exp(u+\tilde{v})$$

where P is a trigonometric polynomial and u and v are continuous and periodic.

9. Essential correlation

Once we have the notion of the largest canonical correlation ρ_1 between past and future, it is natural to ask whether there are associated canonical components π_1 and ϕ_1 with correlation ρ_1, as there are in the finite-dimensional case. We are also interested in knowing whether we can define the second canonical correlation by restricting attention to elements of the past and the future that are uncorrelated with π_1 and ϕ_1 respectively, and so on. Jewell and Bloomfield (1983) discuss the various possibilities. In the first place there may be no pair $\pi_1 \in \mathscr{P}_n$, $\phi_1 \in \mathscr{F}_{n+1}$ whose correlation attains the supremum ρ_1. This is the case in the example $f(\lambda) = \cos \lambda/2$, for which $\rho_1 = 1$ (since the process is not minimal, and hence does not have the positive angle property) and yet there is no pair with correlation 1 (since the process is completely non-deterministic).

Next, it may be possible to find a finite number of principal components by this process, after which the supremum of the correlation of the remaining linear combinations is not attained.

Thirdly, there may be a countable number of such components, but such that the correlations of the remaining variables do not converge to 0.

In all these three situations, there is a residual correlation that cannot be removed by orthogonalizing with respect to canonical components. Jewell and Bloomfield (1983) call this the essential correlation of the process (it is the supremum of the essential spectrum of a certain linear operator associated with the spectrum of the process).

The final case is where there is a countable set of canonical correlations $\{\rho_i\}$, perhaps only finitely many being non-zero, and these in a sense carry all of the correlation between past and future. In this case there is no essential correlation. It was shown by Jewell and Bloomfield (1983) that this is equivalent to $\{x(n)\}$ being completely regular.

Theorem 3. $\{x(n)\}$ is completely regular iff there exist $\{\rho_i\}$, $1 \geq \rho_i \downarrow 0$ as $i \to \infty$, and $\pi_i \in \mathscr{P}_n$, $\phi_i \in \mathscr{F}_{n+1}$ such that

$$\operatorname{corr}(\pi_i, \phi_i) = \rho_i$$

$$= \sup \{\operatorname{corr}(\pi, \phi) : \pi \in \mathscr{P}_n, \pi \perp \pi_j, 1 \leq j < i;$$

$$\phi \in \mathscr{F}_{n+1}, \phi \perp \phi_j, 1 \leq j < i\}.$$

Thus the absence of essential correlation, which is a form of non-singularity, is equivalent to complete regularity, a form of asymptotic independence. Note that such processes may fail to have the positive angle property, but only by having at least one unit canonical correlation that is attained by associated canonical components. In this case the process is also not completely non-deterministic, $\mathscr{P}_n \cap \mathscr{F}_{n+1}$ being a finite-dimensional but non-trivial space. The process is also not minimal.

Thus having no essential correlation and no unit canonical correlations gives a strong type of non-singularity.

10. Information

There is one further notion that needs to be explored. If $\{\rho_i\}$ are the canonical correlations between two sets of Gaussian random variables, the quantity

$$-\tfrac{1}{2}\ln \prod (1-\rho_i^2)$$

is the (Shannon) information of either set concerning the other (Gelfand and Yaglom (1959)). Ibragimov and Rozanov (1978) introduced the following definition.

Definition 9. $\{x(n)\}$ is absolutely regular if the mutual information between $\mathscr{P}_n$ and $\mathscr{F}_{n+u}$ converges to 0 as $u \to \infty$.

The same authors characterized the spectra of these processes.

Proposition 9. $\{x(n)\}$ is absolutely regular iff its spectrum is absolutely continuous and its spectral density is of the form $f = |P|^2 e^u$, where P is a trigonometric polynomial and the Fourier coefficients u_r of u satisfy $\sum |r| u_r^2 < \infty$.

The definition clearly expresses asymptotic independence, and also clearly implies complete regularity. However, there is also a related form of non-singularity. Recall that if two finite-dimensional random variables both have non-singular marginal distributions, then their joint distribution is non-singular iff their canonical correlations satisfy

$$\prod (1-\rho_i^2) > 0,$$

or in other words, iff their mutual information is finite. By analogy, we obtain a particularly strong form of non-singularity by requiring the mutual information of $\mathscr{P}_n$ and $\mathscr{F}_{n+1}$ to be finite. Jewell and Bloomfield (1983) show that there is a close connection with absolute regularity.

Theorem 4. The mutual information of $\mathscr{P}_n$ and $\mathscr{F}_{n+1}$ is finite iff the spectrum of the process is absolutely continuous and the spectral density is of the form e^u where the Fourier coefficients of u satisfy $\sum |r| u_r^2$.

The only strengthening of absolute regularity is the elimination of the polynomial factor that is allowed in Proposition 9. This factor introduces unit canonical correlations which cause the information to be infinite. However, the product of the non-zero $(1 - \rho_i^2)$ converges to a positive quantity.

11. Rational spectra and geometrically dominated autocovariances

It is natural at this point to ask which of the commonly used models satisfy which of these various definitions. It is well known that ARMA models have absolutely continuous spectra, and spectral densities that are rational functions of $\cos \lambda$, and that this is equivalent to there being only a finite number of non-zero correlations between $\mathscr{P}_n$ and $\mathscr{F}_{n+1}$ (Jewell and Bloomfield (1983)). Our strictest definition of asymptotic independence, absolute regularity, is therefore satisfied. The corresponding definition of non-singularity, finiteness of the information, is also satisfied unless there are zeros in the numerator of the rational spectrum (the MA part), as these would introduce unit canonical correlations. Integrability of the spectral density rules out zeros in the denominator.

Thus for processes with rational spectra, the only issue is whether there are zeros in the numerator. If there are none, the process satisfies our strictest definitions. If there are, the process is still purely non-deterministic, but fails all definitions from complete non-determinism on. The same is in fact true of any process for which the autocovariances decay geometrically.

Theorem 5. Suppose that there exist $A > 0$ and B, $0 < B < 1$ such that $|\gamma_r| < AB^r$, or equivalently that $\limsup |\gamma_r|^{1/r} < 1$. Then the process is absolutely regular, and there is finite information in the past concerning the future iff the spectral density is bounded away from 0.

Proof. The Laurent series $\sum \gamma_r z^r$ converges in the annulus $B < |x| < 1/B$, and defines an analytic function $g(z)$ whose restriction to the unit circle $|z| = 1$ is the spectral density function. If g has zeros on the unit circle then $g = ph$ where p is a polynomial with the same zeros and h is 0-free on the unit circle, and hence in an annulus containing the unit circle. Since g is non-negative on the unit circle, both p and h may also be chosen to be non-negative, and hence the restriction of p to the unit circle is of the form $|P|^2$ where P is a trigonometric polynomial. Since h is analytic and non-zero in an annulus, the Fourier coefficients of $\ln h$ are also dominated by a geometric sequence, although the modulus may be larger than B. Thus the summability condition required of the second factor in the characterization of absolute regularity is also satisfied. The polynomial factor, which destroys the finiteness of the information in the past about the future, is present iff the spectral density is not bounded away from 0.

It would be interesting to know whether there are weaker conditions on the autocovariances or the spectrum that similarly reduce the number of different possibilities.

References

BLOOMFIELD, P., JEWELL, N. P. AND HAYASHI, E. (1983) Characterization of completely non-deterministic stochastic process. *Pacific J. Math.* **197**, 307–317.

DELEEUW, K. AND RUDIN, W. (1958) Extreme points and extremum problems in H^1. *Pacific J. Math.* **8**, 467–485.

DOOB, J. L. (1953) *Stochastic Processes*. Wiley, New York.

GELFAND, I. M. AND YAGLOM, A. M. (1959) Calculation of the amount of information about a random function contained in another such function. *Amer. Math. Soc. Translations* **12**, 199–246.

HANNAN, E. J. (1961) The general theory of canonical correlation and its relation to functional analysis. *J. Austral. Math. Soc.* **2**, 229–242.

HANNAN, E. J. (1970) *Multiple Time Series*. Wiley, New York.

HELSON, H. AND SARASON, D. E. (1967) Past and future. *Math. Scand.* **21**, 5–16.

HELSON, H. AND SZEGO, S. (1960) A problem in prediction theory. *Ann. Math. Pura Appl.* **51**, 107–138.

HUNT, R. A., MUCKENHOUPT, B. AND WHEEDEN, R. L. (1973) Weighted norm inequalities for the conjugate function and Hilbert transform. *Trans. Amer. Math. Soc.* **176**, 227–251.

IBRAGIMOV, I. A. AND ROZANOV, Y. A. (1978) *Gaussian Random Processes*. Springer-Verlag, New York.

JEWELL, N. P. AND BLOOMFIELD, P. (1983) Canonical correlations of past and future for time series: definitions and theory. *Ann. Statist.* **11**, 837–847.

KOLMOGOROV, A. N. (1941) Stationary sequences in Hilbert space. *Bull. Moscow State Univ.* **2**, 1–40. Reprinted in translation in *Linear Least Squares Estimation*, ed. T. Kailath, Benchmark Papers in Electrical Engineering and Computer Science, **17**, Dowden, Hutchinson & Ross, Stroudsburg, PA, 66–89.

ROSENBLATT, M. (1956) A central limit theorem and a strong mixing condition. *Proc. Nat. Acad. Sci. USA* **42**, 43–47.

SARASON, D. E. (1972) An addendum to "Past and future." *Math. Scand.* **30**, 62–64.

ZYGMUND, A. (1959) *Trigonometric Series*, Vol. I. Cambridge University Press, Cambridge.

Linear Dynamic Errors-in-Variables Models

M. DEISTLER

Abstract

Linear dynamical systems where both inputs and outputs are contaminated by errors are considered. A characterization of the sets of all observationally equivalent transfer functions is given, the role of the causality assumption is investigated and conditions for identifiability in the case of Gaussian as well as non-Gaussian observations are derived.

LINEAR DYNAMIC SYSTEMS; IDENTIFIABILITY; CAUSALITY

1. Introduction

In the statistical analysis of linear systems, there are two main approaches to modelling the 'stochastic environment' for the system: in the usual approach, all errors are added to the outputs. Models of this kind are called *errors-in-equations* (EE) models: the most important model class (in case of rational transfer functions) are then the ARMAX models. Note that we make no distinction between errors added to the outputs and errors added to the equations, as both kinds of error lead to the same consequences. In the second approach, errors are added to inputs and outputs: this is the errors-in-variables (EV) case.

The EE setting is adequate if our primary aim is to give a good *representation of the observations* by a system, e.g. if we are interested in the prediction of the observed outputs from the observed inputs. On the other hand, the EV setting is appropriate if we are interested in the *analysis of the true system* generating the data, and if we cannot be sure *a priori* that the inputs are observed without errors.

The statistical theory of linear dynamic systems in the EE case, and especially in the ARMAX case, has now reached a certain state of completeness. Thanks in good part to the seminal work of E. J. Hannan, we now understand the problems of parametrization, of maximum likelihood estimation and of order estimation associated with multi-input–multi-output ARMAX systems (see e.g. Hannan and Kavalieris (1984)).

The statistical theory in the EV case, on the other hand, is a long way from reaching this state. This is due to a number of additional complications, arising especially in connection with the problem of identifiability in this context. The problem of identifiability is whether the characteristics of interest are uniquely determined from certain characteristics of the observed processes (see e.g. Deistler and Seifert (1978)). If identifiability cannot be ensured, then the problem is to describe the sets of all observationally equivalent characteristics of interest, i.e. the classes of all characteristics of interest corresponding to given characteristics of the observed processes.

In the EE case, under general conditions, the transfer functions are uniquely determined from the (ensemble) second moments of the observations and the main problem is the parameterization of the (rational) transfer functions. This problem has been treated in three papers by E. J. Hannan (Hannan (1969), (1971), (1976)).

In the EV setting, there is a crucial additional complication: the transfer functions in general are not uniquely determined from the second moments of the observations (see Anderson and Deistler (1984)); furthermore, higher-order moments of the observations may contain additional information about the transfer function. Once the transfer functions are identifiable, of course, as a second step, the problem of the parametrization of these transfer functions has to be solved in the same way as in the EE case.

There is an extensive literature on static EV models (see Madansky (1959), Moran (1971), T. W. Anderson (1984) for surveys). Even for linear static EV models, the multivariate case has not yet been solved completely (Kalman (1982), (1983)). Linear dynamic EV models have been treated in Hannan (1963), Akaike (1966), Maravall (1979), Wegge (1982), Nowak (1983), Aigner et al. (1984), Anderson and Deistler (1984) and B. D. O. Anderson (1985).

In this paper we treat the problems of identifiability and observational equivalence in single-input–single-output linear dynamic EV systems under general assumptions. The paper is organized as follows. In Sections 2 and 3 the characteristics of the observed processes considered are their second moments. In Section 2, for the general case of not necessarily causal and not necessarily rational systems, the sets of observationally equivalent transfer functions are characterized. In Section 3 the effects of an additional causality and rationality assumption are investigated, and conditions for identifiability are given. In Section 4 we derive conditions for identifiability using information from higher-order moments of the observations.

2. The analysis based on second moments of the observations: The general case

In this section the analysis is based on second moments of the observations only. We derive a characterization of the sets of all observationally equivalent characteristics of interest (especially of the transfer function of the underlying system), corresponding to fixed second moments of the observations. These results have essentially been already stated in Anderson and Deistler (1984).

The general form of the linear relation considered is:

$$(2.1) \qquad \lim_{N \to \infty} \sum_{i=-N}^{N} W_i^{(y,\,N)} \hat{y}_{t-i} = \lim_{N \to \infty} \sum_{i=-N}^{N} W_i^{(x,N)} \hat{x}_{t-i}.$$

$(\hat{x}_t)$ and $(\hat{y}_t)$, $t \in \mathbb{Z}$, are therefore the (in general unobserved), true processes considered. Let

$$(2.2) \qquad w^{(\cdot)}(z) = \lim_{N \to \infty} \sum_{i=-N}^{N} W_i^{(\cdot,N)} z^i, \text{ where } \cdot \text{ may be } x \text{ or } y,$$

denote the generating functions, where z denotes a complex variable $z \in \mathbb{C}$ as well as the backward shift operator $z\hat{x}_t = \hat{x}_{t-1}$. The limits in (2.1) are understood in the sense of mean-square convergence of random variables. If the contrary has not been stated explicitly, we consider only the case where $\hat{x}_t$ and $\hat{y}_t$ are real and one-dimensional.

The observed processes (x_t) and (y_t) respectively are given by

$$(2.3) \qquad x_t = \hat{x}_t + u_t$$

$$(2.4) \qquad y_t = \hat{y}_t + v_t$$

where (u_t) are the errors contaminating $(\hat{x}_t)$ and (v_t) are the errors contaminating $(\hat{y}_t)$.

Throughout we assume that

$$(2.5) \qquad \text{all processes considered are (wide-sense) stationary,}$$

$$(2.6) \qquad E\hat{x}_t = Eu_t = Ev_t = 0,$$

$$(2.7) \qquad E\hat{x}_s u_t = E\hat{x}_s v_t = Eu_s v_t = 0 \qquad \forall s, t \in \mathbb{Z},$$

and, for the sake of notational simplicity:

$$(2.8) \qquad \text{all processes have (finite) spectral densities.}$$

The assumption $E\hat{x}_t = 0$ may easily be relaxed. (2.7) gives a definition of the errors, saying that all common effects are due to the system. Although for many cases this is a natural definition, there are applications, where other definitions of errors may be appropriate, e.g. when it is *a priori* known that the frequency ranges of the errors and of true processes are disjoint or when the true processes are known to be non-Gaussian and the errors are Gaussian (or conversely).

Note that (2.1) allows for a completely symmetric treatment of the variables; therefore we have not used words like 'system' or 'inputs and outputs' in this context.

Let

$$f = \begin{pmatrix} f_x & f_{xy} \\ f_{yx} & f_y \end{pmatrix}, \qquad \hat{f} = \begin{pmatrix} f_{\hat{x}} & f_{xy} \\ f_{yx} & f_{\hat{y}} \end{pmatrix}, \qquad \tilde{f} = \begin{pmatrix} f_u & 0 \\ 0 & f_v \end{pmatrix}$$

denote the spectral matrices of $\begin{pmatrix} x_t \\ y_t \end{pmatrix}$, $\begin{pmatrix} \hat{x}_t \\ \hat{y}_t \end{pmatrix}$ and $\begin{pmatrix} u_t \\ v_t \end{pmatrix}$ respectively. Then (2.1)–(2.8) implies

$$(2.9) \qquad\qquad f = \hat{f} + \tilde{f}$$

where $\hat{f}$ is singular (i.e. $\hat{f}(\lambda)$ is singular λ-almost everywhere) and $\tilde{f}$ is diagonal.

The characteristics of interest are $w^{(y)}$, $w^{(x)}$, $f_{\hat{x}}$, f_u, f_v. Given any decomposition (2.9) of f, where $\hat{f}$, $\tilde{f}$ are spectral density matrices (i.e. non-negative-definite Hermitian and integrable matrices $f : [-\pi, \pi] \to \mathbb{C}^{2\times2}$) corresponding to real processes (i.e. $f(\lambda) = f'(-\lambda)$) where $\hat{f}$ is singular and $\tilde{f}$ is diagonal, we can define

$$\bar{w}^{(x)}(\exp(-i\lambda)) = f_{yx}(\lambda)f_x^{-1}(\lambda); \qquad \bar{w}^{(y)}(\exp(-i\lambda)) = 1, \quad \text{if} \quad f_{\hat{x}}(\lambda) \neq 0,$$

$$(2.10) \quad \bar{w}^{(x)}(\exp(-i\lambda)) = 1; \qquad \bar{w}^{(y)}(\exp(-i\lambda)) = f_{xy}(\lambda)f_y^{-1}(\lambda),$$

$$\text{if} \quad f_{\hat{x}}(\lambda) = 0 \quad \text{and} \quad f_{\hat{y}}(\lambda) \neq 0,$$

$$\bar{w}^{(x)}(\exp(-i\lambda)) = \bar{w}^{(y)}(\exp(-i\lambda)) = 0, \quad \text{if} \quad \hat{f}(\lambda) = 0.$$

We can always find a suitable normalization say $n(\exp(-i\lambda)) > 0 \forall \lambda$, where

$$(w^{(x)}(\exp(-i\lambda)), w^{(y)}(\exp(-i\lambda)))$$

$$= n(\exp(-i\lambda))(\bar{w}^{(x)}(\exp(-i\lambda)), \bar{w}^{(y)}(\exp(-i\lambda))),$$

are such that the infinite sums in (2.1) exist.

Thus up to normalization of $w^{(y)}$, $w^{(x)}$, there is a one-to-one relation between $(w^{(y)}, w^{(x)}, f_{\hat{x}}, f_u, f_v)$ and $(\hat{f}, \tilde{f})$. Therefore our problem is to characterize the set of all feasible $\hat{f}$, $\tilde{f}$ in (2.9) for given f.

Once $(w^{(y)}, w^{(x)}, f_x, f_u, f_v)$ are given, then a relation (2.1)–(2.4) satisfying (2.5)–(2.8) can always be found.

Clearly, for given f, a decomposition (2.9) is in one-to-one relation with the pair $(f_{\hat{x}}, f_{\hat{y}})$. Thus given $(f_{\hat{x}}, f_{\hat{y}})$ the characteristics f_u, f_v and $w^{(y)}$, $w^{(x)}$ up to normalization are uniquely determined. From (2.9) we see that every pair $(f_{\hat{x}}, f_{\hat{y}})$ compatible with the given second moments f_x, f_y, f_{xy} of the observations satisfies the following relations:

$$(2.11) \qquad 0 \leq f_{\hat{x}} \leq f_x, \qquad f_{\hat{x}}(\lambda) = f_{\hat{x}}(-\lambda), \quad f_{\hat{x}} \text{ is measurable,}$$

$$(2.12) \qquad 0 \leq f_{\hat{y}} \leq f_y, \qquad f_{\hat{y}}(\lambda) = f_{\hat{y}}(-\lambda), \quad f_{\hat{y}} \text{ is measurable,}$$

and, since $\hat{f}$ is singular

(2.13)
$$|f_{xy}|^2 = f_{\hat{x}} \cdot f_{\hat{y}},$$

and these are the only restrictions for $(f_{\hat{x}}, f_{\hat{y}})$. By the non-negative definiteness of f we have

(2.14)
$$|f_{xy}|^2 \leq f_x \cdot f_y$$

and thus, putting for example

$$f_{\hat{x}} = f_x, \qquad f_{\hat{y}}(\lambda) = \begin{cases} |f_{xy}(\lambda)|^2 f_x^{-1}(\lambda) & \text{if} \quad f_x(\lambda) \neq 0 \\ f_y(\lambda) & \text{if} \quad f_x(\lambda) = 0 \end{cases}$$

for every spectral density matrix f, a decomposition (2.9) or equivalently an EV system exists.

From now on we assume that

(2.15)
$$f_{\hat{x}}(\lambda) > 0 \qquad \forall \lambda \in [-\pi, \pi];$$

(2.1) can then always be written as a *system*

(2.16)
$$\hat{y}_t = w(z)\hat{x}_t$$

with transfer function

(2.17)
$$w(z) = \lim_{N \to \infty} \sum_{i=-N}^{N} W_i^{(N)} z^i$$

and with (true and observed) inputs $\hat{x}_t$ and x_t and (true and observed) outputs $\hat{y}_t$ and y_t respectively.

If $f_{xy}(\lambda) \neq 0$ for all λ then (2.15) is always satisfied; otherwise (2.15) has to be postulated in addition to (2.11)–(2.13) to characterize the set of all feasible $(f_{\hat{x}}, f_{\hat{y}})$.

If $f_{xy} = 0$ then $\tilde{f} = f$ gives a feasible decomposition (2.9); if $f_{xy}(\lambda) \neq 0$ on a set of Lebesgue measure greater than 0, then $\hat{f}$ must be of rank equal to 1 on this set.

Now, from what has been said above we obtain the following result (see Anderson and Deistler (1984)).

Theorem 1. Consider the linear dynamic EV model (2.16), (2.3)–(2.8). Under the additional assumption $f_{\hat{x}}(\lambda) > 0 \, \forall \lambda$, the set of all transfer functions w satisfying

(2.18)
$$|f_{yx}(\lambda)| \cdot f_x^{-1}(\lambda) \leq |w(\exp(-i\lambda))| \leq f_y(\lambda) \cdot |f_{xy}^{-1}(\lambda)|$$

$$w(\exp(-i\lambda)) \cdot |w(\exp(-i\lambda))|^{-1} = f_{yx}(\lambda) \cdot |f_{yx}(\lambda)|^{-1} \quad \text{for} \quad f_{xy}(\lambda) \neq 0$$

and

$$w(\exp(-i\lambda)) = 0 \quad \text{for} \quad f_{xy}(\lambda) = 0,$$

is the set of all transfer functions w corresponding to given second moments of the observations f_x, f_y, f_{xy}.

The corresponding set of the other characteristics of interest $(f_{\hat{x}}, f_u, f_v)$ satisfies the following relations:

$$(2.19) \qquad f_{\hat{x}}(\lambda) = \begin{cases} f_{yx}(\lambda) \cdot w^{-1}(\exp(-i\lambda)) & \text{for} \quad f_{xy}(\lambda) \neq 0 \\ 0 < f_{\hat{x}}(\lambda) \leq f_x(\lambda) & \text{for} \quad f_{xy}(\lambda) = 0 \end{cases}$$

$$(2.20) \qquad\qquad\qquad f_u = f_x - f_{\hat{x}}$$

$$(2.21) \qquad\qquad\qquad f_v = f_y - w(\exp(-i \cdot))f_{xy}.$$

Remark 1. The theorem above shows that the phase of w is uniquely determined from f_x, f_y, f_{xy}, whereas the absolute value of w varies in a band whose boundaries are determined by the two extreme cases where either $u_t = 0$ or $v_t = 0$. This result is analogous to the static case (where w, $f_{\hat{x}}$, f_u, f_v are constant), where the slope of the fitted line varies in an interval whose endpoints correspond to the two regressions, from x_t on y_t and from y_t on x_t (Frisch (1934)). If $f(\lambda)$ is singular (λ-a.e.) then $f = \hat{f}$ gives the only feasible decomposition (2.9) and thus w is unique in this case.

From (2.19)–(2.21) we see that once w is given and $w(\exp(-i\lambda)) \neq 0 \; \forall \lambda$, the characteristics $f_{\hat{x}}$, f_u, f_v are unique. It is also easy to see that, once $f_{\hat{x}}$ is known, w, f_u, f_v are uniquely determined.

3. The causal, rational case

In many cases it is known *a priori* that $w(z)$ is *causal*, i.e. that the summation in (2.17) ranges from 0 to ∞ only:

$$(3.1) \qquad\qquad w(z) = \lim_{N \to \infty} \sum_{i=0}^{N} W_i^{(N)} z^i.$$

In addition, $w(z)$ is often assumed to be rational, i.e.

$$(3.2) \qquad\qquad w(z) = a^{-1}(z) \cdot b(z)$$

where $a(z)$, $b(z)$ are polynomials, and $(\hat{x}_t)$, (u_t), (v_t) are assumed to be ARMA processes. In this section a description of the internal characteristics corresponding to given second moments of the observations under these additional *a priori* assumptions is obtained, and conditions for identifiability are given.

First, let us introduce some additional notation. For a polynomial $p(z) = \sum_{i=0}^{\delta p} P_i z^i$; $P_i \in \mathbb{R}$ $z \in \mathbb{C}$, we denote its degree by δp and by $p^*(z)$ we mean the rational function defined by

$$p^*(z) = p(z^{-1}) = \sum_{i=0}^{\delta p} P_i z^{-i}.$$

Furthermore, by δp_0 we denote the multiplicity of the zero of $p(z)$ at $z = 0$ and by $\tilde{p}(z)$ we denote the polynomial defined by

$$\tilde{p}(z) = p^*(z) \cdot z^{\delta p},$$

where $\tilde{p}(z)$ has degree $\delta p - \delta p_0$. Let $z_1 \cdots z_{\delta p - \delta p_0}$ be the zeros of $p(z)$ which are not equal to 0. Then $z_1^{-1} \cdots z_{\delta p - \delta p_0}^{-1}$ are the zeros of $p^*(z)$. Note that $P_0 \neq 0$ implies $\tilde{\tilde{p}} = p$.

It is well known that every rational function f of the form

$$f(\exp(-i\lambda)) = p_1(\exp(-i\lambda)) \cdot p_2(\exp(-i\lambda))^{-1} p_3^*(\exp(-i\lambda)) \cdot p_4^*(\exp(-i\lambda))^{-1}$$

defined on the unit circle of the complex plane, where p_i, $i = 1, \cdots, 4$, are polynomials, has a unique rational extension to $\mathbb{C}$ given by

$$f(z) = p_1(z) \cdot p_2(z)^{-1} \cdot p_3^*(z) \cdot p_4^*(z)^{-1}.$$

In this section (rational) spectral densities are considered as to be extended in this way and thus to be defined on $\mathbb{C}$ rather than on the unit circle or on $[-\pi, \pi]$. Also, omitting a factor of 2π, we have:

$$\begin{aligned}
w &= a^{-1}b, \\
f_{\hat{x}} &= d^{-1}e\sigma_\varepsilon e^* d^{-1*} \\
f_u &= c^{-1}h\sigma_\mu h^* c^{-1*} \\
f_v &= f^{-1}g\sigma_\mu g^* f^{-1*}
\end{aligned}$$

(3.3)

(where we have omitted the complex variable z) and where

$$a(z) = \sum_{i=0}^{\delta a} A_i z^i; \qquad b(z) = \sum_{i=0}^{\delta b} B_i z^i; \qquad d(z) = \sum_{i=0}^{\delta d} D_i z^i$$

(3.4)

$$e(z) = \sum_{i=0}^{\delta e} E_i z^i; \qquad c(z) = \sum_{i=0}^{\delta c} C_i z^i; \qquad h(z) = \sum_{i=0}^{\delta h} H_i z^i$$

$$f(z) = \sum_{i=0}^{\delta f} F_i z^i; \qquad h(z) = \sum_{i=0}^{\delta h} H_i z^i$$

are polynomials satisfying

$$\begin{aligned}
a(z) \neq 0, \quad |z| \leq 1; & \quad d(z) \neq 0, \quad |z| \leq 1; & \quad e(z) \neq 0, \quad |z| \leq 1, \\
c(z) \neq 0, \quad |z| \leq 1; & \quad h(z) \neq 0, \quad |z| < 1, & \\
f(z) \neq 0, \quad |z| \leq 1; & \quad g(z) \neq 0, \quad |z| < 1, &
\end{aligned}$$

(3.5)

and where σ_ε, σ_μ, σ_v are the respective innovation variances.

$$e(z) \neq 0, |z| = 1 \text{ is then equivalent to } f_{\hat{x}}(z) \neq 0, |z| = 1.$$

To obtain uniqueness of the parameters describing transfer functions and spectra we assume:

(3.6) a, b are relatively prime and so are d, e and c, h and f, g,

(3.7) $a(0) = d(0) = e(0) = c(0) = h(0) = f(0) = g(0) = 1.$

Let

$$b = b^{+} \cdot b^{-} \cdot z^{\delta b_0}$$

where b^{+} has no zeros inside the unit circle, b^{-} has no zeros outside or on the unit circle and where $b^{-}(0) = 1$.

Note that, since our interest is in the transfer function w, we have neither imposed a miniphase assumption on b (i.e. $b = b^{+}$), nor have we assumed $b(0) = 1$.

We maintain all these notations and assumptions (which we call the general assumptions of this section) throughout the section. Unless the contrary has been stated explicitly, we do not assume that the degrees $\delta a \cdots \delta h$ are known *a priori*.

Now, the relation between the second moments of the observations and the characteristics of interest, namely w, $f_{\hat{x}}$, f_u, f_v is of the form:

(3.8) $f_x = d^{-1}e\sigma_\varepsilon e^* d^{-1*} + c^{-1}h\sigma_\mu h^* c^{-1*}$

(3.9) $f_{yx} = a^{-1}bd^{-1}e\sigma_\varepsilon e^* d^{-1*}$

(3.10) $f_y = a^{-1}bd^{-1}e\sigma_\varepsilon e^* d^{-1*}b^* a^{-1*} + f^{-1}g\sigma_\nu g^* f^{-1*}.$

This section consists of two main parts. In the first part we now investigate the information concerning $(w, f_{\hat{x}})$ coming from (3.9).

Under our assumptions there is a one-to-one relation between $(w, f_{\hat{x}})$ and $(a, b, d, e, \sigma_\varepsilon)$, and in this sense we can identify them. We call a 5-tuple $(a, b, d, e, \sigma_\varepsilon)$ satisfying our assumptions and corresponding to a given f_{yx} via (3.9) a *realization* of f_{yx}.

Let $(a, b, d, e, \sigma_\varepsilon)$ be a realization of f_{yx}. By c_e, c_{d+} and c_{d-} respectively we denote the greatest common divisors of a, e, of b^{+}, d and of $\tilde{b}^{-}$, d, respectively. Furthermore, c denotes the greatest common divisor of c_{d+}, c_{d-} and we define

$$c_{+} = c_{d^+} \cdot c^{-1}; \qquad c_{-} = c_{d^-} \cdot c^{-1}.$$

As a norming condition we impose

$$c_e(0) = c_{d^+}(0) = c_{d^-}(0) = c(0) = 1.$$

Then we can write (3.9) as

$$f_{yx} = a^{-1} \cdot b \cdot d^{-1}e\sigma_\varepsilon \tilde{e}\tilde{d}^{-1} \cdot z^{-\delta e + \delta d}$$

(3.11) $$= (a \cdot c_e^{-1})^{-1} \cdot z^{\delta b_0}(b^{+} \cdot c_{d^+}^{-1})(b^{-} \cdot \tilde{c}_{d^-}^{-1})(d \cdot c_{d^+}^{-1})^{-1}$$

$$\times (e \cdot c_e^{-1}) \cdot \sigma_\varepsilon \cdot \tilde{e} \cdot (\tilde{d} \cdot \tilde{c}_{d^-}^{-1})^{-1} z^{-\delta e + \delta d}.$$

Now we define

$$(3.12) \qquad d_N = c_{d^-}^{-1} \cdot d.$$

Note that the zeros of $\tilde{d}_N$ are just the poles of f_{yx} whose absolute values satisfy $0 < |z| < 1$.

Thus, as $d_N(0) = 1$ by assumption, d_N is directly obtained from f_{yx}. From (3.11) we get:

$$(3.13) \qquad f_{yx} d_N \tilde{d}_N = (a \cdot c_e^{-1})^{-1} z^{\delta b_0} (b^+ \cdot c_{d^+}^{-1})(b^- \cdot \tilde{c}_{d^-}^{-1})$$
$$\times c_+ \cdot c_-^{-1} \cdot (e \cdot c_e^{-1}) \sigma_\varepsilon \tilde{e} \cdot z^{-\delta e + \delta d}.$$

Note that $(b^+ \cdot c_{d^+}^{-1})$ and c_- are relatively prime since every common factor f of $(b^+ \cdot c_{d^+}^{-1})$ and c_- is also a common factor of b^+ and of c_-. But since $c_{d^+} \cdot c_-$ is a factor of d, f must be constant since otherwise c_{d^+} would not be a greatest common divisor of b^+ and d. In the same way, it can be shown that $(\tilde{b}^- \cdot \tilde{c}_d^{-1})$ and c_+ are relatively prime.

Thus, if we define

$$(3.14) \qquad a_N = c_- \cdot c_e^{-1} \cdot a$$

then it is easily seen that the zeros of a_N are the poles of $f_{yx} d_N \tilde{d}_N$ which have absolute value greater than 1; and therefore as $a_N(0) = 1$, a_N can be directly calculated from f_{yx}. From (3.13) we obtain

$$(3.15) \qquad \begin{aligned} f_{xy} d_N \tilde{d}_N a_N &= z^{\delta b_0}(b^+ \cdot c_{d^+}^{-1} \cdot b_r^{-1}) \cdot c_+(b^- \tilde{c}_{d^-}^{-1} \tilde{b}_r^{-1}) \tilde{c}_e \\ &\quad \times b_r(ec_e^{-1}) \sigma_\varepsilon \tilde{b}_r(\tilde{e}\tilde{c}_e^{-1}) \cdot z^{-\delta e + \delta d} \end{aligned}$$

where b_r is the greatest common divisor of $(b^+ \cdot c_{d^+}^{-1})$ and $(\tilde{b}^- \cdot c_{d^-}^{-1})$ with the property $b_r(0) = 1$.

We now define

$$b_N^+ = c^{-1} b_r^{-1} b^+$$
$$b_N^- = \tilde{c}_e \tilde{b}_r^{-1} \tilde{c}_{d^-}^{-1} b^-$$
$$(3.16) \qquad \delta b_{N,0} = \delta b_0 - \delta e + \delta d - \delta d_N + \delta e_N = \delta b_0 + \delta c_{d^-} + \delta b_r - \delta c_e$$
$$b_N = z^{\delta b_{N,0}} \cdot b_N^+ b_N^-$$
$$e_N = c_e^{-1} b_r e.$$

Then, for suitably chosen $\sigma_{\varepsilon N} > 0$,

$$(3.17) \quad f_{yx} d_N \tilde{d}_N a_N = z^{\delta b_{N,0}} \cdot b_N^+ b_N^- e_N \sigma_{\varepsilon N} \tilde{e}_N z^{-\delta e_N + \delta d_N} = b_N e_N \sigma_{\varepsilon N} \tilde{e}_N z^{-\delta e_N + \delta d_N}$$

and since $\tilde{b}^- \cdot \tilde{c}_{d^+}^{-1}$ and c_t are relatively prime b_N^+ and $\tilde{b}_N^-$ are also relatively prime; and thus e_N is uniquely defined from all zeros, z_i say, of $f_{yx} d_N \tilde{d}_N a_N$ with the properties that $|z_i| > 1$ and z_i^{-1} are zeros of $f_{yx} d_N \tilde{d}_N a_N$ (i.e. from all zeros of modulus greater than 1 whose reflections on the unit circle are zeros again) and taking into account $e_N(0) = 1$. b_N^- is then obtained from the remaining unreflected) zeros of $f_{yx}\hat{d}_N d_N a_N$ with modulus greater than 0 and less than 1 and taking into account $b_N^-(0) = 1$. b_N^+ is obtained from the remaining zeros of $f_{yx} d_N \tilde{d}_N a_N$ of modulus greater than or equal to 1. b_N^+ is unique only up to multiplication by a positive constant. Note that $\delta b_0 + \delta d - \delta e$ is the multiplicity of the zero of f_{yx} at $z = 0$ and thus $\delta_{N,0}$ is defined from f_{yx} via (3.16). Finally $\sigma_{\varepsilon N}$ is chosen to satisfy (3.18). We call any 5-tuple $(a_N, b_N, d_N, e_N, \sigma_{\varepsilon N})$ obtained in this way from f_{yx} a *standard tuple* for f_{yx}. Note that for such a standard tuple of f_{yx}:

(i) $w_N = a_N^{-1} b_N$ (and thus $f_{\hat{x},N} = d_N^{-1} e_N \sigma_{\varepsilon N} e_N^* d_N^{*-1}$) is a unique up to multiplication by arbitrary positive constants.

(ii) A standard tuple is a realization of f_{yx} if and only if $\delta b_{N,0} = \delta b_0 + \delta c_{d^-} + \delta b_r - \delta c_e \geqq 0$.

Thus we have shown the first part of the following theorem.

Theorem 2. Let the general assumptions of this section hold and let $(a_N, b_N, d_N, e_N, \sigma_{\varepsilon N})$ be a standard tuple of f_{yx}. Then

(i) If $(a, b, d, e, \sigma_\varepsilon)$ is a realization of f_{yx} then

$$
\begin{aligned}
a &= c_e \cdot c_-^{-1} \cdot a_N \\
b &= \nu \cdot c \cdot b_r \tilde{c}_{d^-} \tilde{b}_r \tilde{c}_e^{-1} z^{-\delta c_{d^-} - \delta b_r + \delta c_e} b_N \\
a^{-1}b &= \nu \cdot c_d \tilde{c}_{d^-} b_r \tilde{b}_r c_e^{-1} \tilde{c}_e^{-1} z^{-\delta c_{d^-} - \delta b_r + \delta c_e} a_N^{-1} b_N \\
d &= c_{d^-} \cdot d_N \\
e &= c_e b_r^{-1} e_N \\
d^{-1} e \sigma_\varepsilon e^* d^{-1*} &= \nu^{-1} c_e \tilde{c}_e c_{d^-}^{-1} \tilde{c}_{d^-}^{-1} b_r^{-1} \tilde{b}_r^{-1} z^{-\delta c_e + \delta b_r + \delta c_{d^-}} d_N^{-1} e_N \sigma_{\varepsilon N} e_N^* d_N^{*-1}
\end{aligned}
$$

(3.18)

where $\nu = b^+(0) \cdot b_N^+(0)^{-1}$, where c_{d^-}, c_e and b_r respectively are the greatest common divisors of b^-, d, of a, e and of $(b^+ c_{d^+}^{-1})$, $(\tilde{b}^- \tilde{c}_{d^-}^{-1})$ respectively and where $c_{d^-}(0) = c_e(0) = b_r(0) = 1$. c_e and $c_{d^-} b_r$ are relatively prime.

(ii) If f_1 is a divisor of $\tilde{b}_N^-$ such that $f_1(0) = 1$, if f_2 is a polynomial satisfying

$$
\begin{aligned}
f_2(z) &\neq 1 \qquad |z| \geqq 1 \\
f_2(0) &= 1
\end{aligned}
$$

and

$$
0 \leqq \delta b_N^- + \delta f_2 - \delta f_1 \leqq \delta b_N^- + \delta b_{N,0}
$$

$$
0 \leqq \delta b_{N,0} - \delta f_2 + \delta f_1 \leqq \delta b_N^- + \delta b_{N,0}
$$

and if $\nu > 0$ then by

(3.19)

$$
\begin{aligned}
a^{-1}b &= \nu \cdot f_2 \cdot \tilde{f}_2 \cdot f_1^{-1} \cdot \tilde{f}_1^{-1} z^{\delta f_1 - \delta f_2} a_N^{-1} b_N \\
d^{-1} e \sigma_\varepsilon e^* d^{-1*} &= \nu^{-1} \cdot f_1 \cdot \tilde{f}_1 \cdot f_2^{-1} \cdot \tilde{f}_2^{-1} \cdot z^{-\delta f_1 + \delta f_2} d_N^{-1} e_N \sigma_{\varepsilon N} e_N^* d_N^{*-1}
\end{aligned}
$$

a realization $(a, b, d, e, \sigma_\varepsilon)$ of f_{yx} is given.

We then have

$$
\begin{aligned}
a &= f_1 \cdot f_3^{-1} a_N \\
b &= v \cdot f_5 \cdot f_4 \cdot \tilde{f}_3 \cdot \tilde{f}_5 \cdot \tilde{f}_4 \cdot \tilde{f}_1^{-1} \cdot z^{-\delta f_2 + \delta f_1} b_N \\
d &= f_3 \cdot f_5 \cdot d_n \\
e &= f_1 \cdot f_4^{-1} e_N
\end{aligned}
$$

(3.20)

where $f_2 = f_3 \cdot f_4 \cdot f_5$; f_3 and f_4 respectively are the greatest common divisors of f_2, a_N and of f_2, e_N respectively and where $f_3(0) = f_4(0) = f_5(0) = 1$.

Proof. It remains to prove the second part of the theorem. Let f_1 and f_2 be as required and let $v > 0$ be a constant. Then, as $\delta b_{N,0} + \delta f_1 - \delta f_2 \geqq 0$, (3.19) defines a transfer function $a^{-1}b$ and a spectral density $d^{-1}e\sigma_\varepsilon e^* d^{-1*}$ corresponding to f_{yx} via (3.9) and satisfying the general assumptions of this section. The last statement of the theorem then is easily shown.

Remark 2. Theorem 2 gives a characterization of all realizations of a given f_{yx}. A similar but somewhat different result has been derived by B. D. O. Anderson (1985) for the not necessarily rational case. From Anderson's results, as well as from (3.18), it is clear that $\delta b^- + \delta b_0$, i.e. the number of zeros of b in $|z| < 1$ is an invariant for all realizations of f_{yx}.

Corollary 1. Let the general assumptions of this section hold. If b_N is miniphase, i.e. if $b_N = b_N^+$ then w (and thus $f_{\hat{x}}$) is determined from f_{yx} up to multiplication by an arbitrary positive constant.

Proof. If b_N is miniphase then $\delta b_N^- = \delta b_{N,0} = 0$ and thus from (3.18) we see that $\delta c_e = 0$. But since b is a polynomial, we then have $\delta c_{d^-} + \delta b_r = 0$ and thus $c_e = c_{d^-} = b_r = 1$.

Remark 3. This corollary has also been stated in Anderson (1985) for the more general, not necessarily rational case.

As another example for Theorem 2 consider the case $\delta b_{N,0} = 0$, $\delta b_N^- = 1$. Then the set of all realizations of f_{yx} corresponding to w is either $v a_N^{-1} b_N$, or $v \cdot z \cdot f_1^{-1} \tilde{f}_1^{-1} \cdot a_N^{-1} b_N$ or $v \cdot f_2 \cdot \tilde{f}_2 \cdot f_1^{-1} \cdot \tilde{f}_1^{-1} a_N^{-1} b_N$ where $v \in \mathbb{R}^+$, f_1 is an arbitrary factor of $\tilde{b}_N^-$ of degree equal to 1 and $f_2 = c \cdot (z - z_2)$; $-cz_2 = 1$, $0 < |z_2| < 1$.

Corollary 2. Let the general assumptions of this section hold. If (for all realizations)

(3.21) $d, \tilde{b}^-$ are relatively prime

(3.22) a, e are relatively prime

(3.23) $b^+, \tilde{b}^-$ are relatively prime

then $w = a^{-1}b$ (and thus $f_{\hat{x}} = d^{-1}e\sigma_\varepsilon e^* d^{*-1}$) is determined from f_{yx} up to multiplication by an arbitrary positive constant.

Proof. The result is a direct consequence of (3.18).

In the second main part of this section we investigate the additional information coming from (3.8) and (3.10).

Any 5-tuple $(a, b, d, e, \sigma_\varepsilon)$ satisfying (3.8)–(3.10) and the general assumptions of this section is called a *realization* of f. Clearly, given $(a, b, d, e, \sigma_\varepsilon)$, c, h, σ_μ, f, h and σ_ν are uniquely determined.

Theorem 3. Let the general assumptions of this section hold. Then

(i) If $(a, b, d, e, 1)$ is a realization of f_{yx}, then all 5-tuples $(a, \nu b, d, e, \nu^{-1})$ satisfying

$$(3.24) \qquad\qquad 0 < \nu_{\min} \leqq \nu \leqq \nu_{\max}$$

where $\nu_{\min}$ and $\nu_{\max}$ are given by

$$(3.25) \qquad \min_{|z|=1} (f_x(z) - \nu_{\min}^{-1} d^{-1}(z)e(z)e^*(z)d^{*-1}(z)) = 0$$

and

$$(3.26) \quad \min_{|z|=1} (f_y(z) - \nu_{\max} a^{-1}(z)b(z)d^{-1}(z)e(z)e^*(z)d^{*-1}(z)b^*(z)a^{*-1}(z)) = 0$$

are realizations of f and for all other $\nu(>0)$ $(a, \nu b, d, e, \nu^{-1})$ is not a realization of f.

(ii) If d and c are *a priori* known to be relatively prime, then there is only a finite number of factors f_2 (and f_1) in (3.19) compatible with given f.

(iii) Under the conditions of Corollary 1 or of Corollary 2, each of the following additional conditions:

$$(3.27) \qquad\qquad \delta d > 0 \text{ and } d, c \text{ are relatively prime}$$

$$(3.28) \qquad\qquad \delta ad > 0 \text{ and } a \cdot d, f \text{ are relatively prime}$$

$$(3.29) \qquad\qquad \delta d = 0 \text{ and } \delta e > \delta h - \delta c$$

$$(3.30) \qquad\qquad \delta ad = 0 \text{ and } \delta e + \delta b > \delta g - \delta f$$

guarantees that $(a, b, d, e, \sigma_\varepsilon)$ is uniquely determined from f (and thus we have identifiability).

Proof. (i) is an immediate consequence of Theorem 2, (3.8) and (3.10), using the non-negativity of spectral densities for $|z| = 1$.

(ii) If d and c are relatively prime, then all zeros of d are poles of f_x and thus there is only a finite number of candidates for d and thus for c_d-compatible with given f_x; since b_r divides e_N, by (3.18), there is only a finite number of polynomials $f_2 = c_{d^-} \cdot b_r$ in (3.19) compatible with given f; clearly as $\tilde{c}_e$ divides $b_{\bar{N}}$ there is also only a finite number of f_1 compatible with f_{yx}.

(iii) From (3.8) we have

(3.31) $$df_x d^* = e\sigma_\varepsilon e^* + dc^{-1} h\sigma_\mu h^* c^{-1*} d^*.$$

If (3.27) holds, there is at least one zero of d, z_1 say, and we have

$$df_x d^*(z_1) = e\sigma_\varepsilon e^*(z_1)$$

and from this σ_ε and thus b are uniquely determined. Analogously under (3.28), σ_ε and b are uniquely determined from (3.10).

If (3.29) holds, then (3.8) is of the form

$$f_x = e\sigma_\varepsilon e^* + c^{-1} h\sigma_\mu h^* c^{-1*}$$

and thus c is uniquely obtained from the poles of f_x. Then σ_ε is obtained from a comparison of coefficients of power $\delta e + \delta c$ in

$$cf_x c^* = ce\sigma_\varepsilon e^* c^* + h\sigma_\mu h^*$$

and in the same way we proceed if (3.30) holds.

Remark 4. As has been pointed out in Anderson (1985), it may happen that $\nu_{\min} > \nu_{\max}$; in this case no (positive) multiple of w corresponding to a realization of f_{yx} can correspond to a realization of f. Results similar to Theorem 3 (iii) have been derived by Maravall (1979), Nowak (1983) and Anderson and Deistler (1984). (iii) may be formulated in more general terms as follows. Under each of the assumptions (3.27)–(3.20), for a given set $\{(a, \nu b, d, e, \nu^{-1}\sigma_\varepsilon) \mid \nu \in \mathbb{R}^+\}$, ν is uniquely determined.

Remark 5. If the maximal degrees, $na \cdots ng$ say, of $a, \cdots, g$ have been prescribed *a priori*, then our EV model can be described by a parameter space $\Theta \subset \mathbb{R}^{na + \cdots + ng + 4}$ consisting of all parameters $\theta = (A(1) \cdots A(na), B(0), \cdots, G(ng), \sigma_\varepsilon, \sigma_\mu, \sigma_\nu)$ satisfying the general assumptions of this section. Then Corollary 2 and Theorem 3(iii) show, that if $nd > 0$ or $nad > 0$, then the model is identifiable on a generic subset of Θ. Of course this is not true in general for the static case.

Clearly the assumption $\delta d > 0$ is a requirement for minimal dynamics. (3.21), (3.22) exclude pole-zero cancellations between input spectra and the system.

Remark 6. We do not investigate the not necessarily causal, rational case in great detail here. However, let us mention that under the assumptions of Corollary 2, but if we do not impose *a priori* the causality assumption (3.1) on w, and thus if instead of $a(z) \neq 0$ $|z| \leq 1$ we only assume $a(z) \neq 0$ $|z| = 1$, (in order to ensure the existence of a stationary solution) and under the additional assumptions:

(3.32) $\qquad\qquad a, \tilde{e}$ are relatively prime

(3.33) $\qquad\qquad a^{+}, \tilde{a}^{-}$ are relatively prime

(where a^{+} and a^{-} are defined in the same way as b^{+} and b^{-}) the result of Corollary 2 holds. This is easily seen as then d can be determined from those poles z_i of f_{xy}, where z_i^{-1} is a pole again, in the same way as e is determined from the reflected zeros of f_{yx}.

This shows that under these assumptions (which again are generically fulfilled in Θ) causality can be detected from the data.

Remark 7. Of course, the rational case is by far the most important one. Note however that Theorem 2 and Corollary 2 can be generalized in a straightforward manner to the case where the spectral densities involved have a representation of the form

$$f(z) = \left(\prod_{i=1}^{\infty}(z-z_i)\right)^{-1}\left(\prod_{i=1}^{\infty}(z-w_i)\right)\sigma\left(\prod_{i=1}^{\infty}(z^{-1}-w_i^{-1})\right)\left(\prod_{i=1}^{\infty}(z^{-1}-z_i^{-1})\right)^{-1}.$$

4. Identifiability results in the case of non-Gaussian observations

As is well known, in the static case, identifiability may be obtained for non-Gaussian observations, (using information exceeding the information coming from second moments) whereas the analogous models are not identifiable for Gaussian observations (see Geary (1942), Reiersøl (1950)). Here these results are extended to the dynamic case. The analysis is based on higher-order (i.e. order greater than 2) cumulant spectra. The results obtained here are closely related to the results given in Akaike (1966). One problem associated with this approach in practice is that reliable estimation of higher-order spectra requires a large amount of data.

If $z_1 \cdots z_n$ are (scalar) random variables then their joint nth order cumulant, cum$(z_1 \cdots z_n)$, is defined as the coefficient of $(i)^n t_1 \cdots t_n$ in the Taylor series expansion of $\log E \exp i \sum_{j=1}^{n} z_j t_j$ (see Brillinger (1981)).

We assume the model (2.16), (2.17), (2.3), (2.4) together with the assumptions (2.5)–(2.8) and the additional assumption

$(\hat{x}_t)$, (u_t), (v_t) are strictly stationary processes which are mutually independent, and all moments (up to order n, n sufficiently large) exist, and the cumulants satisfy conditions of the form:

$$(4.1) \qquad \sum_{t_1 \cdots t_{n-1}=-\infty}^{\infty} |\text{cum}\,(\hat{y}_{t_1}^{*} \cdots \hat{y}_{t_r}^{*} \hat{x}_{t_{r+1}}^{*} \cdots \hat{x}_{t_{n-1}}^{*} \hat{x}_0)| < \infty$$

and the same holds for (u_t) and for (v_t).

Note that, whereas the lack of correlation of (x_t), (u_t) and (v_t) is costless, as it is just a definition of the errors, the independence assumption causes a true restriction of generality. Under (4.1) the nth-order cumulant spectra exist (Brillinger (1981)) and are given by

$$f_{\hat{y}(r)\hat{x}(n-r)}(\lambda_1 \cdots \lambda_{n-1})$$

$$(4.2) \qquad = (2\pi)^{-n+1} \sum_{t_1 \cdots t_{n-1}=-\infty}^{\infty} \mathrm{cum}\,(\hat{y}_{t_1}^* \cdots \hat{y}_{t_r}^* \hat{x}_{t_{r+1}}^* \cdots \hat{x}_{t_{n-1}}^* \hat{x}_0^*) \exp\left\{-i \sum_{j=1}^{n-1} \lambda_j t_j\right\}$$

and analogously for the nth-order cumulant spectra $f_{u(n)}$ and $f_{v(n)}$ for (u_t) and (v_t) respectively.

Since the cumulants are linear and continuous with respect to one variable λ_i (when the others are kept constant), we obtain from (2.16) and (4.2)

$$(4.3) \quad f_{\hat{y}(r+1)\hat{x}(n-r-1)}(\lambda_1 \cdots \lambda_{n-1}) = w(\exp\,(-i\lambda_1)) \cdot f_{\hat{y}(r)\hat{x}(n-r)}(\lambda_1 \cdots \lambda_{n-1}).$$

On the other hand, by the properties of the cumulants, the mutual independence of $(\hat{x}_t)$, (u_t) and (v_t) implies

$$(4.4) \qquad\qquad f_{y(r)x(n-r)} = f_{\hat{y}(r)\hat{x}(n-r)} + f_{v(r)u(n-r)}$$

where we have used an evident notation to denote the nth order (cross-) spectra of (y_t) and (x_t) and of (u_t) and (v_t) respectively. Furthermore, the independence of (u_t) and (v_t) implies

$$(4.5) \qquad\qquad f_{v(r)u(n-r)} = 0 \quad \text{for} \quad r > 0, \qquad n - r > 0.$$

Thus we have the following result.

Theorem 4. Consider the linear dynamic EV model (2.16), (2.17), (2.3)–(2.8) and (4.1). Under the additional assumptions:

$$(4.6) \qquad \begin{array}{l} \text{there exist } r > 0;\ n - r > 1 \text{ and } \lambda_2 \cdots \lambda_{n-1} \text{ such that} \\ f_{y(r)x(n-r)}(\lambda_1 \cdots \lambda_{n-1}) \neq 0 \quad \forall \lambda_1 \end{array}$$

w is uniquely determined from

$$(4.7) \quad w(\exp\,(-i\lambda_1)) = f_{y(r+1)x(n-r-1)}(\lambda_1 \cdots \lambda_{n-1}) \cdot f_{y(r)x(n-r)}^{-1}(\lambda_1 \cdots \lambda_{n-1}).$$

Remark 8. Let us assume that $(\hat{x}_t)$ is a regular process with a Wold representation (see e.g. Hannan (1970))

$$\hat{x}_t = w_1(z)\varepsilon_t; \qquad w_1(z) = \sum_{i=0}^{\infty} W_1(i)z^i$$

where we in addition postulate

$$(\varepsilon_t) \text{ is i.i.d. and } \sum_{i=0}^{\infty} |W_1(i)| < \infty$$

then (Brillinger (1981))

$$f_{\hat{x}(n)}(\lambda_1 \cdots \lambda_{n-1}) = (2\pi)^{-n+1} \cdot w_1(\exp(-i\lambda_1)) \cdots w_1(\exp(-i\lambda_{n-1}))$$

(4.8)
$$\times w_1\left(\exp\left\{i \sum_{j=1}^{n-1} \lambda_j\right\}\right) \operatorname{cum}(\varepsilon_0^n).$$

If (ε_t) is non-Gaussian and if all moments exist, then there is an $n > 2$ such that $\operatorname{cum}(\varepsilon_0^n) \neq 0$. If in addition $w(\exp(-i\lambda)) \neq 0$, $w_1(\exp(-i\lambda)) \neq 0$ $\forall \lambda$ then (4.6) is fulfilled due to (4.8), (4.3)–(4.5). Thus in a non-Gaussian situation it is very likely that (4.6) is fulfilled.

Remark 9. Theorem 4 can be generalized in a straightforward way to the multivariable case.

Remark 10. Also the higher-order cumulant spectra of $(\hat{x}_t)$, (u_t) and (v_t) are uniquely determined from w, provided that $w(\exp(-i\lambda)) \neq 0 \, \forall \lambda$. From (4.3)–(4.5) we obtain

$$f_{\hat{x}(n)}(\lambda_1 \cdots \lambda_{n-1}) = w^{-1}(\exp(-i\lambda_1)) \cdot f_{y(1)x(n-1)}(\lambda_1 \cdots \lambda_{n-1})$$

$$f_{u(n)} = f_{x(n)} - f_{\hat{x}(n)}$$

$$f_{v(n)}(\lambda_1 \cdots \lambda_{n-1}) = f_{y(n)}(\lambda_1 \cdots \lambda_{n-1}) - w(\exp(-i\lambda_1)) \cdot f_{y(n-1)x(1)}(\lambda_1 \cdots \lambda_{n-1}).$$

Acknowledgements

I wish to express my deep gratitude for all that I have been able to learn from Ted Hannan.

I am grateful to Professor B. D. O. Anderson of Canberra and Professor R. Schnabl of Vienna for valuable hints.

References

AIGNER, D. J., HSIAO, C., KAPTEYN, A. AND WANSBEEK, T. (1984) Latent variable models in econometrics. In *Handbook of Econometrics*, ed. Z. Grilliches and M. D. Intriligator, North-Holland, Amsterdam.

AKAIKE, H. (1966) On the use of non-Gaussian process in the identification of a linear dynamic system. *Ann. Inst. Statist. Math.* **18**, 269–276.

ANDERSON, B. D. O. (1985) Identification of scalar errors-in-variables models with dynamics. *Automatica.* To appear.

ANDERSON, B. D. O. AND DEISTLER, M. (1984) Identifiability in dynamic errors-in-variables models. *J. Time Series Anal.* **5**, 1–13.

ANDERSON, T. W. (1984) Estimating linear statistical relationships. *Ann. Statist.* **12**, 1–45.

BRILLINGER, D. R. (1981) *Time Series: Data Analysis and Theory.* Holden Day, San Francisco.

DEISTLER, M. AND SEIFERT, H. G. (1978) Identifiability and consistent estimability in dynamic econometric models. *Econometrica* **46**, 969–980.

FRISCH, R. (1934) *Statistical Confluence Analysis by Means of Complete Regression Systems.* Publication No. 5, University of Oslo, Economic Institute.

GEARY, R. C. (1942) Inherent relations between random variables. *Proc. R. Irish Acad.* A **47**, 63–76.

HANNAN, E. J. (1963) Regression for time series with errors of measurement. *Biometrika* **50**, 293–302.

HANNAN, E. J. (1969) The identification of vector mixed autoregressive-moving average systems. *Biometrika* **56**, 223–225.

HANNAN, E. J. (1970) *Multiple Time Series.* Wiley, New York.

HANNAN, E. J. (1971) The identification problem for multiple equation systems with moving average errors. *Econometrica* **39**, 751–765.

HANNAN, E. J. (1976) The identification and parametrization of ARMAX and state space forms. *Econometrica* **44**, 713–723.

HANNAN, E. J. AND KAVALIERIS, L. (1984) Multivariate linear time series models. *Adv. Appl. Prob.* **16**, 492–561.

KALMAN, R. E. (1982) System identification from noisy data. In *Dynamical Systems* II, a University of Florida International Symposium, eds. A. Bednarek and L. Cesari. Academic Press, New York.

KALMAN, R. E. (1983) Identifiability and modeling in econometrics. In *Developments in Statistics* **4**, ed. P. R. Krishnaiah. Academic Press, New York.

MADANSKY, A. (1959) The fitting of straight lines with both variables are subject to error. *J. Amer. Statist. Assoc.* **54**, 173–205.

MARAVALL, A. (1979) *Identification in Dynamic Shock-Errors Models.* Springer-Verlag, Berlin.

MORAN, P. A. P. (1971) Estimating structural and functional relationships. *J. Multivariate Anal.* **1**, 232–255.

NOWAK, E. (1983) Identification of the dynamic shock-error model with autocorrelated errors. *J. Econometrics* **23**, 211–221.

REISERSØL, O. (1950) Identifiability of a linear relation between variables which are subject to error. *Econometrica* **9**, 1–24.

WEGGE, L. (1982) ARMAX-Model parameter identification without and with latent variables. Working Paper. Dept. of Economics, Univ. of California, Davis.

Quantile Spectral Analysis and Long-Memory Time Series

EMANUEL PARZEN[*]

Abstract

An approach to time series model identification is described which involves the simultaneous use of frequency, time and quantile domain algorithms; the approach is called quantile spectral analysis. It proposes a framework to integrate the analysis of long-memory (non-stationary) time series with the analysis of short-memory (stationary) time series.

MODEL IDENTIFICATION; QUANTILE DATA ANALYSIS; HURST EXPONENT; SPECTRAL INDEX OF REGULAR VARIATION

0. Introduction

The casual peruser of this paper may be startled to see that there are not many complicated formulas. The emphasis in this paper is on how to develop a conceptual framework to integrate diverse approaches to time series analysis.

The need to analyse data observed in the form of time series is present in almost every scientific field. The concept of a conventional analysis is not the same in each field. Engineers tend to estimate mean, variance, and spectrum (which may be regarded as a non-parametric signature of models). Economists and forecasters tend to estimate mean, variance, and time domain models such as ARMA or ARIMA (which are parametric models). Spectral and ARMA estimation are not routine procedures; there are many algorithms for spectral estimation and time-domain model identification. In addition, there are critics of spectral- and correlation-based methods of time series analysis who search for alternate methods to model long-range dependence. This paper describes an approach to time series analysis which attempts to use recently developed quantile methods

* This paper was prepared while the author was a Fellow at the Center for Advanced Study in the Behavioral Sciences at Stanford, California. I am grateful for financial support provided by National Science Foundation Grant BNS 76-22943.

The research was supported by the U.S. Office of Naval Research and the U.S. Army Research Office.

of statistical data analysis to integrate spectral and correlation methods with methods for long-memory and/or long-tailed time series.

As diagnostic tools for model identification, we propose quantile analysis of suitable sample spectral densities; these methods are called quantile spectral analysis. Quantile methods also enable the time series analyst to obtain insight into the types of probability distributions that could fit the sample.

Let $Y(t)$, $t = 0, \pm1, \pm2, \cdots$ be a univariate discrete time series of which one has observed a sample $Y(t)$, $t = 1, \cdots, T$. The problem of empirical model identification is to identify (using only the data) models for the probability law of the stochastic process generating the time series. The problem of parameter estimation usually assumes a model which has been identified through the use of model identification methods.

1. Sample spectral density

This paper aims to discuss the following question which should be important to researchers on empirical model identification methods: what is the role to be played by methods of empirical spectral analysis? A time series need not be stationary in order for us to be able to compute and interpret the sample spectral density of a sample $Y(t)$, $t = 1, \cdots, T$, defined by

$$f^{\sim}(\omega) = \left| \sum_{t=1}^{T} \exp(2\pi it\omega) Y(t) \right|^2 \bigg/ \sum_{t=1}^{T} |Y(t)|^2.$$

This function is defined for all ω, but it is usually computed at a grid of frequencies $\omega = k/\text{NFREQ}$, $k = 0, 1, \cdots, \text{NFREQ-1}$. One chooses $\text{NFREQ} > T$; usually $\text{NFREQ} = 2T$.

2. Spectral density

If one assumes that the observed time series $Y(t)$, $t = 1, \cdots, T$ is a sample of a stationary Gaussian time series, the parameters to be estimated are the mean $\mu = E[Y(t)]$, the covariance function $R(v) = \text{Cov}[Y(t+v), Y(t)]$, the correlation function $\rho(v) = R(v)/R(0)$, and the spectral density function

$$f(\omega) = \sum_{v=-\infty}^{\infty} \exp(-2\pi iv\omega)\rho(v), \qquad 0 \leq \omega \leq 1.$$

The spectral density function provides a spectral representation of the correlation function

$$\rho(v) = \int_0^1 \exp(2\pi iv\omega)f(\omega)d\omega, \qquad v = 0, \pm1, \pm2, \cdots.$$

In order to guarantee the existence of the spectral density, one assumes $\rho(v)$ is absolutely summable. One of the goals of empirical spectral analysis is to relax this assumption, and also to provide methods to test the assumption that the random variables $Y(t)$ are Gaussian or have finite second moments.

The variable ω represents frequency and is usually assumed to vary in the interval $-0.5 \leq \omega \leq 0.5$. Only the interval $0 \leq \omega \leq 0.5$ has physical significance when the time series is real-valued. We adopt the interval $0 \leq \omega \leq 1$ to develop an analogy between spectral density functions and quantile density functions (denoted in this paper by $f(\omega)$, $0 \leq \omega \leq 1$, and $q(u)$, $0 \leq u \leq 1$ respectively).

The inverse spectral density, defined by $fi(\omega) = 1/f(\omega)$, obeys the fundamental relation $fi(\omega)f(\omega) = 1$; it enjoys analogous properties to the density quantile function (introduced by Parzen (1979) and denoted by $fQ(u)$, $0 \leq u \leq 1$).

3. Quantile domain functions

To identify and estimate the parameters of probability distributions that fit a data batch $X(1), \cdots, X(n)$ quantile data analysis uses statistical methods based on the properties of functions defined in the quantile domain.

Quantile probability theory is defined for a random variable X with distribution function $F(x) = \Pr[X \leq x]$; we define

$$\text{Quantile function } Q(u) = F^{-1}(u) = \inf\{x : F(x) \geq u\}, \qquad 0 \leq u \leq 1.$$

Given a sample $X(1), \cdots, X(n)$ of X one forms the sample distribution function $F^{\sim}(x)$ and the sample quantile function

$$Q^{\sim}(u) = F^{\sim -1}(u).$$

When $F(x)$ is continuous with probability density function $f(x) = F'(x)$, there is a density for $Q(u)$ which we call the quantile density function and denote $q(u) = Q'(u)$. The density quantile function is $fQ(u) = f(Q(u))$. These two densities form a reciprocal pair in the sense that (differentiating $F(Q(u)) = u$)

$$fQ(u)q(u) = 1.$$

The reciprocal functions $q(u)$ and $fQ(u)$ have direct probability interpretations while the two reciprocal functions $f(\omega)$ and $fi(\omega)$ might appear to be purely mathematically defined concepts. The variables u and ω represent percentiles and frequency respectively. We believe it is remarkable that one can regard $q(u)$ and $f(\omega)$ as having analogous properties (i.e. isomorphic but not in a strict mathematical sense), and similarly $fQ(u)$ and $fi(\omega)$ have analogous properties.

It should be noted that the analogy is not initially perfect since $f(\omega)$ is an even function in the sense that $f(\omega) = f(1 - \omega)$. However this restriction is not necessary if we consider complex-valued zero-mean time series with covariance function defined by $R(v) = E[Y^*(t+v)Y(t)]$. An asterisk $*$ on a complex number z denotes its complex conjugate z^*.

4. Entropy analogy

To understand the type of analogies between a spectral density function $f(\omega)$ and a quantile density function $q(u)$ we start with the formulas for entropy of a continuous random variable X and a Gaussian time series $Y(t)$.

The entropy of X is defined

$$H(f) = \int_{-\infty}^{\infty} \{-\log f(x)\}f(x)\,dx = \int_{0}^{1} -\log fQ(u)\,du = \int_{0}^{1} \log q(u)du.$$

The entropy of $Y(t)$ is defined

$$H(f) = 0.5\left\{1 + \int_{0}^{1} \log f(\omega)d\omega\right\}.$$

The analogy between the density quantile function and the spectral density function is that entropy is essentially equal to the integral of their logarithm.

The concept of maximum entropy representations of spectral density functions and quantile density functions play an important role; autoregressive representations can be justified as maximum entropy representations (for references see Parzen (1982)).

It should be emphasized that entropy concepts play a fundamental role in probability model identification: this was first shown in the pioneering work of Akaike (see Akaike (1977)).

5. Density quantile function representations: regularly varying functions and tail exponents

To understand the nature of the analogies that exist between spectral density functions $f(\omega)$ and density quantile functions $fQ(u)$, it is best to begin with the observation that a density quantile function $fQ(u)$ of a realistic probability distribution has a representation as a regularly varying function:

$$fQ(u) = u^{\alpha_0}L_0(u) \quad \text{and} \quad fQ(u) = (1-u)^{\alpha_1}L_1(u)$$

where $L_0(u)$ is slowly varying as $u \to 0$, and $L_1(u)$ is slowly varying as $u \to 1$. The exponents α_0 and α_1 are called left and right tail exponents.

A tail exponent α provides four types of tail behavior of distributions which we call: super-short ($\alpha < 0$), short ($0 \leq \alpha \leq 1$), medium ($\alpha = 1$), long ($\alpha > 1$). Within medium-tail distributions we may distinguish medium-short, medium-medium, and medium-long.

A distribution $F(x)$ has a left-tail type and a right-tail type. A normal distribution is symmetric, and both its left and right tails are medium-short. An exponential distribution has a short left tail and a medium-medium right tail.

The random variable $X = \cos(2\pi U)$, where U is uniform on 0 to 1, has a distribution with super-short tails on both left and right (one can show that $\alpha_0 = \alpha_1 = -1$).

6. Spectral density representations: index of regular variation

We define a spectral density $f(\omega)$ to have finite dynamic range if it satisfies $0 < c_1 \leqq f(\omega) \leqq c_2 < \infty$ for all ω.

When $f(\omega)$ does not have finite dynamic range, we propose to model its behavior at a frequency ω_0 at which it tends towards zero or infinity as a regularly varying function with the representation

$$f(\omega) = (\omega - \omega_0)^{-\delta} L(\omega)$$

where $L(\omega)$ is slowly varying as $\omega \to \omega_0$. We call δ the index of regular variation at frequency ω_0.

At zero frequency the representation is $f(\omega) = \omega^{-\delta} L(\omega)$ where $L(\omega)$ is slowly varying as $\omega \to 0$. This function is integrable only if $\delta < 1$; the corresponding correlation function decays very slowly to 0: $\rho(v) \sim v^{\delta-1}$.

Note that a finite dynamic range spectral density has $\delta = 0$, but $\delta = 0$ does not imply finite dynamic range since $f(\omega)$ can tend to ∞ as $\omega \to 0$; an example is $f(\omega) \sim (\log \omega)^2$, $\rho(v) \sim (\log v)/v$ as $v \to \infty$.

Long-memory time series models considered by Mandelbrot (1973), Granger and Joyeux (1980), and Geweke and Porter-Hudak (1983) have spectral density $f(\omega)$ satisfying the regular variation representation at $\omega = 0$.

The formula $\delta = 2H - 1$ relates the index δ (of regular variation of the spectral density at $\omega = 0$) to the Hurst exponent H mentioned in Section 14; $0 < \delta < 1$ corresponds to $0.5 < H < 1$.

7. Self-similarity

When a spectral density $f(\omega)$ is regularly varying at $\omega = 0$ it enjoys a property which we call self-similarity at $\omega = 0$, defined by

$$f(y\omega) = y^{-\delta} f(\omega) \quad \text{as} \quad \omega \to 0,$$

in the sense that, for any $y > 0$, $f(y\omega)/y^{-\delta} f(\omega) \to 1$ as $\omega \to 0$. A problem to be investigated is the relation of this type of self-similarity to the concept of asymptotically self-similar processes introduced by Mandelbrot (1982).

8. Short-memory time series

We call a time series short-memory if it can be modeled as a stationary time series with absolutely summable correlation function and spectral density function $f(\omega)$ with finite dynamic range.

A short-memory time series can be modeled as an infinite invertible moving average and autoregression in terms of a white-noise time series $\{e(t)\}$ representing the innovations of $Y(t)$:

$$Y(t) + a(1)Y(t-1) + a(2)Y(t-2) + \cdots = e(t),$$

$$Y(t) = e(t) + b(1)e(t-1) + b(2)e(t-2) + \cdots.$$

A stationary time series is called a white-noise time series if $f(\omega)$ is constant, or equivalently $f(\omega) = 1$, $0 \leq \omega \leq 1$.

We prefer to call a white-noise time series a no-memory time series. To test that a time series is of no-memory type, one tests that its sample spectral density $f^{\sim}(\omega)$ is of no-memory type. We propose that we test this by methods of quantile data analysis, and we therefore obtain the concept of quantile spectral analysis.

9. Role of ARMA models

Identification of the memory type of an observed time series is the essential first step in our approach to time series model identification since (in our view) only when the time series is short-memory is it appropriate to model it approximately by an autoregressive-moving-average, or ARMA (p, q), model of the form

$$Y(t) + a(1)Y(t-1) + \cdots + a(p)Y(t-p) = e(t) + b(1)e(t-1) + \cdots + b(q)e(t-q)$$

where $\{e(t)\}$ is approximately Gaussian white noise; the parameters are estimated by quasi-maximum likelihood estimation procedures. Our conception of a parsimonious ARMA scheme is that the residual time series $e(t)$ just barely is not significantly different from white noise (a no-memory time series). This is tested by comparing the sample spectral density $f^{\sim}(\omega)$ with the theoretical spectral density

$$f_{p,q}(\omega) = \sigma_{p,q}^2 \, |h_q(\exp 2\pi i\omega)|^2 / |g_p(\exp 2\pi i\omega)|^2$$

where $h_q(z) = 1 + b(1)z + \cdots + b(q)z^q, g_p(z) = 1 + a(1)z + \cdots + a(p)z^p$ and $\sigma_{p,q}^2$ is a constant which makes the integral of $f_{p,q}(\omega)$ over the interval $0 \leq \omega \leq 1$ equal to 1. A quantile spectral criterion of ARMA model fit is outlined in Section 22.

In terms of the lag operator L defined by $LY(t) = Y(t-1)$, an ARMA model can be written $g_p(L)Y(t) = h_q(L)e(t)$. The residual time series $e(t)$ is theoretically computed from $Y(t)$ by $e(t) = h_q^{-1}(L)g_p(L)Y(t)$.

An ARMA (p, q) model for $Y(\cdot)$ is usually written

$$g_p(L)(Y(t) - \mu) = h_q(L)e(t).$$

In terms of keeping track of the successive transformations of $Y(t)$ from a long-memory series to a short-memory series $Z(t)$ to a no-memory series $e(t)$, we would write

$$Y(t) - \mu = Z(t),$$

$$g_p(L)Z(t) = h_q(L)e(t).$$

10. Long-memory time series models

A short-memory model is unrealistic for many observed time series which are of a type which we call intuitively 'long-memory' time series; examples of such time series are those with trends, or very slowly decaying correlations, or spectral densities with very large dynamic ranges.

A long-memory time series is modeled parametrically by an operator which transforms it to a short-memory time series. Examples of such operators are non-invertible filters (especially differencing $(I-L)^d$ of degree $d > 0.5$), or representations as the sum of a long-memory signal and a short-memory noise.

The non-invertible filter is often chosen to be a difference operator which transforms $Y(t)$ to

$$Y^\sim(t) = Y(t) + A(1)Y(t-1) + \cdots + A(M)Y(t-M);$$

the operator is chosen so that $Y^\sim(t)$ is just barely a short-memory time series. An example of a non-invertible filter often considered in the modeling of monthly economic time series is $Y^\sim(t) = Y(t) - Y(t-12)$.

An example of a representation model is $Y(t) = S(t) + Y^\sim(t)$ where $Y^\sim(t)$ is a short-memory time series and $S(t)$ is a deterministic function of period 12 months which can be represented as a sum of cosine and sine functions of period 12.

The concept of memory type may be best explained in the context of ARIMA (p, d, q) models. Memory type depends on whether $d > 0$, and whether the transfer functions $g_p(z)$ and $h_q(z)$ have roots close to the unit circle in the complex z-plane.

The orders p and q measure the past lags involved in the model, whereas memory is related to measures of the predictability of future values from past values.

11. Degree of differencing and index of regular variation

When a time series $Y(t)$ is long-memory and can be transformed to a short-memory time series $Z(t)$ (with spectral density $f_Z(\omega)$) by differencing d times, one can think of $Y(t)$ as having a spectral density $f_Y(\omega)$ with representation

$$f_Y(\omega) = |1 - \exp(-2\pi i \omega)|^{-2d} f_Z(\omega).$$

The spectral density $f_Y(\omega)$ is regularly varying at $\omega = 0$ with index $\delta = 2d$. Estimators for δ can provide techniques for estimating d.

Techniques for estimating d have been proposed by Granger and Joyeux (1980), Hosking (1981), Janacek (1982), and Geweke and Porter-Hudak (1983).

12. Zeros in the spectrum and overdifferencing

A zero in the spectral density $f(\omega)$ at a frequency yields a negative index of regular variation there, and consequently classifies the time series as long memory.

The time series $Y(t) = e(t) - e(t - 12)$ obtained by twelfth-differencing white noise $e(t)$ has zeros at frequencies $\omega = 0$ and $\omega = \frac{1}{12}$, and index $\delta = -2$ (for example, at $\omega = 0$, $f(\omega) \sim \omega^2$). If one simulates this time series, and computes the delta estimating sequences δ_k, one learns from the sample that δ is significantly negative.

This technique provides a means of determining how much to difference. Let $Y(t)$ be the celebrated log international airlines passengers time series to which Box and Jenkins (1970) fitted the 'airline model' which takes first and twelfth differences. Parzen (1982) proposes that twelfth-difference suffices as an operation which transforms the original series (which is long-memory) to a new series which is just barely short-memory. Taking first-difference in addition does not lead in this case to an incorrect model. But taking first-difference in addition to twelfth-difference is not in my judgement justified by diagnostic procedures which emphasize the criterion of memory type.

13. Memory as a criterion for model fit

A basic problem of empirical time series analysis is to devise diagnostic tools for identifying the memory of a time series. We propose that a model fitted to an observed time series should have memory characteristics which agree with those estimated from the data by suitable diagnostic tools. Diagnostic tools should be used to recommend 'best' and 'second best' fitting models, rather than a unique model, since the ultimate criterion of fit of a time series model is its scientific (explanatory) goodness of fit in addition to its statistical (curve-fitting) goodness of fit.

14. Hurst exponents and long-memory models

An alternative to spectral analysis for studying dependence in time series is R/S (rescaled adjusted range) analysis (Mandelbrot (1973)). For successive sample sizes T one forms the sample mean μ_T, the sample standard deviation σ_T, and the normalized time series $Y_T(t) = \{Y(t) - \mu_T\}/\sigma_T$, which we call the mean-standard deviation normalization of the time series. Define the cumulative sums $S_T(n) = \sum_{t=1}^{n} Y_T(t)$, $n = 1, \cdots, T$, and their range

$$R_T = \max_{n=1,\cdots,T} S_T(n) - \min_{n=1,\cdots,T} S_T(n).$$

If one plots $\log R_T$ against $\log T$, the slope H of a line fitted to the scatter diagram is called the Hurst exponent. More rigorously the time series is said to have a Hurst exponent H if, as $T \to \infty$, $\log R_T - H \log T$ converges in distribution to a random variable (see Bhattacharya et al. (1983)). Probability theorists have shown that $H = 0.5$ for many models (for example, when $Y(t)$ are independent with finite second moments, or obeying a stable law). An important problem is to determine models that might explain observed values $H > 0.5$. Mandelbrot and Wallis (1968) and Mandelbrot and Taqqu (1979) use stationary time series with extremely long-range dependence. Bhattacharya et al. (1983) show that one can use models which combine trends and weak dependence. The notion of long-memory models is introduced to provide an intuitive framework for relating the facts that both kinds of models have Hurst exponent $H > 0.5$.

15. Quantile spectral analysis

Quantile spectral analysis is a name for the combination of quantile data analysis and spectral analysis techniques to identify models for observed time series.

Quantile spectral analysis is also a name for the study of analogies and isomorphisms between spectral density functions and quantile density functions.

16. Identification-quantile function

An exploratory approach to testing distributional hypotheses such as normality or exponentiality is provided by forming a fully non-parametric estimator of the standardized quantile function $QI(u)$ defined by

$$QI(u) = Q(u) - Q(0.5)/2(Q(0.75) - Q(0.25)).$$

We call $QI(u)$ an identification-quantile function; it equals 0 at $u = 0.5$ and its slope at $u = 0.5$ is approximately equal to 1. It is an easily estimated approximation to the unit-quantile function $Q1(u)$ defined by

$$Q1(u) = \{Q(u) - Q(0.5)\}/q(0.5).$$

The corresponding density function $f1(x)$ is normalized so that the median $= 0$ and $f1(0) = 0$. The unit-density function $f1(x)$ is: $\exp(-\pi x^2)$ for normal distributions; $\exp(-2x)$, $x \geqq -0.5 \log 2$, for exponential distributions.

17. Identification-quantile diagnostics of probability distributions

One can show that the graph of $QI(u)$ provides quick graphical diagnostics of the probability distribution types that might fit a data batch. The tail values of $QI(u)$ at $u = 0.01$, 0.05, 0.10, 0.90, 0.95, 0.99 provide quick arithmetical diagnostics of the tail exponents of the probability distributions (defined in Section 5).

Distributions with a long tail (exponents $\geqq 1.5$) do not have finite second moments; one must then decide if the data needs to be transformed by taking its logarithm, square root, cube root, etc.

18. Identification-quantile normalization

One can always write a time series $Y(\cdot)$ as a sum of a mean $\mu(t) = E[Y(t)]$ and a fluctuation $Z(t) = Y(t) - \mu(t)$; thus, $Y(t) = \mu(t) + Z(t)$. We call $Y(t)$ stationary if $\mu(t) = \mu$, a constant, and $Z(t)$ is a zero-mean stationary time series with variance σ^2 and correlation function $\rho(v)$. If $\rho(v)$ decays slowly then $Z(\cdot)$ is long-memory and $Y(\cdot)$ is long-memory. But even if $Z(\cdot)$ is short-memory we consider $Y(\cdot)$ to be long-memory if μ/σ is large.

One way to transform $Y(\cdot)$ to a short-memory time series is to form $Y(t) - \bar{Y}$ where $\bar{Y}$ is the sample mean. However, this transformation introduces a zero (at $\omega = 0$) in the sample spectral density. One might consider instead forming $Y(t) - Q^\sim(0.5)$, where $Q^\sim(0.5)$ is the sample median of $Y(\cdot)$.

After computing the sample quantile function $Q^\sim(u)$ of the time series sample, we transform $Y(t)$ to $Y_T(t) = \{Y(t) - Q^\sim(0.5)\}/2\{Q^\sim(0.75) - Q^\sim(0.25)\}$, which we call the identification-quantile normalization of the time series.

19. Quantile tests of location-scale parameter models

A location-scale parameter model $F(x) = F_0((x - \mu)/\sigma)$ in the quantile domain assumes the form $Q(u) = \mu + \sigma Q_0(u)$, where $Q_0(u)$ is the quantile function corresponding to $F_0(x)$. Normal and exponential distributions are location-scale parameter families with

$$F_0(x) = \Phi(x), \qquad \Phi(x) = \int_{-\infty}^{x} \phi(x')dx', \qquad Q_0(u) = \Phi^{-1}(u);$$

$$F_0(x) = 1 - \exp(-x), \qquad Q_0(u) = -\log(1-u), \qquad f_0 Q_0(u) = 1 - u.$$

To test the hypothesis $H_0 : Q(u) = \mu + \sigma Q_0(u)$, one could test $H_0 : q(u) = \sigma q_0(u)$, or equivalently $H_0 : f_0 Q_0(u) q(u) = \text{constant}$. To test this hypothesis one forms estimators of the comparison distribution function

$$D(u) = \int_0^u d(u')du', \qquad 0 \le u \le 1,$$

$$d(u) = f_0 Q_0(u) q(u)/\sigma_0, \qquad \sigma_0 = \int_0^1 f_0 Q_0(t) q(t) dt.$$

One tests $H_0 : D(u) = u$, $0 \le u \le 1$. A test statistic $D^{\sim}(u)$ is formed by replacing in the foregoing formulas $q(u)$ by

$$q^{\sim}(u) = \{Q^{\sim}(u+h) - Q^{\sim}(u-h)\}/2h \quad \text{for suitable } h.$$

20. Cumulative weighted spacings function $D^{\sim}(u)$

Usually $Q^{\sim}(u)$ is defined as a piecewise linear function joining the values $Q^{\sim}(j/(n+1)) = X[j]$, where $X[j]$ denotes the jth order statistic when $X(1), \cdots, X(n)$ is written in increasing order as $X[1] < X[2] < \cdots < X[n]$. Then we often define

$$q^{\sim}(j/(n+1)) = \{X[j+1] - X[j-1]\}(n+1)/2.$$

We call $q^{\sim}(u)$ spacings, and $D^{\sim}(u)$ a sample comparison distribution function (or cumulative weighted spacings function).

The quantile approach is particularly powerful for detecting that the data obeys a specified distribution except for the presence of some outliers (extremely large values). Then $D^{\sim}(u)$ appears to be linear up to the value of u corresponding to the percentage of values coming from the specified distribution!

The cumulative weighted spacings function $D^{\sim}(u)$ can provide rigorous (confirmatory) procedures to test distributional hypotheses such as exponentiality or normality. Usually the graph of $D^{\sim}(u)$ readily indicates whether it deviates significantly from the uniform distribution $D(u) = u$.

21. Quantile spectral analysis test of white noise

If $Y(t)$, $t = 0, \pm 1, \pm 2, \cdots$ is a zero-mean stationary time series with short memory, the sample spectral density $f^{\sim}(\omega)$ is not a consistent estimator of

$f(\omega)$, but (for $0<\omega<0.5$) $f^{\sim}(\omega)$ does converge in distribution to an exponential distribution with mean $f(\omega)$: for any $x>0$, $\Pr[f^{\sim}(\omega)>x]\to\exp\{-x/f(\omega)\}$ as $T\to\infty$.

Let the collection $\{f^{\sim}(k/\text{NFREQ}), k=1,\cdots,\text{NFREQ}/2\}$ be denoted by $\{f^{\sim}\}$. We propose to analyse its properties as a data batch, and in particular to form its sample quantile function $Q^{\sim}\{f^{\sim}\}(u)$ and sample distribution function $F^{\sim}\{f^{\sim}\}(x)$. One can show that $F^{\sim}\{f^{\sim}\}(x)$ converges in probability to a limit distribution function denoted $F\%\{f^{\sim}\}$. Further,

$$1-F\%\{f^{\sim}\}(x)=\int_0^{0.5}\exp\{-x/f(\omega)\}d\omega.$$

For a no-memory time series, $F\%\{f^{\sim}\}(x)$ is an exponential distribution. Therefore we propose that to test the hypothesis H_0 that a time series is of no-memory type, we test by quantile spectral analysis the following hypothesis: the sample spectral density data batch $\{f^{\sim}\}$ has a sample quantile function $Q^{\sim}\{f^{\sim}\}$ which can be fitted by an exponential distribution.

It is important to be aware that the sample distribution of the sample spectral density is not asymptotically exponential when the time series does not have finite second moments (see Freedman and Lane (1981) and Rosenblatt (1981)). Therefore we should first determine that the probability distribution of the time series does not have long tails.

22. Function smoothing criterion for model fit

Let
$$d^{\sim}(\omega)=f^{\sim}(\omega)/f_{p,q}(\omega).$$

We can interpret $d^{\sim}(\omega)$ to be an estimator of the sample spectral density $f_e^{\sim}(\omega)$ of the residual time series $e(t)$. A model is said to provide a parsimonious fit if $d^{\sim}(\omega)$ behaves like a sample spectral density which just barely can be smoothed by a function $d^{\wedge}(u)$ which is a constant. In quantile data analysis we use a similar criterion of model fitting; there is a sample quantile density function $q^{\sim}(u)$ and we seek smooth quantile density functions $q^{\wedge}(u)$ such that $d^{\sim}(u)=q^{\sim}(u)/q^{\wedge}(u)$ is just barely fitted by a smooth function $d^{\wedge}(u)$ which is constant. The quantile spectral approach to this smoothing problem is to test how well the sample quantile function of the data batch $\{d^{\sim}\}$ can be fitted by an exponential distribution. Equivalently, we can test how well $\{\log d^{\sim}\}$ can be fitted by an extreme-value distribution.

23. Explicit formula for index of regular variation

The index δ of regular variation of a spectral density $f(\omega)$ at $\omega=0$ can be computed by

$$\delta=\lim_{\omega\to0}\int_0^1\{\log f(y\omega)-\log f(\omega)\}dy.$$

To estimate index δ at a frequency ω_0 from a consistent estimator $f^{\wedge}(k/n)$ of the spectral density at a grid of equi-spaced frequencies, we choose m so that $m/n = \omega_0$ and form a sequence

$$\delta_k = \frac{1}{k} \sum_{j=1}^{k} \log f^{\wedge}\left(\frac{j+m}{n}\right) - \log f^{\wedge}\left(\frac{k+1+m}{n}\right).$$

The convergence properties of the sequence δ_k remain to be investigated. Experience with them on practical examples indicate that they provide useful and interpretable diagnostic tools. However, the convergence of δ_k to the true value of δ is very slow, and we usually use the shape of the curve of δ_k (as a function of k) rather than any of its individual values as the basis for our diagnosis.

24. Summary of quantile spectral analysis model identification procedures

Given a sample $Y(t)$, $t = 1, \cdots, T$, to identify models for the time series we use the following procedures (embodied in our time series analysis computer program ARSPIQ).

(I) To identify heuristically probability distributions that describe the data batch of time series observations, we compute its sample identification-quantile function, and its comparison distribution functions for testing the univariate probability distribution for normality and exponentiality.

(II) To identify heuristically the memory type of the time series we compute the sample spectral density and describe its properties as a data batch by computing the comparison distribution function $D^{\sim}(u)$ which tests for exponentiality. Memory type can be inferred from the shape of $D^{\sim}(u)$. If $D^{\sim}(u)$ fits $D(u) = u$, the series is no-memory (white noise).

(III) When the time series is identified as short-memory it is modelled as an ARMA scheme. To identify heuristically the orders p and q, and significant lags, of the finite-parameter ARMA scheme, one approach is subset regression based on an MA(∞) representation. The coefficients of the MA(∞) representation are computed from an approximating AR scheme, or from sample cepstral correlations (Fourier coefficients of $\log f(\omega)$). A quantile spectral analysis criterion is stated (Section 22) for an ARMA scheme to provide a parsimonious model.

(IV) When the time series is identified as long-memory, insight into the appropriate model for transforming the long-memory time series to a short-memory time series can be obtained by computing estimators δ_k of the spectral index of regular variation at $\omega = 0$ and other specified frequencies. For this purpose one needs to form spectral density estimators. One can use kernel estimators which smooth the sample spectral density and two kinds of AR spectral estimators (depending on whether the AR coefficients used to form AR spectral estimators are computed by the Yule–Walker equations or by least-squares Burg-type algorithms). The order of the AR schemes is determined by AR order-determining criterion functions (such as AIC and CAT).

The approach to time series model identification outlined in this paper can be considered exploratory data analysis. The ultimate validity of this approach can be decided only after additional theoretical refinement of its criteria, and the demonstration of its success in important practical applications.

References

AKAIKE, H. (1977) On entropy maximization principle. In *Applications of Statistics*, ed. P. R. Krishnaiah, North Holland, Amsterdam, 27–41.

BHATTACHARYA, R. N., GUPTA, V. K. AND WAYMIRE, E. (1983) The Hurst effect under trends. *J. Appl. Prob.* **20**, 649–662.

BOX, G. E. P. AND JENKINS, G. M. (1970) *Time Series Analysis, Forecasting, and Control.* Holden Day, San Francisco.

COX, D. R. (1984) Long-range dependence: a review. In *Statistics: An Appraisal*, ed. H. A. David and H. T. David, Iowa State University Press, Ames, 55–74.

FREEDMAN, L. AND LANE, D. (1981) The empirical distribution of the Fourier coefficients of a sequence of independent, identically distributed long-tailed random variables. *Z. Wahrscheinlichkeitsth.* **55**, 21–37.

GEWEKE, J. AND PORTER-HUDAK, S. (1983) The estimation and application of long memory time series models. *J. Time Series Analysis.* **4**, 221–238.

GRANGER, C. W. G. AND JOYEUX, R. (1980) An introduction to long memory time series models and fractional differencing. *J. Time Series Analysis* **1**, 15–29.

HANNAN, E. J. (1970) *Multiple Time Series.* Wiley, New York.

HANNAN, E. J. AND KANTER, M. (1977) Autoregressive processes with infinite variance. *J. Appl. Prob.* **14**, 411–415.

HOSKING, J. R. M. (1981) Fractional differencing. *Biometrika* **68**, 165–176.

JANACEK, G. J. (1982) Determining the degree of differencing for time series via the log spectrum. *J. Time Series Analysis* **3**, 177–183.

MANDELBROT, B. (1973) Statistical methodology for nonperiodic cycles: from the covariance to R/S analysis. *Rev. Econom. Social Measurement*, 259–290.

MANDELBROT, B. (1982) *The Fractal Geometry of Nature.* Freeman, San Francisco.

MANDELBROT, B. AND TAQQU, M. (1979) Robust R/S analysis of long run serial correlation. *42nd Internat. Statist. Inst., Manila*, 1–38.

MANDELBROT, B. AND WALLIS, J. (1968) Noah, Joseph and operational hydrology. *Water Resources Res.* **4**, 909–918.

PARZEN, E. (1979) Nonparametric statistical data modeling (with discussion), *J. Amer. Statist. Assoc.* **74**, 105–131.

PARZEN, E. (1981) Time series model identification and prediction variance horizon. In *Applied Time Series Analysis II*, ed. David F. Findley, Academic Press, New York, 415–447.

PARZEN, E. (1982) ARARMA models for time series analysis and forecasting. *J. Forecasting* **1**, 67–82.

PARZEN, E. (1982) Maximum entropy interpretation of autoregressive spectral densities. *Statist. Prob. Letters*, **1**, 2–6.

ROSENBLATT, M. (1981) Limit theorems for Fourier transforms of functionals of Gaussian sequences. *Z. Wahrscheinlichkeitsth.* **55**, 123–132.

Order Estimation by Accumulated Prediction Errors

JORMA RISSANEN

Abstract

This paper presents a new criterion based on prediction error which allows the estimation of the number of parameters as well as structures in statistical models. The criterion is valid for short and long samples alike. Unlike Akaike's earlier criterion, also based on prediction error, the criterion proposed here appears to produce consistent error estimates in ARMA processes.

ARMA PROCESSES; MINIMUM DESCRIPTION LENGTH; MODEL COST; PARAMETERS; CONSISTENCY

1. Introduction

Based upon the reasoning that since the ultimate use of most models of time series is to provide predictions, there has long been a desire to base the entire estimation procedure on minimization of prediction errors. Indeed, such estimators have been shown to possess desirable properties, and, in fact, in the Gaussian ARMA processes they are comparable with the ML-estimators (Ljung and Caines (1979)). Moreover, as shown by Davisson (1965) and Akaike (1974), if one includes in the prediction error the effect of the estimation errors in the parameters, a criterion results which automatically penalizes the number of parameters in the model. This is an important innovation, for in the past a separate hypothesis testing was required to estimate the number of the parameters.

A meticulous analysis, above all by Shibata (1976), (1980), has revealed that the order estimates minimizing Akaike's AIC criterion, which in the Gaussian ARMA processes is quite equivalent to Davisson's prediction error criterion, have interesting and useful properties, except that they are not consistent even in the case of AR processes. Although one may argue that consistency in itself is not all that important, in particular if the 'true' data-generating system is infinite-dimensional, this author nevertheless feels that the AIC criterion fails an analyzable test of performance. And this, of course, does nothing to increase one's confidence in it when it is to be applied to non-analyzable cases. We do

55

not, incidentally, accept the premise that any physical system is infinite-dimensional. In fact, we regard the finite amount of observed data to *be* the 'true' system, and together with possible prior information the data are all we have about the process; any other 'system' explanation we invent is simply a model, which may well be infinite-dimensional, and as such it will have to compete with other models we care to consider.

Prompted by this shortcoming of the AIC criterion, we proposed in Rissanen (1978), (1983a), (1985) an altogether different principle, the so-called MDL principle (MDL for minimum description length), based on a purely information-theoretic idea: pick the parameters so that the model they define permits a redescription of the observed sequence with the smallest number of binary digits. This principle indeed has been shown to produce consistent order estimates in ARMA processes (Hannan (1980)). But despite the soundness and success of the MDL principle, the failure of such an intuitively attractive principle as one based on prediction errors remained in this author's mind as a puzzling issue, so much so that he suspected that the prediction errors are not fairly and properly represented in the above-mentioned attempts. This, indeed, appears to be the case, and when corrected we do arrive at a criterion based on prediction error, which is not only valid asymptotically like the previous ones but which is also perfectly justified even for short samples. This criterion appears to be asymptotically equivalent to the MDL criterion for Gaussian ARMA processes, and hence their estimates should be consistent.

2. Accumulated prediction error criterion

We consider an observed sample $x = x_1, \cdots, x_n$, where the numbers x_i are delivered to us one after another so that at every time instant $t = 0, 1, \cdots, n-1$ we are given the past sequence $x^t = x_0 x_1 \cdots x_t$. Here, x_0 is a constant, say 0, representing the string of no observations. Suppose now that at each t we are to make a prediction of the next value x_{t+1}, based upon the sequence x^t so far seen. How should we form a measure of the prediction errors? It seems quite natural to define the following accumulated measure:

$$(2.1) \qquad V(k, x) = n^{-1} \sum_0^{n-1} (x_{t+1} - \hat{x}_{t+1})^2,$$

where $\hat{x}_{t+1} = f(x^t, \theta(x^t))$ denotes the prediction made at time t based upon the past sequence x^t with use of parameters estimated in some way, collected in the k-component vector $\theta(x^t)$, which also must depend only on the past observations. Applying the sensible reasoning that we should act on the principle that has worked best in the past (indeed, we cannot think of a better principle for statistical inference!), these estimates should clearly be determined by minimization of the summed past prediction errors. The number of parameters k is then determined so that $V(k, x)$ is minimized.

We illustrate the use of this estimation principle in the case of ARMA models. Suppose that we are willing to model the data as being a sample from some stationary zero-mean one-sided moving-average ARMA (p, q) process:

$$(2.2) \qquad x_t + a_1 x_{t-1} + \cdots + a_p x_{t-p} = e_t + b_1 e_{t-1} + \cdots + b_{t-q} e_{t-q}$$

where $\{e_t\}$ is an uncorrelated, zero-mean process such that the sequence $\{e_t, e_{t-1}, \cdots\}$ spans the same linear space as $\{x_t, x_{t-1}, \cdots\}$. The process x is then defined up to the second moments by the $p+q$ parameters $\theta = (a_1, \cdots, a_p, b_1, \cdots, b_q)$ and the variance σ^2 of the process e_t.

In view of the class of models selected we consider a linear predictor with the prediction error $\varepsilon_{t+1} = x_{t+1} - \hat{x}_{t+1}$ given by $\varepsilon_{t+1} = e_{t+1,t}$, which is determined by $p+q$ parameters $\theta(t) = (a_{1,t}, \cdots, a_{p,t}, b_{1,t}, \cdots, b_{q,t})$ and the data as follows: for $i = 0, \cdots, t+1$, put

$$(2.3) \qquad e_{i,t} + b_{1,t} e_{i-1,t} + \cdots + b_{q,t} e_{i-q,t} = x_i + a_{1,t} x_{i-1} + \cdots + a_{p,t} x_{i-p},$$

where $x_i = e_{i,t} = 0$ for $i \leq 0$. The parameter vector $\theta(t)$, in turn, should be determined so that the criterion

$$(2.4) \qquad s^2(t) = t^{-1} \sum_0^{t-1} e_{i+1,t}^2,$$

is minimized. Finally, the numbers p and q are determined so that the accumulated prediction errors (2.1) are minimized.

An outstanding feature of the criterion (2.1) is that it uses the same given set of observations both as the basis for estimation and as a test of the validity of the estimates. We may compare this with Akaike's criterion AIC, which can also be interpreted in terms of prediction errors. It follows from a result in Davisson (1965), when specialized to a stationary AR (p^0)-process (see also Fuller and Hasza (1981)) that

$$(2.5) \qquad E_{\theta^0}(x_{n+1} - \hat{x}_{n+1})^2 = \sigma^2(1 + p/n) + o(n^{-1})$$

where $\hat{x}_{n+1}$ denotes the predictor obtained with the least squares estimator $\theta(x^n)$ having $p, p \geq p^0$, components. In other words,

$$\hat{x}_{n+1} = -\hat{a}_1 x_n - \cdots - \hat{a}_p x_{n-p+1},$$

where $\hat{a}_i$ are determined from the observed sequence x^n by minimization of the error squares (2.4) for $t = n$. Further, σ^2 denotes the variance of the stationary process $\{e_t\}$ which with the p^0-component θ^0 determines the process x_t. We thus see that the more parameters (above the 'true' number p^0) we pick in the model, the greater the mean prediction error (2.5). The next step is to replace the variance σ^2 by the estimate, the minimized error squares $s^2(n)$ in (2.4). This estimator has a bias, which is asymptotically given by $-\sigma^2 p/n$. By correcting this and substituting the result in (2.5), we get Akaike's AIC criterion after taking the natural logarithm:

$$(2.6) \qquad\qquad 2 \ln s(n) + 2p/n.$$

It seems to us that both applications of the prediction error principle are meaningful, although in (2.6) no specific samples are used as a means of validating the estimates. Instead, the asymptotic mean is taken to provide a sort of validation. However, something appears to be lost in such a substitute validation procedure, which becomes evident in the lack of consistency of the resulting order estimates. Our criterion (2.1), instead, forces a validation after each observation is received, which leads to a greater penalty on the number of parameters used. That this, in turn, should produce consistent order estimates can be seen from the asymptotic analysis carried out in the next section, but its plausibility is easy to see intuitively. It certainly seems reasonable to expect that a model's predictive capability cannot be improved by estimating excessive unnecessary parameters, while an improvement does result if a new relevant parameter is added to the model. Hence, the best predictions are obtained when the model has as many parameters as the—this time imaginary—data-generating system.

In Stone (1977) and Geisser and Eddy (1979) another predictive approach to model selection was described, which also uses the common batch of data both for estimation and validation. However, that approach is not 'honestly' predictive in the same sense as ours, and, in fact, Stone in the case with independence shows that the resulting criterion is asymptotically equivalent to Akaike's criterion. Hence, cross-validation in itself does not seem to guarantee consistency, and indeed why should it?

3. Asymptotic properties

In order to analyze further the proposed criterion and its estimators, suppose that the strings x are generated by a process in the class of Gaussian ARMA processes. We wish to find out how small the mean of the accumulated prediction error criterion (2.1) then can be made. In Rissanen (1984) we proved an asymptotic result, which states the following.

No matter how many parameters we estimate, and no matter how we estimate them, the inequality

$$(3.1) \qquad\qquad E_{\theta^0} V(k, x) \geqq \sigma^2 [1 + ((k^0 - \varepsilon)/n) \ln n],$$

holds for all positive numbers ε and all 'true' parameters θ^0, defining stationary one-sided moving-average processes, except some in a set $A_\varepsilon(n)$ whose volume goes to 0 as $n \to \infty$.

The issue remains whether the right-hand-side bound, this time with $\varepsilon < 0$, can be reached by some estimator, for example, by the ML estimator. It is readily seen from (2.5) that this is true at least for the AR processes. Indeed, if we add the mean prediction errors in (2.5), with n replaced by t, from $t = 0$ to $t = n - 1$ in accordance with (2.1), we get the sum of a harmonic series, and

$$(3.2) \qquad E_{\theta^0} V(p^0, x) = \sigma^2 [\,+ (p^0/n) \ln n\,] + o(n^{-1} \ln n).$$

In other words, the right-hand-side bound in (3.1) with $\varepsilon = 0$ and $k^0 = p^0$, corresponding to the AR processes, is attained asymptotically.

These asymptotic results provide an interesting connection with the MDL criterion (Rissanen (1978), (1984)) which, when expressed in terms of the natural logarithmic unit in the AR case, is as follows:

$$(3.3) \qquad \ln s(n) + \tfrac{1}{2}((p+1)/n) \ln n.$$

If we replace in (3.2) σ^2 by its bias-corrected estimate, as above following (2.5), we get the criterion

$$(3.4) \qquad 2 \ln s(n) + (p/n) \ln n,$$

which differs from (3.3) in having one fewer parameter. The extra parameter in (3.3) is σ^2, whose best estimate is $s(n)$ no matter how the other p parameters are estimated. Hence, for estimating these the two criteria are equivalent. We conjecture that the difference between $V(p, x)$ and (3.4), multiplied by $n(\ln n)^{-1}$, tends to 0. If this is true, then because the order estimates obtained by minimization of (3.3) have been shown to be consistent (Hannan and Quinn (1979), Rissanen (1980)), so should be the estimates obtained by minimization of the accumulated prediction errors (2.1).

4. Simulations

We illustrate the use of the criterion (2.1) by applying it to a sequence of observations generated by a Gaussian ARMA system. We fitted models of type ARMA (p, q) with $(p, q) = (1, 0)$, $(2, 0)$, $(1, 1)$, and $(0, 2)$. Table 1 gives the minimized criterion $V(k, x^n)$ for five different values of n along the single sample of size 600. If we add that models $(2, 2)$ and $(0, 1)$ gave uniformly worse values than the two best models $(1, 0)$ and $(1, 1)$ in the table (we did not calculate the last entry for the two worst models), the reader can conclude that a system $(1, 1)$ was the one that generated the data. Notice, however, that up to the sample size 200 the simpler first-order AR model $(1, 0)$ performed better than the eventual winner $(1, 1)$.

TABLE 1
Minimized criterion values for four models

		Length n				
		50	100	200	300	600
	(1, 0)	1.336	1.276	1.101	1.107	1.015
Model	(2, 0)	1.629	1.385	1.156	1.120	—
(p, q)	(1, 1)	1.505	1.307	1.117	1.096	0.996
	(0, 2)	1.925	1.520	1.221	1.159	—

The data-generating system had the parameters $a_1 = 0.5$ and $b_1 = -0.3$. The 300-sample estimate of the (1, 0)-model was $\hat{a}_1 = 0.59$ while the same estimates for the two parameters in the (1, 1)-model were $\hat{a}_1 = 0.335$ and $\hat{b}_1 = -0.405$. The associated minimized sample variance for the best (1, 1)-model was $s^2(300) = 1.024$, which gives the value $1.024(1 + \log 300)/300 = 1.073$ for the MDL criterion (3.3), where we now used the binary logarithm. This is a little smaller than the table entry 1.096. For the final sample size of 600 we got with the same model $s^2(600) = 0.953$, which gives the value 0.988 for the MDL criterion. This, again, is less than the corresponding table entry 0.996. We conclude that the criterion (2.1) imposes a greater penalty on the system complexity than the MDL criterion. This is particularly noticeable for short samples where the relative model cost is greater. Hence, the criterion (2.1) tends to underestimate the number of parameters, which is, perhaps, just as it ought to be. After all, the 'information' in the data, the word taken in the technical sense as the infinum of the code lengths, cannot be defined without including the estimation of the parameters, and, hence, to achieve the total information only those parameters which 'buy' enough performance should be retained. This means that initially when the sample size is small the optimum model necessarily has only a few parameters and others will be included gradually as more data is received. This is in keeping with our general philosophy that there never is any 'true' system nor a 'true' number of parameters—only an optimum number—and an excessively complex model is bad not only because of practicability reasons in being more difficult and expensive to implement, but because it performs worse.

In conclusion, we point out that the criterion (2.1) ought to give reasonable results even when used to estimate the structure of vector ARMA processes. After all, when a model in a 'bad' structure is selected, the parameters are expressed in a coordinate system with some axes tending to be near parallel, and one may expect large estimation errors and hence large prediction errors. For the estimation of structure with a three-term MDL criterion we refer to Rissanen (1983b).

References

AKAIKE, H. (1974) A new look at the statistical model identification. *IEEE Trans. Automatic Control* **AC-19**, 716–723.

DAVISSON, L. D. (1965) The prediction error of stationary Gaussian time series of unknown covariance. *IEEE Trans. Inf. Theory* **IT-11**, 527–532.

FULLER, W. A. AND HASZA, D. P. (1981) Properties of predictors for autoregressive time series. *J. Amer. Statist. Assoc.* **76**, 155–161.

GEISSER, S. AND EDDY, W. (1979) A predictive approach to model selection. *J. Amer. Statist. Assoc.* **74**, 153–160.

HANNAN, E. J. (1980) The estimation of the order of an ARMA process. *Ann. Statist.* **8**, 1071–1081.

HANNAN, E. J. AND QUINN, B. G. (1979) The determination of the order of an autoregression. *J. R. Statist. Soc.* B **41**, 190–195.

LJUNG, L. AND CAINES, P. (1979) Asymptotic normality of prediction error estimators for approximate system models. *Stochastics* **3**, 29–46.

RISSANEN, J. (1978) Modeling by shortest data description. *Automatica* **14**, 465–471.

RISSANEN, J. (1980) Consistent order estimates of autoregressive processes by shortest description of data. *In Analysis and Optimisation of Stochastic Systems*, ed. O. Jacobs, M. Davis, M. Dempster, C. Harris, P. Parks, Academic Press, New York.

RISSANEN, J. (1983a) A universal prior for integers and estimation by minimum description lengths. *Ann. Statist.* **11**, 416–431.

RISSANEN, J. (1983b) Estimation of structure by minimum description length. *Circuits, Systems, and Signal Processing (Special Issue on Rational Approximations)* **1**, 395–406.

RISSANEN, J. (1984) Universal coding information, prediction, and estimation. *IEEE Trans. Inf. Theory* **IT-30**, 629–636.

RISSANEN, J. (1985) Minimum description length principle. In *Encyclopaedia of Statistical Sciences*, Vol. 5, ed. S. Kotz and N. L. Johnson. Wiley, New York.

SHIBATA, R. (1976) Selection of the order of an autoregressive model by Akaike's information criterion. *Ann. Statist.* **63**, 117–126.

SHIBATA, R. (1980) Asymptotically efficient selection of the order of the model for estimating parameters of a linear process. *Ann. Statist.* **8**, 147–164.

STONE, M. (1977) An asymptotic equivalence of choice of model by cross-validation and Akaike's criterion *J. R. Statist. Soc.* B **39**, 44–47.

Identifiability of Time Series Models with Errors in Variables

VICTOR SOLO

Abstract

Straightforward derivations are provided for some identifiability results for time series models with errors in variables.

Introduction

It is a pleasure to join in this Festschrift for Ted Hannan on the occasion of his sixty-fifth birthday. Ted has shaped present-day time series in many ways. He showed how to formulate problems (in econometrics and delay estimation) for precise asymptotic analysis. He forged many tools for such analysis and perceived so many of the issues that he helped to create an environment where, by now, asymptotic analysis has become almost routine. With all this, though, he has always emphasized the practical aspects of the results: see his books, Hannan (1960), (1970). One of Ted's most impressive contributions to time series analysis is his realization and exploitation of the role of multivariable polynomials in the identifiability problem for multivariate ARMA time series (Hannan (1969), (1979)). Another area that revolves around identifiability is the error in variables problem and Ted was an early worker here too (Hannan (1963)). This essay is concerned with such problems.

The problem

Recently Maravall (1979) has provided a careful exposé of the identifiability of some time series models with errors in variables. The interesting results revealed by Maravall are that often identifiability is not a problem, as it is in the traditional static case. He shows that the identifiability problem differs

---

This work was completed while the author was Visiting Associate Professor in the Department of Statistics, Purdue University.

depending on whether or not the input or exogenous sequence is serially uncorrelated or serially correlated. In any case the results are expressed as simple counting rules.

However, Maravall's argument is very long (developed over 120 pages)—something that he lamented in his preface. In this essay a simple direct development is given of the basic theorems of Maravall. The present discussion clearly reveals the origin of the identifiability conditions. Actually the results obtained include Maravall's in that the exogenous sequence need not be modelled by an ARMA process for some of the results.

Söderström (1980) has independently given some identifiability results for EIV problems, but his model is different from the one used by Maravall. He also models the input sequence as an ARMA process though. Further, Söderström excluded the case that the input sequence is white noise (so carefully discussed by Maravall). Otherwise the two sets of results are more or less the same. The relationship will be discussed below. Recently Anderson and Deistler (1982) have extended Söderström's discussion but again excluded the case of white-noise input.

Section 1 opens the discussion of the single-input–single-output case, pointing out how the white-noise input case is different. In Section 2 the basic results of Maravall are derived in a direct and simple way. In Section 3 Söderström's results are reviewed and extended to cover the case of a white-noise input.

1. Single-input–single-output identifiability with errors in variables: first steps

We begin with a basic model for the observed output (y) and input (z) and unobserved output (Y) and input (Z) as follows:

(1a)
$$y_t = Y_t + \varepsilon_{yt}$$

(1b)
$$A_y(L)Y_t = B_y(L)Z_t + v_t$$

(1c)
$$z_t = Z_t + \varepsilon_{zt}$$

(1d)
$$A_v(L)v_t = B_v(L)\nu_{vt}$$

where $LY_t = Y_{t-1}$ etc.; $A_w(L) = \sum_0^{p_w} a_{w_i}L^i$ etc.; the ε's and ν are white noises and uncorrelated among each other; $A_{y0} = 1 = A_{v0} = B_{v0}$. Also we denote $(a_{y1} \cdots a_{yp_y})'$ by $\boldsymbol{a}_y$ etc.; $B_w(L) = \sum_0^{q_w} b_{w_i}L^i$.

(1e)   All roots of $Z^p A_y(Z^{-1}) = 0$ are <1 in modulus

(1f) Z_t is stationary

(1g) $Z^p A_y(Z^{-1})$ and $Z^p B_y(Z^{-1})$ have no common factors.

To begin a study of the identifiability of model 1 (i.e. Equations (1)) we naturally calculate the following covariances:

(2a) $\qquad\qquad E(y_0 y_\tau) = \gamma_y(\tau) = \gamma_Y(\tau) \qquad \tau \neq 0$

(2b) $\qquad\qquad E(z_0 z_\tau) = \gamma_z(\tau) = \gamma_z(\tau) \qquad \tau \neq 0$

(2c) $\qquad\qquad E(y_0 z_{-\tau}) = \gamma_{yz}(\tau) = \gamma_{YZ}(\tau) \qquad \text{all } \tau.$

The step that now follows naturally is to seek the parameters $\boldsymbol{a}_y, \boldsymbol{b}_y$ by taking cross-covariances in (1b) with $Z_{t+\tau}$ to find

(3a) $\qquad\qquad A_y(L)\gamma_{YZ}(-\tau) = B_y(L)\gamma_Z(-\tau) \quad \text{all } \tau.$

Now a problem is immediately apparent. Suppose Z_t is a white noise then clearly for $-\tau > q_y = \deg B_y(L)$

$$\gamma_{\tilde{Y}Z}(-\tau) = A_y(L)\gamma_{YZ}(-\tau) = 0$$

and we can find $\boldsymbol{a}_y$ by assembling and solving these equations in matrix form. However, we have trouble in finding $\boldsymbol{b}_y$ for the only equations apparently available are for $-\tau = 0, +1, \cdots, q_y$, namely

(4) $\qquad\qquad \gamma_{\tilde{Y}Z}(-\tau) = b_{y\tau}\gamma_Z(0)$

so that only $B_{y\tau} = b_{y\tau}\gamma_Z(0)$ can be found. So we are missing a scale factor.

For this reason it is clear (as Maravall found) that we need two types of result: one when Z_t is white, the other when it is not.

Before continuing the discussion an important simplification with regard to the identifiability of ARMA models is made. In the appendix it is shown that the $p+q+1$ parameters $\boldsymbol{a}, \boldsymbol{b}, \sigma^2$ of an ARMA (p, q) model such as (1d) are equivalent to the $r+1 = p+q+1$ autocovariances $\gamma_0, \gamma, \cdots, \gamma_r$ (i.e. each set may be obtained from the other). Thus identifiability of an ARMA model is established if these autocovariances are found. This equivalence is more than just a theoretical point. The author has recently discussed algorithms for constructing the exact likelihood in both the scalar and multivariate cases parameterised by autocovariances: see Solo (1982), (1983a).

Now let us observe from (2b) that $\gamma_Z(1) \cdots \gamma_Z(r)$ are available for any r. So that the whole covariance sequence of Z_t is available once $\gamma_Z(0)$ is found. Then, if appropriate, an ARMA model may be constructed for Z_t (as observed in the appendix this includes the orders p_Z, q_Z). For these reasons we are henceforth only concerned with the identifiability of $\gamma_Z(0)$.

So far as v_t is concerned we similarly need only show how $\gamma_v(0)$, $\gamma_v(1), \cdots, \gamma_v(r_v)$ may be found. It will still prove convenient, though, to state results in terms of $\boldsymbol{a}_v, \boldsymbol{b}_v, \sigma^2_{vv}$.

The following results will be established in the next section. For other results see Section 3.

Result 1C. If Z_t is serially correlated then

$$[\boldsymbol{a}_y : \boldsymbol{b}_y], [\gamma_Z(0), \sigma_{\varepsilon Z}^2], \boldsymbol{a}_v \text{ are always identified.}$$

Result 1C'. If Z_t is serially correlated then

$$[\boldsymbol{b}_v, \sigma_{vv}^2], \sigma_{\varepsilon y}^2 \text{ are also identified iff } p_y \geqq \max(0, q_v - p_v + 1).$$

Result 1W. If Z_t is a white noise $\boldsymbol{a}_y, \boldsymbol{a}_v$ are always identified.

Result 1W'. If Z_t is a white noise then

$$[\boldsymbol{b}_y, \sigma_{\varepsilon y}^2], [\gamma_Z(0), \sigma_{\varepsilon Z}^2], [\boldsymbol{b}_v, \sigma_{vv}^2] \text{ are also identified iff}$$

$$\min(p_y, q_y) \geqq \max(0, 1 + q_v - p_v)$$

$$\max(p_y, q_y) \geqq 1 + \max(0, 1 + q_v - p_v).$$

Remark 1. Note in the above results that p_y, q_y, p_v can be obtained from the covariances. However in Results 1C', 1W', q_v must be known.

Remark 2. For our Result 1W' cf. Maravall's Theorem 4; for 1C' cf. his Theorem 7. Note though we do not require, as does Maravall, that Z_t be ARMA. Actually Maravall considered multiple but independent inputs and so joined his Theorems 4 and 7 into Theorem 8 (see p. 121 and the table on p. 122). These results are easily established by straightforward extensions of the discussion given below.

Remark 3. The case of correlated inputs is much harder (Maravall only found sufficient conditions) and will be discussed elsewhere.

2. Single-input–single-output identifiability with errors in variables: Details

First we treat the correlated case.

Result 1C. Assemble Equations (3a) for $\tau = 1, 2, \cdots, (r_y + 1)$ in matrix form and solve (via (2b), (2c)) for $(\boldsymbol{a}_y, \boldsymbol{b}_y)$. This is straightforward unless Z_t is MA with $q_Z < q_y$ or ARMA with $p_Z < q_y$ for then the last rows in the matrix are null or linearly dependent on previous ones. We deal with this in a moment.

Next put $\tau = 0$ in (3a) to yield an equation for $\gamma_Z(0)$. In case $b_{y0} = 0$ put $\tau = 1$ etc. Finally from $\gamma_z(0) = \gamma_Z(0) + \sigma_{\varepsilon Z}^2$ we obtain $\sigma_{\varepsilon Z}^2$.

To cover the MA or ARMA case we have only to be a little more judicious. First reduce the ARMA case to the MA. We can (cf. the appendix) find $A_Z(L)$ with $A_Z(L)\gamma_Z(-\tau) = 0$ for $\tau > q_Z$. Thus introducing $\tilde{Z}_t = A_Z(L)Z_t$ we find that (3a) becomes

$$(3b) \qquad\qquad A_y(L)\gamma_{Y\tilde{Z}}(-\tau) = B_y(L)\gamma_{\tilde{Z}}(-\tau)$$

and $\tilde{Z}_t$ is MA (q_Z). We find for $\tau > q_Z$

$$A_y(L)\gamma_{Y\tilde{Z}}(-\tau) = 0$$

which gives (via (2c)) a set of equations for $\boldsymbol{a}_y$. Assumption (1g) ensures $\boldsymbol{a}_y$ is unique. So introduce $\tilde{Y}_t = A_y(L)Y_t$; then (3b) becomes

(3c) $$\gamma_{\tilde{Y}\tilde{Z}}(-\tau) = B_y(L)\gamma_{\tilde{Z}}(-\tau).$$

This yields a set of $q_y + 2q_Z + 1$ equations for the $q_y + 2$ unknowns $\boldsymbol{b}_y$, $\gamma_Z(0)$. By writing out an example or two the reader can see that these equations have a quadrilateral shape. So, by solving from the top down and bottom up, a set of linear equations is obtained for $\boldsymbol{b}_y$, $\gamma_Z(0)$. The argument continues as above for σ_Z^2. Again, (1g) ensures (by contradiction) $\boldsymbol{b}_y$ is unique.

We have $\boldsymbol{a}_v$ left to find. Take covariances in (1b) to see

$$\gamma_v(\tau) = E\{v_0[A_y(L)Y_{-\tau} - B_y(L)Z_{-\tau}]\}$$

(5a) $$= E(v_0 A_y(L)Y_{-\tau})$$

$$= \sum_0^{p_y}\sum_0^{p_y} a_{yi}a_{yj}\gamma_Y(i-j-\tau) - \sum_0^{q_y}\sum_0^{q_y} b_{yi}a_{yj}\gamma_{YZ}(i-j-\tau).$$

This set of equations involves $\gamma_{YZ}(\cdot)$ which is identified and $\gamma_Y(\cdot)$ which is identified except for $\gamma_Y(0)$. However for $\tau > p_y$, $\gamma_Y(0)$ does not appear so we can calculate $\gamma_v(p_y + q_v) \cdots \gamma_v(p_y + r_v)$ and hence (cf. Appendix A) determine the autoregressive (AR) parameters $\boldsymbol{a}_v$. Result 1C is thus established.

Result 1C′. Now to find the other parameters we first reduce v_t to the MA case. Multiply through (1b) by $A_v(L)$ to see

(6) $$A_y'(L)Y_t = B_y'(L)Z_t + v_t'$$

where $A_y'(L) = A_y(L)A_v(L)$; $p_y' = \deg(A_y'(L)) = p_y + p_v$ etc.; v_t' is a MA (q_v) process. Now return to Equation (5a) (with v, p_y, q_y replaced respectively by v_y', p_y', q_y'): call it then (a′).

We need the autocovariance sequence of v_t'. From Equation (6) it is clearly available (via (2a)) once we know $\gamma_Y(0)$.

Now according to assumption $\gamma_{v'}(\tau) = 0$, $\tau > q_v$. In order to determine $\gamma_Y(0)$ from this assumption we must have it appear in (5a′) for a lag $\tau > q_v$. However, it only appears when $\tau \leq p_y'$. We can thus use our assumption to find $\gamma_Y(0)$ if and only if $p_y' > q_v$, i.e. $p_y \geq \max(0, 1 + q_v - p_v)$. Then we find $\gamma_{v'}(\tau)$ by taking covariances in (6) while $\sigma_{\varepsilon y}^2 = \gamma_y(0) - \gamma_Y(0)$. Result 1C′ is thus established.

Now we turn to the white-noise case.

Results 1W, 1W′. Previous argument has established 1W. For 1W′ recall that from (3a) we can only find $b_{y\tau}\gamma_Z(0) - \tau = 0, 1, \cdots, q_y$ and $\gamma_Z(0)$ is still missing. It is Equation (5a) that can rescue us again.

If we substitute (4) into (5a) we see

$$(5b) \qquad \gamma_v(\tau) = \sum_0^{p_y} \sum_0^{p_y} a_{yi} a_{yi} \gamma_Y(i-j-\tau) - \gamma_Z^{-1}(0) \sum_0^{q_y} \gamma_{\bar{Y}Z}(-i) \gamma_{\bar{Y}Z}(i-\tau).$$

Once more we reduce consideration to the case where v_t is MA. Observe that for $\tau > p_y$, $\gamma_Y(0)$ does not appear in (5b) while (in view of (4)) if $\tau > q_y$, $\gamma_Z(0)$ does not appear. Thus for $\tau > \max(p_y, q_y)$ we can directly calculate $\gamma_v(\tau)$ and so find the AR parameters a_v. We then multiply through (1b) by $A_v(L)$ as before to obtain an equation like (6). We can then return to (5b) (now called (5b')) with v, p_y, q_y replaced by v', p'_y, q'_y as before.

According to assumption, $\gamma_{v'}(\tau) = 0$ for $\tau > q_v$. In order to determine $\gamma_Y(0), \gamma_Z(0)$ from this assumption we must have them appear in (5b') for two lags $\tau > q_v$. Now $\gamma_Y(0)$ only appears when $\tau \leq p'_y$ and $\gamma_Z(0)$ only appears when $\tau \geq q'_y$. We then have two possibilities:

(i) $\gamma_Y(0)$ appears in the lag q_v equation (which requires clearly $p'_y > q_v$) and $\gamma_Z(0)$ appears in the lag $(q_v + 1)$ equation (which requires $q'_y > q_v + 1$).

(ii) Vice versa. We can summarize this by saying we need

$$\min(p_y, q_y) \geq \max(0, 1 + q_v - p_v)$$

$$\max(p_y, q_y) \geq 1 + \max(0, 1 + q_v - p_v).$$

Again we find $\sigma^2_{\varepsilon y} = \gamma_y(0) - \gamma_Y(0)$. Result 1W' is thus established. (Since $\gamma_Y(0)$ and $\gamma_Z(0)$ are both available (5b) yields the remaining unknown covariances $\gamma_v(\tau)$.)

3. The transfer function model

The transfer function model used by Söderström is as follows:

$$(7a) \qquad\qquad\qquad Y_t = T(L)Z_t$$

$$(7b) \qquad\qquad\qquad y_t = Y_t + v_t^*$$

$$(1c) \qquad\qquad\qquad z_t = Z_t + \varepsilon_{Zt}$$

$$(7c) \qquad\qquad\qquad A_v^*(L)v_t^* = B_v(L)v_{st}^*$$

where v^* is a white noise; $T(L) = A_y^{-1}(L)B_y(L)$.

$$(1f) \qquad\qquad\qquad Z_t \text{ is stationary}$$

$$(7d) \qquad\qquad\qquad v_t^* \text{ is stationary.}$$

Let us first observe the connexion with the ARMAX model used by Maravall. From (1a) we see

$$Y_t = Y_t^* + A_y^{-1}(L)v_t.$$

Thus we can replace (1a), (1b) by (7a), (7b) where

$$v_t^* = A_y^{-1}(L)v_t + \varepsilon_{yt}.$$

Thus Maravall's scheme is included in Söderström's.

As before we learn from (7) that

(8a) $$\gamma_{Y^*Z}(\tau) = T(L)\gamma_Z(\tau) \qquad \text{all } \tau$$

(8b) $$\gamma_{yz}(\tau) = \gamma_{Y^*}(\tau) \qquad \text{all } \tau$$

(8c) $$\gamma_z(\tau) = \gamma_Z(\tau) \qquad \tau \neq 0.$$

Now in solving (8) for a_y, b_y we meet exactly the same trouble as before if Z_t is a white noise. This point was not discussed by Söderström or Anderson and Deistler.

Note that if Z_t is serially correlated we can as before obtain $[a_y, b_y]$, $[\sigma_{\varepsilon Z}^2, \gamma_Z(0)]$, a_v^*. From (7a) we then get $\gamma_{Y^*}(\tau)$ for all τ and then from (7b) $\gamma_{v^*}(\tau) = \gamma_y(\tau) - \gamma_{Y^*}(\tau)$. Thus we have (cf. Söderström (1980)) the following result.

Result 2C. If Z_t is serially correlated then $[a_y, b_y]$, $[\sigma_{\varepsilon Z}^2, \gamma_z(0)]$, $[a_v^*, b_v^*, \sigma_{vv}^2]$ are identified, i.e. all the parameters are identified.

Now Söderström's results are extended by proving the following.

Result 2W. If Z_t is white noise then

$$[a_y, b_y], [\sigma_{\varepsilon Z}^2, \gamma_Z(0)], [a_v^*, b_v^*, \sigma_{vv}^2] \text{ are identified iff } p_y - q_y < p_v^* - q_v^*.$$

Since Z_t is a white noise we have as before that $\tilde{b}_{y\tau} = b_{y\tau}\gamma_Z(0)$ are identified. We are led then to write (7b) as

(9a) $$y_t' = Y_t' + v_t'$$

(9b) $$\Rightarrow \gamma_t'(\tau) = \gamma_Y'(\tau) + \gamma_v'(\tau)$$

where

(10a) $$y_t' = A(L)y_t = A_v^*(L)A_y(L)y_t$$

(10b) $$Y_t' = \gamma_Z^{-1}(0)A_v^*(L)\tilde{B}_y(L)Y_t$$

(10c) $$v_t' = A_y(L)B_v^*(L)\nu_{vt}$$

so that the dashed quantities are MA processes. Now y_t is an ARMA process in this case (since Z_t is white) so that we can (as in the appendix) identify $A(L)$ and hence (since $A_y(L)$ is identified) $A_v^*(L)$. Thus $\gamma_y'(\tau)$ can be found from $\gamma_y(\tau)$.

The idea is of course to use (9b) to get an equation for $\gamma_Z(0)$. For this we need at least one lag for which $\gamma_Y'(\tau) \neq 0$ yet $\gamma_v'(\tau) = 0$. Now (10b) shows us that

$\gamma_Z(0)$ appears in (9b) for $\tau \leq p_{v\cdot} + q_y$. However, from (10c) we see $\gamma_v'(\tau)$ vanishes for $\tau > p_y + q_v$. We clearly will have an equation for $\gamma_Z(0)$ iff

$$p_{v\cdot} + q_y > p_y + q_v.$$

Then since $\gamma_Z(0)$ is identified, so is $B_y(L)$ and hence via (7a) so is $\gamma_{Y\cdot}(\tau)$. Thus $\gamma_v(\tau)$ is found from (7b). Finally $\sigma_{\varepsilon Z}^2 = \gamma_z(0) - \gamma_Z(0)$. Result 2W is thus verified.

Remark. If Z_t is a white noise and $\gamma_Z(0)$ cannot be identified we still have valuable information. Indeed $T(L)$ is known to within a scale factor $(\gamma_Z(0))$. Further, we can bound $\gamma_Z(0)$ since it is $\leq \gamma_z(0)$. Of course independent knowledge of $\sigma_{\varepsilon Z}^2$ resolves the problem completely. In any case all the phase information (i.e. lagging) is available in $T(L)$: this can of course be extremely useful.

Appendix. Parameterization of ARMA models by autocovariances

It is shown here how the parameters a, b, σ^2 of an ARMA (p, q) model determine and are determined by the $r + 1 = p + q + 1$ autocovariances $\gamma(0), \gamma(1), \cdots, \gamma(r)$. Obtaining the γ's from a, b, σ^2 is simply the question of generating the autocovariances of an ARMA process. The best algorithm for this is due to Hwang (1978) and independently to Wilson (1979). Thus only the reverse transformation is considered here.

Consider the ARMA model $A(L)\omega_t = B(L)\nu_t$ and take cross-covariances with $\omega_{t-\tau}$ for $\tau > q$ to see

$$A(L)\gamma_\tau = 0 \qquad \tau > q$$

and $\gamma_\tau \equiv \gamma_\omega(\tau)$. We can write this in matrix form as

$$-\boldsymbol{H}_p(a_p \cdots a_1)' = (\gamma_{q+1} \cdots \gamma_{r+1})'$$

where $\boldsymbol{H}_p$ is the Hankel matrix (i.e. one whose cross-diagonal entries are equal)

$$\boldsymbol{H}_p = \begin{bmatrix} \gamma_{q-p} & \gamma_{q-p+1} & \cdots & \gamma_q \\ \gamma_{q-p+1} & & & \\ & \cdot & & \\ & \cdot & & \\ & \cdot & & \\ \gamma_q & \cdots \cdots \cdots & & \gamma_{q+p} \end{bmatrix}.$$

From this set of equations a is clearly obtained. Actually p can also be found since it is defined by

$$p \text{ is the smallest index} \ni \text{rank } H_{p+i} = R \quad \forall i$$

(see e.g. Solo (1983a)).

There are several ways to proceed now (cf. Solo (1983b)), but a simple one is to reduce consideration to the MA case. Since by definition $u_t = A(L)\omega_t$ is an MA (q) process we can use $\gamma_0, \gamma_1, \cdots, \gamma_r$ to determine $\gamma_u(0) \cdots \gamma_u(q)$ from the equations.

$$\gamma_u(\tau) = \sum_0^p \sum_0^p a_i a_j \gamma_{i-j+\tau}; \qquad a_0 = 1.$$

Then from the $(q+1)$ MA covariances $\gamma_u(\cdot)$ we can determine σ^2, b by spectral factorization, say for example the iterative procedure of Wilson (1969).

References

ANDERSON, B. D. O. AND DEISTLER, M. (1982) Identifiability in dynamic error in variables models. Technical Report #15, Dept of Systems Engineering, RSSS, The Australian National University.

HANNAN, E. J. (1960) *Time Series Analysis*. Methuen, London.

HANNAN, E. J. (1963) Regression for time series with errors of measurement. *Biometrika* **50**, 293–302.

HANNAN, E. J. (1969) The identification of vector mixed ARMA systems. *Biometrika* **56**, 223–225.

HANNAN, E. J. (1970) *Multiple Time Series*. Wiley, New York.

HANNAN, E. J. (1979) The statistical theory of linear systems. In *Developments in Statistics*, ed. P. R. Krishnaiah, Academic Press, New York, 83–121.

HWANG, (1978) Solutions of complex integrals using the Laurent expansion. *I.E.E.E. Trans. Acoustics Speech and Signal Processing* **26**, 263–266.

MARAVALL, A. (1979) *Identification in Dynamic Shock-Error Models*. Lecture Notes in Economics and Mathematical Systems **165**, Springer-Verlag, Berlin.

SÖDERSTRÖM, T. (1980) Spectral decomposition with application to identification. In *Numerical Techniques for Stochastic Systems*, ed. F. Archetti and M. Cugiani. North-Holland, Amsterdam.

SOLO, V. (1982) ARMA models without MA parameters. Submitted.

SOLO, V. (1983a) The exact likelihood for a multivariate ARMA model. *J. Multivariate Anal.*

SOLO, V. (1983b) Order estimation in time series models by singular value decomposition. Unpublished.

WILSON, G. T. (1969) Factorization of the generating function of a pure MA process. *SIAM J. Numerical Analysis* **6**, 1–11.

WILSON, G. T. (1979) Some efficient computational procedures for high order ARMA models. *J. Statist. Comput. Simul.* **8**, 301–309.

PART 2

ESTIMATION FOR TIME SERIES

The Identification of ARMA Processes

AN HONG-ZHI

CHEN ZHAO-GUO

Abstract

This paper presents a review of recent results for the identification of ARMA processes according to the principles introduced by Akaike, i.e. assuming that the true orders exist and proposing criteria such as AIC and BIC. The development both of these methods and of consistency theory has been led by E. J. Hannan.

AIC; BIC; ESTIMATE OF ARMA ORDER; SUBSET FITTING; STRONG CONSISTENCY

1. Introduction

Consider a stationary process, $y(t)$, generated by

$$(1.1) \qquad \sum_{j=0}^{p} a_j y(t-j) = \sum_{j=0}^{q} b_j \varepsilon(t-j), \qquad a_0 = b_0 = 1$$

$$E\{\varepsilon(t) \mid F_{t-1}\} = 0, \qquad E\{\varepsilon^2(t) \mid F_{t-1}\} = \sigma^2,$$

$$(1.2) \qquad E\varepsilon^4(t) < \infty$$

$$(1.3) \qquad \sum_{j=0}^{p} a_j Z^j \neq 0, \qquad \sum_{j=0}^{q} b_j Z^j \neq 0, \qquad |Z| \leq 1,$$

where F_t is the σ-algebra generated by $\varepsilon(j)$, $j \leq t$. It is also assumed that $\sum a_j Z^j$ and $\sum b_j Z^j$ have no common zero. These conditions ensure that $y(t)$ may be represented as

$$(1.4) \qquad y(t) = \sum_{0}^{\infty} \beta_j \varepsilon(t-j), \qquad \sum_{0}^{\infty} \beta_j Z^j = \left(\sum a_j Z^j\right)^{-1}\left(\sum b_j Z^j\right),$$

where the β_j decrease to 0 at a geometric rate.

Our purpose is to find estimates for the true order (p_0, q_0). A subscript 0 will be used throughout for true quantities, e.g., p_0, q_0, a_{0j} and so on.

The traditional technique for this purpose is using statistical inference based on the central limit theorem for sample autocorrelations and partial correlations (Box and Jenkins (1970); Hannan (1970)). This does not seem to be an effective way of identifying the orders in the case where both $p_0 > 0$ and $q_0 > 0$ hold.

A new type of criterion proposed by Akaike was introduced to identify the orders of ARMA processes, that is FPE (in the AR case) and AIC (Akaike (1969), (1973)). These two criteria are asymptotically equivalent in the AR case and are very powerful in many practical problems. Unfortunately, consistency does not hold (Shibata (1976); Hannan (1980)) although the probability of over-estimation is small (Shibata (1976)). The definition of AIC is given by

$$(1.5) \qquad \text{AIC}(p, q) = \log \hat{\sigma}^2_{pq} + 2(p+q)/T,$$

where T is the number of observations, $\hat{\sigma}^2_{pq}$ is the ML estimate of σ^2 obtained by maximizing the Gaussian likelihood of ARMA (p, q) process (though the Gaussian assumption is not always maintained).

Minimizing AIC (p, q) in the range $p = 1, 2, \cdots, P$, $q = 1, 2, \cdots, Q$, we obtain $(\hat{p}, \hat{q})$ as the estimate of the true order (p_0, q_0), where P, Q are taken *a priori* such that

$$(1.6) \qquad 0 \leq p_0 \leq P, \qquad 0 \leq q_0 \leq Q.$$

The criterion (1.5) may be extended to a more general form

$$(1.7) \qquad H(p, q) = \log \hat{\sigma}^2_{pq} + c(T)(p+q)/T.$$

Then, to obtain a consistent estimate of the orders, it is necessary that

$$(1.8) \qquad c(T) \to \infty, \qquad c(T)/T \to 0, \quad \text{as } T \to \infty.$$

Hannan and Quinn (1979) and Hannan (1980) proved that if we take $c(T)$ as $\log T$, or $c \log \log T$, e.g.

$$(1.9) \qquad \text{BIC}(p, q) = \log \hat{\sigma}^2_{pq} + (p+q) \log T/T$$

$$(1.10) \qquad \varphi(p, q) = \log \hat{\sigma}^2_{pq} + c(p+q) \log \log T/T, \qquad c > 2$$

then the criteria will offer consistent estimates.

There are still problems we have to face, especially in practice.

(1) How do we choose P, Q such that (1.6) holds?

(2) In the case $q_0 > 0$, the procedures for calculation (even though $\hat{\sigma}^2_{pq}$ is an approximate ML estimate) are non-linear; especially in the case $p > p_0$, $q > q_0$, the redundant parameters cause a lot of trouble.

(3) Even if we derive an easier procedure for calculating $\hat{\sigma}^2_{pq}$, could we derive a recursive scheme to find the minimum point of $H(p, q)$ instead of point by point calculation over the lattice (p, q), $p = 1, 2, \cdots, P$, $q = 1, 2, \cdots, Q$?

(4) If the true model is a subset ARMA model, could we give a procedure to obtain a consistent identification of the non-zero parameters' subscripts? This requirement is more severe in the multivariate case.

We shall discuss these problems in the following sections by offering some new methods and theoretical results.

2. The AR model

In this section we discuss the estimate of the order for AR models, i.e. $q_0 = 0$ in (1.1).

Throughout the paper we denote

$$\gamma(j) = Ey(t)y(t+j), \qquad j = 0, \pm 1, \cdots$$

$$\hat{\gamma}(j) = \hat{\gamma}(-j) = (1/T) \sum_{t=1}^{T-j} y(t)y(t+j), \qquad j = 0, 1, \cdots, T-1.$$

It will be convenient to put $\hat{y}(t) = y(t)$, for $t = 1, 2, \cdots, T$; $\hat{y}(t) = 0$, otherwise. Thus we may write

$$\hat{\gamma}(j) = (1/T) \sum_{t=1}^{T} \hat{y}(t)\hat{y}(t+j), \qquad j = 0, \pm 1, \cdots, \pm(T-1).$$

In this case the ML estimate, $\hat{\sigma}_p^2$, of σ^2 is approximately

$$(2.1) \qquad \hat{\sigma}_p^2 = \hat{\gamma}(0) + \sum_{j=1}^{p} \hat{a}_{pj}\hat{\gamma}(j) = \hat{\gamma}(0) - \hat{a}_p \hat{\Gamma}_p \hat{a}_p^\tau$$

where $\hat{a}_p = (\hat{a}_{p1}, \hat{a}_{p2}, \cdots, \hat{a}_{pp})$, matrix $\hat{\Gamma}_p = (\hat{\gamma}(i-j))_{p \times p}$, and $\hat{a}_p$ satisfies

$$(2.2) \qquad \hat{a}_p \hat{\Gamma}_p = -\hat{\gamma}_p, \qquad \hat{\gamma}_p = (\hat{\gamma}(1), \hat{\gamma}(2), \cdots, \hat{\gamma}(p)).$$

$\hat{a}_p$, $\hat{\sigma}_p^2$, $p = 0, 1, \cdots$, can be calculated by Levinson's recursion, i.e.

$$(2.3) \qquad \begin{cases} \hat{a}_{pp} = -\left\{ \sum_{j=0}^{p-1} \hat{a}_{p-1,j}\hat{\gamma}(p-j) \right\} \Big/ \hat{\sigma}_{p-1}^2, \\[2mm] \hat{a}_{pj} = \hat{a}_{p-1,j} + \hat{a}_{pp}\hat{a}_{p-1,p-j}, \qquad j = 1, 2, \cdots, p-1 \\[2mm] \hat{\sigma}_p^2 = (1 - \hat{a}_{pp}^2)\hat{\sigma}_{p-1}^2 = \hat{\gamma}(0) \prod_{j=1}^{p} (1 - \hat{a}_{jj}^2). \end{cases}$$

In the vector case, an AR model is

$$\sum_{j=0}^{p} A_j Y(t-j) = \varepsilon(t), \qquad A_0 = I,$$

$$(2.4) \qquad E\{\varepsilon(t) \,|\, F_{t-1}\} = 0, \qquad E\{\varepsilon(t)\varepsilon(t)^\tau \,|\, F_{t-1}\} = \Sigma > 0,$$

$$\varepsilon(t) = (\varepsilon_1(t), \varepsilon_2(t), \cdots, \varepsilon_v(t))^\tau, \qquad E\varepsilon_j^4(t) < \infty, \qquad j = 1, 2, \cdots, v,$$

(where v is the dimension of $y(t)$),

$$\det\left\{ \sum A_j Z^j \right\} \neq 0, \quad \text{for} \quad |Z| \leq 1.$$

We may get forms similar to (2.1) and (2.2) by replacing $\hat{a}_p$, $\hat{a}_{pj}$ and $\hat{\sigma}_p^2$ by $\hat{A}_p = (\hat{A}_{p1}\,\hat{A}_{p2}\cdots\hat{A}_{pp})$, $\hat{A}_{pj}$ and $\hat{\Sigma}$ respectively with

$$\hat{\gamma}(j) = (1/T) \sum_{t=1}^{T} \hat{Y}(t)\hat{Y}(t+j)^{\tau}.$$

Instead of Levinson's recursion, we use Whittle's recursion in the vector case, i.e.

$$(2.5) \quad \begin{cases} \hat{A}_{pp} = -\hat{\Delta}_{p-1}\hat{\Lambda}_{p-1}^{-1}, \qquad \hat{B}_{pp} = -\hat{\Delta}_{p-1}^{\tau}\hat{\Sigma}_{p-1}, \\[2mm] \hat{\Delta}_p = \sum_0^p \hat{A}_{pj}\hat{\gamma}(j-p-1), \\[2mm] \hat{A}_{pj} = \hat{A}_{p-1,j} + \hat{A}_{pp}\hat{B}_{p-1,p-j}, \qquad \hat{B}_{pj} = \hat{B}_{p-1,j} + \hat{B}_{pp}\hat{A}_{p-1,p-j}, \\[2mm] \hspace{6cm} j = 1, 2, \cdots, p-1, \\[2mm] \hat{\Sigma}_p = (I - \hat{A}_{pp}\hat{B}_{pp})\hat{\Sigma}_{p-1}, \qquad \hat{\Lambda}_p = (I - \hat{B}_{pp}\hat{A}_{pp})\hat{\Lambda}_{p-1} \end{cases}$$

with $\hat{A}_{p0} = \hat{B}_{p0} = I$, $\hat{\Sigma}_0 = \hat{\Lambda}_0 = \hat{\gamma}(0)$.

For the vector case AIC can be defined as follows:

$$\text{AIC}(p) = \log \det \hat{\Sigma}_p + 2v^2 p/T.$$

So minimizing AIC(p) above in the range $p = 0, 1, \cdots, P$, we obtain an estimate of the order p_0.

Theorem 2.1 (Shibata (1976)). In the scalar case with Gaussian $y(t)$, suppose $\hat{p}$ minimizes AIC(p) over $p = 0, 1, \cdots, P$, and suppose $0 \leq p_0 \leq P$, then

$$(2.6) \qquad \lim_{T \to \infty} P(\hat{p} = p) = \begin{cases} 0, & 0 \leq p < p_0 \\ \pi_{p-p_0}\pi'_{P-p}, & p_0 \leq p \leq P \end{cases}$$

where

$$\pi_n = \sum_n^* \left\{ \prod_{i=1}^n \frac{1}{r_i!}\left(\frac{\alpha_i}{i}\right)^{r_i} \right\}$$

$$\pi'_n = \sum_n^* \left\{ \prod_{i=1}^n \frac{1}{r_i!}\left(\frac{1-\alpha_i}{i}\right)^{r_i} \right\}$$

with $\pi_0 = \pi'_0 = 1$, $\alpha_i = P(\chi^2 > 2i)$, χ^2 being a random variable having χ^2 distribution with freedom i, and the summation $\sum^*$ is over all groups $(r_1, r_2, \cdots, r_n)$ of non-negative integers satisfying

$$r_1 + 2r_2 + 3r_3 + \cdots + nr_n = n.$$

Theorem 2.2 (Hannan and Quinn (1979)). In the scalar case with the additional condition of strict stationary on $y(t)$, suppose $\hat{p}$ minimize BIC $(p, 0)$ or

$\varphi(p, 0)$ in range $p = 0, 1, \cdots, P$, and suppose $p_0 \leqq P$, then

(2.7)
$$\lim_{T \to \infty} \hat{p} = p_0, \quad \text{a.s.}$$

'Take P such that $p_0 \leqq P$' is not logical, because we do not know p_0. The following theorem dealt with this problem.

Theorem 2.3 (An, Chen and Hannan (1982)). Under the conditions of Theorem 2.2, put $P(T) = O((\log T))$, and $P(T) \uparrow \infty$ with $T \uparrow \infty$. Suppose $\hat{p}$ minimizes $\text{BIC}(p, 0)$ in the range $p = 0, 1, \cdots, P(T)$, then

$$\lim_{T \to \infty} \hat{p} = p_0, \quad \text{a.s.}$$

In Theorem 2.3, if $y(t)$ is generated by an ARMA (p, q) model and $q_0 > 0$, Hannan and Kavalieris (1983) showed that

(2.8)
$$\hat{p} = \{\log T / 2 \log \lambda\}\{1 + o(1)\}, \quad \text{a.s.}$$

where λ is the modulus of the root of

$$\sum_{j=0}^{p} a_{0j} Z^j$$

nearest to $|Z| = 1$.

For the vector case the result of Theorem 2.3 is also true, but we leave it for Section 4.

3. The ARMA model

In the literature there are many suggested procedures for calculating $\hat{\sigma}_{pq}^2$, the estimates of the variance of the residuals under the assumption of (p, q) orders, see e.g. Anderson (1977), Box and Jenkins (1970), Hannan (1970), (1973). All these estimates originated from the MLE for Gaussian process, though the Gaussian assumption is not always necessary for the final form of the estimates.

Theorem 3.1 (Hannan (1980)). Under the conditions (1.1), (1.2), (1.3), for $\text{AIC}(p, q)$,

(3.1)
$$\begin{cases} \lim_{T \to \infty} P(\hat{p} = p, \hat{q} = q) = \pi_{q-q_0, Q-q}, & (\pi_{q-q_0, Q-q} = 0 \text{ for } q < q_0) \\[2ex] \lim_{T \to \infty} P(\hat{p} < p, \hat{q} = q_0) = 0, & P = p_0; \\[2ex] \lim_{T \to \infty} P(\hat{p} = p, \hat{q} = q_0) = \pi_{p-p_0, P-p}, & (\pi_{p-p_0, P-p} = 0 \text{ for } p < p_0) \\[2ex] \lim_{T \to \infty} P(\hat{p} = p, \hat{q} < q_0) = 0, & Q = q_0. \end{cases}$$

Theorem 3.2 (Hannan (1980)). Under the conditions (1.1), (1.2), (1.3), $\varepsilon(t)$ are independent, then for BIC(p, q) or $\varphi(p, q)$,

$$\text{(3.2)} \qquad \lim_{T \to \infty} \hat{p} = p_0, \qquad \lim_{T \to \infty} \hat{q} = q_0, \quad \text{a.s.}$$

If $\varepsilon(t)$ are not independent but $E|\varepsilon(t)|^\eta < \infty$, $\eta > 4$, then for BIC(p, q), (3.2) is still true.

Theorem 3.3 (Hannan (1980)). Under the conditions (1.1), (1.2), (1.3), for $H(p, q)$ with $c(T)$ satisfying (1.8), $\hat{p}$, $\hat{q}$ are weakly consistent.

As we have pointed out in Section 1, the procedures for calculating $\hat{\sigma}^2_{pq}$ mentioned above are non-linear and computationally difficult, but the work of Hannan and Rissanen (1982) seems to give us a key to a new way of overcoming this difficulty. We call it the H–R procedure and state it as follows.

Step 1. Fit the data with an AR$(P(T))$, $P(T) = O(\log T)^d$, $0 < d < \infty$. Denote the coefficients by $\hat{\alpha}_T = (\hat{\alpha}_{T1}, \cdots, \hat{\alpha}_{T,P(T)})$ which is obtained by solving (cf. (2.2))

$$\hat{\alpha}_T \hat{\Gamma}_{P(T)} = -\hat{\gamma}_{P(T)},$$

and

$$\hat{\sigma}^2_T = \hat{\gamma}(0) - \hat{\alpha}_T \hat{\Gamma}_{P(T)} \hat{\alpha}^\tau_T$$

is obtained simultaneously. Calculate $\hat{\varepsilon}(t)$ recursively by

$$\text{(3.3)} \qquad \begin{cases} \hat{\varepsilon}(t) = \sum_{j=0}^{P(T)} \hat{\alpha}_{Tj} \hat{y}(t-j), & \alpha_{T0} = 1, \qquad t = 1, 2, \cdots, T \\ \hat{\varepsilon}(t) = 0, & \text{otherwise.} \end{cases}$$

Step 2. Form

$$\text{(3.4)} \qquad \begin{cases} \hat{\gamma}^y(j) = (1/T) \sum_{t=1}^{T} \hat{y}(t)\hat{y}(t+j), \\ \\ \hat{\gamma}^{y\hat{\varepsilon}}(j) = (1/T) \sum_{t=1}^{T} \hat{y}(t)\hat{\varepsilon}(t+j) = \hat{\gamma}^{\hat{\varepsilon}y}(-j), \\ \\ \hat{\gamma}^{\hat{\varepsilon}}(j) = (1/T) \sum_{t=1}^{T} \hat{\varepsilon}(t)\hat{\varepsilon}(t+j). \end{cases}$$

Denote

$$\hat{\Gamma}^{y\hat{\varepsilon}}_{pq} = \begin{pmatrix} \hat{\gamma}^{y\hat{\varepsilon}}(0) & \cdots & \hat{\gamma}^{y\hat{\varepsilon}}(1-q) \\ \cdots\cdots\cdots\cdots\cdots\cdots\cdots\cdots \\ \hat{\gamma}^{y\hat{\varepsilon}}(p-1) & \cdots & \hat{\gamma}^{y\hat{\varepsilon}}(p-q) \end{pmatrix} = (\hat{\Gamma}^{\hat{\varepsilon}y}_{qp})^\tau$$

$$\hat{\gamma}^y_p = (\hat{\gamma}^y(1) \cdots \hat{\gamma}^y(p)),$$

$$\hat{\gamma}^{y\hat{\varepsilon}}_p = (\hat{\gamma}^{y\hat{\varepsilon}}(-1) \cdots \hat{\gamma}^{y\hat{\varepsilon}}(-q)),$$

and

$$\hat{\Gamma}_p^y = (\hat{\gamma}(i-j))_{p \times p}.$$

The $q \times q$ matrix $\hat{\Gamma}_q^{\hat{\varepsilon}}$ is defined similarly. Define (we use the notation $\hat{\sigma}_{pq}^2$ again but here it is not the approximate MLE of the variance of the residuals)

$$\hat{\sigma}_{pq}^2 = \min_{a_{pj}, b_{qj}} (1/T) \sum_{t=1}^{T} \left\{ \sum_{0}^{p} a_{pj} \hat{y}(t-j) - \sum_{1}^{p} b_{qj} \hat{\varepsilon}(t-j) \right\}^2$$

(3.5)

$$\simeq \hat{\gamma}^y(0) + \sum_{1}^{p} \hat{a}_{pj} \hat{\gamma}^y(j) - \sum_{1}^{q} \hat{b}_{qj} \hat{\gamma}^{\hat{\varepsilon}y}(j),$$

where $\hat{a}_p = (\hat{a}_{p1} \cdots \hat{a}_{pp})$, $\hat{b}_q = (\hat{b}_{q1} \cdots \hat{b}_{qq})$ satisfy

(3.6) $$(\hat{a}_p \quad \hat{b}_q) \begin{pmatrix} \hat{\Gamma}_p^y & -\hat{\Gamma}_{pq}^{y\hat{\varepsilon}} \\ -\hat{\Gamma}_{qp}^{\hat{\varepsilon}y} & \hat{\Gamma}_p^{\hat{\varepsilon}} \end{pmatrix} = (-\hat{\gamma}_p^y \quad \hat{\gamma}_q^{y\hat{\varepsilon}}).$$

Step 3. Minimize

(3.7) $$\text{BIC}_\delta(p, q) = \log \hat{\sigma}_{pq}^2 + (p+q)(\log T)^{1+\delta}/T, \qquad \delta > 0$$

over $p = 0, 1, \cdots, P$, $q = 0, 1, \cdots, Q$, to obtain $\hat{p}$, $\hat{q}$.

Theorem 3.4 (Hannan and Rissanen (1982), and the correction). Under the conditions (1.1), (1.2), (1.3), with $\hat{p}$, $\hat{q}$ obtained as above, then (3.2) holds.

If we put $p = q$ and rearrange (3.6) in the form

(3.6′) $$(\hat{a}_{p1} \hat{b}_{p1} \cdots \hat{a}_{pp} \hat{b}_{pp}) \hat{\Gamma}_p = -(1 \ 0) \hat{\gamma}_p,$$

where $\hat{\gamma}_p = (\hat{\gamma}(-1) \cdots \hat{\gamma}(-p))$ and $\hat{\Gamma}_p$ has $\hat{\gamma}(i-j)$ as its (i, j)th block,

(3.7′) $$\hat{\gamma}(j) = \begin{pmatrix} \hat{\gamma}^y(j) & -\hat{\gamma}^{y\hat{\varepsilon}}(j) \\ -\hat{\gamma}^{\hat{\varepsilon}y}(j) & \hat{\gamma}^{\hat{\varepsilon}}(j) \end{pmatrix}.$$

Imbed (3.6′) into the matrix equation

(3.8) $$(\hat{A}_{p1} \cdots \hat{A}_{pp}) \hat{\Gamma}_p = -\hat{\gamma}_p, \qquad \hat{A}_{pj} = \begin{pmatrix} \hat{a}_{pj} & \hat{b}_{pj} \\ * & * \end{pmatrix}$$

(the elements $*$ in $\hat{A}_{pj}$ are of no concern), using Whittle's (2.5), we obtain $\hat{\Sigma}_p$ for $p = 0, 1, \cdots, P$. The first element in $\hat{\Sigma}_p$ is $\hat{\sigma}_{pp}^2$. The minimization of $\text{BIC}_\delta(p, p)$ offers a consistent estimate of $r_0 = \max(p_0, q_0)$.

We have to find some ways of giving a P such that $P > r_0$. Wang and Chen (1985) suggested a monitoring criterion by checking

(3.9) $$\det \hat{\Sigma}_p < (\log T/T)^{\frac{1}{2}} \hat{\sigma}_T^4.$$

When we carry on Whittle's recursion along $p = 0, 1, \cdots$ we denote the first p for which (3.9) holds by r. It can be proved that r is a consistent estimate of r_0.

Simulations show good results even for $T = 100$. Then r may serve for both P and Q in Step 3 above.

One may be afraid that underestimation of r_0 by r will happen when T is not too large (in practice usually there are no true orders, so to take such 'first' p as the upper bound of the orders seems unsafe). We can carry on Whittle's recursion further till, say, $s = r + 2$, or $r + 3$ and so on, but we make $\det \hat{\Sigma}_p$ keep away from 0 for all $p \leqq s$ and satisfy (3.11) below. Then we take s as an upper bound for both p_0 and q_0; this is much more plausible than saying 'given P, Q such that (1.6) is satisfied' and it is not too wasteful.

Now at $p = s$, we obtained $\hat{a}_{sj}$, $\hat{b}_{sj}$, $j = 1, 2, \cdots, s$. Calculate $\tilde{\varepsilon}(t)$ recursively by

$$(3.10) \qquad \tilde{\varepsilon}(t) = -\sum_{j=1}^{s} \hat{b}_{sj} \tilde{\varepsilon}(t-j) + \sum_{j=0}^{s} \hat{a}_{sj} \hat{y}(t-j), \qquad \hat{a}_{s0} = 1, \qquad t = 1, \cdots, T,$$

$$\tilde{\varepsilon}(t) = 0, \qquad\qquad\qquad\qquad\qquad\qquad\qquad\qquad \text{otherwise.}$$

Instead of $\tilde{\varepsilon}(t)$ obtained in Step 1 in the H–R procedure, we use $\tilde{\varepsilon}(t)$ and carry on the procedure in Step 2 to obtain $\tilde{\sigma}^2_{pq}$, $\tilde{a}_p$, $\tilde{b}_q$ corresponding to $\hat{\sigma}^2_{pq}$, $\hat{a}_p$, $\hat{b}_q$ respectively in the case of $\hat{\varepsilon}(t)$.

Theorem 3.5 (Wang and Chen (1985)). Suppose $\eta > 0$ is fixed and $a_0(Z) \neq 0$, $b_0(Z) \neq 0$, $|Z| \leqq 1 + \eta$. If $\hat{a}_{sj}$, $\hat{b}_{sj}$, $j = 1, 2, \cdots, s$, satisfy

$$(3.11) \quad \hat{a}_s(Z) = \sum_{j=0}^{s} \hat{a}_{sj} Z^j \neq 0, \qquad \hat{b}_s(Z) = \sum_{j=0}^{s} \hat{b}_{sj} Z^j \neq 0, \qquad |Z| \leqq 1 + \eta$$

and $(\tilde{p}, \tilde{q})$ minimize (1.9) with $\hat{\sigma}^2_{pq}$ replaced by $\tilde{\sigma}^2_{pq}$ in the range $p = 0, 1, \cdots, s$, $q = 0, 1, \cdots, s$ then

$$(3.12) \qquad\qquad\qquad \lim_{T \to \infty} \tilde{p} = p_0, \qquad \lim_{T \to \infty} \tilde{q} = q_0 \quad \text{a.s.}$$

Here η is introduced only for the sake of proving (3.12); in fact η can be taken as small as one likes.

Obviously, minimization of $\text{BIC}(p, p)$ in (1.9) with $\hat{\sigma}^2_{pp}$ replaced by $\tilde{\sigma}^2_{pp}$, $0 \leqq p \leqq s$, offers another consistent estimate of r_0 (we also denote it by r) which is intuitively more convincing than monitoring. We shall use this r at the end of Section 4. Of course, we need a simple feasible procedure for searching for (p, q) in the lattice $p = 0, 1, \cdots, s$, $q = 0, 1, \cdots, s$, not exhausting all the points (p, q). One can easily figure out such a procedure after reading the next section.

4. Subset fitting

Consider the vector AR model (2.4) again. Let

$$(4.1) \qquad\qquad Y(t) = (y_1(t), \cdots, y_v(t))^\tau, \qquad \varepsilon(t) = (\varepsilon_1(t), \cdots, \varepsilon_v(t))^\tau.$$

Put $p = P$ in (2.4) and

$$(4.2) \qquad A_j = \begin{pmatrix} a_{11}(i) & a_{12}(i) & \cdots & a_{1v}(i) \\ a_{21}(i) & a_{22}(i) & \cdots & a_{2v}(i) \\ \vdots & \vdots & & \vdots \\ a_{v1}(i) & a_{v2}(i) & \cdots & a_{vv}(i) \end{pmatrix}, \qquad i = 1, 2, \cdots, P$$

are the parameter matrices satisfying the stationary condition. So we have the following model for $y_1(t)$:

$$(4.3) \qquad \begin{aligned} y_1(t) =\ & a_{11}(1)y_1(t-1) + a_{11}(2)y_1(t-2) + \cdots + a_{11}(P)y_1(t-P) \\ & + a_{12}(1)y_2(t-1) + a_{12}(2)y_2(t-2) + \cdots + a_{12}(P)y_2(t-P) + \cdots \\ & + a_{1v}(1)y_v(t-1) + a_{1v}(2)y_v(t-2) + \cdots + a_{1v}(P)y_v(t-P) \\ & + \varepsilon_1(t). \end{aligned}$$

Suppose the true values of the parameters $a_{1j}(i)$, $i = 1, 2, \cdots, P$, $j = 1, 2, \cdots, v$, are non-zero only for

$$a_{1e}(i_{es}) \neq 0, \qquad s = 1, 2, \cdots, P_e, \quad e = 1, 2, \cdots, v,$$

where $0 \leq p_e \leq P$, and $1 \leq i_{es} \leq P$, but $a_{1e}(i) = 0$ for other values of i.

If the subscripts i_{es} are given for $s = 1, 2, \cdots, p_e$, $e = 1, 2, \cdots, v$, we can fit model (4.3) by using only the variables $y_e(t - i_{es})$, $e = 1, 2, \cdots, v$, $s = 1, 2, \cdots, p_e$. If we do not know which ones to use, we have to determine them first. We fit a similar model for $y_2(t), y_3(t), \cdots, y_v(t)$, by the same method. Combining all the fitted models of $y_1(t), y_2(t), \cdots, y_v(t)$, we get the subset fitting model of (2.4). Thus we only need consider the subset fitting model of (4.3).

Suppose we have n observations $Y(1), Y(2), \cdots, Y(n)$, set

$$\begin{aligned} T = n - P, \qquad K = vP, \qquad z(k) = y_1(P+k), \qquad e(k) = \varepsilon_1(P+k), \\ x_{ip-p+j}(k) = y_i(P+k-j), \qquad j = 1, 2, \cdots, P, \qquad i = 1, 2, \cdots, v, \end{aligned}$$

and then

$$Z = (z(1), z(2), \cdots, z(T))^\tau, \qquad e = (e(1), e(2), \cdots, e(T))^\tau,$$

$$\theta = (\theta_1, \theta_2, \cdots, \theta_K)^\tau,$$

$$X = \begin{pmatrix} x_1(1) & x_2(1) & \cdots & x_K(1) \\ x_1(2) & x_2(2) & \cdots & x_K(2) \\ \vdots & \vdots & & \vdots \\ x_1(T) & x_2(T) & \cdots & x_K(T) \end{pmatrix}.$$

It is easy to see that the model (4.3) for $t = P+1, P+2, \cdots, n$ can be written as the following linear model:

$$(4.4) \quad z(k) = \theta_1 x_1(k) + \theta_2 x_2(k) + \cdots + \theta_K X_K(k) + e(k), \qquad k = 1, 2, \cdots, T,$$

or equivalently the model

$$(4.5) \qquad\qquad\qquad Z = X\theta + e.$$

By the definition of θ the true values of θ_j satisfy

$$\theta_{eP-P+i_{es}} \neq 0, \qquad e = 1, 2, \cdots, v, \quad s = 1, 2, \cdots, p_e$$

$$\theta_j = 0 \qquad \text{for other } j.$$

Put

$$i_{(p_1+p_2+\cdots+p_{e-1})+s} = i_{es} + (e-1)P, \qquad e = 1, 2, \cdots, v, \quad s = 1, 2, \cdots, p_e,$$

and $p_0 = p_1 + p_2 + \cdots + p_v$, $I = \{i_1, i_2, \cdots, i_{p_0}\}$ as the true set of the non-zero parameters' subscripts.

To estimate set I means to select variables in model (4.4), which is similar to a regression model in form. In the regression analysis literature, there are many methods available to select variables. An and Gu (1985) proposed a set of new methods using AIC, BIC to select regression variables and obtained strong consistency theorems for these and other 'criterion methods'. These new methods can be used to identify set I for model (4.4), and some consistency to the true set I can be shown (see Theorem 4.1 below).

Let $J_R = \{1, 2, \cdots, R\}$ with $R = vP$, and $J_s = \{j_1, j_2, \cdots, j_s\}$ be a subset of J_R, and

$$X(J_s) = \begin{pmatrix} x_{j_1}(1) & x_{j_2}(1) & \cdots & x_{j_s}(1) \\ x_{j_1}(2) & x_{j_2}(2) & \cdots & x_{j_s}(2) \\ \vdots & \vdots & & \vdots \\ x_{j_1}(T) & x_{j_2}(T) & \cdots & x_{j_s}(T) \end{pmatrix},$$

$$\theta(J_s) = \begin{pmatrix} \theta_{j_1} \\ \theta_{j_2} \\ \vdots \\ \theta_{j_s} \end{pmatrix},$$

$$(4.6) \qquad \hat{\theta}(J_s) = (X(J_s)^\tau X(J_s))^{-1} X(J_s)^\tau Z,$$

$$(4.7) \quad \begin{aligned} S(J_s) &= \|Z - X(J_s)\hat{\theta}(J_s)\|^2 \\ &= Z^\tau Z - Z^\tau X(J_s)(X(J_s)^\tau X(J_s))^{-1} X(J_s)^\tau Z. \end{aligned}$$

It is well known that $\hat{\theta}(J_s)$ is the least squares estimate of $\theta(J_s)$ in fitting model (4.4) by using variables $x_{j_1}(k), x_{j_2}(k), \cdots, x_{j_s}(k)$, and $S(J_s)$ is the residual sum of squares in fitting model (4.4). We define three random sequences of subset of

J_R as follows:

$\{L_s^{(1)}\}$ is determined by

$$S(L_s^{(1)}) = \min_{J_s \subset J_R} S(J_s), \qquad s = 1, 2, \cdots, R;$$

$\{L_s^{(2)}\}$ is determined by $L_0^{(1)} = \varnothing$ (empty set), and

$$S(L_s^{(2)}) = \min_{j \in J_R \setminus L_{s-1}^{(2)}} S(L_{s-1}^{(2)} \cup \{j\}), \qquad s = 1, 2, \cdots, R;$$

$\{L_s^{(3)}\}$ is determined by $L_R^{(3)} = J_R$, and

$$S(L_{R-s}^{(3)}) = \min_{j \in L_{R-s+1}^{(3)}} S(L_{R-s+1}^{(3)} \setminus \{j\}), \qquad s = 1, 2, \cdots, R,$$

Put

(4.8) $$\mathrm{AIC}_i(s) = \log S(L_s^{(i)}) + 2s/T, \qquad s = 0, 1, \cdots, R,$$

(4.9) $$\mathrm{BIC}_i(s) = \log S(L_s^{(i)}) + s \log T/T, \qquad s = 0, 1, \cdots, R.$$

Let $\mathrm{AIC}_1(\hat{p})$ be the smallest value of $\mathrm{AIC}_1(s)$ over $s = 0, 1, \cdots, R$, then take $L_{\hat{p}}^{(1)} = \hat{I}$ as the estimate of set I. We call this method the AIC_1 method. Similarly we have AIC_i ($i = 2$ or 3), and BIC_i ($i = 1$, 2 or 3).

In An and Gu's paper (1985) $\{L_s^{(i)}\}$ ($i = 2$ or 3) are determined by recursive procedures which are similar to stepwise regression. Take $\{L_s^{(2)}\}$, for example. Put $Q = R + 1$, and

$$B^{(0)} = \begin{pmatrix} X^\tau X & X^\tau Z \\ Z^\tau X & Z^\tau Z \end{pmatrix} = (b_{ij}^{(0)})_{1 \leq i,j \leq Q}.$$

Let $T_k B^{(0)}$ be the transform of matrix $B^{(0)}$, defined by $B' = T_k B^{(0)} = (b'_{ij})$ where

(4.10) $$b'_{ij} = \begin{cases} b_{ij}^{(0)} - b_{ik}^{(0)} b_{kj}^{(0)}/b_{kk}^{(0)}, & \text{for} \quad i \neq k \text{ and } j \neq k, \\ -b_{ik}^{(0)}/b_{kk}^{(0)}, & \text{for} \quad i \neq k \text{ and } j = k, \\ b_{kj}^{(0)}/b_{kk}^{(0)}, & \text{for} \quad i = k \text{ and } j \neq k, \\ 1/b_{kk}^{(0)}, & \text{for} \quad i = k \text{ and } j = k. \end{cases}$$

The following recursive procedure obtains sequence $\{L_s^{(2)}\}$:

(1) Calculate $u_j^{(1)} = (b_{jQ}^{(0)})^2/b_{jj}^{(0)}$, for each $j \in L_R^{(2)} = J_R$; Let $u_{m_1}^{(1)}$ be the biggest one, then take $L_1^{(2)} = \{m_1\}$;

(2) Calculate $T_{m_1} B^{(0)} = B^{(1)} = (b_{ij}^{(1)})$, put $S^{(1)} = b_{QQ}^{(1)}$, $\hat{\theta}^{(1)} = b_{m_1 Q}^{(1)}$;

(3) Calculate $u_j^{(2)} = (b_{jQ}^{(1)})^2/b_{jj}^{(1)}$, for each $j \in J_R \setminus L_1^{(2)}$, let $u_{m_2}^{(2)}$ be the biggest one, then take $L_2^{(2)} = \{m_1, m_2\}$;

(4) Calculate $T_{m_2} B^{(1)} = B^{(2)} = (b_{ij}^{(2)})$, put $S^{(2)} = b_{QQ}^{(2)}$, $\hat{\theta}^{(2)} = (b_{m_1 Q}^{(2)}, b_{m_2 Q}^{(2)})^\tau$;

(5) Similarly, repeat (3) and (4) until we get $L_R^{(2)}$.

It is known that

$$\hat{\theta}^{(j)} = \hat{\theta}(L_j^{(2)}), \qquad S^{(j)} = S(L_j^{(2)}).$$

Theorem 4.1 (Gu and An (1983)). Suppose that $Y(t)$ satisfy the model (2.4), and $P = O(\log n)$, then BIC_1 and BIC_3 are consistent methods, i.e. the estimates $\hat{I}$ given by minimizing (4.9) for $i = 1$ or 3, satisfy

$$(4.11) \qquad\qquad \lim_{n \to \infty} \hat{I} = I, \quad \text{a.s.}$$

and then BIC_2 and AIC_i ($i = 1$, 2 or 3) are overconsistent methods, i.e. the estimates $\hat{I}$ given by minimizing (4.9) for $i = 2$ or minimizing (4.8) for $i = 1$, 2 or 3, satisfy

$$(4.12) \qquad\qquad \liminf_{n \to \infty} \hat{I} \supset I, \quad \text{a.s.}$$

It is easy to see that in order to get sequence $\{L_s^{(1)} : s = 1, 2, \cdots, R\}$, we have to calculate $S(J_s)$ for each s and for each $J_s \subset J_R$, the total number of which is

$$C^1 + C^2 + \cdots + C_R^R = 2^R - 1.$$

But to obtain sequences $\{L_s^{(2)}\}$ or $\{L_s^{(3)}\}$ we only need to calculate $R(R+1)/2$ of $S(J_s)$. In fact the calculations for BIC_i and AIC_i ($i = 2$ or 3) can be greatly reduced again by using recursive procedures as mentioned above.

Now we turn to subset ARMA models as in Section 3. The scalar AR is also a special case of this. We have obtained an estimate of the upper bound of r_0 as well as $\bar{\varepsilon}(t)$ by (3.10), and also a consistent estimate r for r_0 in Section 3. Thus for sufficient large T, $r = r_0$,

$$(4.13) \quad \hat{y}(t) = \sum_{j=1}^{r} \{a_j(-\hat{y}(t-j)) + b_j\bar{\varepsilon}(t-j)\} + e(t), \qquad t = 1, 2, \cdots, T,$$

then this is a special case of (4.4) though the probability structure of (4.13) is different from (4.3) which is also of the form (4.4).

We can use either one of the procedures mentioned above to identify the subscript set I corresponding to coefficients which are non-zero in the true $\mathrm{ARMA}(p_0, q_0)$ model.

Theorem 4.2 (Chen (1985)). Suppose conditions (1.1), (1.2), (1.3) hold, using BIC_3 for (4.13) to obtain $\hat{I}$, then

$$(4.14) \qquad\qquad \lim_{n \to \infty} \hat{I} = I, \quad \text{a.s.}$$

References

AKAIKE, H. (1969) Fitting autoregression models for prediction. *Ann. Inst. Statist. Math.* **21**, 243–247.

AKAIKE, H. (1973) Information theory and estimation of the maximum likelihood principle. In

2nd International Symposium on Information Theory, ed. B. N. Petrov and F. Csaki, Akademiai Kiado, Budapest, 261–281.

An Hong-Zhi, Chen Zhao-Guo and Hannan, E. J. (1982) Autocorrelation, autoregression and autoregressive approximation. *Ann. Statist.* **10**, 926–936.

An Hong-Zhi and Gu Lan (1985) On the selection of regression variables. *Acta Math. Appl. Sinica, English Series* **2** (1).

Anderson, T. W. (1977) Estimation for autoregressive moving average models in the time and frequency domain. *Ann. Statist.* **5**, 842–856.

Box, G. E. P. and Jenkins, G. M. (1970) *Time Series Analysis, Forecasting and Control.* Holden Day, San Francisco.

Chen Zhao-Guo (1985) On the recursive fitting of subset autoregressive-moving average process. *Acta Math. Appl. Sinica B English Series* **2** (1).

Gu Lan and An Hong-Zhi (1983) Fitting AR models by using partial parameters. *Acta Math. Appl. Sinica* A. To appear.

Hannan, E. J. (1970) *Multiple Time Series.* Wiley, New York.

Hannan, E. J. (1973) The asymptotic theory of linear time series models. *J. Appl. Prob.* **10**, 130–145.

Hannan, E. J. (1980) The estimation of the order of an ARMA series. *Ann. Statist.* **8**, 1071–1081.

Hannan, E. J. and Kavalieris, L. (1983) The convergence of autocorrelations and auto-regressions. *Austral. J. Statist.* **25**, 287–297.

Hannan, E. J. and Quinn, B. G. (1979) The determination of the order of an autoregression. *J. R. Statist. Soc.* B **41**, 190–195.

Hannan, E. J. and Rissanen, J. (1982) Recursive estimation of mixed autoregressive-moving average order. *Biometrika* **69**, 81–94.

Shibata, R. (1976) Selection of the order of an autoregressive model by Akaike's information criterion. *Biometrika* **63**, 117–126.

Wang Shou-Ren and Chen Zhao-Guo (1985) Estimation of the order of ARMA model by linear procedures. *Chinese Ann. Math.* B **6**, 53–70.

Whittle, P. (1963) On the fitting multivariate autoregression and the approximate canonical factorization of a spectral density matrix. *Biometrika* **50**, 129–134.

Time Series Regression with Unequally Spaced Data

RICHARD H. JONES

Abstract

Regression analysis with stationary errors is extended to the case when observations are not equally spaced. The errors are modelled as either a discrete-time ARMA process with missing observations, or as a continuous-time autoregression with observational error observed at arbitrary times. Using a state-space representation, a Kalman filter is used to calculate the exact likelihood. The linear regression coefficients are separated out of the likelihood so non-linear optimization is required only with respect to the parameters modelling the error structure.

CORRELATED ERRORS; STATE SPACE; KALMAN FILTER

1. Introduction

Hannan (1963), following the work of Grenander and Rosenblatt (1957), developed asymptotically efficient methods of estimating regression coefficients with a stationary error structure. These frequency domain methods are illustrated in Hamon and Hannan (1963). Harvey and Phillips (1979) used a state-space approach to obtain exact maximum likelihood estimates of regression coefficients with autoregressive–moving-average (ARMA) errors (see also Harvey (1981)). By including the linear regression coefficients in the state vector, and concentrating them out of the likelihood function, non-linear optimization need only be carried out with respect to the ARMA parameters. This paper extends these results to include missing and unequally spaced data, and uses a different method of separating the linear regression coefficients which avoids the initialization problems at the beginning of the data. Instead of including the linear regression coefficients in the state vector, the Kalman recursion is applied to the regression equation giving a transformed regression equation which is solved for the residual sum of squares at the end of each iteration.

Using a state-space representation, the Kalman filter can be used to calculate the exact likelihood function. If a discrete-time process is modelled as a stationary ARMA process, missing observations pose no difficulty at all when estimating the parameters (Jones (1980)). If observations are taken at arbitrary times, with no basic underlying sampling interval, it is necessary to use a continuous-time model for the process. A continuous-time autoregression where the process satisfies a stable linear differential equation driven by white noise is one possibility (Jones (1981)). The state-space representation allows the addition of observational error to these models increasing the flexibility in model fitting; however, for a discrete-time ARMA(p, q) process, the observational error will be confounded with the other parameters if $q \geqq p$. For a discrete-time process, a continuous-time autoregression of order p, CAR(p), with observational error, may be a more parsimonious representation than the corresponding discrete-time ARMA(p, p) process (Jones (1985)).

2. The method

The model to be considered is the usual linear regression equation,

$$(2.1) \qquad\qquad\qquad y = X\beta + \varepsilon$$

where y is an $n \times 1$ vector of observations on the dependent variable, X is an $n \times m$ matrix of observations on the independent variables, assumed to be observed without error, β is an $m \times 1$ vector of regression coefficients, assumed to be unknown constants to be estimated, and ε is an $n \times 1$ vector of random errors. It will be assumed that the ε's are a zero-mean stationary Gaussian process observed at equal or unequal intervals of time. If the sampling times are multiples of some basic time interval, the errors will be modelled either as an ARMA process, possibly with observational error, or as a continuous-time autoregression with observational error. If the observation times are not multiples of some basic time interval, the continuous-time autoregression with observational error will be used.

Using the principle of the separability of linear and non-linear parameters in estimation (Golob and Pereyra (1973)), assume for the purpose of each iteration consisting of one pass of the Kalman filter through the data, that the parameters of the stationary error structure are fixed. The parameter values determine the initial state covariance matrix for the Kalman filter, as well as the complete covariance recursion. Given these covariances, the Kalman filter is simply a linear recursion involving the current and past observations and can be represented as an $n \times n$ lower-triangular matrix operating on the data. This is similar to the procedure used by Duncan and Jones (1966), except that the transformation used in their study was a Fourier transformation. The lower-triangular matrix based on the Kalman filter can be constructed in such a way that the result is the normalized innovation at each time point, i.e. the innovation divided by its standard deviation determined to within a constant, σ^2,

$$(2.2) \qquad\qquad K\varepsilon = \tilde{\varepsilon}$$

where K is an $n \times n$ lower-triangular matrix representing the Kalman filter, and $\tilde{\varepsilon}$ is the vector of innovations. K is effectively a Cholesky decomposition (Wecker and Ansley (1983)). It is not necessary to write a closed-form expression for K since it will usually be generated using a recursion. The variance of each element of the innovation vector, before normalization, enters into the likelihood function as in weighted least squares when the weights are varied. The variance of $\tilde{\varepsilon}_i$ will be denoted by $\sigma^2 V_i$.

The innovation vector will consist of uncorrelated random variables only if the stationary error structure is properly modelled with the correct values of the parameters. Given a trial model and initial guesses for the parameters, the Kalman filter is used to calculate the exact likelihood. This is embedded in a non-linear optimization routine, where a function evaluation consists of one pass through the data, and maximum likelihood estimates are obtained for the parameters. The whiteness of the transformed errors or innovations can then be checked by some method such as a cumulative periodogram as an indication of whether the assumed model is appropriate.

This same procedure can be used in a regression situation by premultiplying the regression equation by the matrix K representing the Kalman filter,

$$(2.3) \qquad \begin{aligned} Ky &= KX\beta + K\varepsilon \\ \tilde{y} &= \tilde{X}\beta + \tilde{\varepsilon} \end{aligned}$$

where

$$(2.4) \qquad \tilde{y} = Ky, \qquad \tilde{X} = KX \quad \text{and} \quad \tilde{\varepsilon} = K\varepsilon.$$

This transformation simply indicates that the Kalman filter operates, not only on the error, but on y and each element of X. It does not take much more computer effort to apply the recursion to $m + 1$ vectors than it does to apply it to one vector, since the covariance recursions need only be calculated once.

An efficient way to calculate the transformation is to augment the X matrix by the y vector and accumulate the components of the transformed normal equations during the pass through the recursion,

$$(2.5) \qquad\qquad [\tilde{X}'\tilde{X} \,|\, \tilde{X}'\tilde{y}],$$

where $'$ denotes the transposed matrix. The regression sum of squares for these weighted normal equations can be simply calculated by applying a Cholesky factorization routine to (2.5) which replaces it by

$$(2.6) \qquad\qquad [T \,|\, (T')^{-1}\tilde{X}'\tilde{y}]$$

where T is $m \times m$ upper-triangular and

$$(2.7) \qquad\qquad T'T = \tilde{X}'\tilde{X}$$

(Graybill (1976)). The weighted regression sum of squares is then the sum of squares of the elements of the right augmented vector of (2.6). If $\tilde{y}'\tilde{y}$, the weighted total sum of squares, is also accumulated, the weighted residual sum of squares for the given values of the ARMA parameters is the difference between the weighted total sum of squares and the weighted regression sum of squares. The weighted residual sum of squares will be denoted RSS.

To complete the calculation of the likelihood, assuming Gaussian errors, it is necessary to account for the innovation variances which were used to normalize the residuals. This is the determinant term in the Gaussian likelihood function and must be included when weights or variances are modified during optimization. If the errors contain the usual scale factor, σ^2, $-2 \ln$ likelihood is calculated as

$$(2.8) \qquad\qquad l = n \ln \text{RSS} + \sum_{i=1}^{n} \ln V_i$$

where the V_i are the innovation variances calculated by the Kalman filter. The scale factor is estimated as

$$(2.9) \qquad\qquad \tilde{\sigma}^2 = \frac{1}{n} \text{RSS}$$

after convergence.

3. A simple special case

In order to demonstrate this procedure by a simple example, consider the case when the errors are a first-order stationary autoregression,

$$(3.1) \qquad\qquad \varepsilon_i = \alpha \varepsilon_{i-1} + \eta_i \qquad -1 < \alpha < 1$$

where η_i are independent Gaussian random variables with zero mean and variance σ^2. For equally spaced observations, the error covariance matrix has the following Toeplitz structure:

$$(3.2) \qquad V = \frac{\sigma^2}{1-\alpha^2} \begin{bmatrix} 1 & \alpha & \alpha^2 & \cdots & \alpha^{n-1} \\ \alpha & 1 & \alpha & \cdots & \alpha^{n-2} \\ \alpha^2 & \alpha & 1 & \cdots & \alpha^{n-3} \\ \vdots & \vdots & \vdots & \vdots & \vdots \\ \alpha^{n-1} & \alpha^{n-2} & \alpha^{n-3} & \cdots & 1 \end{bmatrix}.$$

The inverse of this matrix is tridiagonal,

$$(3.3) \qquad V^{-1} = \frac{1}{\sigma^2} \begin{bmatrix} 1 & -\alpha & 0 & \dots & 0 & 0 & 0 \\ -\alpha & 1+\alpha^2 & -\alpha & \dots & 0 & 0 & 0 \\ 0 & -\alpha & 1+\alpha^2 & \dots & 0 & 0 & 0 \\ \vdots & \vdots & \vdots & \vdots & \vdots & \vdots & \vdots \\ 0 & 0 & 0 & \dots & 1+\alpha^2 & -\alpha & 0 \\ 0 & 0 & 0 & \dots & -\alpha & 1+\alpha^2 & -\alpha \\ 0 & 0 & 0 & \dots & 0 & -\alpha & 1 \end{bmatrix}$$

and has a Cholesky factorization that is lower-triangular with ones on the main diagonal except in the upper left-hand corner, $-\alpha$ on the sub-diagonal, and zeros everywhere else,

$$(3.4) \qquad K = \frac{1}{\sigma} \begin{bmatrix} (1-\alpha^2)^{\frac{1}{2}} & 0 & 0 & \dots & 0 & 0 \\ -\alpha & 1 & 0 & \dots & 0 & 0 \\ 0 & -\alpha & 1 & \dots & 0 & 0 \\ \vdots & \vdots & \vdots & \vdots & \vdots & \vdots \\ 0 & 0 & 0 & \dots & 1 & 0 \\ 0 & 0 & 0 & \dots & -\alpha & 1 \end{bmatrix}.$$

For a given value of α, premultiplication of the error vector, ε by K multiplies the first element by $(1-\alpha^2)^{\frac{1}{2}}$ and replaces the other elements by one-step prediction residuals or innovations. This is the proper weighting for calculating the exact likelihood since it is based on the square root of the exact inverse of the error covariance matrix. Premultiplying the regression equation (2.3) by K performs the same operation on Y and each column of X. Solving the transformed normal equations (2.5) allows the likelihood to be calculated for the chosen value of α. By varying α until $-2 \ln$ likelihood (2.8) is minimized, maximum likelihood estimates of both α and β are obtained. In this case

$$(3.5) \qquad V_1 = 1/(1-\alpha^2), \qquad V_i = 1, \qquad i > 1,$$

so

$$(3.6) \qquad \sum_{i=1}^{n} \ln V_i = -\ln (1-\alpha^2).$$

This term is important in the likelihood function for a stationary first-order autoregression since it gets large as $|\alpha| \to 1$. If α is estimated using a non-linear optimization routine, the logistic-type transformation used by Jones (1980) can be used to keep α in the proper range,

$$(3.7) \qquad \alpha = \frac{1 - e^{-u}}{1 + e^{-u}} \qquad u = \ln\left[\frac{1 + \alpha}{1 - \alpha}\right].$$

As u varies from $-\infty$ to ∞, α varies from -1 to 1.

A generalized likelihood ratio test for the significance of α can be obtained by first calculating $-2\ln$ likelihood for $\alpha = 0$ and testing the change in $-2\ln$ likelihood at this value and the value at the minimum as chi square with one degree of freedom. If the maximum likelihood adjustment for serial correlation is used, it is felt that the usual t and F tests for the regression coefficients, although now approximate rather than exact, can still be used if the denominator degrees of freedom are reduced by 1 to allow for the estimation of α.

4. ARMA$(1, 1)$ errors

An alternate parameterization of an ARMA$(1, 1)$ process is as an AR(1) process with observational error (Box and Jenkins (1976); Jones (1984)). This is a two-parameter model with sufficient flexibility to approximately model the error structure of many regression situations, especially when the sample size is not too large. It also allows the use of a scalar Kalman filter to transform the regression equation, and can be used to demonstrate the general case of ARMA(p, q) errors.

The model for the regression errors consists of a state equation,

$$(4.1) \qquad w_i = \alpha w_{i-1} + \eta_i,$$

and an 'observation' equation

$$(4.2) \qquad \varepsilon_i = w_i + v_i.$$

Here the term observation equation is used rather loosely since it is the dependent variable, y_i, and the corresponding row of the X matrix, that are actually observed. Equations (4.1) and (4.2) are simply an alternate way of parameterizing an ARMA$(1, 1)$ process. It is assumed that the η_i are independent $N(0, \sigma^2)$, and v_i are independent $N(0, \sigma^2 R)$, and independent of η_i. The additional parameter to be estimated in this example is the scalar R.

For given values of α and R, the Kalman recursion starts by specifying the initial state as

$$(4.3) \qquad w(0 \mid 0) = 0,$$

with variance equal to the unconditional variance of w_i divided by σ^2,

$$(4.4) \qquad P(0 \mid 0) = 1/(1 - \alpha^2).$$

The recursion for the Kalman filter proceeds as follows:

[1] Predict the state,

$$(4.5) \qquad w(i \mid i-1) = \alpha w(i-1 \mid i-1).$$

[2] Calculate the variance of the prediction,

$$(4.6) \qquad P(i \mid i-1) = \alpha^2 P(i-1 \mid i-1) + 1.$$

This assumes that σ^2 has been concentrated out of the likelihood and will be estimated at the end.

[3] Using the next observation, calculate the innovation,

$$(4.7) \qquad I(i) = \varepsilon_i - w(i \mid i-1),$$

[4] and the innovation variance,

$$(4.8) \qquad V_i = P(i \mid i-1) + R.$$

At this stage the term

$$(4.9) \qquad I^2(i)/V_i$$

is accumulated into RSS, and

$$(4.10) \qquad \ln V_i$$

is accumulated for the determinant contribution to $-2 \ln$ likelihood.

[5] The state estimate is updated,

$$(4.11) \qquad w(i \mid i) = w(i \mid i-1) + P(i \mid i-1)I(i)/V_i,$$

[6] as is its variance,

$$(4.12) \qquad P(i \mid i) = P(i \mid i-1)R/(P(i \mid i-1) + R).$$

The preceding recursion uses ε_i as its data because it is being applied to the error vector in a regression equation. For the actual calculations in a regression situation, ε_i is replaced by y_i, and each of the x_{ij}.

5. Unequally spaced data

For unequally spaced data, the error structure can be modelled as a continuous-time autoregression with observational error as in Jones (1981). To illustrate this situation, details will be given for a continuous-time AR(1), or CAR(1), process. Generalization to a CAR(p) process with observational error is straightforward, and the FORTRAN code given in Jones (1981) is easily modified to handle regression.

A stationary CAR(1) process is

$$(5.1) \qquad dz(t) + \gamma z(t)dt = dW(t),$$

where $\gamma > 0$ and $W(t)$ is a Wiener process, i.e. $dW(t)$ is continuous-time zero-mean Gaussian 'white noise' with instantaneous variance

$$(5.2) \qquad\qquad E\,|dW(t)|^2 = Q\,dt.$$

Process (5.1) has covariance function at lag τ of

$$(5.3) \qquad\qquad C(\tau) = \frac{Q}{2\gamma}\exp\left(-\gamma\,|\,\tau\,|\right),$$

and spectral density

$$(5.4) \qquad\qquad s(f) = \frac{Q}{\gamma^2 + (2\pi f)^2}$$

where f is frequency in cycles per unit time.

If sampled at equally spaced time points, with sampling interval h, the resulting sampled process is a discrete-time AR(1) process with parameter

$$(5.5) \qquad\qquad \alpha = \exp\left(-\gamma h\right)$$

and variance of the random input

$$(5.6) \qquad\qquad \sigma^2 = \frac{Q}{2\gamma}\left(1 - \exp\left(-2\gamma h\right)\right).$$

The problem considered here is when observations are taken at unequally spaced time points, $t_1 < t_2 < \cdots < t_n$, and the process is observed with error,

$$(5.7) \qquad\qquad \varepsilon_i = z(t_i) + v_i,$$

where the observational error, v_i, is assumed to be zero-mean Gaussian with variance QR. As in the previous section where the observational error was assumed to have variance $\sigma^2 R$ so that when σ^2 was concentrated out of the likelihood, R remained to be estimated, here Q will be concentrated out leaving R to be estimated. Actually, R represents the ratio of the observational error variance to the input noise variance.

The recursion given in Section 4 can now be simply modified to allow for unequally spaced data. Before any observations are taken, the initial state of the process is

$$(5.8) \qquad\qquad z(0\,|\,0) = 0$$

with variance

$$(5.9) \qquad\qquad P(0\,|\,0) = \frac{1}{2\gamma}.$$

For a given time step of length

$$(5.10) \qquad\qquad \delta_i = t_i - t_{i-1},$$

let

(5.11) $$\alpha(\delta_i) = \exp(-\gamma\delta_i).$$

[1] The first step of the recursion (4.5) becomes

(5.12) $$z(t_i \mid t_{i-1}) = \alpha(\delta_i) z(t_i \mid t_{i-1}),$$

[2] with variance

(5.13) $$P(t_i \mid t_{i-1}) = \alpha(\delta_i)^2 P(t_{i-1} \mid t_{i-1}) + \frac{1}{2\gamma}(1 - \alpha^2(\delta_i)).$$

[3] The innovation is

(5.14) $$I(t_i) = \varepsilon_i - z(t_i \mid t_{i-1}),$$

[4] with innovation variance

(5.15) $$V(t_i) = P(t_i \mid t_{i-1}) + R.$$

(5.16) $$I(t_i)^2 / V(t_i) \quad \text{and} \quad \ln V(t_i)$$

are accumulated to determine $-2 \ln$ likelihood.

[5] The updated state estimate is

(5.17) $$z(t_i \mid t_i) = z(t_i \mid t_{i-1}) + P(t_i \mid t_{i-1}) I(t_i) / V(t_i),$$

[6] with variance

(5.18) $$P(t_i \mid t_i) = P(t_i \mid t_{i-1}) R / (P(t_i \mid t_{i-1}) + R).$$

6. Example

Davis and Nihan (1984) have used the procedures in this paper to study traffic flow in Interstate highway I-5 just north of Seattle, Washington. This was an intervention analysis since during the study period a median improvement project blocked half of the median lane. The dependent variable was the peak 15-minute volume each weekday. Of the 107 weekdays, 21 were coded as missing because of detector failures and other problems. Variables considered in the regression were an indicator variable for the intervention, an indicator variable denoting whether there was an accident or other incident that day, and a covariate giving the traffic flow in another part of the freeway system not affected by the blockage. The purpose of the study was to determine if the improvement project affected the traffic flow in the area. Using ordinary regression, the intervention effect was significant at the 5 per cent level; however, there was significant lag-1 correlation in the residuals. When the errors of the regression were modelled as an MA(1) process using the techniques in this paper for handling missing observations, the intervention effect was no longer significant.

7. Conclusion

It is well known that serial correlation in the errors of a regression analysis often causes false significant results. If this error structure is approximately modelled, the statistical tests on regression coefficients are much more realistic. This paper has concentrated on two-parameter models for stationary error structure, an ARMA(1, 1) model for equally spaced data with missing observations, and a continuous-time AR(1) process with observational error for unequally spaced data. While these simple models are often sufficient for relatively short data series, the more general ARMA(p, q) or continuous-time AR(p) process with observational error can be used if the error correlation structure is more complicated.

References

BOX, G. E. P. AND JENKINS, G. M. (1976) *Time Series Analysis: Forecasting and Control.* revised edn. Holden-Day, San Francisco.

DAVIS, G. A. AND NIHAN, N. L. (1984) Using time series designs to estimate changes in freeway level of service, despite missing data. *Transportation Res.* **18**A, 431–438.

DUNCAN, D. B. AND JONES, R. H. (1966) Multiple regression with stationary errors. *J. Amer. Statist. Assoc.* **61**, 917–928.

GOLOB, G. H. AND PEREYRA, V. (1973) The differentiation of pseudoinverses and non-linear least squares problems whose variables separate. *SIAM J. Numerical Analysis* **10**, 413–452.

GRAYBILL, F. A. (1976) *Theory and Application of the Linear Model.* Duxbury Press, North Scituate, Massachusetts.

GRENANDER, U. AND ROSENBLATT, M. (1957) *Statistical Analysis of Stationary Time Series.* Wiley, New York.

HAMON, B. V. AND HANNAN, E. J. (1963) Estimating relations between time series. *J. Geophysical Res.* **68**, 6033–6041.

HANNAN, E. J. (1963) Regression for time series. In *Time Series Analysis*, ed. M. Rosenblatt, Wiley, New York, 17–37.

HARVEY, A. C. (1981) *Time Series Models.* Wiley, New York.

HARVEY, A. C. AND PHILLIPS, G. D. A. (1979) Maximum likelihood estimation of regression models with autoregressive-moving average disturbances. *Biometrika* **66**, 49–58.

JONES, R. H. (1980) Maximum likelihood fitting of ARMA models to time series with missing observations. *Technometrics* **22**, 389–395.

JONES, R. H. (1981) Fitting a continuous time autoregression to discrete data. In *Applied Time Series Analysis* II, ed. D. F. Findley, Academic Press, New York, 651–682.

JONES, R. H. (1985) Time series analysis with unequally spaced data. In *Handbook of Statistics, Volume 5: Time Series in the Time Domain*, ed. E. J. Hannan, P. R. Krishnaiah and M. M. Rao. North-Holland, Amsterdam.

WECKER, W. E. AND ANSLEY, C. F. (1983) The signal extraction approach to nonlinear regression and spline smoothing. *J. Amer. Statist. Assoc.* **78**, 81–89.

The 'Derived' Moving-Average Model and its Role in Causality

J. H. W. PENM
R. D. TERRELL

Abstract

In the situations where restrictions on the multivariate subset AR model are known, we propose methods of providing suitable standard errors of estimate and prediction which assist in assessing the importance of the coefficients appearing the 'derived' moving-average (MA) model. The coefficient patterns of the derived moving-average model are proposed as an alternative basis for detecting Granger-causality.

GRANGER-CAUSALITY; SUBSET AUTOREGRESSION; RESTRICTED SUBSET AUTOREGRESSION; 'DERIVED' MOVING-AVERAGE REPRESENTATION; ASYMPTOTIC STANDARD ERROR OF ESTIMATE

1. Introduction

Investigation of the cause and effect relationship, or simply the causality which exists among economic and social variables, is usually based on procedures employing economic and econometric model building. Granger (1963), (1969) suggested a definition of causality that makes no explicit use of economic laws to provide *a priori* restrictions on the structure; he also fitted a multivariate autoregression (AR) for empirical model building to detect Granger-causality. Subsequently, Sims (1972), (1980), Hsiao (1979), (1982), Caines et al. (1981), and Tjøstheim (1981) proposed procedures to test Granger-causality in a class of multivariate autoregressive (AR) models. Geweke (1978), (1981), (1982) showed how these notions relate to exogeneity in the context of a complete dynamic simultaneous equation model, and Penm and Terrell (1982), (1984a,b) have proposed algorithms to study the multivariate restricted subset AR models, which allow possible zero entries in the non-zero coefficient matrices of AR models: the optimal AR model is then used as a basis for detecting causality.

However, to better understand the dynamic behaviour of the chosen model, Sims (1972), (1980) has advocated expressing the model in its moving-average

(MA) representation. This latter representation has the added advantage that it may be easily comprehended as it describes the system's response to typical random shocks. Hsiao (1982) also defined possible indirect causality and spurious causality in terms of both AR and MA representation. An additional reason for the development of the derived moving average representation is seen when we consider multivariate models of the type discussed in Hsiao (1982). To illustrate this issue we consider a model:

$$\begin{pmatrix} u_{1t} \\ u_{2t} \\ u_{3t} \end{pmatrix} = \begin{Bmatrix} 1 & 0.5L & 0 \\ 0 & 1 & 0.5L^4 \\ 0 & 0 & 1 \end{Bmatrix} \begin{pmatrix} \varepsilon_{1t} \\ \varepsilon_{2t} \\ \varepsilon_{3t} \end{pmatrix}$$

where $(\varepsilon_{1t}, \varepsilon_{2t}, \varepsilon_{3t})$ are mutually orthogonal disturbances generated by an independent Gaussian process. The associated vector autoregressive process is simply found by inversion to be

$$\begin{Bmatrix} 1 & -0.5L & 0.25L^5 \\ 0 & 1 & -0.5L^4 \\ 0 & 0 & 1 \end{Bmatrix} \begin{pmatrix} u_{1t} \\ u_{2t} \\ u_{3t} \end{pmatrix} = \begin{pmatrix} \varepsilon_{1t} \\ \varepsilon_{2t} \\ \varepsilon_{3t} \end{pmatrix}.$$

If attention is focused only on which coefficients in the above restricted subset autoregression are zero or non-zero then the causal nature of the underlying process will be misunderstood. As the vector autoregressive model is much easier to use to obtain a model which fits the data well, we advocate the derivation of the associated moving-average representation and the relevant variance–covariance matrix for these derived estimates.

In the full-order AR model, Baillie (1981), (1982) provided the asymptotic standard error of estimate for the coefficients in the AR equivalent MA representation. However, in situations where it is necessary to estimate a restricted AR model, his proposed methods need slight modification; we therefore extend his work to provide the asymptotic prediction mean squared error for the multivariate restricted AR model and the asymptotic standard error of estimate for those coefficients appearing in an MA representation *derived* from a restricted AR. The detailed account is presented in Section 4.

As an alternative to the method based on an autoregression we consider the MA representation to assess Granger-causality. After we identify either the optimum subset or restricted subset AR model (see Penm and Terrell (1982), (1984b)), we consider the derived MA representation and with suitable standard errors of estimate for the MA representation we assess from the non-zero coefficients the Granger-causal patterns. Section 2 briefly introduces restricted multivariate AR models. Section 3 comments on assessment of patterns of Granger-causality. Section 4 supplies the required theoretical developments on the standard errors of estimate and the mean square prediction error of a restricted AR model and Section 5 gives an empirical example. Conclusions are in Section 6.

2. Multivariate restricted autoregression

In AR modelling, let $\boldsymbol{u}(t)$ be a zero-mean, wide-sense-stationary time series of dimension m. We consider the multivariate AR(q) model of the form

$$(2.1) \qquad B^q(L)\boldsymbol{u}(t) = \boldsymbol{u}(t) + \sum_{\tau=1}^{q} B_\tau \boldsymbol{u}(t-\tau) = \varepsilon(t),$$

where $B^q(L) = I + \sum_{\tau=1}^{q} B_\tau L^\tau$, L is the lag operator, $L\boldsymbol{u}(t) = \boldsymbol{u}(t-1)$, $B_0 = I$, B_τ, $\tau = 1, 2, \cdots, q$ are $m \times m$ parameter matrices, $b_\tau(i, j)$ is the i, jth entry of B_τ, and $\varepsilon(t)$ is an $m \times 1$ stationary vector process with $E\{\varepsilon(t)\} = 0$ and with

$$E\{\varepsilon(t)\varepsilon'(t-\tau)\} = \begin{cases} G, & \tau = 0 \\ 0, & \tau > 0. \end{cases}$$

Conventionally, researchers proposed procedures to identify q, the order of AR model, in (2.1). After q is identified, then every entry $b_\tau(i, j)$ of B_τ, $\tau = 1, \cdots, q$ may be assumed to be non-zero, i.e. they may neglect the possible AR (q) model with possible zero entries $b_\tau(i, j)$ in the coefficient matrices, B_τ. In this paper we consider two kinds of restricted autoregression, subset autoregression and restricted subset autoregression. In subset autoregressions, there exist possible zero B_τ matrices in (2.1) for $\tau < q$. In restricted subset autoregression, there exist possible zero entries $b_\tau(i, j)$ in the non-zero coefficient matrices, B_τ, of the subset autoregression.

McClave (1975) developed an algorithm to identify scalar subset AR Unfortunately most economic time series analysis needs multivariate modelling. Penm and Terrell (1982), (1984b) presented recursive methods and search methods to identify *multivariate subset* AR *models*. We also needed to assess whether elements belonging to these non-zero coefficient matrices should be included within the optimal subset model, so we further augmented the search methods by employing a block Choleski decomposition to fit *restricted* multivariate subset AR. In conjunction with model selection criteria we then selected the optimum restricted subset AR. The optimal AR model is then used as a basis for detecting causality.

3. Causality testing

For ease of exposition, we shall partition $\boldsymbol{u}(t)$ of (2.1) into $[\boldsymbol{u}_1(t), \boldsymbol{u}_2(t), \boldsymbol{u}_3(t)]'$. The AR representation of (2.1) is then

$$(3.1) \qquad \begin{pmatrix} I + B_{11}(L) & B_{12}(L) & B_{13}(L) \\ B_{21}(L) & I + B_{22}(L) & B_{23}(L) \\ B_{31}(L) & B_{32}(L) & I + B_{33}(L) \end{pmatrix} \begin{pmatrix} \boldsymbol{u}_1(t) \\ \boldsymbol{u}_2(t) \\ \boldsymbol{u}_3(t) \end{pmatrix} = \varepsilon(t),$$

where

$$B_{ij}(L) = \sum_{\tau=1}^{q} B_{ij,\tau} L^{\tau}.$$

It is also well known that (3.1) also admits an MA representation

(3.2)

$$\begin{pmatrix} u_1(t) \\ u_2(t) \\ u_3(t) \end{pmatrix} = A(L)\varepsilon(t) = A_0\varepsilon(t) + A_1\varepsilon(t-1) + \cdots$$

$$= \begin{pmatrix} I + A_{11}(L) & A_{12}(L) & A_{13}(L) \\ A_{21}(L) & I + A_{22}(L) & A_{23}(L) \\ A_{31}(L) & A_{32}(L) & I + A_{33}(L) \end{pmatrix} \varepsilon_t$$

where $A_0 = I$. A natural way of providing a basic causal ordering which links AR and MA models may then be developed from the work of Sims (1972) and Hsiao (1982).

Case 1. A sufficient condition for $u_3(t)$ not to Granger-cause $u_1(t)$ in the form (3.1) and (3.2) is that the following conditions hold:
   1.a: The autoregressive operator $B(L)$ is lower block triangular,
   1.b: The moving average operator $A(L)$ is lower block triangular.

Consider the AR model of the form (3.1) with the autoregressive operator of the form

$$\begin{pmatrix} I + B_{11}(L) & 0 & 0 \\ B_{21}(L) & I + B_{22}(L) & B_{23}(L) \\ B_{31}(L) & B_{32}(L) & I + B_{33}(L) \end{pmatrix},$$

where, although $B_{23}(L) \neq 0$, $u_3(t)$ does not Granger-cause $u_1(t)$ and so $B(L)$ is not a lower block triangular matrix in the sense indicated by the dotted lines, and therefore clearly Condition 1.a is not a necessary condition for $u_3(t)$ not to Granger-cause $u_1(t)$. Similarly we could show that Condition 1.b is not a necessary condition for $u_3(t)$ not to Granger-cause $u_1(t)$. However, we notice that if we consider a matrix polynomial with the form

$$\begin{pmatrix} I + B_{11}(L) & 0 & 0 \\ B_{21}(L) & I + B_{22}(L) & B_{23}(L) \\ B_{31}(L) & B_{32}(L) & I + B_{33}(L) \end{pmatrix}$$

then for this particular lower block-triangular matrix, which is the least restrictive of the possible forms, the Conditions 1.a and 1.b are alternative necessary and sufficient conditions for $u_3(t)$ not to Granger-cause $u_1(t)$. Analogously a sufficient condition for $u_1(t)$ not to Granger-cause $u_3(t)$ in the form (3.1) and (3.2) is that either of the following conditions hold:
   1.c: The autoregressive operator $B(L)$ is upper block triangular,
   1.d: The moving average operator $A(L)$ is upper block triangular.

There are clearly different forms of block triangularity when there are more than two groups of variables in $\boldsymbol{u}(t)$ and so the actual nature of the block triangularity is only relevant when defining necessary and sufficient conditions.

More general causal ordering can be treated using definitions suggested by Hsiao (1982). Consider the pattern

$$\boldsymbol{u}_1(t) \longleftarrow \boldsymbol{u}_2(t)$$
$$\uparrow$$
$$\boldsymbol{u}_3(t)$$

which indicates $\boldsymbol{u}_3(t)$ causing $\boldsymbol{u}_1(t)$ but only through $\boldsymbol{u}_2(t)$. We call this situation indirect causality from $\boldsymbol{u}_3(t)$ to $\boldsymbol{u}_1(t)$, and we have the following.

Case 2a. A sufficient condition for $\boldsymbol{u}_3(t)$ not to Granger-cause $\boldsymbol{u}_1(t)$ directly, but to cause $\boldsymbol{u}_1(t)$ indirectly in the AR model (3.1) is that

$$B_{13,\tau} = 0 \quad \text{for all } \tau, \quad \text{and} \quad B_{12,\tau} \neq 0, \quad B_{23,\tau} \neq 0 \quad \text{for some } \tau.$$

Case 2b. A sufficient condition for $\boldsymbol{u}_3(t)$ not to Granger-cause $\boldsymbol{u}_1(t)$ directly, but to cause $\boldsymbol{u}_1(t)$ indirectly in the MA model of the form (3.2) is that

$$A_{13,\tau} = 0 \quad \text{for all } \tau, \quad \text{and} \quad A_{12,\tau} \neq 0, \quad A_{23,\tau} \neq 0 \quad \text{for some } \tau.$$

If we multiply $\varepsilon(t)$ of (3.1) by P, where $P\Omega P' = I$, then (3.1) has an MA representation:

$$(3.3) \qquad \boldsymbol{u}(t) = \Phi_0 a(t) + \sum_{j=1}^{\infty} \Phi_j a(t-j),$$

where

$$a(t) = P\varepsilon(t), \qquad E\{a(t)a(r-\tau)\} = \begin{cases} I, & \tau = 0 \\ 0, & \tau \neq 0, \end{cases}$$

and Φ_0 is a lower-triangular matrix. With this representation, Condition 1.b has to be reconstructed. We can similarly transform (3.1) to have an MA representation where Φ_0 is a lower-triangular matrix, and Condition 1.d has then to be modified too. As an illustration, we consider an MA representation of (3.2) with the form

$$(3.4) \qquad \begin{pmatrix} \boldsymbol{u}_1(t) \\ \boldsymbol{u}_2(t) \\ \boldsymbol{u}_3(t) \end{pmatrix} = \varepsilon(t) + \begin{pmatrix} p & 0 & 0 \\ 0 & q & s \\ 0 & r & t \end{pmatrix} \varepsilon(t-1) + \begin{pmatrix} u & 0 & 0 \\ 0 & v & w \\ 0 & x & y \end{pmatrix} \varepsilon(t-2) + \cdots$$

where $\boldsymbol{u}_1(t)$ does not Granger-cause $\boldsymbol{u}_3(t)$. Now we transform the form (3.4) into (3.3) and we have

$$(3.5) \qquad \begin{pmatrix} \boldsymbol{u}_1(t) \\ \boldsymbol{u}_2(t) \\ \boldsymbol{u}_3(t) \end{pmatrix} = \begin{pmatrix} a & 0 & 0 \\ b & c & 0 \\ d & e & f \end{pmatrix} a(t) + \begin{pmatrix} pa & 0 & 0 \\ qb+sd & qc+se & sf \\ rb+td & rc+te & tf \end{pmatrix} a(t-1) + \cdots.$$

In assessing Granger-causality from the form (3.5), it is not at all straight-forward to recognize that $u_1(t)$ does not Granger-cause $u_3(t)$, which illustrates the inconvenience of detecting causality patterns from an MA-representation of the form (3.3), and indicates our reasons for preferring to use the canonical (the reduced form) MA representation with $A_0 = I$.

4. The theoretical development

In this section, for a situation where the constraints which make up the restricted AR model are all known, we derive the asymptotic prediction mean squared error for multivariate restricted autoregression, and the asymptotic standard error of estimate of those coefficients appearing in its MA representation. Following Baillie's method (1982), it is convenient to express the full AR model of (2.1) by means of the companion form representation

$$
\begin{pmatrix} u(t) \\ u(t-1) \\ \cdot \\ \cdot \\ \cdot \\ u(t-q+1) \end{pmatrix} = \begin{pmatrix} -B_1 & -B_2 & \cdot & \cdot & \cdot & -B_q \\ I & 0 & \cdot & \cdot & \cdot & 0 \\ 0 & I & \cdot & \cdot & \cdot & 0 \\ \cdot & \cdot & \cdot & \cdot & \cdot & \cdot \\ \cdot & \cdot & \cdot & \cdot & \cdot & \cdot \\ 0 & 0 & \cdot & \cdot & I & 0 \end{pmatrix} \begin{pmatrix} u(t-1) \\ u(t-2) \\ \cdot \\ \cdot \\ \cdot \\ u(t-q) \end{pmatrix} + \begin{pmatrix} \varepsilon(t) \\ 0 \\ \cdot \\ \cdot \\ \cdot \\ 0 \end{pmatrix}
$$

which is written as

$$(4.1) \qquad\qquad U(t) = BU(t-1) + \eta(t).$$

We premultiply (4.1) by N' which gives

$$(4.2) \qquad\qquad u(t) = N'BU(t-1) + \varepsilon(t),$$

where $N' = [I_m : 0]$, which is of dimension $m \times mq$, and the coefficients of the multivariate AR model of (4.2) can be represented by a row vector γ', where

$$(4.3) \qquad\qquad \gamma' = R(N'B),$$

and R is the row stacking operator.

Consider the $c_1, c_2, \cdots, c_k$th elements of γ' are constrained to 0, where $1 \leq c_1 \leq c_2 \leq \cdots \leq c_k \leq m^2q$, then the formula (4.3) becomes

$$\gamma'(C_k) = R(N'B(C_k)),$$

where C_k is an integer set which contains $c_1, c_2, \cdots, c_k$, and the Equation (4.2) is

$$(4.4) \qquad\qquad u(t) = N'B(C_k)U(t-1) + \eta(t).$$

The equivalent MA representation of the restricted AR model with the form

(4.1) is

$$U(t) = \sum_{j=0}^{\infty} B^j(C_K)\eta(t-j), \qquad B_0 = I$$

and the equivalent MA representation of the restricted AR model with the form (2.1) is

(4.5)
$$u(t) = \sum_{j=0}^{\infty} A_j\varepsilon(t-j), \quad \text{where} \quad A_j = N'B^j(C_k)N,$$

which gives an expression for the multivariate restricted time series model. At time $n+l$, the model of the form (4.4) becomes

$$u(n+l) = N' \sum_{j=0}^{l-1} B^j(C_k)\eta(n+l-j) + N'B^l(C_k)U(n),$$

therefore the minimum mean square error for a prediction made at time n, l periods ahead, is

(4.6)
$$u_{n,l} = N'B^l(C_k)U(n).$$

Once the model has been estimated, (4.6) includes estimates of $B^l(C_k)$ and becomes the predictor

$$\hat{u}_{n,l} = N'\hat{B}^l(C_k)U(n),$$

where the l-step prediction error is

(4.7)
$$e_{n,l} = N' \sum_{j=0}^{l-1} B^j(C_k)\eta(n+l-j) - [N'\hat{B}^l(C_k) - N'B^l(C_k)]U(n).$$

Hence,

(4.8)
$$e'_{n,l} = \sum_{j=0}^{l-1} \eta'(n+l-j)B'^j(C_k)N - R[N'\hat{B}^l(C_k) - N'B^l(C_k)][I_m \otimes U(n)]$$

where I_m is an $m \times m$ identity matrix.

Using a first-order Taylor expansion with higher-order terms neglected we have

(4.9)
$$R[N'\hat{B}^l(C_k)] \simeq R[N'B^l(C_k)] + [\hat{\gamma}'(C_k) - \gamma'(C_k)]Q_l,$$

where

$$Q_l = \frac{\partial R[N'B^l(C_k)]}{\partial \gamma(C_k)}.$$

Now we define $\gamma'_r(C_k) = R[B(C_k)]$, then

$$\frac{\partial \gamma_r(C_k)}{\partial \gamma(C_k)} = I_{m^2q}(C_k)[N' \otimes I_{mq}],$$

where $(c_1, c_1), (c_2, c_2), \cdots, (c_k, c_k)$th diagonal element of $I_{m^2q}(C_k)$ has the value 0.

Using the relation $A_j = N'B^j(C_k)N$, and modifying the result due to Yamamoto (1976), Schmidt (1974) and Baillie (1982), we have

$$Q_l = I_{m^2q}(C_k)[N \otimes I_{mq}]\frac{\partial R[N'B^l(C_k)]}{\partial \gamma_r(C_k)}$$

(4.10)
$$= I_{m^2q}(C_k)[N' \otimes I_{mq}]\left(\sum_{j=0}^{l-1} B''^{j}(C_k) \otimes B^{l-1-j}(C_k)\right)[N \otimes I_{mq}]$$

$$= I_{m^2q}(C_k) \sum_{j=0}^{l-1} A_j' \otimes B^{l-1-j}(C_k),$$

which appropriately modifies M_l proposed in Baillie's (1982) work.

Next, to combine the formulas (4.7), (4.8), (4.9) and (4.10) we can obtain $E[e_{n,l}e_{n,l}']$, i.e. the asymptotic mean square error of an l-step prediction is then given by

(4.11)
$$\text{amse}(\hat{u}_{n,l}) = N' \sum_{j=0}^{l-1} B^j(C_k)NGN'B''^{j}(C_k)N$$
$$+ (I_m \otimes U_n)'Q_l'K(C_k)Q_l(I_m \otimes U_n)$$

where

$$K(C_k) = E\{[\hat{\gamma}(C_k) - \gamma(C_k)][\hat{\gamma}'(C_k) - \gamma'(C_k)]\}.$$

The last term in Equation (4.11) is purely due to the sample error in the parameter estimation.

Consider the asymptotic standard error of estimate of those coefficients appearing in the MA representation, (4.5). We define

(4.12)
$$\alpha_j = R(A_j) = R[N'B^j(C_k)][I_m \otimes N],$$

and

(4.13)
$$\hat{\alpha}_j = R[N'\hat{B}^j(C_k)][I_m \otimes N].$$

Using the relations in (4.9), (4.10), (4.12) and (4.13), we can show that

(4.14)
$$E[(\alpha_j - \hat{\alpha}_j)(\alpha_j' - \hat{\alpha}_j')] = [I_m \otimes N']Q_j'K(C_k)Q_j[I_m \otimes N].$$

Now we consider the calculation of $K(C_k)$. In a full-order autoregression, we have

(4.15)
$$K = G \otimes \Gamma^{-1} = \begin{pmatrix} g_{11}\Gamma^{-1} & g_{12}\Gamma^{-1} & \cdots & g_{1m}\Gamma^{-1} \\ \cdot & \cdot & \cdots & \\ \cdot & \cdot & \cdots & \\ \cdot & \cdot & \cdots & \\ g_{m1}\Gamma^{-1} & g_{m2}\Gamma^{-1} & \cdots & g_{mm}\Gamma^{-1} \end{pmatrix}$$

where the (i, j)th block matrix of (4.15) is

$$(4.16) \qquad g_{ij}(\Gamma)^{-1}(\Gamma)(\Gamma)^{-1} = g_{ij}\Gamma^{-1},$$

and where g_{ij} is the (i, j)th entry of G, and $\Gamma = E[\boldsymbol{u}(t)\boldsymbol{u}'(t)]$. In a restricted AR the (i, j)th block matrix of $K(C_k)$ becomes

$$g_{ij}[I_{mq}(C_i)\Gamma I_{mq}(C_i)]^{-1}[I_{mq}(C_i)\Gamma I_{mq}(C_j)][I_{mq}(C_j)\Gamma I_{mq}(C_j)]^{-1}$$

where C_i is an integer set which contains $c^1, c^2, \cdots, c^a$, and the $c^1, c^2, \cdots, c^a$th entry of the ith row of $N'B$ has been constrained to 0. Similarly C_j is an integer set which contains $c_1, c_2, \cdots, c_b$ and the $c_1, c_2, \cdots, c_b$th entry of the jth row of $N'B$ has been constrained to 0.

5. Empirical example

In conventional AR modelling, model-builders often ignore the possible AR models with the intermediate orders constrained to 0 and zero entries in the coefficient matrix. To test overall causality based on such sub-optimal models may lead to a higher probability of incorrect causal patterns. Following Sim's view that a multivariate economic system is better interpreted in the MA form we could also choose to detect the Granger-causality pattern after an MA form has been derived from an estimated AR model. With this approach, we would therefore identify the subset or restricted subset autoregression, and establish suitable standard errors of estimate for the coefficients derived in the equivalent MA representation, and so detect Granger-causality. This approach provides an alternative to use of the restricted subset autoregression as a basis for detecting causality patterns.

In practice, the restrictions on the autoregression may not be all known and so the need to search for an appropriate restricted model may bias our inference. It is, however, our purpose at this stage to illustrate how to use the methods of Section 4 to provide a linkage between a restricted AR and its MA representation and in particular to provide as well the associated errors of estimates. We consider the final subset autoregression estimated by ordinary least squares in conjunction with HC (see Hannan (1981) and Hannan and Quinn (1979)) on the quarterly seasonally adjusted series for Canadian income (M_2), bank rate (BR), and money (GNP) for the period 1955 I to 1977 IV. The variables actually used are $\boldsymbol{u}_1 = \Delta \log M_2$, $\boldsymbol{u}_2 = \log \mathrm{BR}$, and $\boldsymbol{u}_3 = \Delta \log \mathrm{GNP}$; where Δ is the difference operator, and the final subset autoregression involves lag 1 and lag 3. The interested reader is referred to Penm and Terrell (1984a). The first twelve coefficient matrices, i.e. $A_j, j = 1, 2, \cdots, 12$, of the MA representation have been assessed by the proposed method in Section 4, and the coefficient matrices with the lag number j larger than 6 are effectively zero

TABLE 1

The optimum restricted subset AR selected for HC

$$u_t = [\Delta \log M_{2t}, \log BR_t, \Delta \log GNP_t]'$$

Criterion	HC
The order of the optimum AR	1, 3

The type of AR coefficient matrices estimated

$$\hat{B}_1 = \begin{pmatrix} -0.323 & 0.013 & 0 \\ (0.1019) & (0.0052) & \\ 0 & -0.741 & -4.11 \\ & (0.0716) & (1.309) \\ 0 & 0 & 0 \end{pmatrix}$$

$$\hat{B}_3 = \begin{pmatrix} 0 & 0 & 0 \\ -3.05 & 0 & 0 \\ (1.3616) & & \\ 0 & 0.019 & -0.184 \\ & (0.0054) & (0.0984) \end{pmatrix}$$

Patterns of causality

$$\begin{array}{c} \Delta \log GNP \\ \Updownarrow \qquad \diagdown \\ \log BR \Leftrightarrow \Delta \log M_2 \end{array}$$

(a) $x \Leftrightarrow y$: feedback relation and instantaneous causation between x and y.

$x - y$: instantaneous causation between x and y.

(b) The values in parentheses are the standard error of estimate of the non-zero coefficients.

TABLE 2

The subset AR model

$$u_t = [\Delta \log M_{2t}, \log BR_t, \Delta \log GNP_t]'$$

Criterion	HC
The order of the conventional AR	1, 3

The type of AR coefficient matrices estimated

$$\hat{B}_1 = \begin{pmatrix} -0.3254 & 0.0145 & 0.0500 \\ (0.1009) & (0.006) & (0.0903) \\ -0.4754 & -0.7853 & -4.059 \\ (1.4543) & (0.085) & (1.302) \\ 0.0786 & 0.0012 & -0.0604 \\ (0.1119) & (0.0066) & (0.1002) \end{pmatrix}$$

$$\hat{B}_3 = \begin{pmatrix} -0.0301 & -0.0069 & 0.0076 \\ (0.10371) & (0.0059) & (0.0091) \\ -2.3761 & 0.0083 & 0.30923 \\ (1.4942) & (0.0086) & (1.3221) \\ -0.0079 & 0.0016 & -0.18799 \\ (0.1150) & (0.0066) & (0.10179) \end{pmatrix}$$

matrices, therefore only the information in the MA representation up to lag 6 is considered in detecting Granger-causality. We compare the results based on conversion of a subset AR to the MA model in Table 1. We also show the results using only the restricted subset AR model in Table 2 for comparison.

In looking for the causal relations among the three variables in this case, the restricted subset autoregression in Table 2 supports an AR model with $\hat{B}_{13}(L)$ and $\hat{B}_{31}(L)$ zero entries, and the remaining $\hat{B}_{ij}(L)$ are non-zero entries. The MA representation derived from the subset autoregression has $\hat{A}_{13}(L)$ and $\hat{A}_{31}(L)$ with zero entries and has the remaining $\hat{A}_{ij}(L)$ non-zero entries in the form (3.2).

Therefore both the *restricted subset* autoregression and the equivalent MA representative of the *subset* autoregression (Table 3) indicate that feedback relations exist between the pair $\Delta \log \text{GNP}$ and $\log \text{BR}$ and the pair $\Delta \log M_2$ and $\log \text{BR}$, and instantaneous causality exists among all three variables.

TABLE 3

Equivalent MA representation for subset AR models in Table 2.

The maximum lag of the MA representation	7

The type of MA coefficient matrices estimated when

$\hat{A}_0 = I$

$$\hat{A}_1 = \begin{pmatrix} 0.325 & 0.015 & 0.050 \\ (0.100) & (0.006) & (0.090) \\ 0.475 & 0.785 & 4.059 \\ (1.454) & (0.086) & (1.301) \\ 0.0786 & -0.001 & 0.006 \\ (0.112) & (0.006) & (0.100) \end{pmatrix} \qquad \hat{A}_5 = \begin{pmatrix} -0.049 & 0.0002 & -0.027 \\ (0.054) & (0.005) & (0.038) \\ 3.118 & -0.057 & 1.034 \\ (1.454) & (0.128) & (1.010) \\ -0.006 & -0.012 & -0.062 \\ (0.043) & (0.004) & (0.032) \end{pmatrix}$$

$$\hat{A}_2 = \begin{pmatrix} 0.103 & -0.016 & -0.042 \\ (0.008) & (0.007) & (0.005) \\ 0.847 & 0.605 & 3.236 \\ (1.52) & (0.121) & (1.081) \\ 0.025 & -0.002 & -0.0008 \\ (0.042) & (0.005) & (0.031) \end{pmatrix} \qquad \hat{A}_6 = \begin{pmatrix} -0.046 & 0.004 & -0.029 \\ (0.043) & (0.005) & (0.037) \\ 2.234 & -0.136 & 0.016 \\ (1.222) & (0.119) & (0.952) \\ -0.039 & -0.010 & -0.016 \\ (0.061) & (0.005) & (0.058) \end{pmatrix}$$

$$\hat{A}_3 = \begin{pmatrix} 0.053 & -0.007 & -0.137 \\ (0.108) & (0.007) & (0.090) \\ 3.194 & 0.376 & 2.208 \\ (1.821) & (0.126) & (1.426) \\ 0.087 & -0.018 & 0.181 \\ (0.112) & (0.006) & (0.098) \end{pmatrix} \qquad \hat{A}_7 = \begin{pmatrix} -0.027 & 0.005 & 0.009 \\ (0.356) & (0.004) & (0.025) \\ 1.227 & -0.159 & -0.342 \\ (1.013) & (0.101) & (0.750) \\ -0.059 & -0.005 & -0.054 \\ (0.047) & (0.004) & (0.035) \end{pmatrix}$$

$$\hat{A}_4 = \begin{pmatrix} -0.18 & -0.004 & -0.038 \\ (0.080) & (0.006) & (0.056) \\ 3.595 & 0.119 & 2.184 \\ (1.704) & (0.127) & (1.213) \\ 0.034 & -0.015 & -0.072 \\ (0.064) & (0.005) & (0.062) \end{pmatrix}$$

6. Conclusion

In this paper, for a situation where the restrictions on the AR or the subset AR model are known, we provide methods to provide suitable standard errors of estimate and prediction which assist in assessing the imporance of the coefficients appearing in the equivalent MA representation and so to detect Granger-causality. This provides another approach to evaluating the behaviour of a vector economic time series and if desired an assessment of causal links.

References

BAILLIE, R. T. (1981) Prediction from the dynamic simultaneous equation model with vector autoregression errors. *Econometrica* **49**, 1331–1337.

BAILLIE, R. T. (1982) Asymptotic standard errors for moving average representation coefficients. Working Paper, University of Birmingham.

CAINES, P. E., KENG, C. W. AND SETHI, S. P. (1981) Causality analysis and multivariate autoregressive modelling with an application to supermarket sales analysis. *J. Economic Dynamics and Control* **3**, 267–298.

GEWEKE, J. (1978) Testing the exogeneity specification in the complete dynamic simultaneous equation model. *J. Econometrics* **7**, 163–186.

GEWEKE, J. (1981) Inference and causality in economic time series models. In *Handbook of Econometrics*, ed. Z. Griliches and M. Intrilligator. North-Holland, Amsterdam.

GEWEKE, J. (1982) Measurement of linear dependence and feedback between multiple time series. *J. Amer. Statist. Assoc.* **77**, 304–313.

GRANGER, C. W. J. (1963) Economic processes involving feedback. *Information and Control* **6**, 28–48.

GRANGER, C. W. J. (1969) Investigating causal relations by econometric models and cross-spectral methods. *Econometrica* **37**, 424–438.

HANNAN, E. J. (1981) Estimating the dimension of a linear system. *J. Multivariate Anal.* **11**, 474–497.

HANNAN, E. J. AND QUINN, B. G. (1979) The determination of the order of an autoregression. *J. R. Statist. Soc.* B **41**, 190–195.

HATANAKA, M. AND ODAKI, M. (1983) Policy analyses with and without a priori conditions. Mimeo, Osaka University.

HSAIO, C. (1979) Autoregressive modelling of Canadian money and income data. *J. Amer. Statist. Assoc.* **74**, 553–560.

HSAIO, C. (1982) Autoregressive modelling and causal ordering of economic variables. *J. Economic Dynamics and Control* **4**, 243–259.

McCLAVE, J. (1975) Subset autoregression. *Technometrics* **17**, 213–220.

PENM, J. H. W. AND TERRELL, R. D. (1982) On the recursive fitting of subset autoregressions. *J. Time Series Analysis* **3**, 43–59.

PENM, J. H. W. AND TERRELL, R. D. (1984a) Multivariate subset autoregressive modelling with zero constraints for detecting Granger-causality at all. *J. Econometrics* No. 3, 311–330.

PENM, J. H. W. AND TERRELL, R. D. (1984b) Multivariate subset autoregression. *Commun. Statist.–Simulations*, **13**, 449–461.

SCHMIDT, P. (1974) The asymptotic distribution of forecasts in the dynamic simulation of an econometric model. *Econometrica* **42**, 303–309.

SCHWARZ, G. (1978) Estimating the dimension of a model. *Ann. Statist.* **6**, 461–464.

SIMS, C. A. (1972) Money, income and causality. *Amer. Econom. Rev.* **62**, 540–552.

SIMS, C. A. (1977) Comment. *J. Amer. Statist. Assoc.* **72**, 23–34.

SIMS, C. A. (1980) Macro-economics and reality. *Econometrica* **48**, 1–48.

TJØSTHEIM, D. (1981) Granger-causality in multiple time series. *J. Econometrics* **17**, 157–176.

YAMAMOTO, T. (1976) Asymptotic mean square prediction error for an autoregressive model with estimated coefficients. *J. R. Statist. Soc.* C **25**, 123–127.

On a Model for a Time Series of Cross-Sections

P. M. ROBINSON

Abstract

Dynamic stationary models for mixed time series and cross-section data are studied. The models are of simple, standard form except that the unknown coefficients are not assumed constant over the cross-section; instead, each cross-sectional unit draws a parameter set from an infinite population. The models are framed in continuous time, which facilitates the handling of irregularly-spaced series, and observation times that vary over the cross-section, and covers also standard cases in which observations at the same regularly-spaced times are available for each unit. A variety of issues are considered, in particular stationarity and distributional questions, inference about the parameter distributions, and the behaviour of cross-sectionally aggregated data.

RANDOM COEFFICIENT MODELS: CONTINUOUS-TIME PROCESSES: ASYMPTOTIC EXPANSIONS

1. Introduction

Statistical data frequently occur in the form of a cross-section of time series; examples in economics, the social sciences, biology and growth and reliability studies come to mind. Such data sets vary greatly in form. In the standard situation, observations are recorded on each cross-sectional unit at the same equally-spaced times; however, even here there are great variations in both the size of cross-section and length of time series, from the very small to the very large, calling for different methods of statistical inference. In other cases each unit can be observed at the same irregular time points; or there can be missing of observations on some of the units at certain time points, owing for example to non-response in a repeated sample survey, rotation of sampled units, or because of the extinction of a cross-sectional unit during the course of the experimental or observational period; there can even be continuous-time records, as in the data set described by Lenz and Robinson (1983).

A variety of models and procedures has been developed for analyzing mixed time series and cross-section data. Models differ, in particular, in the ways in which attributes of individual cross-sectional units are represented. Considerable interest has been shown in error-components models, wherein the residual for a particular observation contains additive stochastic effects that are common to all units and all times respectively, plus an effect due to that observation only (see e.g. Balestra and Nerlove (1966), Wallace and Hussain (1969)). Another theme has been linear regression models that permit the regression coefficients to vary across units by treating each unit's coefficients as a sample from a population (see e.g. Swamy (1970); Hsiao (1975)). Most of the literature has emphasized a regression approach, and relatively little attention has been given to the dynamic modelling of mixed time series and cross-section data, or even to a satisfactory treatment of serial correlation in the residuals from regression models for such data. Most workers have assumed the autocorrelation structure to be identical across realizations (see e.g. Azzalini (1981)). Given long enough time series, separate time series models could of course be estimated for each unit, though some common features would be expected and it might be desirable to investigate and exploit these, and in much longitudinal data the time series will be too short to contemplate standard time series modelling. Motivated by such considerations, Robinson (1978a) proposed a first-order autoregressive (AR) model in which the AR coefficient varies across units, each unit sampling a coefficient value from an infinite population. The philosophy is like that of Swamy (1970) and Hsiao (1975), suggesting a similar fundamental structure for all units, but permitting variations of a possibly general nature in fine-grain behaviour.

In this paper we again consider a random coefficient approach. However, our models are framed in continuous time, because of the applicability of such models to the relatively non-standard types of mixed time series and cross-section data described in the first paragraph above, as well as to standard data sets. Cross-sectional units are modelled by first- and second-order stochastic differential equations, with coefficients varying randomly over units. Soong ((1973), pp. 222–228) suggested that a similar model would be of interest in engineering and biomedical applications, though the multivariate first-order model which he studied is, unlike our models, deterministic conditional on coefficient values and initial conditions.

Discrete equally-spaced observations on the first-order differential equation model studied in Section 2 satisfy the model of Robinson (1978a), and in fact we discuss for our models the type of issues studied by Robinson (1978a) for the discrete AR (1). However, we deal also with other aspects. We consider a second-order differential equation model having either real or complex conjugate roots. We allow to some extent for irregularly-spaced and cross-sectionally varying observation times. We allow for a random scale parameter. We study the distributional structure of the data. We consider the stationarity and autocorrelation structure of cross-sectionally aggregated data.

This paper is dedicated to Professor E. J. Hannan on his sixty-fifth birthday.

2. First-order model

2.1. *The model.* We consider a population of unobservable, independent, real-valued homogeneous random processes $\{X_\omega(t), -\infty < t < \infty; \ \omega = 1, 2, \cdots\}$, each having zero mean and orthonormal increments, that is, for almost all ω,

$$
(1) \quad
\begin{cases}
E\{X_\omega(s) - X_\omega(t)\} = 0, \quad \text{all} \quad s, t; \\[4pt]
E[\{X_\omega(s) - X_\omega(t)\}\{X_\omega(u) - X_\omega(v)\}] = s - t, \quad s = u > t = v, \\[4pt]
\hphantom{E[\{X_\omega(s) - X_\omega(t)\}\{X_\omega(u) - X_\omega(v)\}]} = 0, \quad s > t \geqq u > v,
\end{cases}
$$

see Hannan (1970). Consider a process $Y_\omega(t)$ defined by

$$
(2) \qquad \frac{dY_\omega(t)}{dt} = \alpha_\omega Y_\omega(t) + \beta_\omega \frac{dX_\omega(t)}{dt}, \qquad -\infty < t < \infty,
$$

where α_ω and β_ω are real random variables that are mutually independent, and independent of $X_\omega(t)$, for all t. We imagine that we have some time series and cross-section observations on $Y_\omega(t)$ though these may be the residuals from a regression model for the raw data, or the results of applying some other type of filtering procedure. For given α_ω and β_ω, irregularly-spaced observations on $Y_\omega(t)$ satisfy a time-varying first-order AR model with heteroscedastic innovations (Robinson (1977)), reducing to a standard first-order AR in the equally-spaced case (Hannan (1970), p. 406).

The probability distribution functions of α_ω and β_ω are denoted F_α and F_β respectively. We assume throughout that

$$
(3) \qquad\qquad\qquad F_\alpha(0-) = 1.
$$

When $\alpha_\omega = \alpha$ almost surely (a.s), (3) reduces to the usual stability requirement $\alpha < 0$. It is convenient to define the raw moments

$$
(4) \quad
\begin{cases}
\mu_r = \displaystyle\int x^r dF_\alpha(x), \\[10pt]
\nu_r = \displaystyle\int x^r dF_\beta(x),
\end{cases}
$$

when these exist; throughout the paper all integrals are over $(-\infty, \infty)$ except where otherwise indicated.

2.2. *Stationarity requirement.* We note that (1) has solution

$$
(5) \qquad\qquad Y_\omega(t) = \beta_\omega \int_{-\infty}^{t} \exp\{\alpha_\omega(t - s)\} dX_\omega(s)
$$

a.s. and so $E\{Y_\omega(t) \mid \alpha_\omega, \beta_\omega\} = 0$, a.s., and thus the unconditional expectation $E\{Y_\omega(t)\}$ is 0 also, for all t. We call $\{Y_\omega(t); \ \omega = 1, 2, \cdots\}$ covariance stationary if the unconditional autocovariance function $E\{Y_\omega(s) Y_\omega(s + t)\}$ depends only on t. This concept is perhaps of most interest in its implications for cross-sectionally aggregated data. Consider a cross-section of size N at time t, $Y_1(t), \cdots, Y_N(t)$, and suppose that what is observed is

$$(6) \qquad Z(t) = \sum_{\omega=1}^{N} Y_\omega(t).$$

Then by independence of $X_\omega(t)$ across ω

$$E\{N^{-1}Z(s)Z(t)\} = E\{Y_\omega(s)Y_\omega(t)\},$$

so our concept of stationarity of $Y_\omega(t)$ implies stationarity of $Z(t)$.

For covariance stationarity it is necessary to strengthen (3) by restricting the probability mass of X_ω near the origin. Specifically a necessary and sufficient condition for covariance stationarity is

$$(7) \qquad \int \frac{dF_\alpha(x)}{x} > -\infty,$$

$$(8) \qquad \nu_2 < \infty.$$

If (7) and (8) are true we have the unconditional autocovariance function

$$(9) \qquad \gamma(t) = E\{Y_\omega(s)Y_\omega(s+t)\} = -\tfrac{1}{2}\nu_2 \int \frac{\exp(|t|\,x)}{x} dF_\alpha(x).$$

The latter two assertions follow from the easy consequence of (1), (3) and (5) that

$$E\{Y_\omega(s)Y_\omega(s+t) \mid \alpha_\omega, \beta_\omega\} = -\frac{1}{2}\frac{\beta_\omega^2}{\alpha_\omega}\exp(\alpha_\omega\,|t|) \quad \text{a.s.}$$

By independence of α_ω and β_ω the latter expression has expectation (9), which is finite for all t if and only if $\gamma(0) < \infty$, that is, (7) and (8) hold.

Bearing in mind that if $\alpha_\omega = 0$ in a realization of (2), that realization is non-stationary, (7) implies that nearly non-stationary sequences are rare in the population indexed by ω. Note that if the $X_\omega(t)$ are Brownian motions, (7) and (8) imply strict stationarity also.

2.3. *Deduction of distributional properties of* α_ω. When F_α is assumed absolutely continuous then the probability density function of α_ω is obtainable in terms of $\gamma(t)$ by inverse Laplace transformation, in view of the form of (9). Alternatively, expressions are available for those moments of α_ω that exist. Now for $t \geqq 0$ it follows from (9) and Fubini's theorem that

$$\gamma(t) = -\int_{-\infty}^{0} \int_{t}^{\infty} e^{tx} dt dF_\alpha(x) = -\int_{t}^{\infty} m(s)ds,$$

where

$$m(t) = \int e^{tx} dF_\alpha(x)$$
$$= \frac{-2}{\nu_2} \frac{d\gamma(t)}{dt}, \qquad t \geqq 0,$$

where the derivative at $t = 0$ is one-sided. Thus when the $(t+1)$th one-sided derivative of $\gamma(t)$ exists at $t = 0$ we have

$$(10) \qquad E(\alpha_\omega^r) = \frac{-2}{\nu_2} \frac{d^{j+1}\gamma(t)}{dt} \bigg|_{t=0}.$$

We can substitute for ν_2 in terms of $\gamma(t)$ or $\gamma(0)$ and F_α,

$$(11) \qquad \nu_2 = \frac{-2d\gamma(t)}{dt} \bigg|_{t=0} = -2\gamma(0) \bigg/ \int \frac{dF_\alpha(x)}{x},$$

see (9) and (10).

When $Y_\omega(t)$ is not continuously observable the derivatives of $\gamma(t)$ cannot be directly estimated. However, if a given parametric form for F_α is proposed, the unknown parameters can be identified in terms of estimable functions of $\gamma(t)$. A mathematically convenient and reasonably flexible class of distributions for $-\alpha_\omega$ is the Pearson Type III, which has density function

$$(12) \qquad \frac{\theta^\lambda}{\Gamma(\lambda)} x^{\lambda-1} e^{-\theta x}, \qquad \theta > 0, \qquad \lambda > 1, \qquad x \geqq 0.$$

Our stipulation that $\lambda > 1$ is in order to satisfy (7). Note that if we take $\theta = \lambda/\zeta$ then (12) converges as $\lambda \to \infty$ to a delta function with spike at ζ; thus we can use (12) to represent various degrees of approximation to the case of constant α_ω across units. We straightforwardly deduce from (9) that

$$(13) \qquad \gamma(t) = \frac{\nu_2 \theta}{2(\lambda - 1)} \left(1 + \frac{|t|}{\theta}\right)^{1-\lambda}.$$

For a given value of λ, for example, we can determine θ and ν_2 in terms of $\gamma(0)$ and $\gamma(t)$,

$$\nu_2 = \frac{2(\lambda - 1)\gamma(0)}{\theta}, \qquad \theta = \frac{t}{\{\gamma(0)/\gamma(t)\}^{1/(1-\lambda)} - 1}, \qquad t \neq 0.$$

(Note that these derivations reflect the choice of time scale.) Likewise we can so determine $E(\alpha_\omega) = -\lambda/\theta$, $\mathrm{Var}\,(\alpha_\omega) = \lambda/\theta^2$ and so on.

2.4. *Deduction of distributional properties of β_ω.* When $X_\omega(t)$ is symmetrically distributed, we cannot identify the odd moments of β_ω from $Y_\omega(t)$. However, the raw second moment of β_ω is given by (11) and higher-order even moments are expressible in terms of higher-order moments of $Y_\omega(t)$. For integer r, a necessary and sufficient condition for the $2r$th-order stationarity of $Y_\omega(t)$ is seen to be the finiteness of the $2r$th cumulants of the increments of $X_\omega(t)$ and

$$\left| \int \frac{dF_\alpha(x)}{x^r} \right| < \infty,$$

implying increasingly strong restrictions on the presence of nearly non-stationary units, with increasing r. Indeed when $X_\omega(t)$ is Brownian motion

$$E\{Y_\omega(t)^{2r}\} = E\left[\beta_\omega \int_{-\infty}^{t} \exp\{\alpha_\omega(t-s)\}dX_\omega(s) \right]^{2r}$$

$$= v_{2r} \frac{(2r)!}{2^r r!} E\left[\int_{-\infty}^{t} \exp\{\alpha_\omega(t-s)\}ds \right]^{r}$$

$$= v_{2r} \frac{(-1)^r (2r)!}{2^{2r} r!} \int \frac{dF_\alpha(x)}{x^r}, \qquad r \geq 1.$$

2.5. *Estimation of distributional properties.* We suppose that we have observations on $Y_\omega(t)$ at times $t_{\omega j_\omega}$ for units $\omega = 1, \cdots, N$. Under (7) an unbiased estimator of $\gamma(t)$ is

$$\hat{\gamma}(t) = \frac{1}{M} \sum_{\omega=1}^{N} \sum_{j(\omega)}' Y_\omega(t_{\omega j}) Y_\omega(t_{\omega j} + t),$$

where the primed sum runs over all $j = 1, \cdots, T_\omega$ such that both $t_{\omega j}$ and $t_{\omega j} + t$ are among $t_{\omega 1}, \cdots, t_{\omega T_\omega}$ and $M = \sum_{\omega=1}^{N} \sum_{j(\omega)}' 1$.

Consider the behaviour of $\hat{\gamma}(t)$ as N increases. If $M \to \infty$ as $N \to \infty$ then, by independence of units, it follows by the Kolmogorov strong law of large numbers that (7) and (8) are necessary and sufficient for the strong consistency of $\hat{\gamma}(t)$ for $\gamma(t)$. Moreover if the increments of $X_\omega(t)$ have finite fourth cumulant, then the central limit theorem for independent identically distributed random variables implies that as $M \to \infty$

$$M^{\frac{1}{2}}\{\hat{\gamma}(t_1) - \gamma(t_1), \cdots, \hat{\gamma}(t_p) - \gamma(t_p)\}$$

converges to a p-variate normal distribution if and only if

$$\int \frac{dF_\alpha(x)}{x^2} < \infty.$$

Similar properties can readily be established for estimators of higher-order moments, such as $E\{Y_\omega(t)^{2r}\}$ occurring in the previous section. Thus, insertion of estimators of moments and lag-moments of $Y_\omega(t)$ into smooth relationships between such moments and distributional properties of α_ω and β_ω produces consistent and asymptotically normal estimators of the latter.

2.6. *Unconditional distribution of $Y_\omega(t)$.* When $X_\omega(t)$ is Brownian motion then, conditional on α_ω and β_ω,

$$(14) \qquad Y_\omega(t) \sim N\left(0, -\frac{1}{2}\frac{\beta_\omega^2}{\alpha_\omega}\right).$$

It is of some interest to determine the nature of the unconditional distribution of $Y_\omega(t)$. The probability density of $Y_\omega(t)$ conditional on β_ω is

$$f_Y(y \mid \beta_\omega) = (\pi\beta_\omega^2)^{-\frac{1}{2}}\int (-x)^{\frac{1}{2}} \exp\,(xy^2/\beta_\omega^2)dF_\alpha(x).$$

Suppose $-\alpha_\omega$ has the distribution (12). Then

$$(15) \qquad f_Y(t \mid \beta_\omega) = \{(\theta\beta_\omega^2)^{\frac{1}{2}}B(\lambda, \tfrac{1}{2})(1+y^2/\theta\beta_\omega^2)^{\lambda+\frac{1}{2}}\}^{-1},$$

a Pearson Type VII distribution. When the scale factor is constant over units, i.e. $\beta_\omega = \beta$ a.s., (15) is the unconditional density function of $Y_\omega(t)$; because moments of degree only $2\lambda - \varepsilon$ (any $\varepsilon > 0$) exist, the presence of nearly non-stationary realizations leads to a longer-tailed unconditional distribution than the conditional distribution (14). (However, it may be shown that if there exists $b < 0$ such that $\alpha_\omega \leqq b$ a.s., the unconditional distribution decays at least as fast as (14).) Note that when $\beta_\omega = \beta$ is given it is possible, using (15), to estimate θ and λ by maximum likelihood on the basis of only a single observation on each of a number of units.

We can obtain an asymptotic approximation for the unconditional distribution of $Y_\omega(t)$ when β_ω has a proper distribution. Starting from (15) and using the approximation

$$(16) \qquad (1+x)^{-c} = 1 + \sum_{j=1}^{p-1} \frac{(-1)^j x^j}{jB(j, c)} + O(x^p)$$

for $c > 0$, $x > 0$, we have as an approximation to the unconditional density of $Y_\omega(t)$ as $|y| \to \infty$,

$$f_Y(\lambda) \sim \frac{1}{\pi^{\frac{1}{2}}\Gamma(\lambda)} \sum_{j=0}^{p-1} \frac{(-1)^j \Gamma(\lambda + \frac{1}{2} + j)}{\Gamma(j+1)}\, \frac{\theta^{\lambda+2j}\nu_{2(\lambda+j)}}{|y|^{2(\lambda+j)+1}} + O(|y|^{-2(\lambda+p)-1})$$

if $\nu_{2(\lambda+p)} < \infty$.

2.7. *Unconditional autocovariance function and spectral density.* As well as making inferences concerning the random parameters from knowledge of $\gamma(t)$, as discussed in Sections 2.3–2.6 above, we may consider what the distributional properties of α_ω and β_ω imply for the behaviour of $\gamma(t)$. This is particularly of interest when we observe only the 'macro' variables (6) for which we wish to choose an approximate time series model on the bases of postulated 'micro' behaviour. Under the Pearson Type III distribution (12) for $-\alpha_\omega$ we have already found that $\gamma(t)$ is given by (13) which decays only as a power law, rather than exponentially. Nearly non-stationary micro realizations have led here to a 'self-similar' macro process (for asymptotic inference covering such processes see e.g. Robinson (1978b)).

Similar inferences can be drawn even when an explicit form of F_α is not specified. Suppose F_α is absolutely continuous, its density function f_α having support (a, b). Then if in a neighbourhood $(b - \varepsilon, b)$, $\varepsilon > 0$, f_α has the series expansion

$$f_\alpha(x) = \sum_{j=0}^{p-1} c_j (x - b)^{d_j} + O(|x|^{d_p}),$$

where $d_0 < d_1 < \cdots < d_p$, we have by Watson's lemma (see e.g. Bleistein and Handelsman (1975))

$$\gamma(t) = e^{b|t|} \sum_{j=0}^{p-1} c_j \Gamma(d_j + 1) |t|^{-d_j} + O(|t|^{-d_p}).$$

It follows that if $b < 0$, so that the realizations of $Y_\omega(t)$ are bounded away from non-stationarity, $\gamma(t)$ decays to 0 somewhat faster than $e^{b|t|}$. However if f_α has a zero of degree d_0 at $b = 0$, then $\gamma(t)$ decays only like $|t|^{-d_0}$.

The unconditional frequency-domain properties of $Y_\omega(t)$, and thus of $Z(t)$, are also of interest. The spectral density function of $Z(t)$, when it exists, is

$$S(\xi) = \frac{1}{2\pi} \int \gamma(t) e^{-it\xi} dt$$

$$= \frac{\nu_2}{2\pi} \int \frac{dF_\alpha(x)}{x^2 + \xi^2}, \qquad -\infty < \xi < \infty.$$

In case $-\alpha_\omega$ has density (12), the condition $\lambda > 2$ is sufficient to ensure the existence and continuity of $S(\xi)$. As for the tail behaviour of $S(\xi)$, we deduce via (16) the asymptotic expansion

$$S(\xi) = \frac{\nu_2}{2\pi} \sum_{j=0}^{p-1} \frac{(-1)^j \mu_{2j}}{|\xi|^{2(j+1)}} + O(|\xi|^{-2(p+1)}), \quad \text{as} \quad |\xi| \to \infty,$$

when $\mu_{2p} < \infty$, the μ_j having been defined in (4).

3. Second-order model

3.1. *The model.* We consider the second-order differential equation model

$$\frac{d^2 Y_\omega(t)}{dt} = \alpha_{1\omega} \frac{dY_\omega(t)}{dt} + \alpha_{2\omega} Y_\omega(t) + \beta_\omega \frac{dX_\omega(t)}{dt}$$

$$(17) \qquad \qquad + \beta_{2\omega} \frac{d^2 X_\omega(t)}{dt^2}, \qquad -\infty < t < \infty,$$

where $\alpha_{i\omega}$, $\beta_{i\omega}$, $i = 1, 2$, are real random variables. We assume throughout that $(\alpha_{1\omega}, \alpha_{2\omega})$ is independent of $(\beta_{1\omega}, \beta_{2\omega})$. An important special case of (17) is that when we know $\beta_{2\omega} = 0$ a.s., that is (17) is a second-order stochastic differential equation driven by white noise, but it costs little extra to treat the more general case. Conditional on $\alpha_{i\omega}$, $\beta_{i\omega}$, $i = 1, 2$, a discrete equally-spaced realization of $Y_\omega(t)$ generated by (17) has an autocovariance function that coincides with that of an autoregressive moving average model with autoregressive order 2 and moving-average order 1 (see Hannan (1970), p. 406).

Let $\phi_{1\omega}$ and $\phi_{2\omega}$ be the zeros of $x^2 - \alpha_{1\omega}x - \alpha_{2\omega}$. Now $\phi_{1\omega}$ and $\phi_{2\omega}$ are either both real or they are complex conjugate, and we treat these cases separately below. It seems unnecessary to discuss all the topics we considered in Section 2 in relation to model (2), because in many respects the extensions are obvious. However, it is of interest to examine the conditions for unconditional stationarity of $Y_\omega(t)$ (and thus the aggregated process $Z(t)$), as well as the behaviour for large t of the autocovariance function $\gamma(t)$. Because higher-order real stochastic differential equations will contain a mixture of real and complex conjugate roots, our approach could be extended to such models, at the cost of some algebraic complexity.

3.2. *Complex conjugate roots: stationarity requirement.* We reparameterize complex conjugate $\phi_{i\omega}$ as

$$\phi_{1\omega} = \rho_\omega \exp(i\psi_\omega), \qquad \phi_{2\omega} = \rho_\omega \exp(-i\psi_\omega),$$

where ρ_ω and ψ_ω are real. We assume that ρ_ω and ψ_ω are independently distributed, their distribution functions F_ρ and F_ψ satisfying

$$F_\rho(0) = 0,$$

$$F_\psi(\tfrac{1}{2}\pi) = 0, \qquad F_\psi(\pi) = 1.$$

These conditions imply that $\phi_{1\omega}$ and $\phi_{2\omega}$ have negative real parts a.s., corresponding to the usual stationarity requirement. We allow the possibility $\psi_\omega = \pi$, when $\phi_{1\omega}$ and $\phi_{2\omega}$ are real and equal.

Conditional on $(\alpha_{1\omega}, \alpha_{2\omega}, \beta_{1\omega}, \beta_{2\omega})$, the lag-$t$ autocovariance of $Y_\omega(t)$ may be written as

$$(18) \quad \frac{\exp(|t|\,\rho_\omega \cos\psi_\omega)}{2 \sin 2\psi_\omega}$$

$$\times \left\{ \frac{\beta_{1\omega}^2 \sin(|t|\,\rho_\omega \sin\psi_\omega - \psi_\omega)}{\rho_\omega^3} - \frac{\beta_{2\omega}^2 \sin(|t|\,\rho_\omega \sin\psi_\omega + \psi_\omega)}{\rho_\omega} \right\},$$

when $\tfrac{1}{2}\pi < \psi_\omega < \pi$, and as

$$\frac{\exp(-|t|\,\rho_\omega)}{4} \left\{ \frac{\beta_{1\omega}^2}{\rho_\omega^3}(1 + |t|\,\rho_\omega) + \frac{\beta_{2\omega}^2}{\rho_\omega}(1 - |t|\,\rho_\omega) \right\}$$

when $\psi_\omega = \pi$. When $t = 0$ both forms are covered by the expression

$$-\tfrac{1}{4} \sec \psi_\omega \left\{ \frac{\beta_{1\omega}^2}{\rho_\omega^3} + \frac{\beta_{2\omega}^2}{\sigma_\omega} \right\}.$$

The requirements for unconditional covariance stationarity are thus

$$(19) \qquad \int \frac{dF_\rho(x)}{x^3} < \infty,$$

$$(20) \qquad \int \sec x \, dF_\psi(x) > -\infty,$$

$$(21) \qquad \int x^2 \, dF_i(x) < \infty, \qquad i = 1, 2,$$

where F_i is the distribution function of $\beta_{i\omega}$, $i = 1, 2$. Conditions (19) and (21) correspond to (7) and (8) respectively for the first-order model, though (19) is more stringent than (7), requiring $\lambda > 3$ when β_ω is a Pearson Type III variable (see (12)). A ψ_ω close to $\tfrac{1}{2}\pi$ also indicates near non-stationarity, and condition (20) restricts the chances of this event. For example, denoting by I the indicator of the interval $(\tfrac{1}{2}\pi, \pi)$, we see that (20) is not true if ψ_ω has density $f_\omega(x) = I \sin x$ (or, *a fortiori*, if ψ_ω is uniform on I) but it is satisfied by $f_\omega(x) = -I \sin 2x$ and $f_\psi(x) = -I \cos x$.

3.3. *Complex conjugate roots: autocovariance function.* For convenience, assume ρ_ω has the Pearson Type III density function (12) with $\lambda > 3$. Using (18), where $\tfrac{1}{2}\pi < \psi_\omega < \pi$, the lag-$t$ autocovariance of $Y_\omega(t)$ conditional on ψ_ω is found to be

$$
\begin{aligned}
(22) \quad \frac{\theta^\lambda}{2\Gamma(\lambda)\sin 2\psi_\omega} &\left[\frac{\chi_1 \Gamma(\lambda - 3)}{(t^2 - 2|t|\,\theta \cos \psi_\omega + \theta^2)^{\frac{1}{2}(\lambda - 3)}} \right. \\
&\times \sin \left\{ (\lambda - 3) \arctan \left(\frac{|t| \sin \psi_\omega}{\theta - |t| \cos \psi_\omega} \right) - \psi_\omega \right\} \\
&- \frac{\chi_2 \Gamma(\lambda - 1)}{(t^2 - 2|t|\,\theta \cos \psi_\omega + \theta^2)^{\frac{1}{2}(\lambda - 1)}} \\
&\left. \times \sin \left\{ (\lambda - 1) \arctan \left(\frac{|t| \sin \psi_\omega}{\theta - |t| \cos \psi_\omega} \right) + \psi_\omega \right\} \right],
\end{aligned}
$$

using formulae given, e.g., by Erdélyi et al. (1954), and with χ_i denoting the left-hand side of (21). This expression decays to 0 like $|t|^{-(\lambda - 3)}$ as $|t| \to \infty$, or as $|t|^{-(\lambda - 1)}$ if $\chi_1 = 0$ (that is if $\beta_{1\omega} = 0$ a.s.), though in a possibly oscillatory fashion.

For example when $\lambda = 5$ and $\beta_{2\omega} = 0$ a.s. (22) is

$$
(23) \qquad \frac{\chi_1 \theta^\lambda}{48 \cos \psi_\omega} \cdot \frac{t^2 - (2\,|t|\cos\psi_\omega - \theta)^2}{(t^2 - 2\,|t|\,\theta \cos\psi_\omega + \theta^2)^2}\,,
$$

which is negative for $|t| > \theta/(2\cos\psi_\omega + 1)$, when $\frac{1}{2}\pi < \psi_\omega < \frac{2}{3}\pi$.

If we now average (22) over an arbitrary ψ_ω population satisfying (20), we deduce after some algebra, via series expansions in powers of $|t|^{-1}$, the unconditional autocovariance

$$
\begin{aligned}
(24) \qquad \gamma(t) = \frac{\chi_1 \theta^\lambda}{2\Gamma(\lambda)} &\left[\frac{\Gamma(\lambda-3)}{|t|^{\lambda-3}} \int \frac{\sin\{(\lambda-3)\pi - (\lambda-2)x\}dF_\psi(x)}{\sin 2x} \right. \\
&\left. + \frac{\theta\Gamma(\lambda-2)}{|t|^{\lambda-2}} \int \frac{\sin\{(\lambda-3)\pi - (\lambda-1)x\}dF_\psi(x)}{\sin 2x} \right] \\
&+ O(|t|^{-(\lambda-1)}), \quad \text{as} \quad |t| \to \infty.
\end{aligned}
$$

(The presence of a $\beta_{2\omega}$ that is not a.s. 0 affects only the $O(|t|^{-\lambda-1})$ remainder.) No additional requirement beyond (20) is needed to justify (24). When ψ_ω has density $-I \sin 2x$, (24) is

$$
\begin{aligned}
(25) \qquad \gamma(t) = \frac{\chi_1 \theta^\lambda}{2\Gamma(\lambda)} &\left[\frac{\Gamma(\lambda-3)}{|t|^{\lambda-3}(\lambda-2)} \left\{1 + \cos\left(\frac{\lambda\pi}{2}\right)\right\} - \frac{\theta\Gamma(\lambda-2)}{|t|^{\lambda-2}(\lambda-1)} \left\{1 - \sin\left(\frac{\lambda\pi}{2}\right)\right\} \right] \\
&+ O(|t|^{-(\lambda-1)}), \quad \text{as} \quad |t| \to \infty.
\end{aligned}
$$

Thus when $\lambda = 5$ (cf. (23)) we have

$$
\gamma(t) = \frac{\chi_1 \theta^5}{144 t^2} + O(t^{-4}),
$$

so the $|t|^{-3}$ term is annihilated, Likewise (25) implies that when $\lambda = 6$ the leading term of $\gamma(t)$ is $O(t^{-4})$, rather than $O(|t|^{-3})$.

The lag-t autocovariance conditional on $\psi_\omega = \pi$ is

$$
\frac{\theta^\lambda}{4\Gamma(\lambda)} \left[\chi_1 \left\{ \frac{\Gamma(\lambda-3)}{(|t|+\theta)^{\lambda-3}} + \frac{|t|\,\Gamma(\lambda-2)}{(|t|+\theta)^{\lambda-2}} \right\} + \chi_2 \left\{ \frac{\Gamma(\lambda-1)}{(|t|+\theta)^{\lambda-1}} - \frac{|t|\,\Gamma(\lambda)}{(|t|+\theta)^\lambda} \right\} \right].
$$

3.4. *Real roots: stationarity requirement.* For real $\phi_{i\omega}$ we adopt the parameterization

$$
\phi_{1\omega} = -\rho_\omega, \qquad \phi_{2\omega} = -\rho_\omega \sigma_\omega,
$$

where ρ_ω and σ_ω are independent non-negative random variables, so that the usual stationarity requirement is a.s. satisfied for each realization. Formally the case of equal real roots is again included ($\sigma_\omega = 1$).

Conditional on $(\alpha_{1\omega}, \alpha_{2\omega}, \beta_{1\omega}, \beta_{2\omega})$ and $\sigma_\omega \neq 1$, the lag-t autocovariance of

$Y_\omega(t)$ is

$$(26) \quad \frac{1}{2(1-\sigma_\omega^2)}\left[\frac{\beta_{1\omega}^2}{\rho_\omega^3}\left\{\frac{\exp\left(-|t|\,\sigma_\omega\rho_\omega\right)}{\sigma_\omega}-\exp\left(-|t|\,\rho_\omega\right)\right\}\right.$$
$$\left.+\frac{\beta_{2\omega}^2}{\rho_\omega}\left\{\exp\left(-|t|\,\rho_\omega\right)-\sigma_\omega\exp\left(-|t|\,\sigma_\omega\rho_\omega\right)\right\}\right].$$

On putting $t = 0$ and taking expectations we find that the covariance stationarity conditions are (19), (21) and

$$\int\frac{dF_\sigma(x)}{x}<\infty,$$

where F_σ is the distribution function of σ_ω.

3.5. *Real roots: autocovariance function.* Again we let β_ω have density (12), where $\lambda > 3$. The lag-t autocovariance of $Y_\omega(t)$ conditional on $\sigma_\omega(\neq 1)$ is then deduced from (26) to be

$$\frac{\theta^\lambda}{2\Gamma(\lambda)(1-\sigma_\omega^2)}\left[\chi_1\left\{\frac{\Gamma(\lambda-3)}{\sigma_\omega(\theta+|t|\,\sigma_\omega)^{\lambda-3}}-\frac{\Gamma(\lambda-2)}{(\theta+|t|)^{\lambda-3}}\right\}\right.$$
$$\left.+\chi_2\left\{\frac{\Gamma(\lambda-1)}{(\theta+|t|)^{\lambda-1}}-\frac{\sigma_\omega\Gamma(\lambda-1)}{(\theta+|t|\,\sigma_\omega)^{\lambda-1}}\right\}\right].$$

If we now employ (16) we have an asymptotic expansion for the unconditional autocovariance function,

$$\gamma(t)=\frac{\theta^\lambda}{\Gamma(\lambda)}\left[\frac{\chi_1\Gamma(\lambda-3)\tau_{\lambda-2}}{|t|^{\lambda-3}}-\frac{\chi_1\theta\Gamma(\lambda-2)\tau_{\lambda-1}}{|t|^{\lambda-2}}\right.$$
$$\left.+\sum_{j=2}^{p}\frac{\Gamma(\lambda+j-3)\theta^{j-2}}{|t|^{\lambda+j-3}}\cdot\left\{\frac{\chi_1\theta^2\tau_{\lambda+j-2}}{\Gamma(j+1)}-\frac{\chi_2\tau_{\lambda+j}}{\Gamma(j-1)}\right\}\right]+O(|t|^{-(\lambda+p-2)})$$

as $|t|\to\infty$, where

$$\tau_\lambda=\int\frac{(1-x^\lambda)dF_\sigma(x)}{x^\lambda(1-x^2)}$$

and it is assumed that

$$\int\frac{dF_\sigma(x)}{x^{\lambda+p+1}}<\infty.$$

Thus, as in the case of complex roots, we can obtain an expansion for $\gamma(t)$ with leading term $O(|t|^{-(\lambda-3)})$, but only if σ_ω has a zero at the origin of sufficiently high order, a more detailed representation requiring further restrictions on F_σ.

References

AZZALINI, A. (1981) Replicated observations of low order autoregressive time series. *J. Time Series Analysis* **2**, 63–70.

BALESTRA, P. AND NERLOVE, M. (1966) Pooling cross section and time series data in the estimation of a dynamic model: the demand for natural gas. *Econometrica* **34**, 585–612.

BLEISTEIN, N. AND HANDELSMAN, R. A. (1975) *Asymptotic Expansions of Integrals*. Holt, Rinehart and Winston, New York.

ERDÉLYI, A., MAGNUS, W., OBERHETTINGER, F. AND TRICOMI, F. G. (1954) *Tables of Integral Transforms*, Vol. 1. McGraw-Hill, New York.

HANNAN, E. J. (1970) *Multiple Time Series*. Wiley, New York.

HSIAO, C. (1975) Some estimation methods in a random coefficient model. *Econometrica* **43**, 305–325.

LENZ, H.-J. AND ROBINSON, P. M. (1983) Sampling of cross-sectional time series. In *Recent Trends in Statistics*. ed. S. Heiler. Vandenhoek and Ruprecht, Göttingen.

ROBINSON, P. M. (1977) Estimation of a time series model from unequally spaced data. *Stoch. Proc. Appl.* **6**, 9–24.

ROBINSON, P. M. (1978a) Statistical inference for a random coefficient autoregressive model. *Scand. J. Statist.* **5**, 163–168.

ROBINSON, P. M. (1978b) Alternative models for stationary stochastic processes. *Stoch. Proc. Appl.* **8**, 141–152.

SOONG, T. T. (1973) *Random Differential Equations in Science and Engineering*. Academic Press, New York.

SWAMY, P. A. V. B. (1970) Efficient inference in a random coefficient regression model. *Econometrica* **38**, 311–323.

WALLACE, T. D. AND HUSSAIN, A. (1969) The use of error components models in combining cross section with time series data. *Econometrica* **37**, 55–72.

Consistency of Model Selection and Parameter Estimation

RITEI SHIBATA

Abstract

The relationship between consistency of model selection and that of parameter estimation is investigated. It is shown that the consistency of model selection is achieved at the cost of a lower order of consistency of the resulting estimate of parameters in some domain. The situation is different when selecting autoregressive moving average models, since the information matrix becomes singular when overfitted. Some detailed analyses of the consistency are given in this case.

AUTOREGRESSIVE PROCESSES; AKAIKE INFORMATION CRITERION; FINITE PARAMETER ESTIMATION

1. Introduction

This paper aims to clarify a relation between the consistency of model selection and that of parameter estimation after a model is selected. Although many types of procedure have been proposed for selecting a statistical model, some of them are consistent while others are not, unless separated models are considered. For example, procedures such as the minimum BIC (Schwarz [10]) or the minimum HQ (Hannan and Quinn [7]) are consistent, while others such as the minimum AIC (Akaike [2]), or the minimum FPE (Akaike [1]) or the minimum C_p (Mallows [9]) are not consistent. Therefore, a natural question arises as to whether the consistency of model selection is really needed or not.

In this paper, we first demonstrate that if model selection is consistent, then the least order of consistency of the parameter estimate becomes lower than $\sqrt{n}$, which is the order of consistency of the original parameter estimate. This fact was partly pointed out by Shibata [13]. We shall show that the consistency of model selection is achieved at the cost of a lower order of consistency of the resulting estimate of parameters in some parameter domain.

Next, in Section 4, we consider the case where the consistency of model selection seems more important than the consistency of the parameter estimate. An example is in the case of selecting an autoregressive–moving-average model ARMA(p, q). Since a lower-order model ARMA$(p-1, q-1)$ is not identifiable in a higher-order model ARMA(p, q), the use of an inconsistent procedure such as AIC might be troublesome. For such a case we propose a modification of AIC or FPE. With this modification the procedure correctly selects the lower model if a common root exists, but otherwise the behaviour remains unchanged and the order of consistency is retained as $\sqrt{n}$.

2. Consistency of model selection

To simplify our discussion, consider only two models, one of which, Model 1, $\{f(x, \theta); \theta \in \Theta\}$, is a family of density functions parametrised by θ, where Θ is an open subset of $\boldsymbol{R}^1$, and the other, Model 0, $\{f(x, \theta_0)\}$, consists of a density function specified by a parameter θ_0 in Θ, which is nested in Model 1. Since model selection here is very simple, it is enough to consider the likelihood ratio testing of the null hypothesis,

$$\mathrm{H}_0; \theta = \theta_0.$$

Given n independent samples $\boldsymbol{x}_n = (x_1, \cdots, x_n)$, the above selection is denoted by a random variable

$$\hat{m} = \begin{cases} 0, & \boldsymbol{x}_n \in A_n \\ 1, & \boldsymbol{x}_n \notin A_n, \end{cases}$$

where

$$A_n = \left\{ \boldsymbol{x}_n; T = 2 \sum_{i=1}^{n} \log \frac{f(x_i, \hat{\theta})}{f(x_i, \theta_0)} < \alpha_n \right\},$$

and $\hat{\theta}$ denotes the maximum likelihood estimate of θ in Model 1. By $\hat{m}$ we can denote various selection procedures, the minimum AIC with $\alpha_n = 2$, the minimum HQ with $\alpha_n = c \log \log n$ for some $c > 2$, and the minimum BIC with $\alpha_n = \log n$. We call $\hat{m}$ weakly consistent if $\hat{m}$ converges in probability to 0 under the null hypothesis H_0, and to 1 under the alternatives. We call it strongly consistent if the convergence is almost sure. We assume the following.

The Fisher information $I(\theta)$ per sample is finite and not 0, and the following condition holds true for the Kullback–Leibler information $I(\theta, \theta_0)$:

$$\inf_{\theta} \frac{I(\theta, \theta_0)}{|\theta - \theta_0|^2 \, I(\theta)} \geqq C_0 > 0.$$

Next assume that the asymptotic distribution of $\{nI(\theta)\}^{\frac{1}{2}} |\hat{\theta} - \theta|$ does not degenerate uniformly in θ, or, more explicitly, that

$$\lim_{M \to \infty} \lim_{n \to \infty} \sup \sup_{\theta} P_\theta [\{nI(\theta)\}^{\frac{1}{2}} |\hat{\theta} - \theta| > M] = 0,$$

and

$$\lim_{M \to 0} \lim_{n \to \infty} \inf \inf_{\theta} P_\theta [\{nI(\theta)\}^{\frac{1}{2}} |\hat{\theta} - \theta| > M] = 1.$$

We further assume that the likelihood ratio

$$T = 2 \sum_{i=1}^{n} \left\{ \log \frac{f(x_i, \hat{\theta})}{f(x_i, \theta)} + \log \frac{f(x_i, \theta)}{f(x_i, \theta_0)} \right\}$$

is well approximated by

(2.1)
$$nI(\theta)(\hat{\theta} - \theta)^2 + 2nI(\theta, \theta_0)$$

for large enough n when $\mathbf{x}_n$ is generated from a population specified by a density function $f(x, \theta)$. An example, for which the above assumptions all hold true, is the normal family with location parameter θ. The following proposition is easily obtained by noting that

$$\left(\frac{nI(\theta_0)}{2 \log \log n} \right)^{\frac{1}{2}} (\hat{\theta} - \theta_0)$$

stays between -1 and 1 for large enough n under the null hypothesis H_0, which follows from the law of the iterated logarithm (see Hannan [6]).

Proposition 2.1. A selection $\hat{m}$ is weakly consistent if and only if

$$\liminf_{n \to \infty} \alpha_n = \infty \quad \text{and} \quad \limsup_{n \to \infty} \alpha_n/n = 0,$$

and is strongly consistent if and only if

$$\liminf_{n \to \infty} \alpha_n/(2 \log \log n) > 1 \quad \text{and} \quad \limsup_{n \to \infty} \alpha_n/n = 0.$$

For uniform consistency, the following proposition holds true.

Proposition 2.2. For any α_n,

$$\liminf_{n \to \infty} \inf_{\{\theta \,;\, 2nI(\theta, \theta_0) \geq \alpha_n\}} P_\theta(\hat{m} = 1) = 1.$$

If $\liminf_{n \to \infty} \alpha_n = \infty$, then for any $\varepsilon > 0$

$$\limsup_{n \to \infty} \sup_{\{\theta \,;\, 2nI(\theta, \theta_0) \leq \alpha_n(1-\varepsilon)\}} P_\theta(\hat{m} = 1) = 0,$$

otherwise, if α_n is bounded,

$$\liminf_{n \to \infty} \inf_{\theta \in \Theta} P_\theta(\hat{m} = 1) > 0.$$

3. Consistency of parameter estimate

In this section, our main concern is the order of consistency of the parameter estimate resulting from the model selection $\hat{m}$:

$$\hat{\theta}(\hat{m}) = \begin{cases} \theta_0 & \text{if} \quad \hat{m} = 0, \\ \hat{\theta} & \text{if} \quad \hat{m} = 1. \end{cases}$$

We call $\hat{\theta}(\hat{m})$ consistent if it converges to θ in probability. For such an estimate, it is necessary to introduce the concept of non-uniform order of consistency. For a consistent estimate $\tilde{\theta}$, consider a sequence of functions $c_n(\theta)$ defined on $\Theta_n \subset \Theta$ such that

$$(3.1) \qquad \lim_{M \to \infty} \limsup_{n \to \infty} \sup_{\theta \in \Theta_n} P_\theta(c_n(\theta)\,|\tilde{\theta} - \theta| > M) = 0.$$

The limit distribution of $c_n(\theta)\,|\tilde{\theta} - \theta|$ is then not degenerate at ∞. We call $c_n^*(\theta)$ an order of consistency of the estimate $\tilde{\theta} = \hat{\theta}(\hat{m})$, if

$$\liminf_{n \to \infty}\left\{ \inf_{\theta \in \Theta_n} \frac{c_n^*(\theta)}{c_n(\theta)} \right\} > 0$$

for any other $c_n(\theta)$ which satisfies (3.1). The order of consistency $c_n^*(\theta)$ defined here is then unique in the following sense. For any $c_n(\theta)$ which satisfies (3.1),

$$0 < \liminf_{n \to \infty} \inf_{\theta \in \Theta_n} \frac{c_n^*(\theta)}{c_n(\theta)} \leqq \limsup_{n \to \infty} \sup_{\theta \in \Theta_n} \frac{c_n^*(\theta)}{c_n(\theta)} < \infty.$$

Theorem 3.1. The order of consistency $c_n^*(\theta)$ of $\hat{\theta}(\hat{m})$ depends on the rate of divergence of α_n to ∞ in the following manner.

If α_n is bounded, then $c_n^*(\theta) = \{nI(\theta)\}^{\frac{1}{2}}$ for all $\theta \in \Theta$. If $\liminf_{n \to \infty} \alpha_n = \infty$ and $\limsup_{n \to \infty} \alpha_n/n = 0$, then $c_n^*(\theta)$ is defined only on $\Theta_n = \{\theta \in \Theta;\, 2nI(\theta, \theta_0) \geqq \alpha_n$ or $0 < 2nI(\theta, \theta_0) \leqq \alpha_n(1 - \varepsilon)\}$ for any $\varepsilon > 0$, and

$$c_n^*(\theta) = \begin{cases} \{nI(\theta)\}^{\frac{1}{2}} & \text{if} \quad 2nI(\theta, \theta_0) \geqq \alpha_n, \\[2mm] \dfrac{1}{|\theta - \theta_0|} & \text{if} \ 0 < 2nI(\theta, \theta_0) \leqq \alpha_n(1 - \varepsilon). \end{cases}$$

If $\liminf_{n \to \infty} \alpha_n/n \neq 0$, then $\hat{\theta}(\hat{m})$ is not consistent.

Proof. For the case when α_n is bounded, we have

$$P_\theta[\{nI(\theta)\}^{\frac{1}{2}}\,|\hat{\theta}(\hat{m}) - \theta| > M]$$
$$= P_\theta[\{nI(\theta)\}^{\frac{1}{2}}\,|\theta - \theta_0| > M,\ \hat{m} = 0] + P_\theta[\{nI(\theta)\}^{\frac{1}{2}}\,|\hat{\theta} - \theta| > M,\ \hat{m} = 1]$$
$$\leqq P_\theta[\{nI(\theta)\}^{\frac{1}{2}}\,|\theta - \theta_0| > M,\ T < \alpha_n] + P_\theta[\{nI(\theta)\}^{\frac{1}{2}}\,|\hat{\theta} - \theta| > M].$$

From (2.1), the first term on the right-hand side of the above inequality is bounded by

$$P_\theta(C_0 M^2 < \alpha_n) + P_\theta[\{nI(\theta)\}^{\frac{1}{2}}\,|\hat{\theta} - \theta| > M].$$

Therefore we see that $c_n^*(\theta) = \{nI(\theta)^{\frac{1}{2}}$ satisfies (3.1). On the other hand, if there exists a function $c_n(\theta)$ such that $\lim_{n \to \infty} \{nI(\theta_n)\}^{\frac{1}{2}}/c_n(\theta_n) = 0$ for a sequence θ_n in

Θ, then

$$P_{\theta_n}[c_n(\theta_n)\,|\hat{\theta}(\hat{m})-\theta_n|>M] \geqq P_{\theta_n}\left[\{nI(\theta_n)\}^{\frac{1}{2}}\,|\hat{\theta}-\theta_n|>\frac{M\{nI(\theta_n)\}^{\frac{1}{2}}}{c_n(\theta_n)},\hat{m}=1\right]$$

$$\geqq P_{\theta_n}(\hat{m}=1)-P_{\theta_n}\left[\{nI(\theta_n)\}^{\frac{1}{2}}\,|\hat{\theta}-\theta_n|\leqq\frac{M\{nI(\theta_n)\}^{\frac{1}{2}}}{c_n(\theta_n)}\right]$$

which converges to 1 as $n\to\infty$ for any M. Therefore, such a $c_n(\theta)$ does not satisfy (3.1). This implies that $c_n^*(\theta)=\{nI(\theta)\}^{\frac{1}{2}}$ is the order of consistency.

Next, consider the case when $\liminf_{n\to\infty}\alpha_n=\infty$ and $\limsup_{n\to\infty}\alpha_n/n=0$. If $2nI(\theta,\theta_0)\geqq\alpha_n$, then

$$(3.2)\quad P_\theta[\{nI(\theta)\}^{\frac{1}{2}}\,|\hat{\theta}(\hat{m})-\theta|>M]\leqq P_\theta(\hat{m}=0)+P_\theta[\{nI(\theta)\}^{\frac{1}{2}}\,|\hat{\theta}-\theta|>M].$$

If $0<2nI(\theta,\theta_0)\leqq\alpha_n(1-\varepsilon)$,

$$(3.3)\quad P_\theta\left\{\frac{1}{|\theta-\theta_0|}\,|\hat{\theta}(\hat{m})-\theta_0|>M\right\}\leqq P_\theta(1>M,\hat{m}=0)+P_\theta(\hat{m}=1).$$

From Proposition 2.2 together with (3.2) and (3.3), we see that $c_n^*(\theta)$ satisfies (3.1).

To show $c_n^*(\theta)$ to be the order of consistency, suppose that there exists a sequence $c_n(\theta_n)$ for which $\lim_{n\to\infty}c_n^*(\theta_n)/c_n(\theta_n)=0$. Then for any θ_n such that $2nI(\theta_n,\theta_0)>\alpha_n$,

$$P_{\theta_n}(c_n(\theta_n)\,|\hat{\theta}(\hat{m})-\theta_0|>M)$$

is bounded away from

$$P_{\theta_n}\left[\{nI(\theta_n)\}^{\frac{1}{2}}\,|\hat{\theta}-\theta_n|>M\frac{c_n^*(\theta_n)}{c_n(\theta_n)},\hat{m}=1\right],$$

and, for any θ_n such that $0<2nI(\theta_n,\theta_0)\leqq\alpha_n(1-\varepsilon)$ it is bounded away from

$$P_{\theta_n}\left\{1>M\frac{c_n^*(\theta_n)}{c_n(\theta_n)},\hat{m}=0\right\}+P_{\theta_n}(\hat{m}=1).$$

Therefore, from Proposition 2.2 we have

$$\lim_{M\to\infty}\liminf_{n\to\infty}P_{\theta_n}(c_n(\theta_n)\,|\hat{\theta}(\hat{m})-\theta_0|>M)=1.$$

The function $c_n(\theta)$ then does not satisfy (3.1).

If $\liminf_{n\to\infty}\alpha_n/n\neq0$, then we can choose a small δ that

$$0<I(\theta,\theta_0)<\liminf_{n\to\infty}\alpha_n/2n\quad\text{and}\quad|\theta-\theta_0|>\delta.$$

Since $P_\theta(\hat{m}=0)$ goes to 1 for such θ, from the inequality

$$P_\theta(|\hat{\theta}(\hat{m})-\theta|>\delta) \geqq P_\theta(|\theta-\theta_0|>\delta, \hat{m}=0),$$

we see that $\hat{\theta}(\hat{m})$ is not consistent.

Theorem 3.1 implies that if $\alpha_n \to \infty$ and $\alpha_n/n \to \infty$ then for θ in $\{\theta; 0<2nI(\theta,\theta_0)\leqq\alpha_n(1-\varepsilon)\}$ the order of consistency of $\hat{\theta}(\hat{m})$ is lower than the order $\{nI(\theta)\}^{\frac{1}{2}}$ of consistency of the maximum likelihood estimate $\hat{\theta}$. To demonstrate this explicitly, we shall further assume that $I(\theta,\theta_0)$ is well approximated by $|\theta-\theta_0|^2 I(\theta_0)$ in the neighbourhood of θ_0. Then for θ such that $2nI(\theta,\theta_0)=\alpha_n(1-\varepsilon)$, $c_n^*(\theta)=1/|\theta-\theta_0|$ is approximated by

$$\{2/(1-\varepsilon)\}^{\frac{1}{2}}\{nI(\theta_0)/\alpha_n\}^{\frac{1}{2}},$$

and the corollary follows.

Corollary 3.1. The estimate $\hat{\theta}(\hat{m})$ is consistent if and only if $\liminf_{n\to\infty}\alpha_n/n=0$. If, in addition, for any $\varepsilon>0$ there exists a δ such that

$$\left|\frac{I(\theta,\theta_0)}{I(\theta_0)|\theta-\theta_0|^2}-1\right|<\varepsilon \quad \text{for} \quad 0<|\theta-\theta_0|<\delta,$$

then the least order of consistency, $\inf_{\theta\in\Theta_n} c_n^*(\theta)$, of $\hat{\theta}(\hat{m})$ is $\{nI(\theta_0)\}^{\frac{1}{2}}/\alpha_n^{\frac{1}{2}}$.

For example, if $\alpha_n=\log\log n$ as in the HQ criterion the least order of consistency is $\{n/(\log\log n)\}^{\frac{1}{2}}$, provided that $I(\theta)$ is bounded below 0.

4. Selection of an autoregressive–moving-average model

As was seen in the previous section, the procedure specified by a bounded α_n, such as the AIC or the FPE, is safer than others specified by a divergent α_n, from the viewpoint of the consistency of the estimate $\hat{\theta}(\hat{m})$. This also holds true for the case of multidimensional θ. However, the same discussion does not follow if the Fisher information matrix $I(\theta)$ is singular at $\theta=\theta_0$. A typical example is in the case of selecting an autoregressive–moving-average model ARMA(p,q) of orders p and q. If the autoregressive and moving-average operators have a common root, the Fisher information matrix $I(\theta)$ is singular, so that the behaviour of the maximum likelihood estimates under the ARMA(p,q) is not standard. In this case, a lower-order model ARMA$(p-1,q-1)$ may yield a better estimate. From such a point of view, Hannan and Rissanen [8] and Hannan [4], [5], [6] used a consistent selection procedure such as the HQ (Hannan and Quinn [7]) or the BIC (Schwarz [10]). In what follows, we shall explicitly analyse the asymptotic behaviour of the parameter estimates when a common root exists. The results support the result of Hannan [6] that AIC and FPE are not consistent selection procedures. However, to protect ourselves in other cases from a possible decrease of the order of consistency of parameter estimates, we propose a modification of AIC and FPE.

To simplify our discussion, we first consider the case of selecting one of the models ARMA$(1, 1)$ and ARMA$(0, 0)$.

Let us consider the ARMA$(1, 1)$ model

$$z_t - \alpha z_{t-1} = a_t - \beta a_{t-1},$$

where $|\alpha| < 1$, $|\beta| < 1$ and $\{a_t\}$ is a sequence of independent and identically normally distributed random variables with mean 0 and variance σ^2. Clearly, if $\alpha = \beta$ the above model degenerates to the ARMA$(0, 0)$ model. To get estimates of α, β and σ^2 under the model ARMA$(1, 1)$, we use the following algorithm (see Hannan and Rissanen [8]). As a convention, define $z_t = 0$ for $t \leq 0$, for given observations $z_1, \cdots, z_n$. The time t runs from 1 to n in summation unless otherwise stated.

(1) Fit a long-lag autoregressive model AR(k_n), and obtain an estimate of a_t by

$$\tilde{a}_t = \sum_{l=0}^{k_n} \hat{\varphi}_l z_{t-l}$$

for $t \geq 1$. Here, $\hat{\varphi}_0 = 1$ and

$$\sum_{l=1}^{k_n} \hat{\varphi}_l \left(\sum z_{t-1-l} z_{t-1-m} \right) = -\sum z_{t-1} z_{t-1-m}$$

for $1 \leq m \leq k_n$. The order k_n should be taken as diverging rapidly to ∞ with n but with k_n^2/n converging to 0.

(2) Obtain a pair of initial estimates $\tilde{\alpha}$ and $\tilde{\beta}$ by minimising

$$\sum (z_t - \alpha z_{t-1} + \beta \tilde{a}_{t-1})^2.$$

(3) Form

$$\hat{a}_t = \tilde{\beta} \hat{a}_{t-1} + z_t - \tilde{\alpha} z_{t-1} \quad \text{for} \quad t \geq 1$$

initialising with $\hat{a}_0 = 0$. Next, form

$$\eta_t = \tilde{\alpha} \eta_{t-1} + \hat{a}_t$$

and

$$\xi_t = \tilde{\beta} \xi_{t-1} + \hat{a}_t \quad \text{for} \quad t \geq 1,$$

initialising with $\eta_0 = \xi_0 = 0$. Regress $\hat{a}_t$ on η_{t-1} and ξ_{t-1}. Let $\tilde{\delta}$ and $\tilde{\varepsilon}$ be the δ and the ε which minimise

$$\sum_t (\hat{a}_t - \delta(\eta_{t-1} - \xi_{t-1}) + \varepsilon \xi_{t-1})^2.$$

Our estimates of α and β are then

$$(4.1) \qquad \hat{\alpha} = \tilde{\delta} + \tilde{\alpha} \quad \text{and} \quad \hat{\beta} = \tilde{\delta} + \tilde{\varepsilon} + \tilde{\beta},$$

respectively.

It is known that the estimates $\hat{\alpha}$ and $\hat{\beta}$ are asymptotically equivalent to the corresponding maximum likelihood estimates, provided that $\alpha \neq \beta$. The proof is partly given on page 92 of Hannan and Rissanen [8]. The full proof can be found in Shibata [15] or in Chen [3]. We now proceed to evaluate the asymptotic behaviour of those estimates in the case where $\alpha = \beta$. The following proposition is for the initial estimates $\tilde{\alpha}$ and $\tilde{\beta}$.

Proposition 4.1. If $z_1, \cdots, z_n$ are generated from an ARMA$(0, 0)$ process, then $\sqrt{n}(\tilde{\alpha} - \tilde{\beta})$ is asymptotically normally distributed with mean 0 and variance 1. The initial estimates $\sqrt{k_n}\,\tilde{\alpha}$ and $\sqrt{k_n}\,\tilde{\beta}$ are asymptotically equivalent to each other, and distributed normally with mean 0 and variance 1.

Proof. Let $\Delta_t = \tilde{a}_t - z_t$, then

$$\Delta_{t-1} = \sum_{l=0}^{k_n} \hat{\varphi}_l z_{t-1-l} - z_{t-1}$$

$$= \sum_{l=1}^{k_n} \hat{\varphi}_l z_{t-1-l},$$

so that

$$\sum \Delta_{t-1}^2 = \sum \left[\sum_{l,m=1}^{k_n} \hat{\varphi}_l \hat{\varphi}_m z_{t-1-l} z_{t-1-m} \right]$$

$$= \sum_{l,m} \hat{\varphi}_l \hat{\varphi}_m \left[\sum z_{t-1} z_{t-1-m} \right]$$

$$= -\sum_l \hat{\varphi}_l \left[\sum z_{t-1} z_{t-1-l} \right]$$

$$= -\sum \Delta_{t-1} z_{t-1}.$$

We have the following explicit expression of $\tilde{\alpha}$ and $\tilde{\beta}$ by solving the least-squares equation

$$\begin{bmatrix} \alpha \\ -\beta \end{bmatrix} = \frac{1}{D} \begin{bmatrix} \sum \tilde{a}_{t-1}^2 & -\sum z_{t-1} \tilde{a}_{t-1} \\ -\sum z_{t-1} \tilde{a}_{t-1} & \sum z_{t-1}^2 \end{bmatrix} \begin{bmatrix} \sum z_t z_{t-1} \\ \sum z_t \tilde{a}_{t-1} \end{bmatrix}.$$

Here D is the determinant of the matrix on the right-hand side of the above equation.

$$\tilde{\alpha} = -\frac{\sum z_t \Delta_{t-1}}{\sum \Delta_{t-1}^2}$$

$$\tilde{\beta} = \frac{-\sum z_t z_{t-1} - \left[\dfrac{\sum z_{t-1}^2}{\sum \Delta_{t-1}^2}\right](\sum z_t \Delta_{t-1})}{\sum z_{t-1}^2 - \sum \Delta_{t-1}^2},$$

and

$$\tilde{\alpha} - \tilde{\beta} = \frac{\sum z_t z_{t-1} + \sum z_t \Delta_{t-1}}{\sum z_{t-1}^2 - \sum \Delta_{t-1}^2}.$$

From the assumption that $z_t = a_t$, we see that

(4.2)
$$\sum \Delta_{t-1}^2 = \sum_{l,m} \sqrt{n}\,\hat{\varphi}_l \sqrt{n}\,\hat{\varphi}_m \left[\frac{1}{n}\sum a_{t-1-l}a_{t-1-m}\right]$$
$$= \sigma^2 \sum_{l=1}^{k_n} (\sqrt{n}\,\hat{\varphi}_l)^2 \left[1 + O_p\left(\frac{1}{\sqrt{n}}\right)\right].$$

Since it is known (Shibata [11], [12]) that $\sqrt{n}\,\hat{\varphi}_1, \cdots, \sqrt{n}\,\hat{\varphi}_{k_n}$ are asymptotically distributed as independent normal random variables with mean 0 and variance 1 under the assumption that $z_t = a_t$, the right-hand side of (4.2) is asymptotically distributed as $\sigma^2 \chi_{k_n}^2$. In the same manner, from the evaluation

$$\sum z_t \Delta_{t-1} = \sigma^2 \sum_{l=1}^{k_n} \sqrt{n}\,\hat{\varphi}_l \sqrt{n}\,\hat{\varphi}_{l+1}\left[1 + O_p\left(\frac{1}{\sqrt{n}}\right)\right],$$

with $\hat{\varphi}_{k_n+1} = \dfrac{1}{n}\sum z_{t-k_n-2}z_{t-1}$, we see that $(\sum z_t \Delta_{t-1})/(\sqrt{k_n}\,\sigma^2)$ is asymptotically normally distributed with mean 0 and variance 1. The first part of the theorem then follows from

(4.3)
$$\sqrt{n}\,(\tilde{\alpha} - \tilde{\beta}) = \frac{\dfrac{1}{\sqrt{n}}\sum a_t a_{t-1}}{\dfrac{1}{n}\sum a_{t-1}^2}\{1 + O_p((k_n/n)^{\frac{1}{2}})\}.$$

For the latter part, it is enough to note that

$$\sqrt{k_n}\,\tilde{\beta} = -\sqrt{k_n}\,\frac{\sum z_t \Delta_{t-1}}{\sum \Delta_{t-1}^2}\{1 + O_p((k_n/n)^{\frac{1}{2}})\}$$

$$= -\frac{\dfrac{1}{\sqrt{k_n}}\sum_{l=1}^{k_n} \sqrt{n}\,\hat{\varphi}_l \sqrt{n}\,\hat{\varphi}_{l+1}}{\dfrac{1}{k_n}\sum_{l=1}^{k_n} (\sqrt{n}\,\hat{\varphi}_l)^2}\{1 + O_p((k_n/n)^{\frac{1}{2}})\}.$$

It is worth noting that the proof of consistency of $\tilde{\alpha}$ and $\tilde{\beta}$ in Proposition 4.1 owes to the error structure of the autoregressive parameter estimates $\hat{\varphi}_1, \cdots, \hat{\varphi}_{k_n}$. The consistency does not necessarily follow if other estimates are used for obtaining an estimate of a_t, or if such errors are not taken into consideration (see Hannan [5], [6]).

Proposition 4.2. If $z_1, \cdots, z_n$ are generated from an ARMA$(0, 0)$ process, then $\tilde{\delta}$ is asymptotically distributed as Cauchy. On the other hand, $\tilde{\varepsilon}$ is of the order $n^{-\frac{1}{2}}k^{-\frac{1}{2}}$ in probability, so that $\sqrt{n}(\hat{\alpha} - \hat{\beta})$ is asymptotically equivalent to $\sqrt{n}(\tilde{\alpha} - \tilde{\beta})$ and both $\hat{\alpha}$ and $\hat{\beta}$ are distributed as Cauchy.

Proof. Since

$$\eta_t = \sum_{l=0}^{t} \tilde{\alpha}^l a_{t-l} \quad \text{and} \quad \xi_t = \sum_{l=0}^{t} \tilde{\beta}^l \hat{a}_{t-l}$$

we obtain the following evaluations:

$$\sum (\eta_{t-1} - \xi_{t-1})^2 = (\tilde{\alpha} - \tilde{\beta})^2 \left(\sum \hat{a}_{t-2}^2 \right) \{1 + O_p(k_n^{-\frac{1}{2}})\},$$

$$\sum \xi_{t-1}^2 = \left(\sum a_{t-1}^2 \right) \{1 + O_p(k_n^{-\frac{1}{2}})\},$$

$$\sum \hat{a}_t (\eta_{t-1} - \xi_{t-1}) = (\tilde{\alpha} - \tilde{\beta}) \left(\sum a_t a_{t-2} \right) \{1 + O_p(k_n^{-\frac{1}{2}})\},$$

$$\sum \xi_{t-1} (\eta_{t-1} - \xi_{t-1}) = (\tilde{\alpha} - \tilde{\beta}) \left\{ \tilde{\beta} \sum a_{t-2}^2 + O_p(k_n^{\frac{1}{2}}) \right\},$$

and

$$\sum \hat{a}_t \xi_{t-1} = \left\{ \sum a_t a_{t-1} + (\tilde{\beta} - \tilde{\alpha}) \sum a_{t-1}^2 \right\} + O_p(k_n^{-\frac{1}{2}}n^{\frac{1}{2}})$$
$$= O_p(k_n) + O_p(k_n^{-\frac{1}{2}}n^{\frac{1}{2}})$$
$$= O_p(k_n^{-\frac{1}{2}}n^{\frac{1}{2}})$$

from the formula (4.3). By solving the least-squares equation for the step (3) in the algorithm, we have, as in the proof of Proposition 4.1,

$$\tilde{\delta} = \frac{(\sum a_{t-1}^2)(\sum a_t a_{t-2}) + \tilde{\beta}(\sum a_{t-2}^2)O_p(k_n^{-\frac{1}{2}}n^{\frac{1}{2}})}{(\tilde{\alpha} - \tilde{\beta})(\sum a_{t-2}^2)(\sum a_{t-1}^2)} \{1 + O_p(k_n^{-\frac{1}{2}})\}$$
$$= \frac{\sum a_t a_{t-2} + O_p(k_n^{-\frac{1}{2}}n^{\frac{1}{2}})}{\sum a_t a_{t-1}} \{1 + O_p(k_n^{-\frac{1}{2}})\}$$
$$= \frac{\sum a_t a_{t-2}}{\sum a_t a_{t-1}} \{1 + O_p(k_n^{-\frac{1}{2}})\}.$$

Therefore the asymptotic distribution of $\tilde{\delta}$ is Cauchy. On the other hand, $\tilde{\varepsilon}$ can be evaluated as

$$\tilde{\varepsilon} = \frac{\tilde{\beta}\,(\sum a_{t-2}^2)(\sum a_t a_{t-2}) - (\sum a_{t-2}^2)O_p(k_n^{-\frac{1}{2}}n^{\frac{1}{2}})}{(\sum a_{t-2}^2)(\sum a_{t-1}^2)}\{1 + O_p(k_n^{-\frac{1}{2}})\}$$
$$= O_p(k_n^{-\frac{1}{2}}n^{-\frac{1}{2}}).$$

This proposition implies that if $\alpha = \beta$, $\hat{\alpha}$ and $\hat{\beta}$ are not consistent but their asymptotic distribution is Cauchy, while the initial estimates $\tilde{\alpha}$ and $\tilde{\beta}$ are consistent with order $k_n^{\frac{1}{2}}$. This does not coincide with the result of Hannan [6] that the maximum likelihood estimates of α and β almost surely converge to ± 1, but this is because our estimates are not necessarily equivalent to the exact maximum likelihood estimates when $\alpha = \beta$.

Now let us consider the two criteria AIC and FPE. It is already known that these criteria are asymptotically equivalent to each other in the case of autoregressive models. But this does not hold for the selection of autoregressive–moving-average models. While AIC is likelihood-oriented, FPE is prediction-error-oriented. AIC for an ARMA(p, q) is, by definition,

$$\text{AIC}(p, q) = \log \hat{\sigma}_{\text{ML}}^2(p, q) + 2(p + q)/n,$$

where $\hat{\sigma}_{\text{ML}}^2(p, q)$ is the maximum likelihood estimate of σ^2 under the model ARMA(p, q). But FPE is different from AIC.

To obtain an explicit form of FPE for the ARMA$(1, 1)$ case, consider another ARMA$(1, 1)$ process $\{z_t^*\}$ which is independent of $\{z_t\}$ but which has the same probabilistic structure as $\{z_t\}$. Since the best predictor of z_t^* is $\bar{z}_t^* = \alpha z_{t-1}^* - \beta a_{t-1}^*$, the error of the estimated predictor

$$\hat{z}_t^* = \hat{\alpha} z_{t-1}^* - \hat{\beta} a_{t-1}^*$$

is given by

(4.4)
$$E^*(\hat{z}_t^* - z_t^*)^2 = E^*(\bar{z}_t^* - z_t^*)^2 + E^*(\hat{z}_t^* - \bar{z}_t^*)^2$$
$$= \sigma^2\{1 + (\hat{\theta} - \theta)'Q(\hat{\theta} - \theta)\}.$$

Here E^* denotes the expectation with respect to $\{z_t^*\}$, $\hat{\theta}' = (\hat{\alpha}, \hat{\beta}')$, $\theta' = (\alpha, \beta')$,

$$Q = \begin{bmatrix} \gamma_0/\sigma^2 & -1 \\ -1 & 1 \end{bmatrix}$$

and $\gamma_k = E^*(z_t^* z_{t-k}^*) = E(z_t z_{t-k})$. From the fact that the underlying model is an ARMA$(1, 1)$, the variance γ_0 of $\{z_i\}$ is written as

$$\gamma_0 = \sigma^2\left\{1 + \frac{(\alpha - \beta)^2}{1 - \alpha^2}\right\}.$$

Since $\hat{\theta}$ is asymptotically equivalent to the corresponding maximum likelihood estimates unless $\alpha = \beta$, the estimation error $\sqrt{n}(\hat{\theta} - \theta)$ is asymptotically normally distributed with the mean 0 and the covariance matrix J^{-1}, where

$$J = \begin{bmatrix} 1/(1-\alpha^2) & -1/(1-\alpha\beta) \\ -1/(1-\alpha\beta) & 1/(1-\beta^2) \end{bmatrix}.$$

The expectation of the right-hand side of (4.3) is then asymptotically equal to

$$\sigma^2\{1 + \operatorname{tr}(J^{-1}Q)/n\} = \sigma^2\{1 + 2(1-\alpha\beta)/n\}.$$

Replacing σ^2 by a bias-corrected estimate $\hat\sigma^2_{\mathrm{ML}}(1,1)/(1-2/n)$, α by $\hat\alpha$, and β by $\hat\beta$, we obtain the FPE for ARMA$(1,1)$,

$$\mathrm{FPE}(1,1) = \frac{\hat\sigma^2_{\mathrm{ML}}(1,1)\{1 + 2(1-\hat\alpha\hat\beta)/n\}}{1-2/n}.$$

The difference between AIC$(1,1)$ and FPE$(1,1)$ is significant unless $\hat\alpha$ and $\hat\beta$ converge to 0.

Now we proceed to evaluate the behaviour of AIC and FPE. We first consider the case where

$$(4.5) \qquad \hat\sigma^2(1,1) = \frac{1}{n}\sum (z_t - \hat\alpha z_{t-1} + \hat\beta\hat a_{t-1})^2$$

is used in place of $\hat\sigma^2_{\mathrm{ML}}(1,1)$. We should note that the above estimate is based on $\hat\alpha$ and $\hat\beta$, but is not itself asymptotically efficient even when $\alpha \neq \beta$.

Let us define

$$\hat\sigma^2 = \frac{1}{n}\sum a_t^2, \qquad \hat\rho_1 = \left(\sum a_t a_{t-1}\right)\Big/\left(\sum a_t^2\right)$$

and $\hat\rho_2 = (\sum a_t a_{t-2})/(\sum a_t^2)$, then from Propositions 4.1 and 4.2, if $\alpha = \beta$,

$$\hat\beta - \hat\alpha = -2\hat\rho_1(1 + O_p(k_n^{-\frac{1}{2}})), \qquad \hat\beta = \frac{\hat\rho_2}{\hat\rho_1}(1 + O_p(k_n^{-\frac{1}{2}}))$$

and

$$\hat a_{t-1} - a_{t-1} = (\tilde\beta - \tilde\alpha)a_{t-2}(1 + O_p(k_n^{-\frac{1}{2}}))$$
$$= \hat\rho_1 a_{t-2}(1 + O_p(k_n^{-\frac{1}{2}})).$$

Therefore, if $\alpha = \beta$, then $z_t = a_t$ and

$$\hat\sigma^2(1,1) = \frac{1}{n}\sum (a_t - \hat\alpha a_{t-1} + \hat\beta\hat a_{t-1})^2$$

$$= \frac{1}{n}\sum \{a_t + (\hat\beta - \hat\alpha)a_{t-1} + \hat\beta(\hat a_{t-1} - a_{t-1})\}^2$$

$$= \frac{1}{n}\sum [a_t - (\hat\rho_1 a_{t-1} + \hat\rho_2 a_{t-2})\{1 + O_p(k_n^{-\frac{1}{2}})\}]^2$$

$$= \hat\sigma^2 - \hat\sigma^2(\hat\rho_1^2 + \hat\rho_2^2 + 2\hat\rho_1^2\hat\rho_2)\{1 + O_p(k_n^{-\frac{1}{2}})\}.$$

Since $\hat{\sigma}^2_{\mathrm{ML}}(0, 0) = \hat{\sigma}^2$, from the definition, $\mathrm{AIC}(0, 0) = \log \hat{\sigma}^2_{\mathrm{ML}}(0, 0)$ and $\mathrm{FPE}(0, 0) = \hat{\sigma}^2_{\mathrm{ML}}(0, 0)$, we have

$$\mathrm{AIC}(1, 1) - \mathrm{AIC}(0, 0) = \log\{1 - (\hat{\rho}_1^2 + \hat{\rho}_2^2 + 2\hat{\rho}_1^2\hat{\rho}_2)(1 + O_p(k_n^{-\frac{1}{2}}))\} + 4/n$$

$$= \frac{1}{n}\{4 - n\hat{\rho}_1^2 - n\hat{\rho}_2^2 + O_p(k_n^{-\frac{1}{2}})\},$$

and

$$\mathrm{FPE}(1, 1) - \mathrm{FPE}(0, 0) = \frac{\hat{\sigma}^2}{1 - 2/n}[(1 - \hat{\rho}_1^2 - \hat{\rho}_2^2)\{1 + 2(1 - \hat{\rho}_2^2/\hat{\rho}_1^2 - 2\hat{\rho}_2)/n\}] - \hat{\sigma}^2$$

$$= \frac{\hat{\sigma}^2}{n}\{4 - n\hat{\rho}_1^2 - n\hat{\rho}_2^2 - 2\hat{\rho}_2^2/\hat{\rho}_1^2 + O_p(k_n^{-\frac{1}{2}})\}.$$

Therefore, under the assumption that $\alpha = \beta$

$$\lim_{n\to\infty} P(\mathrm{AIC}(0, 0) < \mathrm{AIC}(1, 1)) = \lim P(n\hat{\rho}_1^2 + n\hat{\rho}_2^2 < 4)$$

$$= P(X < 4) = 0.86466,$$

and

$$\lim_{n\to\infty} P(\mathrm{FPE}(0, 0) < \mathrm{FPE}(1, 1)) = \lim P(n\hat{\rho}_1^2 + n\hat{\rho}_2^2 + 2\hat{\rho}_2^2/\hat{\rho}_1^2 < 4)$$

$$= P(X + 2Y < 4) = 0.45120,$$

where X and Y are independent, and distributed as χ^2 with degree of freedom 2, and as Cauchy, respectively. It is obvious that if $\alpha \neq \beta$, then $\mathrm{AIC}(1, 1) < \mathrm{AIC}(0, 0)$ and $\mathrm{FPE}(1, 1) < \mathrm{FPE}(0, 0)$ hold true in probability for large enough n, and we have the following theorem.

Theorem 4.1. Suppose that the competing models are only $\mathrm{ARMA}(1, 1)$ and $\mathrm{ARMA}(0, 0)$. If $\hat{\sigma}^2(1, 1)$ is used in place of the exact maximum likelihood estimate, then the probability of correct selection tends to 1 if $\alpha \neq \beta$, otherwise it tends to 0.86466 and 0.45120, respectively, for AIC and FPE.

It is worth noting that in spite of non-regularity, the error probability of AIC is the same as that in the regular case, for example, in the case of $\mathrm{AR}(p)$ fitting (Shibata [12]), while that of FPE is different and higher.

Next we consider the case where

$$(4.6) \qquad \tilde{\sigma}^2(1, 1) = \frac{1}{n}\sum (z_t - \tilde{\alpha}z_{t-1} + \tilde{\beta}\hat{a}_{t-1})^2$$

is used instead of $\hat{\sigma}^2(1, 1)$.

Under the assumption that $\alpha = \beta$, it follows that

$$\tilde{\sigma}^2(1, 1) - \tilde{\sigma}^2(0, 0) = \frac{1}{n} \sum \hat{a}_t^2 - \frac{1}{n} \sum a_t^2$$

$$= (\tilde{\beta} - \tilde{\alpha})^2 \frac{1}{n} \sum a_{t-1}^2 + 2(\tilde{\beta} - \tilde{\alpha}) \frac{1}{n} \sum a_t a_{t-1}$$

$$= \hat{\sigma}^2 \{(\tilde{\beta} - \tilde{\alpha})^2 + 2(\tilde{\beta} - \tilde{\alpha})\hat{\rho}_1\}$$

$$= -\hat{\sigma}^2 \hat{\rho}_1^2.$$

By similar discussion as in the proof of Theorem 4.1, we have

$$\lim_{n \to \infty} P(\text{AIC}(0, 0) < \text{AIC}(1, 1)) = P(\chi_1^2 < 4) = 0.9545,$$

provided that $\alpha = \beta$.

Theorem 4.2. Suppose that the competing models are only ARMA$(1, 1)$ and ARMA$(0, 0)$. If $\tilde{\sigma}^2(1, 1)$ is used in place of the exact maximum likelihood estimate in the definition of AIC$(1, 1)$, then the probability of correct selection tends to 1 if $\alpha \neq \beta$, otherwise it tends to 0.9545.

Therefore we can see that $\tilde{\sigma}^2(1, 1)$ is better than $\hat{\sigma}^2(1, 1)$ as a replacement of $\hat{\sigma}_{\text{ML}}^2(1, 1)$ in the definition of AIC$(1, 1)$. However, there still remains a small probability of selecting the overfitted model ARMA$(1, 1)$, where both $\hat{\alpha}$ and $\hat{\beta}$ are not consistent. On the other hand, we have already seen that if we require consistency of selection as in HQ and BIC, then the least order of consistency becomes lower than $\sqrt{n}$. An alternative is to modify AIC as follows. For simplicity, consider the models ARMA(p, q) $0 \leq p, q \leq 1$. Select ARMA$(0, 0)$ if $|\tilde{\delta}| > \alpha_n / \sqrt{n}$, otherwise select the model which minimises AIC or FPE from those models except ARMA$(0, 0)$. Here, α_n is a divergent sequence with n as used in the definition of the HQ or the BIC, which diverges to ∞ with $\alpha_n / \sqrt{n}$ converging to 0. By this modification, AIC and FPE correctly select ARMA$(0, 0)$ when $\alpha = \beta$, because $\tilde{\delta}$ is distributed as Cauchy. In this paper, we do not give any further proof, but it can be easily understood that by such a modification the order of consistency goes down to $O((n/\alpha_n)^{\frac{1}{2}})$ only in a domain like

$$\{(\alpha, \beta); (\alpha_n/n)^{\frac{1}{2}}(1 - \varepsilon) < |\alpha - \beta| < (\alpha_n/n)^{\frac{1}{2}}\},$$

while that of BIC or HQ also goes down in a domain like

$$\{(\alpha, \beta); (\alpha_n/n)^{\frac{1}{2}}(1 - \varepsilon) < |\alpha| < (\alpha_n/n)^{\frac{1}{2}}\}$$

or

$$\{(\alpha, \beta); (\alpha_n/n)^{\frac{1}{2}}(1 - \varepsilon) < |\beta| < (\alpha_n/n)^{\frac{1}{2}}\}.$$

More detailed analysis and extension remain for the future.

Acknowledgements

The author would like to express his sincere thanks to the referee. His suggestions led to the correction of some serious errors in the original manuscript; and also improved the readability of the paper. Theorem 4.2 is due to the referee.

References

[1] AKAIKE, H. (1970) Statistical predictor identification. *Ann. Inst. Statist. Math.* **22**, 203–217.

[2] AKAIKE, H. (1973) Information theory and an extension of the maximum likelihood principle. In *2nd Int. Symposium on Information Theory*, ed. B. N. Petrov and F. Csáki, Akadémia Kiado, Budapest, 267–281.

[3] CHEN, ZHAO-GUO (1984) The asymptotic efficiency of a linear procedure of estimation for ARMA models. *J. Time Series Anal.*

[4] HANNAN, E. J. (1980) The estimation of the order of an ARMA process *Ann. Statist.* **8**, 1071–1081.

[5] HANNAN, E. J. (1982) Testing for autocorrelation and Akaike's criterion. In *Essays in Statistical Science: Papers in Honour of P. A. P. Moran*, ed. J. M. Gani and E. J. Hannan, Applied Probability Trust, Sheffield, 403–412.

[6] HANNAN, E. J. (1982) Fitting multivariate ARMA models. In *Statistics and Probability, Essays in Honour of C. R. Rao*, ed. G. Kallianpur, P. R. Krishnaiah and J. K. Ghosh, North-Holland, Amsterdam, 307–316.

[7] HANNAN, E. J. AND QUINN, B. G. (1979) The determination of the order of an autoregression. *J. R. Statist. Soc.* B **41**, 190–195.

[8] HANNAN, E. J. AND RISSANEN, J. (1982) Recursive estimation of mixed autoregressive-moving average order. *Biometrika* **69**, 81–94.

[9] MALLOWS, C. L. (1973) Some comments on C_p. *Technometrics* **15**, 661–675.

[10] SCHWARZ, G. (1978) Estimating the dimension of a model. *Ann. Statist.* **6**, 461–464.

[11] SHIBATA, R. (1976) Selection of the order of an autoregressive model by Akaike's information criterion. *Biometrika* **63**, 117–126.

[12] SHIBATA, R. (1977) Convergences of least squares estimates of autoregressive parameters. *Austral. J. Statist.* **19**, 226–235.

[13] SHIBATA, R. (1980) Asymptotically efficient selection of the order of the model for estimating parameters of a linear process. *Ann. Statist.* **8**, 147–164.

[14] SHIBATA, R. (1983) A theoretical view of the use of AIC. In *Time Series Analysis: Theory and Practice* **4**, ed. O. D. Anderson, Elsevier, Amsterdam, 237–244.

[15] SHIBATA, R. (1984) Identification and selection of ARMA models. Tech. Rep. RT-MAE-8406, Institute of Mathematics and Statistics, University of São Paulo.

Band-Limited Spectral Estimation of Autoregressive–Moving-Average Processes

P. J. THOMSON

Abstract

Consider an autoregressive–moving-average process of given order where it is known that a number of moving-average roots are of unit modulus. Such a situation might arise, for example, when a time series has been differenced to induce stationarity by removing a non-stationary polynomial or seasonal trend. A band-limited spectral estimation procedure is proposed for estimating the coefficients of such a process and the asymptotic properties of the estimators investigated. The asymptotic theory is illustrated with reference to simulated and real data. A preliminary investigation of the use of Akaike's AIC criterion and this procedure to determine the number of roots of unit modulus (in the case where this is unknown) is also carried out by means of simulation.

The proposed band-limited spectral estimation procedure can also be used to take account of other possible effects met in practice. These include, for example, the band-limited response of a recording device or trend-contaminated low-frequency components.

ARMA MODEL; TIME SERIES INFERENCE; FOURIER METHODS; FILTERING; ASYMPTOTIC RESULTS

1. Introduction

Let $\{X_n\}$ be an autoregressive–moving-average process satisfying the difference equation

$$(1.1) \qquad \sum_{j=0}^{p} a_j X_{n-j} = \sum_{j=0}^{q} b_j \varepsilon_{n-j} \quad (a_0 = b_0 = 1; \, n = 0, \pm 1, \cdots)$$

where $\{\varepsilon_n\}$ is a sequence of uncorrelated random variables each with zero mean and variance σ^2. Moreover $A(z) = \sum a_j z^i$, $B(z) = \sum b_j z^i$ are required to have no zeros inside the unit circle and $A(z)$ to have no zeros on it. We are concerned with the situation where d of the q roots of $B(z)$ have unit modulus.

This model might arise in a number of ways. Consider, for example, a time series consisting of a zero-mean stationary error process superimposed on a non-linear trend. If the trend is a polynomial or can be approximated piecewise by low-order polynomials then this time series, appropriately differenced, may well be modelled by (1.1) above with moving-average roots of unity. If the trend is periodic or seasonal with period s (s an integer), then differencing observations s time intervals apart will again yield a time series which can be modelled by (1.1), but now there will be s complex moving-average roots of unit modulus. In either case, if the error structure comprises random-walk-type components then the degree of differencing required to induce stationarity in the time series will exceed the resulting number of roots of $B(z)$ with unit modulus. In this context it is worth pointing our that, for the autoregressive integrated moving-average process of Box and Jenkins (1976), the degree of differencing required to induce stationarity in the trend never exceeds that required to induce stationarity in the error structure.

In practice the model (1.1) will arise following differencing operations such as those just described. However, in many instances, the resulting number of moving-averaging roots of unit modulus may well be unknown due, perhaps, to over-differencing or the unknown nature of the error structure. In this context it is of interest to determine the number of moving-average roots of unit modulus.

The situations described in the previous paragraph arise through pre-filtering the data using the difference filter. Although this example is central to the paper it is appropriate to extend the treatment to cover the case where the data has been pre-filtered by a filter with transfer function $\Phi(z) = C^{-1}(z)D(z)$. Here $C(z) = \sum c_i z^i$, $D(z) = \sum d_i z^i$ are finite-order polynomials with $C(z)$, $D(z)$ having no roots inside and $C(z)$ having no roots on the unit circle. Thus, in general, we consider the model (1.1) where p and q are known, $A^{-1}(z)B(z) = \Phi(z)G(z)$, $G(z) = \alpha^{-1}(z)\beta(z)$ and $\alpha(z) = \sum \alpha_i z^i$, $\beta(z) = \sum \beta_i z^i$ are finite-order polynomials with $\alpha(z)$, $\beta(z)$ having no roots inside and $\alpha(z)$ having no roots on the unit circle. The parameters α_i, β_i and σ^2 are unknown. Hence $\Phi(z)$ is a known function of z, whereas $G(z)$ involves the unknown parameters. In Section 2 an estimation criterion is proposed yielding estimators of the α_i, β_i and σ^2. The asymptotic properties of these estimators are also established.

It is noted that the model (1.1) where $B(z)$ has exactly one root of unity is considered by Pham-Dinh (1978) using methods different to those given here. See also Dzhaparidze (1977) and Davies (1983) for related results.

The estimation procedure proposed in Section 2 is a band-limited version of the procedures due to Hannan (1969) and Akaike (1973) with the data adjusted for the known effects of the transfer function $\Phi(z)$. This procedure is also appropriate in many other practical situations where one might wish to fit autoregressive–moving-average spectral models over only a part of the frequency range. One such example is the case where the recording device used has a band-limited response. Another example concerns a trend-plus-error model where a band of low-frequency spectral components has been contaminated by trend or long-term effects. Excluding these frequencies from the analysis, autoregressive–moving-average models can be fitted to the remaining frequencies to explain the error structure or short-term effects. See also Robinson (1977), where the band-limited spectral estimation of more general finite-parameter models is discussed.

This paper owes much to Professor Hannan's considerable research in this area of time series and, in particular, builds directly on his influential paper Hannan (1973).

2. An estimation criterion

Consider (1.1) where $\{X_n\}$ is a zero-mean, ergodic, stationary process. Given a finite sample $X_1, \cdots, X_N$ we introduce the finite Fourier transforms

$$(2.1) \quad W(\omega_k) = \frac{1}{\sqrt{2\pi N}} \sum_{n=1}^{N} X_n \exp(in\omega_k) \quad \left(\omega_k = \frac{2\pi k}{N}, -\tfrac{1}{2}N \le k \le [\tfrac{1}{2}N] \right)$$

where $[\tfrac{1}{2}N]$ denotes the integral part of $\tfrac{1}{2}N$. As is well known, the $W(\omega_k)$ can be computed extremely rapidly if N is highly composite. $W(\lambda)$ is, apart from a constant of proportionality, essentially the sample analogue of $dZ(\lambda)$ where $Z(\lambda)$ is the (complex) spectral measure underlying the spectral representation of X_n. As expected, the asymptotic covariance properties of $W(\lambda)$ parallel those of $dZ(\lambda)$. Moreover, under certain general conditions, the $W(\omega_k)$ can be regarded as asymptotically independent complex normal random variables each with zero mean and variance $f(\omega_k)$. (See Hannan (1970) and Brillinger (1975).) Here $f(\lambda)$, the spectral density of X_n satisfies

$$E(X_m X_{m+n}) = \int_{-\pi}^{\pi} \exp(in\lambda) f(\lambda) d\lambda \quad (n = 0, \pm 1, \cdots)$$

and, adopting the notation of Section 1,

$$f(\lambda) = \frac{\sigma^2}{2\pi} |\Phi(\exp(i\lambda))|^2 |G(\exp(i\lambda))|^2 \quad (|\lambda| \le \pi).$$

For convenience we shall henceforth write $\phi(\lambda)$ for $|\Phi(\exp(i\lambda))|^2$ and $g(\lambda)$ for $|G(\exp(i\lambda))|^2$. Now, one obvious approach to estimating the α_j, β_j and σ^2 is that of maximum likelihood treating the limiting distribution of the $W(\omega_k)$ as the likelihood of the observations. Of course such a procedure will not, in general, yield maximum likelihood estimators. (Hannan (1974) refers to this approximate likelihood as the quasi-likelihood.) Nevertheless, such procedures have been widely used and generally provide sound estimators in practice. (See Hannan (1974) and Brillinger (1974).)

The function $\phi(\lambda)$ can have at most a finite number of zeros. To remove the effects of these singularities and to take account of other possible effects (such as the band-limited response of a recording device or trend-contaminated low-frequency components, for example) we now consider a fixed band of frequencies $\mathscr{B}$ on which $\phi(\lambda)$ and $g(\lambda)$ are strictly positive. This means, in particular, that we must know the (potential) zeros of $g(\lambda)$ *a priori*. With this in mind the (approximate) log-likelihood of the relevant $W(\omega_k)$ is, apart from a constant,

$$(2.2) \qquad -\sum_{\mathscr{B}} \log\left\{\frac{\sigma^2}{2\pi}\phi(\omega_k)g(\omega_k;\boldsymbol{\theta})\right\} - \frac{2\pi}{\sigma^2}\sum_{\mathscr{B}}\frac{|W(\omega_k)|^2}{\phi(\omega_k)g(\omega_k;\boldsymbol{\theta})}.$$

Here $\boldsymbol{\theta}$ denotes the vector of parameters α_j and β_j and $g(\lambda;\boldsymbol{\theta})$ has been written in place of $g(\lambda)$ reflecting the fact that it is through this function alone that the (approximate) likelihood depends on $\boldsymbol{\theta}$. Let $\boldsymbol{\theta}_0$, σ_0^2 denote the true values of the parameters.

Maximising (2.2) with respect to σ^2 yields

$$(2.3) \qquad \hat{\sigma}_N^2(\boldsymbol{\theta}) = \frac{2\pi}{N'}\sum_{\mathscr{B}}\frac{|W(\omega_k)|^2}{\phi(\omega_k)g(\omega_k;\boldsymbol{\theta})}$$

where N' is the number of frequencies of the form $2\pi k/N$ in $\mathscr{B}$. Substituting back into (2.2), we choose to estimate $\boldsymbol{\theta}_0$ by $\hat{\boldsymbol{\theta}}_N$ and σ_0^2 by $\hat{\sigma}_N^2(\hat{\boldsymbol{\theta}}_N)$ where $\hat{\boldsymbol{\theta}}_N$ minimises

$$(2.4) \qquad \hat{q}_N(\boldsymbol{\theta}) = \log\hat{\sigma}_N^2(\boldsymbol{\theta}) + \frac{1}{N'}\sum_{\mathscr{B}}\log g(\omega_k;\boldsymbol{\theta})$$

over some suitable parameter space Θ.

3. Strong consistency of the estimators and a central limit theorem

In summary, $\{X_n\}$ is assumed to be a zero-mean, ergodic, stationary process satisfying (1.1) with spectral density

$$f(\lambda) = \frac{\sigma^2}{2\pi}\phi(\lambda)g(\lambda;\boldsymbol{\theta}) = \frac{\sigma^2}{2\pi}|\Phi(\exp(i\lambda))|^2 |G(\exp(i\lambda))|^2 \quad (|\lambda|\leq\pi).$$

The transfer functions $\Phi(z)$, $G(z)$ are given by $C^{-1}(z)D(z)$, $\alpha^{-1}(z)\beta(z)$ respectively where $\alpha(z)$, $\beta(z)$, $C(z)$, $D(z)$ are finite-order polynomials with $\alpha(z)$, $C(z)\neq 0$ for $|z|\leq 1$ and $\beta(z)$, $D(z)\neq 0$ for $|z|<1$. In particular $\phi(\lambda) = |\Phi(\exp(i\lambda))|^2$ is known *a priori*, as are the orders of $\alpha(z)$, $\beta(z)$. For identification purposes it is assumed that $\alpha(z)$, $\beta(z)$ have no common factors, $\alpha_0 = \beta_0 = 1$ and, letting r, s denote the respective orders of $\alpha(z)$, $\beta(z)$, that α_r, β_s are not both 0. The vector of parameters is denoted by $\boldsymbol{\theta} = (\boldsymbol{\alpha}', \boldsymbol{\beta}')'$ where the prime denotes vector transpose and $\boldsymbol{\alpha} = (\alpha_1, \cdots, \alpha_r)'$, $\boldsymbol{\beta} = (\beta_1, \cdots, \beta_s)'$. The true values of the parameters are denoted by σ_0^2, $\boldsymbol{\theta}_0$ and $\mathscr{B}$, the band of frequencies of interest, is assumed to be a compact subset of $(-\pi, \pi]$, symmetric about $\lambda = 0$ with $\phi(\lambda)$, $g(\lambda; \boldsymbol{\theta}_0) > 0$, $\lambda \in \mathscr{B}$. We are concerned with minimising (2.4) over the parameter space Θ where $\Theta = \{\boldsymbol{\theta} : g(\lambda; \boldsymbol{\theta}) \geqq \delta > 0, \lambda \in \mathscr{B}\}$ some $\delta > 0$. The conditions of this paragraph will be referred to collectively as 'Conditions A'.

Theorem 3.1. If $\{X_n\}$ is an autoregressive–moving-average process satisfying (1.1) then, under Conditions A, $\hat{\boldsymbol{\theta}}_N \to \boldsymbol{\theta}_0$ a.s. and $\hat{\sigma}_N^2(\hat{\boldsymbol{\theta}}_N) \to \sigma_0^2$ a.s. as $N \to \infty$.

Proof. Now $\hat{\boldsymbol{\theta}}_N$ minimises $\hat{q}_N(\boldsymbol{\theta})$ or, equivalently,

$$\tilde{q}_N(\boldsymbol{\theta}) = \log \hat{\sigma}_N^2(\boldsymbol{\theta}) - \frac{1}{N'} \sum_{\mathscr{B}} \log \frac{g(\omega_k; \boldsymbol{\theta}_0)}{g(\omega_k; \boldsymbol{\theta})} \, .$$

Suppose $\hat{\boldsymbol{\theta}}_N$ does not converge to $\boldsymbol{\theta}_0$. Then there exists a subsequence $\tilde{\boldsymbol{\theta}}_N (\tilde{\boldsymbol{\theta}}_N \in \Theta)$ converging to $\boldsymbol{\theta}' \in \Theta$ where $\boldsymbol{\theta}' \neq \boldsymbol{\theta}_0$. Note that

$$(3.1) \qquad \limsup \tilde{q}_N(\tilde{\boldsymbol{\theta}}_N) \leq \limsup \tilde{q}_N(\boldsymbol{\theta}_0) = \log \sigma_0^2.$$

This result follows in a straightforward way from the proof of Lemma 1 of Hannan (1973). Moreover, for arbitrary $\varepsilon \in (0, 1)$ and N large enough,

$$\tilde{q}_N(\tilde{\boldsymbol{\theta}}_N) \geqq \log \left\{ \frac{2\pi}{N'} \sum_{\mathscr{B}_\varepsilon} \frac{|W(\omega_k)|^2}{\phi(\omega_k) g(\omega_k; \tilde{\boldsymbol{\theta}}_N)} \right\} - \frac{1}{N'} \sum_{\mathscr{B}_\varepsilon} \log \frac{g(\omega_k; \boldsymbol{\theta}_0)}{g(\omega_k; \tilde{\boldsymbol{\theta}}_N)}$$

where $\mathscr{B}_\varepsilon = \{\lambda : g(\lambda; \boldsymbol{\theta}_0) \geqq \varepsilon g(\lambda; \boldsymbol{\theta}')\} \cap \mathscr{B}$. Thus

$$\limsup \tilde{q}_N(\tilde{\boldsymbol{\theta}}_N) \geqq \log \sigma_0^2 + \log \left\{ \frac{1}{b} \int_{\mathscr{B}_\varepsilon} \frac{g(\lambda; \boldsymbol{\theta}_0)}{g(\lambda; \boldsymbol{\theta}')} \, d\lambda \right\} - \frac{1}{b} \int_{\mathscr{B}_\varepsilon} \log \frac{g(\lambda; \boldsymbol{\theta}_0)}{g(\lambda, \boldsymbol{\theta}')} \, d\lambda$$

where b denotes the Lebesgue measure of $\mathscr{B}$. This again follows from the proof of Lemma 1, Hannan (1973). Letting b_ε denote the Lebesgue measure of $\mathscr{B}_\varepsilon$, note that

$$\log \left\{ \frac{1}{b_\varepsilon} \int_{\mathscr{B}_\varepsilon} \frac{g(\lambda; \boldsymbol{\theta}_0)}{g(\lambda; \boldsymbol{\theta}')} \, d\lambda \right\} \geqq \frac{1}{b_\varepsilon} \int_{\mathscr{B}_\varepsilon} \log \frac{g(\lambda; \boldsymbol{\theta}_0)}{g(\lambda; \boldsymbol{\theta}')} \, d\lambda$$

with equality iff $g(\lambda; \boldsymbol{\theta}_0) = g(\lambda; \boldsymbol{\theta}')$, $\lambda \in \mathscr{B}_\varepsilon$; i.e. iff $\boldsymbol{\theta}_0 = \boldsymbol{\theta}'$. Now take ε sufficiently small so that

$$\log \left\{ \frac{1}{b} \int_{\mathscr{B}_\varepsilon} \frac{g(\lambda; \boldsymbol{\theta}_0)}{g(\lambda; \boldsymbol{\theta}')} \, d\lambda \right\} > \frac{1}{b} \int_{\mathscr{B}_\varepsilon} \log \frac{g(\lambda; \boldsymbol{\theta}_0)}{g(\lambda; \boldsymbol{\theta}')} \, d\lambda.$$

Then

$$\liminf \tilde{q}_N(\tilde{\boldsymbol{\theta}}_N) > \log \sigma_0^2,$$

which contradicts (3.1). Thus $\hat{\boldsymbol{\theta}}_N \to \boldsymbol{\theta}_0$ a.s. as $N \to \infty$.

Finally, since $\hat{\sigma}_N^2(\boldsymbol{\theta})$ converges uniformly to

$$\frac{\sigma_0^2}{b} \int_{\mathcal{B}} \frac{g(\lambda; \boldsymbol{\theta}_0)}{g(\lambda; \boldsymbol{\theta})} \, d\lambda$$

in any closed subset of Θ that contains $\boldsymbol{\theta}_0$, we conclude that $\hat{\sigma}_N^2(\hat{\boldsymbol{\theta}}_N) \to \sigma_0^2$ a.s. as $N \to \infty$.

To establish the asymptotic distributional properties of the estimators additional assumptions are needed. Let $\mathcal{F}_n$ be the σ-algebra generated by the ε_m for $m \le n$. Following Hannan (1976) we assume that the quantities

$$E\{\varepsilon_n^j \mid \mathcal{F}_{n-1}\} \quad (j = 1, 2, 3, 4)$$

are almost surely constants. Moreover, $g(\lambda; \boldsymbol{\theta}_0)$ is now assumed to be strictly positive on $(-\pi, \pi]$. Thus Θ becomes $\Theta = \{\boldsymbol{\theta}; g(\lambda; \boldsymbol{\theta}) \ge \delta > 0, |\lambda| \le \pi\}$. These conditions will be referred to as 'Conditions B'.

Let $\partial \hat{q}_N(\boldsymbol{\theta})/\partial \boldsymbol{\theta}$ denote the vector with typical element $\partial \hat{q}_N(\boldsymbol{\theta})/\partial \theta_i$ and $\partial^2 \hat{q}_N(\boldsymbol{\theta})/\partial \boldsymbol{\theta} \partial \boldsymbol{\theta}'$ the matrix with typical element $\partial^2 \hat{q}_N(\boldsymbol{\theta})/\partial \theta_i \partial \theta_j$. The quantities $\partial g(\lambda; \boldsymbol{\theta})/\partial \boldsymbol{\theta}$, $\partial \log g(\lambda; \boldsymbol{\theta})/\partial \boldsymbol{\theta}$, $\partial^2 g(\lambda; \boldsymbol{\theta})/\partial \boldsymbol{\theta} \partial \boldsymbol{\theta}'$ are defined analogously.

Theorem 3.2. If $\{X_n\}$ is an autoregressive–moving-average process satisfying (1.1) then, under Conditions A and B, $N^{\frac{1}{2}}(\hat{\boldsymbol{\theta}}_N - \boldsymbol{\theta}_0)$ is asymptotically normal with zero-mean vector and covariance matrix $4\pi \Omega^{-1}$ where

$$\Omega = \int_{\mathcal{B}} \frac{\partial \log g(\lambda; \boldsymbol{\theta}_0)}{\partial \boldsymbol{\theta}} \frac{\partial \log g(\lambda; \boldsymbol{\theta}_0)'}{\partial \boldsymbol{\theta}} \, d\lambda$$

$$- b^{-1} \left\{ \int_{\mathcal{B}} \frac{\partial \log g(\lambda; \boldsymbol{\theta}_0)}{\partial \boldsymbol{\theta}} \, d\lambda \right\} \left\{ \int_{\mathcal{B}} \frac{\partial \log g(\lambda; \boldsymbol{\theta}_0)}{\partial \boldsymbol{\theta}} \, d\lambda \right\}'.$$

Proof. Expanding $\partial \hat{q}_N(\boldsymbol{\theta})/\partial \boldsymbol{\theta}$ in a Taylor series about $\hat{\boldsymbol{\theta}}_N$ yields

$$(3.2) \qquad -N^{\frac{1}{2}} \frac{\partial \hat{q}_N(\boldsymbol{\theta}_0)}{\partial \boldsymbol{\theta}} = \left\{ \frac{\partial^2 \hat{q}_N(\bar{\boldsymbol{\theta}}_N)}{\partial \boldsymbol{\theta} \partial \boldsymbol{\theta}'} \right\} N^{\frac{1}{2}}(\hat{\boldsymbol{\theta}}_N - \boldsymbol{\theta}_0)$$

where $\|\bar{\boldsymbol{\theta}}_N - \boldsymbol{\theta}_0\| \le \|\hat{\boldsymbol{\theta}}_N - \boldsymbol{\theta}_0\|$ and $\|\cdot\|$ denotes the standard Euclidean norm. Moreover $\partial^2 \hat{q}_N(\bar{\boldsymbol{\theta}}_N)/\partial \boldsymbol{\theta} \partial \boldsymbol{\theta}'$ converges almost surely to $b^{-1}\Omega$. This follows from Theorem 3.1 and the proof of Lemma 1 of Hannan (1973).

Now

$$(3.3) \qquad -N^{\frac{1}{2}} \frac{\partial \hat{q}_N(\boldsymbol{\theta}_0)}{\partial \boldsymbol{\theta}} = \frac{2\pi N^{\frac{1}{2}}}{N' \hat{\sigma}_N^2(\boldsymbol{\theta}_0)} \sum \frac{|W(\omega_k)|^2}{\phi(\omega_k) g(\omega_k; \boldsymbol{\theta}_0)} \boldsymbol{\psi}_N(\omega_k)$$

where

$$\boldsymbol{\psi}_N(\lambda) = \frac{\partial \log g(\lambda; \boldsymbol{\theta}_0)}{\partial \boldsymbol{\theta}} - \frac{1}{N'} \sum_{\omega_k \in \mathcal{B}} \frac{\partial \log g(\omega_k; \boldsymbol{\theta}_0)}{\partial \boldsymbol{\theta}}$$

for $\lambda \in \mathcal{B}$ and is 0 otherwise. However, (3.3) can be written as

$$\frac{1}{N^{\frac{1}{2}}\hat{\sigma}_N^2(\boldsymbol{\theta}_0)} \sum_{m=1}^{N} \sum_{n=1}^{N} X_m X_n \left\{ \frac{1}{N'} \sum \frac{\exp i(m-n)\omega_k}{\phi(\omega_k)g(\omega_k;\boldsymbol{\theta}_0)} \psi_N(\omega_k) \right\}.$$

Note that the last bracketed factor differs by a term of $O(N^{-1})$ from

$$\frac{1}{b} \int \frac{\exp i(m-n)\lambda}{\phi(\lambda)g(\lambda;\boldsymbol{\theta}_0)} \psi(\lambda)\, d\lambda$$

where

$$\psi(\lambda) = \frac{\partial \log g(\lambda;\boldsymbol{\theta}_0)}{\partial \boldsymbol{\theta}} - \frac{1}{b} \int_{\mathcal{B}} \frac{\partial \log g(\omega;\boldsymbol{\theta}_0)}{\partial \boldsymbol{\theta}}\, d\omega$$

for $\lambda \in \mathcal{B}$ and is 0 otherwise. Hence, since $\hat{\sigma}_N^2(\boldsymbol{\theta}_0)$ converges to σ_0^2 a.s., the ε_n have finite fourth moment and $\psi(\lambda)$ integrates to 0, (3.3) differs from

$$(3.4) \qquad \frac{2\pi N^{\frac{1}{2}}}{\sigma_0^2 b} \int \left\{ \frac{|W(\lambda)|^2}{\phi(\lambda)g(\lambda;\boldsymbol{\theta}_0)} - \frac{1}{2\pi N} \sum_{1}^{N} \varepsilon_n^2 \right\} \psi(\lambda)\, d\lambda$$

by terms which converge in probability to 0. Now replace $\psi(\lambda)$ by $\tilde{\psi}(\lambda)$ which is continuous on $(-\pi, \pi]$ and differs from $\psi(\lambda)$ only on a set of measure δ, some arbitrarily small $\delta > 0$. In turn, $\tilde{\psi}(\lambda)$ is approximated by $\tilde{\psi}_M(\lambda)$ where the latter is the Cesaro sum to M terms of the Fourier series of $\tilde{\psi}(\lambda)$ and M is sufficiently large that $\sup_\lambda \|\tilde{\psi}(\lambda) - \tilde{\psi}_M(\lambda)\| < \delta$. Then, from Bernstein's lemma (see Hannan (1970), p. 242) and the development on p. 138 of Hannan (1973), we are reduced to considering

$$\frac{2\pi N^{\frac{1}{2}}}{\sigma_0^2 b} \int \left(|W_\varepsilon(\lambda)|^2 - \frac{1}{2\pi N} \sum_{1}^{N} \varepsilon_n^2 \right) \tilde{\psi}_M(\lambda)\, d\lambda$$

where $W_\varepsilon(\lambda)$ is given by (2.1) with X_n replaced by ε_n. The fact that this is asymptotically normal follows from pp. 138–139 of Hannan (1973) and Hannan (1976). Moreover (3.5) has zero mean and covariance matrix which converges to $4\pi b^{-2}\Omega$ as $\delta \to 0$. This completes the proof.

In many situations the band of frequencies $\mathcal{B}$ will differ only marginally from $(-\pi, \pi]$. For such situations, versions of Theorems 3.1 and 3.2 with b large would be useful. To this end we define $\{\mathcal{B}_M\}$ to be a sequence of compact subsets of $(-\pi, \pi]$, symmetric about 0 with $\mathcal{B}_M \uparrow (-\pi, \pi]$. In the result that follows we have written $\hat{\boldsymbol{\theta}}_N^{(M)}$ in place of $\hat{\boldsymbol{\theta}}_N$ to indicate dependence of $\hat{\boldsymbol{\theta}}_N$ on the band of frequencies $\mathcal{B}_M$.

Corollary 3.2. Under the above conditions and the conditions of Theorems 3.1 and 3.2 there exists a sequence M_N increasing with N such that $\hat{\boldsymbol{\theta}}_N^{(M_N)} \to \boldsymbol{\theta}_0$

a.s., $\hat{\sigma}_N^2(\hat{\boldsymbol{\theta}}_N^{(M_N)}) \to \sigma_0^2$ a.s. and $N^{\frac{1}{2}}(\hat{\boldsymbol{\theta}}_N^{(M_N)} - \boldsymbol{\theta}_0)$ is asymptotically normal with zero-mean vector and covariance matrix

$$4\pi\left\{\int_{-\pi}^{\pi} \frac{\partial \log g(\lambda; \boldsymbol{\theta}_0)}{\partial \boldsymbol{\theta}} \frac{\partial \log g(\lambda; \boldsymbol{\theta}_0)'}{\partial \boldsymbol{\theta}}\right\}^{-1}.$$

Proof. This result follows from Lemmas 1 and 2 of Hannan and Thomson (1973), and the fact that

$$\int_{-\pi}^{\pi} \frac{\partial \log g(\lambda; \boldsymbol{\theta}_0)}{\partial \boldsymbol{\theta}} \, d\lambda = 0.$$

The results given in this section can be readily extended to cover the case of a more general class of finite-parameter models for $g(\lambda; \boldsymbol{\theta})$ and more general transfer functions $\phi(\lambda)$. Such extensions would require additional regularity conditions such as those given in Hannan (1973).

4. Numerical studies

A number of simulation studies were carried out to investigate the adequacy of the asymptotic results given in Section 3. The three models chosen for the simulation were simple examples of autoregressive–moving-average processes satisfying (1.1) with parameters given in Table 1. On the interval $[0, \pi]$ the spectral densities at Models I, II and III are unimodal with the spectral peaks for Models I and II located at $\lambda = 0$ and $\lambda = \pi$ respectively, and the spectral peak for Model III located near $\lambda = \pi/3$. In each case 20 independent Gaussian realisations were generated for sample sizes $N = 128$ and $N = 256$.

Taking p, q as known *a priori*, the relevant models were fitted over frequency bands of the form

$$B(a, b) = \{\lambda \mid 2\pi a \leqq |\lambda| \leqq 2\pi b\}.$$

The three pairs of values chosen for (a, b) were $(0, 0.5)$, $(0.01, 0.5)$ and $(0.05, 0.4)$, which utilise 100%, 98% and 70% of the available frequency range respectively. The latter models might correspond to the situation where low-frequency components have been excluded, for example, due to contamination by a trend. As the bandwidth shrinks the parameter estimates will become less

TABLE 1

Model	p	q	Parameters
I	0	1	$b_1 = -0.7, \sigma = 1$
II	1	0	$a_1 = -0.5, \sigma = 1$
III	2	0	$a_1 = -0.5, a_2 = 0.25, \sigma_1 = 1$

precise at a rate that depends on the sample size N and the variation of the spectral density over the band $B(a, b)$. In addition, for each of the cases where $B(a, b) \neq [-\pi, \pi]$, the data was differenced using the difference filter and the appropriate model fitted using (2.4) and (2.3) with

$$(4.1) \qquad \phi(\lambda) = |1 - \exp(i\lambda)|^2 = 4 \sin^2 \tfrac{1}{2}\lambda.$$

This corresponds to the situation where the observations comprise a non-stationary trend plus autoregressive–moving-average errors and the data has been differenced to remove the trend.

The results of the simulation are given in Tables 2, 3 and 4. For these tables

TABLE 2

Simulation results for Model I ($b_1 = -0.7$)

Band	$B(0, 0.5)$	$B(0.01, 0.5)$		$B(0.05, 0.4)$	
Differencing	—	—	D	—	D
$\bar{x}$	−0.732	−0.719	−0.454	−0.712	−0.605
s_x	0.076	0.090	0.190	0.182	0.211
$N = 128$ σ_x	0.063	0.072	0.072	0.152	0.152
MSE	0.007	0.008	0.095	0.032	0.051
MSED	—	0.1033		0.0281	
$\bar{x}$	−0.691	−0.677	−0.494	−0.720	−0.650
s_x	0.044	0.051	0.143	0.176	0.163
$N = 256$ σ_x	0.045	0.051	0.051	0.108	0.108
MSE	0.002	0.003	0.062	0.030	0.028
MSED	—	0.0568		0.0199	

TABLE 3

Simulation results for Model II ($a_1 = -0.5$)

Band	$B(0, 0.5)$	$B(0.01, 0.5)$		$B(0.05, 0.4)$	
Differencing	—	—	D	—	D
$\bar{x}$	−0.464	−0.472	−0.516	−0.504	−0.511
s_x	0.062	0.074	0.075	0.107	0.106
$N = 128$ σ_x	0.077	0.082	0.082	0.127	0.127
MSE	0.005	0.006	0.006	0.011	0.011
MSED	—	0.0058		0.0007	
$\bar{x}$	−0.488	−0.499	−0.525	−0.531	−0.537
s_x	0.052	0.055	0.051	0.084	0.082
$N = 256$ σ_x	0.054	0.058	0.058	0.090	0.090
MSE	0.003	0.003	0.003	0.008	0.008
MSED	—	0.0015		0.0003	

TABLE 4(a)
Simulation results for Model III ($a_1 = -0.5$)

Band	$B(0, 0.5)$	$B(0.01, 0.5)$		$B(0.05, 0.4)$	
Differencing	—	—	D	—	D
$\bar{x}$	−0.468	−0.477	−0.502	−0.450	−0.457
s_x	0.087	0.087	0.089	0.126	0.104
$N = 128$ σ_x	0.086	0.087	0.087	0.108	0.108
MSE	0.008	0.008	0.007	0.018	0.012
MSED	—		0.0014		0.0017
$\bar{x}$	−0.517	−0.520	−0.533	−0.516	−0.520
s_x	0.053	0.053	0.045	0.074	0.068
$N = 256$ σ_x	0.061	0.061	0.061	0.077	0.077
MSE	0.003	0.003	0.003	0.005	0.005
MSED	—		0.0006		0.0003

TABLE 4(b)
Simulation results for Model III ($a_2 = 0.25$)

Band	$B(0, 0.5)$	$B(0.01, 0.5)$		$B(0.05, 0.4)$	
Differencing	—	—	D	—	D
$\bar{x}$	0.229	0.220	0.200	0.190	0.196
s_x	0.081	0.082	0.084	0.143	0.147
$N = 128$ σ_x	0.086	0.087	0.087	0.115	0.115
MSE	0.007	0.007	0.009	0.023	0.023
MSED	—		0.0009		0.0015
$\bar{x}$	0.240	0.237	0.230	0.217	0.213
s_x	0.062	0.064	0.069	0.094	0.096
$N = 256$ σ_x	0.061	0.061	0.061	0.081	0.081
MSE	0.004	0.004	0.005	0.009	0.010
MSED	—		0.0003		0.0002

the $\bar{x}$, s_x and σ_x rows give the sample means, sample standard deviations and population standard deviations respectively of the 20 independent parameter estimates. The σ_x row is calculated using Theorem 3.2. The MSE row gives the (sample) mean squared errors of the estimates and the MSED row gives the mean squared errors of the (paired) differences between the estimates obtained with and without differencing. The simulations for the various models and sample sizes were generated independently. However, the five estimation procedures indicated by the columns in the tables were always applied to the same set of 20 independent realisations. The D column indicates that the model has been fitted to the differenced data using (2.4) and (2.3) with $\phi(\lambda)$ given by (4.1).

On the whole the simulations agree reasonably well with the theoretical results given in Section 3. The notable exception concerns the frequency band $B(0.01, 0.5)$ and the case where the parameters are estimated after correcting for the difference filter. In this situation the spectral ordinates $|W(\omega_k)|^2$ of the differenced process will be close to 0 near $\lambda = 0.02\pi$. Dividing these through by $\phi(\lambda) = 4 \sin^2 \frac{1}{2}\lambda$, which is also close to 0 near $\lambda = 0.02\pi$, has clearly introduced bias into the estimation procedure. This is particularly evident in the case of the moving-average given by Model I, but it is also present to a lesser degree in the other two models.

Bearing in mind the location of the spectral mass for Models I, II and III it might be expected that, as the bandwidth shrinks, the estimates for Models I and II would become less precise relative to those for Model III. Moreover, since the smallest band $B(0.05, 0.4)$ excludes more high frequencies than low frequencies this effect should be more marked for Model I than Model II since the spectral density for Model I has its peak at $\lambda = \pi$. The simulation results reported in Tables 2–4 confirm these expectations.

The band-limited spectral estimation method described in Section 2 was applied to the viscosity readings given in series D of the appendix to Box and Jenkins (1976). The graph of the data and the various analyses found in Box and Jenkins (1976) suggest that the data might reasonable well be described by a model that consists of a (slight) linear trend plus autoregressive error. To eliminate the trend the data was differenced and then the autoregression fitted using (2.4) and (2.3) with $\phi(\lambda) = 4 \sin^2 \frac{1}{2}\lambda$ and $p = 1$, $q = 0$. For the frequency band $(0.01, 0.5)$ the autoregressive parameter estimate obtained was $\hat{a}_1 = -0.84$ with estimated standard deviation 0.043. The innovation variance was estimated as 0.089, which is slightly lower than those listed for the models fitted to this data given in Box and Jenkins (1976).

Finally, as indicated in the introduction to this paper, one might wish to use band-limited autoregressive–moving-average spectral estimation and Akaike's AIC criterion (see Akaike (1973)) to determine the number of moving-average roots of unity. This might arise if the observed process consists of a non-stationary trend plus error and the degree of differencing required to induce stationarity in the trend exceeds that required to induce stationarity in the error structure. Although the validity of the AIC criterion in this particular situation has not been established, it is of interest to consider the results of a limited number of simulations using AIC in this way.

The simulations considered generated 20 independent realisations for each of three moving-average models. The latter had parameters $b_1 = -1$ (corresponding to differenced white noise and a moving-average root of unity),

TABLE 5

Proportion of times (out of 20) that AIC
selected $b_1 = -1$

True b_1	-1	-0.8	-0.6
Proportion	0.95	0.55	0.10

$b_1 = -0.8$ and $b_1 = -0.6$. In all cases Gaussian data were generated with $\sigma^2 = 1$, $N = 256$ and the estimation restricted to $B(0.01, 0.5)$. For each case a moving average of order 1 was fitted and compared, using AIC to the moving average with unit root $(b_1 = -1)$. Table 5 gives the proportion of times that the AIC criterion selected the differenced model $(b_1 = -1)$ in each of the three situations considered. If AIC is used in this way then these results indicate that there is roughly a 45% chance of successfully discriminating between a moving average with a root on the unit circle $(b_1 = -1)$ and one whose root is near the unit circle $(b_1 = -0.8)$. Consider applying the technique to the concentration readings given in series A of the appendix to Box and Jenkins (1976). There a first-order moving-average model was fitted to the differenced data yielding an estimated parameter of -0.70 and an innovation variance of 0.101. For the frequency band $B(0.01, 0.5)$ a moving-average model was fitted to the differenced data using (2.3) and (2.4) with $\phi(\lambda) = 1$. The parameter b_1 was estimated as -0.65 with standard deviation 0.08 and innovation variance 0.104. For the same frequency band a moving-average model with unit root $(b_1 = -1)$ was fitted using (2.3) and (2.4) with $\phi(\lambda) = 4 \sin^2 \frac{1}{2}\lambda$ and $p = q = 0$. The estimated innovation variance was 0.111. Based on these figures the AIC criterion favours the model with no unit root, although the difference involved is not great.

The use of the AIC criterion in this way and the results quoted are somewhat speculative at this stage. Further theoretical and practical work is needed to ascertain the utility or otherwise of this particular technique.

References

AKAIKE, H. (1973) Information theory and an extension of the maximum likelihood principle. *2nd Internat. Symp. Information Theory*, ed. B. N. Petrov and F. Csaki, Akademiai Kiado, Budapest, 267–281.

BOX, G. E. P. AND JENKINS, G. M. (1976) *Time Series Analysis: Forecasting and Control* (Revised edn). Holden-Day, San Francisco.

BRILLINGER, D. R. (1974) Fourier analysis of stationary processes. *Proc. IEEE* **62**, 1628–1643.

BRILLINGER, D. R. (1975) *Time Series Data Analysis and Theory*. Holt, Rinehart and Winston, New York.

DAVIES, R. B. (1983) Optimal inference in the frequency domain. In *Handbook of Statistics*, Vol. 3, ed. D. R. Brillinger and P. R. Krishnaiah, North-Holland, Amsterdam, 73–92.

DZHAPARIDZE, K. O. (1977) Estimation of parameters of a spectral density with fixed zeroes. *Theory Prob. Appl.* **22**, 708–729.

HANNAN, E. J. (1969) The estimation of mixed autoregressive moving average models. *Biometrika* **56**, 579–593.

HANNAN, E. J. (1970) *Multiple Time Series.* Wiley, New York.

HANNAN, E. J. (1973) The asymptotic theory of linear time-series models. *J. Appl. Prob.* **10**, 130–145.

HANNAN, E. J. (1974) Time series analysis. *IEEE Trans. Autom. Control* AC-**19**, 706–715.

HANNAN, E. J. (1976) The asymptotic distribution of serial covariances. *Ann. Statist.* **4**, 396–399.

HANNAN, E. J. AND THOMSON, P. J. (1973) Estimating group delay. *Biometrika* **60**, 241–253.

PHAM-DINH, T. (1978) Estimation of parameters in the ARMA model when the characteristic polynomial of the MA operator has a unit zero. *Ann. Statist.* **6**, 1369–1389.

ROBINSON, P. M. (1977) The construction and estimation of continuous time models and discrete approximations in econometrics. *J. Econometrics* **6**, 173–197.

PART 3

HYPOTHESIS TESTING AND DISTRIBUTION THEORY FOR TIME SERIES

The Comparison of Transient Responses to Stimuli

MURRAY A. CAMERON

Abstract

In some experiments an 'observation' is a time series and the 'treatments' are stimuli which elicit changes in the nature of the time series for only a short period. One approach to analysing data from such experiments is presented here. Tests are given for detecting responses in such experiments, and for comparing responses elicited under different treatment regimes. The tests are based on frequency-domain estimates of the prediction variance for stationary time series. If several stimuli are applied within each observed time series at regular intervals, then the method of analysis may be designed so that the estimates of the different responses are orthogonal and most of the calculations can be performed using standard analysis of variance programs.

ANALYSIS OF POWER; FOURIER METHODS; TIME SERIES; TRANSIENTS

1. Introduction

Transients occur frequently in time series data: outliers, earthquakes in geophysics, short regime changes in economic series, evoked potentials from the nervous system in neurophysiology, are all examples. Transients are often of fundamental importance in data, but few general methods are available for the statistical analysis of transients. Although Jones et al. (1970) discussed the detection of transients using an outlier identification technique, most methods that have been suggested may be derived by supposing that the transient is the response of a linear filter to a pulse. In time-domain applications, Box and Tiao (1975) and Tanaka (1983) used such a model and estimated the parameters of the models for the filters.

In the frequency domain, Brillinger (1973a) considered methods of removing the effects of transients when estimating spectra, while Cameron and Hannan (1979) estimated parameters which describe the change in phase of a transient signal received by a number of sensors at different locations and Brillinger (1981) suggested a linear model in the frequency domain for the comparison of transients. That linear model is considered further in this paper, and overall tests for the presence of transients and for their equality are presented. The tests are based on spectral estimates of the prediction error variance for stationary time series, and are analogous to F-tests which arise in the analysis of variance.

These techniques should be useful for the analysis of some data from electroencephalograms and electrocardiograms in particular. For example, experiments are often conducted to examine the response of, say, heart rate, to stimuli. Stimuli in this case may be odours, sounds, lights, physical pressure or electrical shocks. Typically the experiments will have been performed to examine whether there are observable differences between the responses to different stimuli, or whether the responses alter if the stimuli are presented to the subject repeatedly. An alternative approach to this last problem is given by Möcks et al. (1984). In the next section, the basic problem is described in detail and some notation introduced. Section 3 contains the details of a general method for estimating the Fourier coefficients of transients, while in Section 4 some test statistics for the detection and comparison of transients are given. The appendix gives some conditions on the transients and the noise processes under which the test statistics are asymptotically normally distributed.

Time-domain methods could be devised for the problems described here by extending the work of either Jones et al. (1970) or Box and Tiao (1975) and Tanaka (1983). In some circumstances (for example, if the length of each time series is short) time-domain methods may be preferable. However, the frequency-domain approach has some advantages. Firstly, there are many time series, each of several thousand observations, in many of the experiments of the type described above. This does not greatly increase the effort of analysis required in the frequency domain, but in the time domain it often means that many parameters need to be estimated, usually with non-linear methods, and for each series many models with similar properties have to be compared. Secondly, the structure of the complete analysis is easily displayed using frequency-domain methods, and this provides a basis for similar time-domain approaches. Finally, the computations are often more straightforward in the frequency-domain and allow the use of standard computer programs. Thus frequency-domain methods are often useful for an analysis of data, even when time-domain methods are also to be used later.

Although it is the comparison of transients which is of interest in this paper, spectral analysis of a transient is often required; evoked potentials, for example, often have most of their energy concentrated in one or two narrow frequency bands. Bennett (1979) suggested fitting long autoregressions to such data. The approach taken here may be useful for separating at least some additive stationary noise from the transient signal, thereby allowing the two components to be analysed and modelled separately.

2. Description of the problem

Suppose that some phenomenon is observed at regular time intervals. Relevant examples of such phenomena are a heart rate or a measure of electrical activity at some site in the brain. Suppose that in the 'normal' state the sequence of observations is indistinguishable from a realization of a stationary time series, but that at certain known times a stimulus is applied which could lead to transitory changes in at least one of the mean and variance of the observations. A complete sequence of observations will be referred to as a record and the Ith record will be denoted $y_I(n)$, $n = 0, 1, \cdots, N-1$. For data from experiments of the types mentioned above, N may be several thousand. Suppose that $y_I(n) = t_I(n) + x_I(n)$, where $x_I(n)$ is the observation in the 'normal' state and $t_I(n)$ is the change caused by the stimulus. The first aim is to test the null hypothesis that $t_I(n) = 0$, $n = 0, 1, \cdots, N-1$.

In experiments where such data are collected there are often several subjects (and thus several sequences of observations). There may also be different stimuli, and they may be applied several times in the one record. In these cases, the hypotheses of importance are usually not the detection of a response but rather the measurement of differences between responses. These may be caused by subject or stimulus differences, or perhaps by a learning effect as the subject becomes more accustomed to the stimuli. A general approach is here given for estimating transient responses, both within a record and between records. To simplify the discussion, it is assumed that the times of onset of the transients are known. As has been discussed in Cameron and Hannan (1979), this assumption is not essential and the onset times may be estimated, although that estimation could be difficult if one transient began before the previous one had effectively died away. In many experiments it will be the case that the times at which the stimuli are applied will be known, and each response will begin virtually immediately after its stimulus. Suppose that the onset time of the ath transient in a record is τ_a. In the case where there is more than one record, the times of onset of the transients and even the number of transients may vary from record to record. It will be assumed here that all records are of the same length, that there are the same number (A, say) of transients in each record, and that the ath transient begins at the same time, τ_a, in each record. The more general situation may be analysed using the same basic principles as are described here. When there is more than one observed record, the different series may be classified according to some design. Two simple examples are: (i) the same stimulus is presented to several different subjects, and (ii) several stimuli are each presented to the same group of subjects, but in each record only one type of stimulus is presented, perhaps several times. Of course, far more complicated classification schemes are possible; the subjects may themselves be classifiable into subgroups, or each subject may receive only a specific subset of the stimuli according to some balanced or partially balanced experimental design. The extension of the methods to these more complicated cases is straightforward.

Suppose $\mathscr{I}$ is a set of indices and $I \in \mathscr{I}$. Then the observations on the Ith record will be represented by

$$(1) \qquad y_I(n) = t_I(n) + x_I(n), \qquad n = 0, 1, \cdots, N-1$$

and

$$(2) \qquad t_I(n) = \sum_a t_I(n; a) = \sum_a t_I^0(n - \tau_a; a).$$

Here $t_I^0(n; a) = t_I(n + \tau_a; a)$ denotes the translation in time of the ath transient so that it begins at time $n = 0$. That is, $t_I^0(n; a) = 0$, $n < 0$. Let $\Omega_N = \{\omega_k : \omega_k = 2\pi k/N, 0 < k < N/2\}$. The methods to be used in this paper will be based on the Fourier coefficients of the data, computed at the frequencies $\omega \in \Omega_N$. To this end, suppose that

$$Y_I(\omega) = N^{-\frac{1}{2}} \sum_{n=0}^{N-1} y_I(n) \exp(in\omega); \qquad \omega \in \Omega_N,$$

denotes the discrete Fourier transform of $y_I(n)$ and let $T_I(\omega)$, $T_I(\omega; a)$, $T_I^0(\omega; a)$ and $X_I(\omega)$ be similarly defined from $t_I(n)$, $t_I(n; a)$, $t_I^0(n; a)$ and $x_I(n)$ respectively.

It will be assumed that the noise processes are stationary, Gaussian, incoherent with each other and with the signals, and with a common 'smooth' spectral density. The methods to be used depend upon the Fourier coefficients of the transient being relatively smooth, as a function of frequency. This is equivalent to the translated signals, $t_I^0(n; a)$, having most of their energy concentrated close to the beginning of the record: see Cameron and Hannan (1979). The precise requirements for the asymptotic theory of the test statistics to be derived are given in the Appendix.

3. Estimating the Fourier coefficients of transients

Let

$$\Omega_{N,j,m} = \{\omega : \omega \in \Omega_N, |\omega - \omega_j| \leq 2\pi m/N\}.$$

Consider the Fourier coefficients of a sequence of observations for the band of fundamental frequencies $\omega \in \Omega_{N,j,m}$. It follows from (1) that

$$Y_I(\omega) = \sum_a T_I(\omega; a) + X_I(\omega).$$

If τ_a is not too large compared with N, so that the transient has died away before the end of the record, this model is approximately equivalent to

$$(3) \qquad Y_I(\omega) = \sum_a \exp(i\tau_a\omega) T_I^0(\omega_j; a) + X_I(\omega), \qquad \omega \in \Omega_{N,j,m}$$

provided that the band is sufficiently narrow for $T_I^0(\omega; a)$ to be approximately constant in the band. This is the basic model to be used here for estimation and inference.

If the spectrum of $x_I(n)$ is approximately constant over $\Omega_{N,j,m}$, then (3) is just a linear regression model for complex-valued variables. Brillinger (1981) set down the model (3) and gave the least-squares regression formulae for estimating the coefficients $T_I^0(\omega_j; a)$. If there is only one transient in the record, then the estimation is particularly simple and also provides insight into how the width of the band of frequencies used in estimation should related to the spread, in time, of the transient.

Let

$$\tilde{Y}_I(\omega; a) = \exp(-i\tau_a\omega) Y_I(\omega).$$

When there is just one transient the regression estimate of $T_I^0(\omega_j; 1)$ is

$$(4) \qquad \hat{T}_I^0(\omega_j; 1) = (2m+1)^{-1} \sum_k \tilde{Y}_I(\omega_{j+k}; 1)$$

where, here and below, the summation over k means summation over the frequencies $\omega_{j+k} \in \Omega_{N,j,m}$. Since $T_I^0(\omega_j; 1)$ varies (slowly) with frequency, it may be desirable to down-weight the frequencies further from ω_j and a weighted estimator is $\sum_k w_k \tilde{Y}_I(\omega_{j+k}; 1)$. Now $\tilde{Y}_I(\omega; 1)$ is the discrete Fourier transform of a variable, $\tilde{y}(n; 1)$ say, which is approximately the series $y(n)$ translated in time so that the transient begins at the beginning of the observed record. (When τ_a is an integer, $\tilde{y}(n; 1)$ is merely a cyclic permutation of $y(n)$.) Thus,

$$(5) \qquad \sum_k w_k \tilde{Y}_I(\omega_{j+k}; 1) = N^{-\frac{1}{2}} \sum_{n=0}^{N-1} \tilde{y}(n; 1) W_m(\omega_n) \exp(in\omega_j)$$

where $W_m(\omega_n) = \sum w_k \exp(ik\omega_n)$ and so when there is only one transient, the estimator of $T_I^0(\omega_j; 1)$ is simply the Fourier coefficient of a time-windowed version of the data. If this is to be a good estimator the time-windowed series should match the transient signal closely, and so the time-window should be large near the peak of the transient and should die away so as to be small when the effects of the transient are thought to be negligible. If the window is too narrow then some of the signal will be measured as noise and biases will result. In fact

$$E \sum_n [t_I(n) - W_m(n) y_I(n)]^2$$

is minimized if, for each n,

$$(6) \qquad W_m(n) = t_I(n)^2/[t_I(n)^2 + \sigma^2],$$

where $\sigma^2 = Ex(n)^2$. Thus, as seems reasonable qualitatively, when the signal to noise ratio is high a good time-window $W_m(n)$ will be large and as the signal to noise ratio decreases so should the time-window.

Disregarding the details of the shape of the transient, the duration of the transient gives an indication of the width of the time-window that should be used or, equivalently, the width of the frequency-domain window over which Fourier coefficients should be smoothed. Since a narrow time-domain window corresponds to a broad window in the frequency domain, the shorter the duration of the transient, the larger may m be chosen.

For example, in the simplest case when $w_k = (2m+1)^{-1}$ for $k = -m, \cdots, m$, then

$$(7) \qquad W_m(\omega_n) = \sin[(2m+1)\omega_n/2]/[(2m+1)\sin(\omega_n/2)],$$

the Dirichlet kernel. This has zeros when $\omega_n = 2\pi k/(2m+1)$ for $k = 1, 2, \cdots$, that is, when $n = kN/(2m+1)$. The first zero occurs at $n = N/(2m+1)$ and if the transient is not negligible until a time n_0 say, then to avoid bias, m must be chosen so that the first zero occurs after n_0, i.e. $m < (N - n_0)/(2n_0)$. Note that from (6) the optimal time window is non-negative so that the Dirichlet window is suboptimal. However, it will be used as an example in later sections because it is notationally and statistically simple.

If there is more than one transient in the record then the Fourier coefficients of all transients may be estimated simultaneously with a regression. However, if the transients are sufficiently well separated then it is possible to design data windows so that the contribution to the estimated Fourier coefficients of one transient from any of the other transients is negligible. One example arises when the transients are equally spaced and well separated. In this case there are (suboptimal) time windows that have zeros at the times of onset (or maximum amplitude) of the separate transients and which are easily implemented in the frequency domain. An example is given by Tanaka (1983); if the transients are equispaced, the Dirichlet kernel is used as a time window and if the total length of the observation record, N, is a divisor of $(2m+1)(\tau_b - \tau_a)$ for all a and b then a set of windows may be defined so that each member of the set has its maximum peak over the beginning of a different transient and has its zeros at the onsets of the other transients. In this case the matrix of regressors in (3) has typical element $\exp(i\tau_a\omega_{j+k})$ and $X'X$ matrix is diagonal. Note that N is a divisor of $(2m+1)(\tau_b - \tau_a)$ if $2m+1$ is a multiple of the number of transients. The multiple to be chosen in practice should depend on the fraction of the time between successive transients during which the transient is above the level of noise. If the number of frequencies is restricted to be an integer multiple of A, that is $2m + 1 = kA$, and if the first transient does not die out before time n_0 then, just as in the case of one transient, a restriction on the bandwidth can be calculated. In this case $kA < N/n_0$.

In the most general case, where the coefficients are estimated by a linear regression, the window shape may be varied by performing a weighted rather than an unweighted regression.

4. The detection and comparison of transients

As mentioned in the previous section, the fundamental model is given by (3) and the Fourier coefficients of the transients may be estimated by linear regression of complex-valued random variables. A test for the null hypothesis of no transient is obtained by testing if any of the estimated Fourier coefficients of the possible transient are significantly different from 0. Likewise, the hypothesis of equality of response is tested by testing the null hypothesis $T_I^0(\omega_j; a) = T_I^0(\omega_j; a')$ for all j. Brillinger (1981) has set out this general model and given the formulae for the estimation and testing of individual Fourier coefficients. The tests that he describes are of the hypotheses that specified coefficients are either 0 or equal. The test statistics are the usual F-statistics calculated in linear regression. In this section a method of calculating overall test statistics from these F-statistics is presented and the application in three particular cases is described. In the appendix it is shown that the overall test statistics are asymptotically normally distributed.

To simplify the exposition and to highlight the link with classical analysis of variance, the Fourier coefficients of the transients at the mid-band frequencies will be estimated by unweighted linear regressions, corresponding to the use of the Dirichlet kernel time window.

4.1. *Detection of a transient response.* Consider first the case of a single transient in a single record. Since the time of onset of the transient is known it may be assumed without loss of generality that the transient begins at the beginning of the record. Thus from (3), $T_I^0(\omega_j; 1)$ may be estimated by the mean of the $2m + 1$ observed Fourier coefficients with frequencies closest to ω_j. If $f_I(\omega)$ is the spectral density of $x_I(n)$ then $\hat{T}_I^0(\omega_j; 1) = (2m + 1)^{-1} \sum_k Y_I(\omega_{j+k})$ has variance $2\pi f_I(\omega_j)/(2m + 1)$ and an estimate of $f_I(\omega_j)$ is

$$\hat{f}_I(\omega_j) = [4m\pi]^{-1} \sum_k |Y_I(\omega_{j+k}) - \hat{T}_I^0(\omega_j; 1)|^2.$$

For large N and small m this estimate of $f_I(\omega_j)$ is approximately distributed as $f_I(\omega_j)\chi^2_{4m}/4m$. Thus if $F(\omega_j) = (2m + 1)|\hat{T}_I^0(\omega_j; 1)|^2/[2\pi\hat{f}_I(\omega_j)]$, a test for a non-zero Fourier coefficient of the transient at frequency ω_j compares $F(\omega_j)$ with an F-statistic having 2 and $4m$ degrees of freedom.

If there are A (>1) transients then from (3) their Fourier coefficients may be estimated by a complex-valued linear regression of $Y_I(\omega)$ on the variables $\{\exp(i\tau_a\omega), a = 1, \cdots, A\}$ and a test of the null hypothesis that all the responses are 0 is the F-statistic computed from the regression and residual sums of squares. This statistic has $2A$ and $2(2m+1-A)$ degrees of freedom. As in standard regression, tests of individual coefficients (responses) are readily constructed. The computations for the regression may be performed using a real variable regression program. See Brillinger (1975), p. 222, for details of the computations.

These tests are valid for a single frequency band, but to test for the presence of a transient, information from all frequency bands should be combined. This raises the problem of multiple tests and one approach is to perform simultaneous tests using the Bonferroni inequality to set the marginal significance levels for the individual, 'almost' independent tests. In a similar context, Brillinger (1973b) suggested plotting the relevant F-statistics against frequency. This would show the frequency ranges in which the signal to noise ratio is largest. (If different bandwidths are used over different parts of the frequency range then the probability integral transforms of the F-statistics should be plotted against frequency to make the plotted points directly comparable.) This is a useful exploratory and diagnostic technique. However, a single test statistic may be required in some circumstances either for formal testing of the null hypothesis or as a summary. If there is a weak transient present, with energy spread over most of the frequency range then the single test statistic may be more powerful than the consideration of the statistics from each band separately. However, if the energy of the transient is concentrated in a small number of frequency bands, the overall test may be less powerful.

For the situation where one stationary process is to be regressed on possibly lagged values of other stationary processes, Cameron (1978) suggested a test for non-zero regression coefficients based on spectral estimates of the prediction error variance. Under certain general conditions on the form of the transients, that test statistic can be used to test for the presence of a transient in the observed series and the asymptotic distribution theory remains valid. Details are given in the appendix. An equivalent statistic is obtained by averaging the logarithms of the individual F-statistics over frequency bands. If there are M bands of width $2m+1$ and if $F_D(\omega_j)$ is the F-statistic for the jth band then under the null hypothesis of no transient response,

$$N^{\frac{1}{2}}[M^{-1}\sum \log F_D(\omega_j) + \log A - \log(2m+1-A) - \psi(A) + \psi(2m+1-A)]$$

is asymptotically normally distributed with mean zero and variance

$$(4m+2)[\psi'(A) + \psi'(2m+1-A)]$$

where $\psi(\cdot)$ and $\psi'(\cdot)$ are the digamma and trigamma functions respectively.

4.2. *Comparing responses within a record.* When several transients within a record are produced by similar stimuli, a statistic to test the hypothesis that the responses are the same may be required. For a single frequency band, the statistic suggested by the approximate likelihood ratio is the ratio of two residual sums of squares; one being that obtained by fitting the full model (3) by regression, and the other being that obtained by regressing $Y_I(\omega)$ on $\sum_a \exp[i\tau_a \omega]$. If these sums of squares are denoted $S_1(\omega_j)$ and $S_2(\omega_j)$ respectively then

$$F_C(\omega_j) = (2m+1-A)[S_2(\omega_j) - S_1(\omega_j)]/[(A-1)S_1(\omega_j)]$$

has, asymptotically, under the hypothesis of equal responses, an F-distribution with $2A-2$ and $2(2m+1-A)$ degrees of freedom. As in the previous case, information from all frequency bands may be combined by calculating the average of the logarithms of the F-statistics, the result being a statistic which is normally distributed in large samples under the null hypothesis of equal responses. In particular,

$$N^{\frac{1}{2}}[M^{-1}\sum \log F_C(\omega_j) + \log(A-1) - \log(2m+1-A)$$
$$- \psi(A-1) + \psi(2m+1-A)]$$

is normally distributed with mean 0 and variance $(4m+2)[\psi'(A-1) + \psi'(2m+1-A)]$.

If the estimates of the different transients are orthogonal then the calculations are much simpler. Let

$$\tilde{Y}_I(\omega; a) = \exp(-i\tau_a \omega) Y_I(\omega).$$

Then,

$$\hat{T}_I^0(\omega_j; a) = (2m+1)^{-1} \sum_k \tilde{Y}_I(\omega_{j+k}; a).$$

Further, the variables $\{\exp(i\tau_a \omega), \omega \in \Omega_{N,j,m}\}$ indexed by a are orthogonal over the band of frequencies and therefore the best estimate of the responses under the assumption that they are equal is just $A^{-1}\sum_a \hat{T}_I^0(\omega_j; a)$. Hence the calculations of the estimates (and test statistics) may be derived from those of a complex-valued analysis of variance: from the Fourier coefficients $Y_I(\omega)$ are derived the A sets of observations $\tilde{Y}_I(\omega; a)$, and a one-way analysis of variance with A groups and $2m+1$ replicates in each group may be performed separately on each of the real and imaginary parts of the $\tilde{Y}_I(\omega; a)$'s. Test statistics can be derived by pooling corresponding degrees of freedom and sums of squares in the two ANOVA tables. When the calculations are performed this way, the pooled 'total sum of squares' is $A\sum_k |Y_I(\omega_{j+k})|^2$ and so $(A-1)\sum_k |Y_I(\omega_{j+k})|^2$ must be subtracted to get the correct 'analysis of power' table. The same quantity must also be subtracted from the 'residual sum of squares'. After this adjustment the F-statistics testing for the presence of a transient and for the equality of responses may be calculated in the usual way. The average log F-statistics may then be calculated to provide a simultaneous test over all frequencies.

The advantages of using orthogonal estimates of the transient responses are twofold. Firstly, the calculations outlined above are very easy to perform. Secondly, various other graphical and robust techniques which have been developed for the analysis of tables (Tukey (1977)) and which are available in widely available statistical packages can be applied to provide a thorough data analysis beyond the formal calculation of test statistics without further extensive computer programming. However, as pointed out by Brillinger (1981), the experimenter may lose information if the stimuli are widely spaced to ensure orthogonal estimates.

4.3. *Comparing responses between records.* If there is just one transient response in each record then, after the Fourier coefficients in each record have been multiplied by $\exp[-i\tau_a\omega]$, the problem reduces to an analysis of variance for each frequency band, much as in the 'orthogonal response estimates' case in (4.2), except that in this case the 'residual SS' and 'total SS' produced by an analysis of variance program are correct.

When there is more than one response per record and there are also several records then various models are possible for the $T_I^0(\omega_j; a)$'s, although the simplest is the linear additive model of the form

$$T_I^0(\omega; a) = \mu(\omega) + \alpha_I(\omega) + \beta_a(\omega) + \gamma_{Ia}(\omega)$$

with the usual need for constraints of the form

$$\sum_I \alpha_I(\omega) = \sum_a \beta_a(\omega) = \sum_I \gamma_{Ia}(\omega) = \sum_a \gamma_{Ia}(\omega) = 0.$$

In general the possible null hypotheses

$$\gamma_{Ia}(\omega) = 0; \qquad \beta_a(\omega) = 0; \qquad \alpha_I(\omega) = 0$$

correspond to regression lines being parallel and may be tested by comparing the residual mean squares computed by fitting the appropriate models by regression. Again the information from the different frequency bands may be combined by finding the average log F-statistic.

As in the previous case, the calculations are performed more easily if the responses are equispaced and the bandwidth chosen so that the estimates of the Fourier coefficients are orthogonal. In this case the computation is much as for an analysis of variance. Thus,

$$\tilde{Y}_I(\omega; a) = \sum_b \exp\left(i(\tau_b - \tau_a)\omega\right) T_I^0(\omega; b) + \tilde{X}_I(\omega; a)$$

$$= \sum_b \exp\left(i(\tau_b - \tau_a)\omega\right)[\mu + \alpha_I + \beta_b + \gamma_{Ib}] + \tilde{X}_I(\omega; a).$$

Then,

$$(2m+1)^{-1} \sum_k \tilde{Y}_I(\omega_{j+k}; a) = \tilde{Y}_I(\omega_{j+.}; a) = \mu + \alpha_I + \beta_a + \gamma_{Ia} + \tilde{X}_I(\omega_{j+.}; a)$$

where $\tilde{X}_I(\omega_{j+.})$ is the mean of $\tilde{X}_I(\omega_{j+k}; a)$ over the $2m+1$ values of k from $-m$ to m. Because the functions $\exp[-i\tau_a\omega_{j+k}]$ are orthogonal over the band, the $\tilde{X}_I(\omega_{j+.}; a)$'s are approximately uncorrelated with mean 0 and variance $(2m+1)^{-1}2\pi f_I(\omega_j)$. The procedures for estimation and inference are then straightforward.

5. Conclusion

The methods described here will be useful in a variety of situations. Only the simplest time-window has been used to describe the calculations since it illustrates the methods and since the means and variances of the asymptotic distributions of the test statistics are slightly easier to calculate. It seems likely that, provided the width of the window is chosen appropriately, the benefits gained by varying the shape of the window will be slight. In the material presented here it has been assumed that the spectrum of the noise is estimated over the same band of frequencies as is used for the estimation of the Fourier coefficients of the transients. However, two different bandwidths could be used. There is no reason, other than algebraic convenience, why the amount of smoothing applied to the noise spectrum should be related to the width of the time-window used to estimate the transients.

Finally, it may be that as a result of applying the tests described in the previous section, it is decided that the responses to certain stimuli are not significantly different and that a time domain model of the common response should be fitted. The methods of Tanaka (1983) can then be used on the appropriate estimates of the Fourier coefficients of the common transient and the corresponding noise spectrum estimate. This may be done easily using a program written to perform the Tanaka estimation.

Appendix: Some asymptotic theory

The test statistics given in this paper are similar to some given in Cameron (1978). To derive their asymptotic distributions, the theorem of that paper is shown to hold when the independent variable is a complex exponential rather than a stationary process, and the 'regression coefficients' are, in fact, a transient signal. In this appendix appropriate conditions on the transient are given for that theorem to hold. The asymptotic distributions of the test statistics then follow.

From (1) and (2),

$$(A1) \qquad y_I(n) = \sum_{a=1}^{A} t_I^0(n - \tau_a; a) + x_I(n)$$

where

$$t_I^0(n; a) = 0 \quad \text{for} \quad n < 0.$$

In order to produce a sensible asymptotic theory for transients, the total energy in the transients must be of the same order as the total energy in the noise and thus a sequence of transients must be defined, with the total energy in each transient increasing with N but being concentrated in a decreasing fraction of the total record length. To emphasise the dependence on N, the observations and their Fourier coefficients will be denoted $t_I^0(n; a, N)$ and $T_I^0(\omega; a, N)$ respectively for the remainder of the appendix.

Suppose that for the jth frequency band, $S_I(j)$ is the 'residual sum of squares' obtained by regressing $Y_I(\omega)$ on $\exp[i\tau_a\omega]$ for $a = 1, \cdots, A$ and $\omega \in \Omega_{N,j,m}$. The theorem below provides the asymptotic distribution of a class of statistics of the form

$$(A2) \qquad \theta_N = M^{-1} \sum_{j=0}^{M-1} \phi[S_I(j)].$$

The theorem is proved assuming that the following four conditions hold.

Condition 1. For each $I \in \mathcal{I}$, $x_I(n)$, $n = 0, 1, \cdots, N-1$ is a realization of a discrete-time stationary Gaussian process with absolutely continuous spectrum and spectral density $f_I(\omega)$ which satisfies a Lipschitz condition of order α, $\alpha > \frac{1}{2}$. Further, $x_I(n)$ is independent of $x_{I'}(n)$ for every $I, I' \in \mathcal{I}$.

Condition 2. The observed series $y_I(n)$ are sums of finite numbers of delayed transients, the ath of which satisfies

$$N^{-1} \sum_n |t_I^0(n; a, N)| \, n < \infty.$$

Condition 3. For $-\pi \leqq \omega \leqq \pi$, $\phi(F)$ satisfies a Lipschitz condition of order 1 for F in the range of $f_I(\omega)$.

Condition 4. The expected value of $\phi[S_I(j)]$ is $\phi[f_I(\omega_j)] + O(N^{-\alpha})$ and all its moments exist under the false assumption that the Fourier coefficients $X_I(\omega_{j+k})$

for $k = -m, \cdots, m$, are independent, identically distributed with mean zero and variance $2\pi f_I(\omega_j) + O(N^{-\alpha})$.

Theorem. Suppose $y_I(n)$ is given by (A1), θ_N by (A2) and that Conditions 1, 2, 3 and 4 all hold. Then for fixed m, as $N \to \infty$, θ_N converges almost surely to

$$\theta_0 = \int_{-\pi}^{\pi} \phi[f_I(\omega)] d\omega,$$

and the moments of $N^{\frac{1}{2}}(\theta_N - \theta_0)$ converge to the same limits as would arise if the assumption in Condition 4 were true with variance $2\pi f_I(\omega_j)$.

Proof. Now,

$$Y_I(\omega) = N^{-\frac{1}{2}} \sum_{a=1}^{A} \sum_{n=0}^{N-1} t_I^0(n - \tau_a; a) \exp(in\omega) + X_I(\omega),$$

and for each a and for $\omega \in \Omega_{N,j,m}$,

(A3)
$$N^{-\frac{1}{2}} \sum_{n=0}^{N-1} t_I^0(n - \tau_a; a) \exp(in\omega)$$
$$= T_I^0(\omega_j; a) \exp(i\tau_a\omega) + [T_I^0(\omega; a) - T_I^0(\omega_j; a)] \exp(i\tau_a\omega)$$
$$- N^{-\frac{1}{2}} \sum_{n=0}^{\tau_a} t_I^0(n + N - \tau_a; a) \exp(i(n+N)\omega).$$

By comparing this with Equation (2.2) of Cameron (1978) and examining the proof of the theorem there, it can be seen that the proof there will continue to hold here if the last two terms on the right of Equation (A3) above are both $O(N^{-\frac{1}{2}})$. Of the two, the first will be 'small' if the Fourier coefficients of $t_I^0(n)$ are sufficiently smooth, and the second will be 'small' if an adequate approximation for the Fourier coefficients of $t_I^0(n; a)$ may be derived from the Fourier coefficients of the series $t_I(n; a)$. since for each a $t_I^0(n; a)$ is transient, both terms may be made small by imposing reasonable conditions on the speed at which the transients die away.

For fixed k and for $\omega \in \Omega_{N,j,m}$,

$$|T_I^0(\omega; a, N) - T_I^0(\omega_j; a, N)| \leq 2k\pi N^{-\frac{3}{2}} \sum_n |t_I^0(n; a, N)| \, n$$
$$= O(N^{-\frac{1}{2}}) \quad \text{if} \quad N^{-1} \sum |t_I^0(n; a, N)| \, n < \infty.$$

If τ_a is fixed then the final term in (A3) can easily be seen to be $O(N^{-\frac{1}{2}})$ under the above condition. However the more realistic situation is when τ_a increases with N. For example, if there are A equispaced transients, $\tau_a = (a-1)N/A$. In this case for the Ath transient,

$$\left| \sum_{n=0}^{(A-1)N/A} t_I^0(n + N/a; A, N) \right| < A \sum_{n=N/A}^{N} |t_I^0(n; A, N)| \, n/N < \infty$$

and thus the last term in (A3) is also $O(N^{-\frac{1}{2}})$ if only $N^{-\frac{1}{2}}\sum_n |t_I^0(n; a, N)| < \infty$. Thus the theorem holds.

Example. One model for transients supposes that a stationary process is passed through a time-window. Koopmans (1974), p. 113 gives references. In an example given in Cameron and Hannan (1979) the window is made dependent on N to exemplify asymptotic theory. There it is suggested that the window is of magnitude $(N/\alpha_N)^{\frac{1}{2}}$ for $0 < n < \alpha_N$ and 0 otherwise, where $\alpha_N \to \infty$ but $\alpha_N/N \to 0$. If $N^{-1}\sum |t_I^0(n; a, N)|\, n < \infty$ then $\alpha_N = O(N^{\frac{1}{3}})$.

Once the theorem holds, Corollary 4 of Cameron (1978) holds and the asymptotic distributions of the various average log F-statistics introduced in this paper are easily derived.

References

BENNETT, W. F. (1979) The analysis of transient spectral components with the autoregressive spectral estimator. *Appl. Statist.* **28**, 1–13.

BOX, G. E. P. AND TIAO, G. C. (1975) Intervention analysis with applications to economic and environmental problems. *J. Amer. Statist. Assoc.* **70**, 70–79.

BRILLINGER, D. R. (1973a) A power spectral estimate which is insensitive to transients. *Technometrics* **15**, 559–562.

BRILLINGER, D. R. (1973b) The analysis of time series collected in an experimental design. In *Multivariate Analysis* III, ed. P. R. Krishnaiah. Academic Press, New York, 241–256.

BRILLINGER, D. R. (1975) *Time Series: Data Analysis and Theory.* Holt, Rinehart and Winston, New York.

BRILLINGER, D. R. (1981) The general linear model in the design and analysis of evoked response experiments. *J. Theoret. Neurobiol.* **1**, 105–119.

CAMERON, M. A. (1978) The prediction variance and related statistics for stationary time series. *Biometrika* **65**, 283–296.

CAMERON, M. A. AND HANNAN, E. J. (1979) Transient signals. *Biometrika* **66**, 243–258.

JONES, R. H., CROWELL, D. H. AND KAPUNIAI, L. E. (1970) Change detection model for serially correlated multivariate data. *Biometrics* **26**, 269–280.

KOOPMANS, L. H. (1974) *The Spectral Analysis of Time Series.* Academic Press, New York.

MÖCKS, J., PHAM DINH TUAN AND GASSER, T. (1984) Testing for homogeneity of noisy signals evoked by repeated stimuli. *Ann. Statist.* **12**, 193–209.

TANAKA, K. (1983) Estimation for transients in the frequency domain. *J. Amer. Statist. Assoc.* **78**, 718–724.

TUKEY, J. W. (1977) *Exploratory Data Analysis.* Addison-Wesley, Reading, Mass.

Approximate Distributions of Student's *t*-Statistics for Autoregressive Coefficients Calculated from Regression Residuals

J. DURBIN

Abstract

We consider a multiple regression model in which the regressors are Fourier cosine vectors. These regressors are intended as approximations to 'slowly changing' regressors of the kind often found in time series regression applications. The errors in the model are assumed to be generated by a special type of autoregressive model defined so that the regressors are eigenvectors of the quadratic forms occurring in the exponent of the probability density of the errors. This autoregression is intended as an approximation to the usual stationary autoregression. Both approximations are adopted for the sake of mathematical convenience.

Student's *t*-statistics are constructed for the autoregressive coefficients in a manner analogous to ordinary regression. It is shown that these statistics are distributed as Student's *t* to the first order of approximation, that is with errors in the density of order $T^{-\frac{1}{2}}$, where T is the sample size, while the squares of the statistics are distributed as the square of Student's *t* to the second order of approximation, that is with errors in the density of order T^{-1}.

AUTOREGRESSION; AUTOCORRELATION; PARTIAL AUTOCORRELATION; REGRESSION WITH AUTOCORRELATED ERRORS; ASYMPTOTIC EXPANSIONS

Consider the regression

$$(1) \qquad y_t = \beta_1 x_{1t} + \cdots + \beta_k x_{kt} + u_t, \qquad t = 1, \cdots, T$$

where u_t follows the mth-order autoregression with density

$$p(u_1, \cdots, u_T \mid A_0, \cdots, A_m) = C(A_0, \cdots, A_m) \exp\left[-\frac{1}{2\sigma^2} \sum_0^m A_j u' C_j u \right]$$

where

$$A_j = \frac{1}{\sigma^2} \sum_{i=0}^{m-j} \alpha_i \alpha_{i+j} \qquad (\alpha_0 = 1)$$

and

$$u'C_0u = \sum_1^T u_t^2, \qquad u'C_1u = \tfrac{1}{2}u_1^2 + \tfrac{1}{2}u_T^2 + \sum_1^{T-1} u_t u_{t+1},$$

$$u'C_2u = u_1 u_2 + u_{T-1}u_T + \sum_1^{T-2} u_t u_{t+2}$$

and so on, using the matrices C_j defined in Durbin (1980). We choose this form of the autoregression because the matrices C_j commute, a property that is essential for the theoretical development, and because it gives a density that we may reasonably expect to be closer to that of the stationary mth-order autoregressions than other forms based on commutative matrices such as the circular autoregression. Since the matrices commute they have the same eigenvectors. We assume that $x_r = [x_{r1}, \cdots, x_{rT}]'$ is the rth eigenvector of C_j with eigenvalue λ_{jr} for $j = 1, \cdots, m$ and $r = 1, \cdots, k$. Substituting in (1) we obtain for the density of $y_1, \cdots, y_T$

$$p(y_1, \cdots, y_T \mid A_0, \cdots, A_m, \beta_1, \cdots, \beta_k)$$

(2)
$$= C(A_0, \cdots, A_m) \exp\left[-\frac{1}{2\sigma^2} \sum_{j=0}^m A_j c_j - \frac{1}{2\sigma^2} \sum_{r=1}^k (b_r - \beta_r)^2 \sum_{t=1}^T x_{rt}^2 \sum_{j=0}^m \lambda_{jr} A_j \right]$$

where $c_j = (y - b_1 x_1 - \cdots - b_k x_k)' C_j (y - b_1 x_1 - \cdots - b_k x_k)$ in which b_r is the least-squares estimator of β_r, i.e. $b_r = \sum_1^T x_{rt} y_t / \sum_1^T x_{rt}^2$. This result follows because $x_1, \cdots, x_k$ are eigenvectors of the C_j's.

As stated in Durbin (1980), these eigenvectors are the Fourier cosine vectors with elements $x_{rt} = \cos\left[(2t-1)(r-1)/(2T)\right]$ for $t = 1, \cdots, T$ and $r = 1, 2, \cdots$. Since as functions of t these oscillate with low frequency when r is small, results obtained by assuming that these are the regressors give good approximations for general regressors when the regressors are 'slowly changing'.

It turns out that we may put $\sigma^2 = 1$ without loss, so from now on we shall assume that this has been done.

Our objective is to investigate the asymptotic distributions of Student's t-statistics for estimates of the autoregressive coefficients $\alpha_1, \cdots, \alpha_m$. We see from the form of (1) that $c_0, \cdots, c_m, b_1, \cdots, b_k$ are sufficient for $A_0, \cdots, A_m$, $\beta_1, \cdots, \beta_m$. On transforming to $c_0, \cdots, c_m, b_1, \cdots, b_k$ and $T - k - m$ other variables and integrating out these other variables we obtain for the density of c, b

(3)
$$f(c, b \mid A_0, A, \beta) = p(y_1, \cdots, y_T \mid A_0, A, \beta) h(c, b)$$

where c, b, A, β stand for the sets $c_0, \cdots, c_m, b_1, \cdots, b_k, A_1, \cdots, A_m$, $\beta_1, \cdots, \beta_k$ and where h does not depend on A, β. From (3) we have

(4)
$$f(c, b \mid 1, 0, 0) = p(y_1, \cdots, y_T \mid 1, 0, 0) h(c, b).$$

Dividing (3) by (4) we obtain

$$f(c, b \mid A_0, A, \beta)$$

(5)
$$= \frac{C(A_1, \cdots, A_m) \exp\left[-\tfrac{1}{2}\sum_j A_j c_j - \tfrac{1}{2}\sum_r (b_r - \beta_r)^2 \sum_t x_{rt}^2 \sum_j \lambda_{jr} A_j\right]}{C(1, 0, \cdots, 0) \exp\left[-\tfrac{1}{2}c_0 - \tfrac{1}{2}\sum_r b_r^2\right]}$$

$$\times f(c, b, \mid 1, 0, 0).$$

Now transform from $c_0, \cdots, c_m$ to $c_0, r_1, \cdots, r_m$ where $r_j = c_j/c_0$. The Jacobian is

$$\frac{\partial(c_0, \cdots, c_m, b_1, \cdots, b_k)}{\partial(c_1, r_1, \cdots, r_m, b_1, \cdots, b_k)} = c_0.$$

Denote the set $r_1, \cdots, r_m$ by r and the density of c_0, r and b by $g(c_0, r, b \mid A_0, A, \beta)$. Then $f(c, b \mid A_0, A, \beta) = c_0 g(c_0, r, b \mid A_0, A, \beta)$. Substituting in (5) we obtain

(6) $$g(c_0, r, b \mid A_0, A, \beta) = C \frac{\exp\left[-\tfrac{1}{2}\sum A_j c_j \text{ etc. as in (5)}\right]}{\exp\left[-\tfrac{1}{2}c_0 - \tfrac{1}{2}\sum b_r^2\right]} g(c_0, r, b \mid 1, 0, 0)$$

where from now on C will denote any appropriate normalising constant. Now r is independent of c_0 in the $1, 0, 0$ case, and it is obviously also independent of b. It follows that

$$g(c_0, r, b \mid 1, 0, 0) = C c_0^{\frac{1}{2}(T-k-2)} \exp\left[-\tfrac{1}{2}c_0 - \tfrac{1}{2}\sum b_r^2\right] l(r)$$

where $l(r)$ is the density of r. Substituting in (6) we obtain

(7) $$g(c_0, r, b \mid A_0, A, \beta) = C \exp\left[\text{as in numerator of (6)}\right] c_0^{\frac{1}{2}(T-k-2)} l(r).$$

First let us consider the joint distribution of the estimates $a_1, \cdots, a_m$ of the autoregression coefficients $\alpha_1, \cdots, \alpha_m$ which, as remarked above, satisfy $\sum_{i=0}^{m-j} \alpha_i \alpha_{i+j} = A_j$, $j = 0, \cdots, m$ ($\alpha_0 = 1$). Integrating out $b_1, \cdots, b_k$ we obtain for the density $g(c_0, r)$ of c_0, r (from now on we use g as a generic symbol for density, as we are running out of symbols)

$$g(c_0, r \mid A_0, A) = C c_0^{\frac{1}{2}(T-k-2)} \exp\left[-\frac{c_0}{2}\left(A_0 + \sum_1^m A_j r_j\right)\right] l(r).$$

Integrating out c_0 we obtain for the density $g(r)$ of r

(8) $$g(r \mid A_0, A) = \frac{C l(r)}{Q^{\frac{1}{2}(T-k)}(\alpha)}$$

where

$$Q(\alpha) = A_0 + \sum_1^m A_j r_j = \sum_0^m \alpha_j^2 + 2 \sum_0^{m-1} \alpha_j \alpha_{j+1} r_1 + \cdots + 2\alpha_m r_m \quad \text{and}$$

$$\alpha = [\alpha_1, \cdots, \alpha_m]'.$$

Now we transform to the vector $a = [a_1, \cdots, a_m]'$ of sample autoregressive coefficients computed from the Yule–Walker equations $Ra = -r$ where

$$(9) \qquad R = \begin{bmatrix} 1 & r_1 & r_2 & \cdots & r_{m-1} \\ r_1 & 1 & r_1 & \cdots & r_{m-2} \\ \vdots & & & & \\ r_{m-1} & \cdots\cdots\cdots\cdots & 1 \end{bmatrix}.$$

We observe that

$$(10) \qquad \begin{aligned} Q(\alpha) &= 1 + 2\alpha' r + \alpha' R\alpha \\ &= 1 - a' Ra + a' Ra - 2\alpha' Ra + \alpha' R\alpha \\ &= 1 - a' Ra + (a - \alpha)' R(a - \alpha). \end{aligned}$$

This is of course just the analogue in the present context of the usual partitioning of the total SS into residual SS + regression SS.

We now use an argument which appears roundabout but which is nevertheless valid and convenient. We want to show that $1 - a'Ra = \prod_{j=1}^m (1 - r_{j.}^2)$ where $r_{j.}$ is the jth partial autocorrelation between z_t and z_{t+j} keeping $z_{t+1}, \cdots, z_{t+j-1}$ fixed, and where z_t is the least-squares residual $y_t - b_1 x_{1t} - \cdots - b_k x_{kt}$. We note that $r_1, \cdots, r_m$ is a set of values which give a non-negative definite matrix of the form (9) with $m - 1$ replaced by m. As such, they could just as easily have come from a sample from a circular autoregressive process, since in either case any set of values $r_1, \cdots, r_m$ which gives a non-negative autocorrelation matrix is permissible. Let $z_1^*, \cdots, z_T^*$ be a set of values generated by a circular autoregression such that $r_j = \sum_1^T z_t^* z_{t+j}^* / \sum_1^T z_t^{*2}$ $(z_t^* = z_{t+T}^*)$. Then by standard least-squares regression theory

$$(11) \qquad \frac{\sum_1^T (z_t^* - a_1 z_{t-1}^* - \cdots - a_k z_{t-m}^*)^2}{\sum_1^T z_t^{*2}} = 1 - R^{*2} = \prod_{j=1}^m (1 - r_{j.}^2)$$

where $a_1, \cdots, a_k$ and $r_{1.}, \cdots, r_{m.}$ are exactly the same functions of $r_1, \cdots, r_k$ as in the non-circular case since they are calculated in the same way in both cases, and where R^* is the multiple correlation coefficient. By standard regression theory

$$(12) \qquad \sum_1^T (z_t^* - a_1 z_{t-1}^* - \cdots - a_k z_{t-k}^*)^2 = \sum_1^T z_t^{*2} - a' Ma$$

where

$$M = \begin{bmatrix} \sum_1^T z_t^{*2} & \sum z_t^* z_{t-1}^* & \cdots & \sum z_t^* z_{t-m+1}^* \\ \sum z_t^* z_{t-1}^* & \sum z_t^{*2} & & \\ \vdots & & \ddots & \\ \sum z_t^* z_{t-m+1}^* & & \cdots & \sum_1^T z_t^{*2} \end{bmatrix} = \sum_1^T z_t^{*2} R.$$

This, combined with (11) and (12), proves that $1 - a'Ra = \prod_{j=1}^m (1 - r_{j.}^2)$.
Now let

$$s^2 = \frac{1}{T-k-m}[1 - a'Ra].$$

Substituting in (8) and using (10) we have

$$(13) \qquad g(r \mid A_0, A) = \frac{Cl(r)}{\sum_1^T (1 - r_{j.}^2)^{\frac{1}{2}(T-k)}\left[1 + \dfrac{(a-\alpha)'R(a-\alpha)}{(T-k-m)s^2}\right]^{\frac{1}{2}(T-k)}}.$$

This is the density of r when $r_{j.}$ and a_j are expressed in terms of $r_1, \cdots, r_m$.
From Durbin (1980) we have for the density of $r_. = [r_{1.}, \cdots, r_{j.}]'$

$$g(r_.) = C \prod_{j \text{ odd}} (1 - r_{j.}^2)^{k-1}(1 - r_{j.})^{\frac{1}{2}(T-k)} \prod_{j \text{ even}} (1 - r_{j.})^{k+1}(1 - r_{j.}^2)^{\frac{1}{2}(T-k)-1}[1 + O(T^{-\frac{3}{2}})].$$

Substituting in (13) we have

$$(14) \quad g(r \mid A_0, A) = \frac{C \prod_{j \text{ odd}} (1 - r_{j.})^{k-1} \prod_{j \text{ even}} (1 - r_{j.})^{k+1}(1 - r_j^2)^{-1} J(r_., r)}{\left[1 + \dfrac{(a-\alpha)'R(a-\alpha)}{(T-k-m)s^2}\right]^{\frac{1}{2}(T-k)}}[1 + O(T^{-\frac{3}{2}})]$$

where $J(r_., r)$ is the Jacobian of the transformation from $r_.$ to r, i.e.

$$J(r_., r) = \left| \frac{\partial(r_{1.}, \cdots, r_{m.})}{\partial(r_1, \cdots, r_m)} \right|.$$

To get the density of a we have to multiply (14) by $J(r, a)$ where $J(r, a)$ is the
Jacobian of the transformation from r to a. Now

$$J(r_., r)J(r, a) = J(r_., a) = \left| \frac{\partial(r_{1.}, \cdots, r_{m.})}{\partial(a_1, \cdots, a_m)} \right|.$$

Daniels (1956), p. 183 showed that

$$(15) \qquad J(r_., a) = \prod_{j \text{ odd}} (1 - r_{i.}^2)^{\frac{1}{2}(j-1)} \prod_{j \text{ even}} (1 - r_{j.})(1 - r_j^2.)^{\frac{1}{2}j-1}.$$

Combining (14) and (15) we have

$$(16) \qquad g(a \mid A_0.A) = \frac{C \prod\limits_{j \text{ odd}} (1 - r_{j.})^{k-1}(1 - r_{i.}^2)^{\frac{1}{2}(j-1)} \prod\limits_{j \text{ even}} (1 - r_{j.})^{k+2}(1 - r_j^2.)^{\frac{1}{2}(j-2)}}{\left[1 - \dfrac{(a - \alpha)'R(a - \alpha)}{(T - k - m)s^2}\right]^{\frac{1}{2}(T-k)}}$$
$$\times [1 + O(T^{-\frac{3}{2}})].$$

Now let $t_j = (a_j - \alpha_j)/s$, $j = 1, \cdots, m$, and let $t = [t_1, \cdots, t_m]'$. Since we have already used the symbol t as the time suffix it might appear confusing to use it again with a different interpretation. However, it seems convenient to follow standard practice and employ it to denote a vector of Studentised statistics. Since we shall use it only with this interpretation in the remainder of the paper, there should be no confusion. The Jacobian of this transformation is complicated, since s is a function of $r_.$ and hence of a. Its value does not matter for our purpose, however, since it does not depend on T. Denote the product of this Jacobian and the numerator of (16), when this is expressed in terms of $t_1, \cdots, t_m$, by $K(t \mid \alpha, T)$. We write $K(t \mid \alpha, T)$ since in general K depends on α and T as well as t. However, it is important to note that the factor depending on T does not depend on $t_1, \cdots, t_m$. We have therefore found the density of t to be

$$(17) \qquad g(t \mid \alpha) = \frac{K(t \mid \alpha, T)}{\left[1 + \dfrac{t'Rt}{T - k - m}\right]^{\frac{1}{2}(T-k)}} [1 + O(T^{-\frac{3}{2}})].$$

We have written $g(t \mid \alpha)$ here since A_0, A depends only on α.

Now we want to relate this to the multi-t distribution obtained for an analogous problem in ordinary least-squares regression. Suppose we have the regression model

$$(18) \qquad y_i = \beta_1 x_{1i} + \cdots + \beta_m x_{mi} + \beta_{m+1} x_{m+1,i} + \cdots + \beta_{m+k} x_{m+k,i} + \varepsilon_i,$$
$$i = 1, \cdots, T$$

where $\varepsilon_1, \cdots, \varepsilon_T$ are i.i.d. $N(0, \sigma^2)$ and where all the x's are fixed. Let

$$X = \begin{bmatrix} x_{11} & \cdots & x_{m1} \\ x_{12} & \cdots & x_{m2} \\ \vdots & & \vdots \\ x_{1T} & & x_{mT} \end{bmatrix}$$

and suppose that the vectors $x_j = [x_{j1}, \cdots, x_{jT}]'$ for $j = m+1, \cdots, m+k$ are orthonormal and are orthogonal to all the columns of X. The density of $y = [y_1, \cdots, y_T]$ is

$$(19) \qquad g(y \mid \beta, B, \sigma^2) = \frac{1}{(2\pi\sigma^2)^{\frac{1}{2}T}} \exp\left(-\frac{1}{2\sigma^2} \sum_1^T \varepsilon_i^2\right)$$

where $\beta = [\beta_1, \cdots, \beta_m]$, $B = [\beta_{m+1}, \cdots, \beta_{m+k}]$ and where ε_i is expressed in terms of the y's using (18). Let $b = [b_1, \cdots, b_m]'$ and $\hat{B} = [b_{m+1}, \cdots, b_{m+k}]$ be the least-squares estimators of β, B. Then

$$\sum_1^T \varepsilon_i^2 = \sum_1^T z_i^2 + (b-\beta)'X'X(b-\beta) + \sum_{m+1}^{m+k} (b_j - \beta_j)^2$$

where z_i's are the least-squares residuals. Putting

$$s^2 = \frac{1}{T-m-k} \sum_1^T z_i^2$$

and transforming to b, $\hat{B}$ and s^2 we obtain for the density of the latter,

$$g(b, \hat{B}, s^2 \mid \beta, B, \sigma^2)$$

$$= C \exp\left[-\frac{1}{2\sigma^2}\{(T-m-k)s^2 + (b-\beta)'X'X(b-\beta) + \sum_{m+1}^{m+k}(b_j-\beta_j)^2\right]$$

$$\times (s^2)^{\frac{1}{2}(T-m-k)-1}.$$

Integrating out $b_{m+1}, \cdots, b_{m+k}$ we obtain for the density of b, s^2,

$$(20) \quad g(b, s^2 \mid \beta, \sigma^2) = C \exp\left[-\frac{(T-m-k)s^2}{2\sigma^2}\left\{1 + \frac{(b-\beta)'X'X(b-\beta)}{(T-m-k)s^2}\right\}\right]$$

$$\times (s^2)^{\frac{1}{2}(T-m-k)-1}.$$

Put $t_j = (b_j - \beta_j)/s$ for $j = 1, \cdots, m$. The Jacobian is s^m. Substituting in (20) we have

$$g(t, s^2 \mid \beta, \sigma^2) = C \exp\left[-\frac{(T-m-k)s^2}{2\sigma^2}\left\{1 + \frac{t'X'Xt}{T-m-k}\right\}\right](s^2)^{\frac{1}{2}(T-k)-1}.$$

Integrating out s^2 we have for the density of $t = [t_1, \cdots, t_m]'$

$$(21) \qquad g(t) = \frac{C}{\left[1 + \dfrac{t'X'Xt}{T-k-m}\right]^{\frac{1}{2}(T-k)}}$$

noting that dependence on β and σ^2 has disappeared.

We note the obvious resemblance of (17) to (21) and ask the question, 'To what extent does the multi-t distribution (21) derived from ordinary least-squares theory provide a good approximation to (17)?' We also want to consider the relation of our results to the Mann–Wald (1943) theorem which effectively (for the non-regression case) asserts that

$$t \xrightarrow{\mathscr{D}} N(0, \tilde{R})$$

where

$$\tilde{R} = \begin{bmatrix} 1 & \rho_1 & \rho_2 & \cdots & \rho_{m-1} \\ \rho_1 & 1 & \rho_1 & \cdots & \rho_{m-2} \\ \vdots & & & & \\ \rho_{m-1} & & \cdots & & 1 \end{bmatrix}.$$

First we note that when $K(t\,|\,\alpha, T)$ is expanded in powers of $t_1, \cdots, t_m$ the linear terms are $O(T^{-\frac{1}{2}})$. This follows since if it had been expressed in terms of a not t it could obviously have been expanded in powers of $a - \alpha$ about 0 and $a - \alpha = st = tO(T^{-\frac{1}{2}})$ since, with the definition of s^2 we have used, $s = O(T^{-\frac{1}{2}})$. Thus we have from (17)

$$(22) \qquad g(t, \alpha) = \frac{K(0\,|\,\alpha, T)}{\left[1 + \dfrac{t'Rt}{T - k - m}\right]^{\frac{1}{2}(T-k)}} [1 + O(T^{-\frac{1}{2}})].$$

Since $K(0\,|\,\alpha, T)$ does not depend on t we see that the dominant term of (22) is the same as (21) with R replaced by $X'X$, possibly with an adjustment to the constant $K(0\,|\,\alpha, T)$ to ensure that the first term of the density integrates to unity. We have therefore shown that as a first approximation, autoregression behaves like ordinary least-squares regression.

Now let $T \to \infty$ in (22) and we have obviously

$$g(t, \alpha) \to C \exp\left[-\tfrac{1}{2}t'\tilde{R}t\right].$$

This is effectively the Mann–Wald result. Now

$$\frac{C}{\left[1 + \dfrac{t'Rt}{T - k - m}\right]^{\frac{1}{2}(T-k)}} = C \exp\left[-\tfrac{1}{2}t'\tilde{R}t\right][1 + O(T^{-1})].$$

(The C's differ by a term $O(T^{-1})$!.) Substituting in (22) we have

$$(23) \qquad g(t, \alpha) = C \exp\left[-\tfrac{1}{2}t'\tilde{R}t\right][1 + O(T^{-\frac{1}{2}})].$$

Thus the asymptotic order of difference between the true density and the multi-t distribution is the same as that between the true distribution and the multi-normal. Intuitively it would seem likely that the t-distribution would give a better approximation than the normal, but this would have to be settled by simulation.

Now we consider the marginal distribution of a single t-statistic, say t_1. Reverting to the ordinary regression case (18), it is well known that if in (21) $t_2, \cdots, t_m$ are integrated out successively then the marginal density of t_1 is

$$(24) \qquad g(t_1) = \frac{C}{\left[1 + \dfrac{t_1^2}{(T-m-k)d_1}\right]^{\frac{1}{2}(T-m-k+1)}}$$

where d_1 is the top left-hand element of $(X'X)^{-1}$. This must be so since it is well known that $V(b_1) = \sigma^2 d_1$, so if we put $\bar{t}_1 = (b_1 - \beta_1)(s\sqrt{d_1})^{-1}$, then $\bar{t}_1$ has Student's t-distribution with $T-m-k$ degree of freedom, i.e., has density

$$g(\bar{t}_1) = \frac{C}{\left[1 + \dfrac{\bar{t}_1^2}{T-m-k}\right]^{\frac{1}{2}(T-m-k+1)}} \cdot$$

But this is what we get if we put $\bar{t}_1 = t_1 d_1^{-\frac{1}{2}}$ in (24). Reverting to the autoregression problem, it follows that if we put $\bar{t}_1 = t_1 (R^{11})^{-\frac{1}{2}}$ where R^{11} is the top left-hand element of R^{-1}, and integrate $t_2, \cdots, t_m$ out from density (17) then the marginal density of $\bar{t}_1$ is

$$(25) \qquad g(\bar{t}_1 \mid \alpha) = \frac{K^*(\bar{t}_1 \mid \alpha, T)}{\left(1 + \dfrac{\bar{t}_1^2}{T-m-k}\right)^{\frac{1}{2}(T-m-k+1)}} [1 + O(T^{-\frac{3}{2}})],$$

where K^* is a new constant such that $K^*(\bar{t}_1 \mid \alpha, T) = K^*(0 \mid \alpha, T)[1 + O(T^{-\frac{1}{2}})]$. This means that for testing hypotheses about α_1 or setting confidence intervals for α_1, the use of the $\bar{t}_1$ statistic as if it had been derived from an ordinary regression model is accurate to the first order of approximation.

Now consider the distribution of $|\bar{t}_1|$, the absolute value of $\bar{t}_1$. As before, we can expand K^* about 0 in powers of $\bar{t}_1$, giving $K^*(\bar{t}_1 \mid \alpha, T) = K_0 + \gamma \bar{t}_1 T^{-\frac{1}{2}} + O(T^{-1})$ where $K_0 = K^*(0 \mid \alpha, T)$ and γ is independent of $\bar{t}_1$ and T. Substituting in (25) we have for the density of $|\bar{t}_1|$

$$g(|\bar{t}_1| \mid \alpha) = \left[\frac{K_0 + \gamma |\bar{t}_1|}{\left(1 + \dfrac{\bar{t}_1^2}{T-m-k}\right)^{\frac{1}{2}(T-m-k+1)}} + \frac{K_0 - \gamma |\bar{t}_1|}{\left(1 + \dfrac{\bar{t}_1^2}{T-m-k}\right)^{\frac{1}{2}(T-m-k+1)}} \right]$$

$$\times [1 + O(T^{-1})]$$

$$= \frac{C}{\left(1 + \dfrac{\bar{t}_1^2}{T-m-k}\right)^{\frac{1}{2}(T-m-k+1)}} [1 + O(T^{-1})], \qquad |\bar{t}_1| \geqq 0.$$

This shows that for symmetric two-sided tests or symmetric confidence intervals, the use of Student's t as if it had been calculated from an ordinary regression is accurate to the second order of approximation. It suggests that for two-sided use, the use of the t-distribution in autoregression should be more accurate than the normal approximation though once again this would have to be tested by simulation. It follows that $\bar{t}_1^2$ is distributed as the square of Student's t to the second order of approximation, that is with an error of order T^{-1}. A further consequence is that F-statistics calculated for several coefficients are distributed as Fisher's F to the same order of approximation.

In order to illuminate the foregoing, let us consider the case $m = k = 1$. Since the first-eigenvector is a vector of constants this means we are considering the fitting of a first-order autoregression from residuals from means. Here,

$$-a = r_1 = r_{1.} = r = \left(\tfrac{1}{2}y_1^2 + \tfrac{1}{2}y_T^2 + \sum_1^{T-1} y_t y_{t+1} - T\bar{y}^2\right)\bigg/ \sum_1^T (y_t^2 - T\bar{y}^2).$$

Density (8) reduces to

$$(26) \qquad g(r \mid \alpha) = \frac{C(1-r^2)^{\frac{1}{2}(T-1)}}{(1+\alpha^2+2\alpha r)^{\frac{1}{2}(T-1)}}[1+O(T^{-\frac{3}{2}})]$$

$$(27) \qquad\qquad = \frac{C}{\left(1+\dfrac{(r+\alpha)^2}{1-r^2}\right)^{\frac{1}{2}(T-1)}}[1+O(T^{-\frac{3}{2}})].$$

Put

$$s^2 = \frac{1}{T-2}(1-r^2) \quad\text{and}\quad t = \frac{r+\alpha}{s} = \frac{\sqrt{T-2}\,(r+\alpha)}{(1-r^2)^{\frac{1}{2}}}$$

so

$$t(1-r^2)^{\frac{1}{2}} = \sqrt{T-2}\,(r+\alpha)$$

$$\frac{dt}{dr}(1-r^2)^{\frac{1}{2}} + \frac{\sqrt{T-2}\,(r+\alpha)}{(1-r^2)^{\frac{1}{2}}}\left(-\frac{r}{(1-r^2)^{\frac{1}{2}}}\right) = \sqrt{T-2}$$

$$\frac{dt}{dr} = \frac{\sqrt{T-2}}{(1-r^2)^{\frac{1}{2}}}\left[1+\frac{r(r+\alpha)}{1-r^2}\right] = \frac{\sqrt{T-2}}{(1-r^2)^{\frac{3}{2}}}(1+r\alpha).$$

Substituting in (27) we therefore have for the density of t

$$g(t \mid \alpha) = \frac{C(1-r^2)^{\frac{3}{2}}}{(1+r\alpha)}\frac{1}{\left[1+\dfrac{t^2}{T-2}\right]^{\frac{1}{2}(T-1)}}[1+O(T^{-\frac{3}{2}})]$$

where r is expressed in terms of t.

For the case of a circular autoregression Leipnik (1947) obtained analogous to (26) the approximate density

$$g(r \mid \alpha) = \frac{C(1-r^2)^{\frac{1}{2}(T-1)}}{(1+\alpha^2+2\alpha r)^{\frac{1}{2}T}}$$

and Quenouille (1949) succeeded in expressing this as the product of a Student's t-density

$$\frac{C}{\left(1+\dfrac{t^2}{T+1}\right)^{\frac{1}{2}(T+2)}}$$

times one plus an odd function of t. Thus the distribution of Quenouille's t^2 is distributed as the square of Student's t with $T+1$ degrees of freedom to the second order of approximation. Furthermore, Quenouille was able to show how the effect of the odd function could be calculated, thus effectively obtaining the distribution of his t-statistic itself to the second order of approximation. In our case the absence of the extra factor $(1+\alpha^2+2\alpha r)^{\frac{1}{2}}$ in the denominator prevents us from obtaining an explicit formula for the approximate density of our t-statistic analogous to Quenouille's. Thus for the case $\alpha \neq 0$ consideration of the special case $m = k = 1$ has not led to any further results beyond what we have already obtained for the general case.

Further refinements can, however, be achieved for the $m = k = 1$ case when $\alpha = 0$. Substituting in (27) gives

$$g(r \mid \alpha) = \frac{C}{\left(1+\dfrac{r^2}{1-r^2}\right)^{\frac{1}{2}(T-1)}}[1+O(T^{-\frac{3}{2}})].$$

Putting $\tilde{t} = \sqrt{T+1}\, r/(1-r^2)^{\frac{1}{2}}$ gives $d\tilde{t}/dr = \sqrt{T+1}\,(1-r^2)^{-\frac{3}{2}}$. We therefore obtain for the density of $\tilde{t}$

$$g(\tilde{t}) = \frac{C}{\left(1+\dfrac{\tilde{t}^2}{T+1}\right)^{\frac{1}{2}(T+2)}}[1+O(T^{-\frac{3}{2}})].$$

Thus $\tilde{t}$ is distributed as Student's t with $T+1$ degrees of freedom to the third order of approximation.

This approximation is essentially identical to Hannan's ((1960), p. 85) approximation to the distribution of the lag-1 coefficient

$$r_1 = \frac{\frac{1}{2}(y_1-\bar{y})^2+\frac{1}{2}(y_T-\bar{y})^2+2\sum_{1}^{T-1}(y_t-\bar{y})(y_{t+1}-\bar{y})}{\sum_{1}^{T}(y_t-\bar{y})^2}$$

for a series of independent $N(\mu, \sigma^2)$ variables. From a knowledge of the exact significance points of r_1, Hannan suggested that r_1 should be approximately distributed as the ordinary product-moment correlation coefficient calculated from a sample of $T+3$ independent observations of a pair of independent normal variables. Denoting such a coefficient by r, it is well known that $\sqrt{T+1}\, r/(1-r^2)^{\frac{1}{2}}$ is distributed as Student's t with $T+1$ degrees of freedom. The fact that for this particular case it turns out to be appropriate to use $T+1$ in place of $T-2$ suggests that it might be preferable to use $T-m-k+3$ for the number of degrees of freedom in the general case in place of $T-m-k$, but I have not investigated this further.

Now let us see how far we can obtain t-tests from this approach for the coefficients $\beta_1, \cdots, \beta_k$ of the original model (1). Let us return to (7), which we write as

$$(28) \qquad g(c_0, r, b \mid A_0, A, \beta) = C \exp\left[-\tfrac{1}{2}\sum A_j c_j - \tfrac{1}{2}\sum_r (b_r - \beta_r)^2 \nu_r\right] c_0^{(T-k-2)} l(r),$$

giving

$$g(c_0, r, \hat{t} \mid A_0, A, \beta)$$

$$(29) \qquad = C \exp\left[-\frac{c_0}{2}\left(A_0 + \sum_1^m A_j r_j + \frac{1}{(T-m-k)}\sum_{j=m+1}^{m+k} \hat{t}_j^2\right)\right] c^{\frac{1}{2}(T-2)} l(r)$$

where $\nu_r = \sum_{t=1}^T x_{rt}^2 \sum_{j=0}^m \lambda_{jr} A_j$ and $\hat{t} = [\hat{t}_{m+1}, \cdots, \hat{t}_{m+k}]$ where

$$\hat{t}_j = \sqrt{\frac{(T-m-k)\nu_j}{c_0}}\,(b_i - \beta_i)$$

for $j = m+1, \cdots, m+k$. Integrating out c_0 we obtain corresponding to (8)

$$(30) \qquad\qquad g(r, \hat{t} \mid A_0, A) = \frac{Cl(r)}{\tilde{Q}^{\frac{1}{2}T}(\alpha)}$$

where

$$(31) \qquad \tilde{Q}(\alpha) = \sum_0^m \alpha_j^2 + 2\sum_{s=1}^m \sum_{j=0}^{m-s} \alpha_j \alpha_{j+s} r_s + \frac{1}{T-m-k}\sum_{m+1}^{m+k} \hat{t}_j^2.$$

The first two terms are reduced exactly as before, and we obtain

$$\tilde{Q}(\alpha) = \prod_1^m (1 - r_{j.}^2)\left[1 + \frac{1}{T-m-k}\left(t'Rt + \sum_{j=m+1}^{m+k} t_j^2\right)\right]$$

where $t_j^* = \tilde{t}_j / \prod_{s=1}^m (1 - r_s^2)^{\frac{1}{2}}$ for $j = m+1, \cdots, m+k$. Finally, denoting the vector $t_1, \cdots, t_m, t_{m+1}^*, \cdots, t_{m+k}^*$ by t^*, we have for its density, analogously to (17)

$$(32) \qquad g(t^* \mid \alpha) = \frac{K^*(t^* \mid \alpha, T)}{\left[1 + \dfrac{1}{T-k-m}\left(t'Rt + \sum_{m+1}^{m+k} t_j^2\right)\right]^{\frac{1}{2}T}}.$$

As before, if we integrate all the t^*'s except one, say t_{m+1}^*, we obtain for its density

$$(33) \qquad g(t^*_{m+1} \mid \alpha) = \frac{K^*(t^*_{m+1} \mid \alpha, T)}{\left(1 + \dfrac{t^{*2}_{m+1}}{T-m-k}\right)^{(T-m-k+1)}} \, .$$

Note that we do not have to change from t^*_{m+1} to $\bar{t}^*_{m+1}$ since the appropriate element of the inverse matrix is now unity. Using exactly the same arguments as for the autoregression coefficients, we find that the distribution of each of $t^*_{m+1}, \cdots, t^*_{m+k}$ is the Student's t-distribution as if we had ordinary least-squares regression with $T - m - k$ degrees of freedom with errors of $O(T^{-\frac{1}{2}})$ in the one-sided case and $O(T^{-1})$ in the symmetric two-sided case. However, there is one point of difference from the autoregressive coefficients and that is that the value of t^*_j for $j = m + 1$ to $m + k$ depends on unknown parameters. In fact we have

$$t^*_r = \frac{(b_r - \beta_r)\sqrt{v_r}}{\bar{s}}$$

where $\bar{s}^2$ is the usual least-squares estimates of residual variance and

$$v_r = \sum_{t=1}^{T} x_{rt}^2 \sum_{j=0}^{m} \lambda_{jr} A_j.$$

This depends on the unknown $A_0, \cdots, A_m$. In practice one would need to compute $\hat{t}^*_r = (b_r - \beta_r)\hat{v}^{\frac{1}{2}}_r/s$ where $\hat{v}_r$ is the obvious estimator of v_r. It turns out that the derivation of approximations to the distribution of $\hat{t}^*_r$ seems intractable using the methods of this paper and that another approach is necessary. One such approach is indicated in Durbin (1983), (1984).

References

DANIELS, H. E. (1956) The approximate distribution of serial correlation coefficients. *Biometrika* **43**, 169–185.

DURBIN, J. (1980) The approximate distribution of partial serial correlation coefficients calculated from residuals from regression on Fourier series. *Biometrika* **67**, 335–349.

DURBIN, J. (1983) The price of ignorance of the autocorrelation structure of the errors of a regression model. *Studies in Econometrics, Time Series and Multivariate Statistics*, ed. S. Karlin, T. Amemiya and L. A. Goodman. Academic Press, New York.

DURBIN, J. (1984) Time series analysis. *J. R. Statist. Soc. A* **147**, 161–173.

HANNAN, E. J. (1960) *Time Series Analysis*. Methuen, London.

LEIPNIK, R. B. (1947) Distribution of the serial correlation coefficient in a circularly correlated universe. *Ann. Math. Statist.* **18**, 80–87.

MANN, H. B. AND WALD, A. (1943) On the statistical treatment of linear stochastic difference equations. *Econometrica* **11**, 173–220.

QUENOUILLE, M. H. (1949) Approximate tests of correlation in time series 3. *Proc. Camb. Phil. Soc.* **45**, 483–484.

A Simultaneous Test in the Presence of Nested Alternative Hypotheses

YUZO HOSOYA

Abstract

This paper considers the generalized likelihood ratio (GLR) test or its modification dealing with nested models. The algorithm for evaluating the critical values and the error-rates for the canonical tests are provided; a table of critical values of a class of GLR tests is also given. The test proposed in the paper has applications in time-series model selection.

HYPOTHESIS TESTING; TIME-SERIES ANALYSIS; LIKELIHOOD RATIO TEST; ASYMPTOTIC THEORY; NUMERICAL INTEGRATION; NESTED MODELS; MODEL SELECTION

0. Introduction

This paper explores a generalization of the likelihood-ratio test in order to deal with testing a hypothesis against nested alternatives. Though such a testing situation is very common in the practice of time-series analysis (take for instance testing the hypothesis of white noise against the autoregressive hypotheses of various orders), there does not seem to be an established general method of dealing with it, except probably for the one which chooses the most complicated as the alternative, which is not necessarily a reasonable choice when a simpler alternative is true. The paper proposes a simultaneous use of several likelihood-ratio statistics which is termed a generalized likelihood ratio (GLR) test and a use of a transformation method of such statistics in cases where an algorithmic impasse with respect to the evaluation of simultaneous distribution arises.

In Section 1, the large-sample canonical form of the GLR test is derived for the general linear time-series model. Section 2 first deals with the distributional problem of the canonical tests when χ^2-statistics involved are of even degrees of freedom, giving the computation procedures for the probability error rates

Research supported in part by the Ministry of Education and Science Grant-in-Aid for Scientific Research No. 58530010.

and the critical values; and then it deals with the general case. One of the main obstacles to developing a general theory of simultaneous inference is that, except for special cases, the joint distributions of the statistics involved are often too complicated even in large samples to be tractable for practical use. This distributional impasse might be one of the reasons which the simultaneous inference approach is not fully explored even though the importance is recognized. The GLR test meets the same difficulty in general; in order to deal with it and to give the tests general applicability, a transformation method is proposed in Section 2, which consists basically of transforming χ^2-increments of odd degrees of freedom into those of even ones. Section 3 exhibits a table of critical values for an important class of GLR tests where the dimension of the parameter spaces is low and the χ^2-increments are all of one degree of freedom. Also the computation method is provided.

In this paper, the following notations and symbols are used. A real Euclidean n-space is denoted by R^n; degree of freedom is abbreviated as d.f.; A' and A^* are the transpose and the conjugate transpose of a matrix A, respectively; I_s denotes the identity matrix of order s; if θ is a p-vector, $\partial f(\theta^0)/\partial\theta$ denotes the column p-vector whose elements are $\partial f(\theta^0)/\partial\theta_j, j = 1, \cdots, p$, which are the partial derivatives evaluated at $\theta = \theta^0$.

1. The generalized likelihood ratio test

Suppose a nested model is given as follows. Let $q_0, q_1, \cdots, q_p$ be non-negative integers such that $q_0 < q_1 < \cdots < q_p$. Let Ω_p be a non-empty open region of R^{q_p} and let $\Omega_l = \{\alpha \in \Omega_p; \alpha_{q_l+1} = 0, \cdots, \alpha_{q_p} = 0\}$ where α_{q_i} denotes the q_ith coordinate of q_p-vector α, where all Ω_l are assumed non-empty ($0 \leq l \leq p - 1$). Let $\Omega_l^* = \{\alpha \in \Omega_l; \alpha \notin \Omega_{l-1}\}$ for $l = 1, \cdots, p$ and let $\Omega_0^* = \Omega_0$. Denote by $\alpha(l)$ a q_l-vector $(l = 0, \cdots, p)$ and sometimes $\alpha \in \Omega_p$ such that $\alpha_i = \alpha_i(l), i = 1, \cdots, q_l$ and $\alpha_i = 0$ for $i > q_l$ is identified with $\alpha(l)$. Denote by $\mathbf{y}$ a real random m-vector to be observed and let H_i $(i = 0, 1, \cdots, p)$ be respectively the composite hypothesis that $\mathbf{y}$ has a probability density $f(y, \alpha)$ with respect to a common σ-finite measure on R^m and that $\alpha \in \Omega_i^*$. Denote by $\hat{\alpha}(l)$ $(l = 0, 1, \cdots, p)$ the maximum-likelihood estimate of α under the hypothesis H_l (assuming their existence) and set

$$(1.1) \qquad L_{i,j} = \log f(\mathbf{y}, \hat{\alpha}(i)) - \log f(\mathbf{y}, \hat{\alpha}(j)).$$

As a test of H_0 against the alternative $H_1, \cdots, H_p$, the generalized likelihood ratio (GLR) test is defined to be any test in such a form that H_0 is rejected if and only if

$$(1.2) \qquad L_{0,j} < c_j \quad \text{for some} \quad j, \qquad j = 1, \cdots, p.$$

Let $\chi_i, i = 1, \cdots, p$, be independent χ^2-random variables with d.f. $(q_i - q_{i-1})$

respectively. Then as a natural extension of the known fact that for $p = 1$ the test (1.2) becomes asymptotically a χ^2-test, it can be shown for many cases where $\mathbf{y}$ is generated by a regular stationary process as well as where it consists of i.i.d. random variables that the joint distribution of $\{L_{01}, \cdots, L_{0p}\}$ is the same as that of $-(\frac{1}{2})\sum_{j=1}^{i} \chi_j$, $i = 1, \cdots, p$ and the GLR test (1.2) is reduced for a large sample to the canonical test of the following set-up. Let $\chi_1, \cdots, \chi_p$ be observable independent χ^2-random variables with $(q_i - q_{i-1})$ d.f., $i = 1, \cdots, p$ respectively, then H_0 is rejected if and only if

$$(1.3) \qquad \sum_{j=1}^{i} \chi_j > d_i \quad \text{for some} \quad i, \qquad i = 1, \cdots, p.$$

The reduction is exhibited below for a general linear process under regularity conditions. The case of i.i.d. observations can be dealt with in a rather similar way by means of the assumptions of Wald (1943) and with the assumption that the parameter space has the nested structure $\Omega_0 \subset \Omega_1 \subset \cdots \subset \Omega_p$ given in the above.

The linear process $\{\mathbf{z}(t) : t = \cdots, -1, 0, 1, \cdots\}$ which is most commonly used in time-series analysis often has a representation such as

$$(1.4) \qquad \mathbf{z}(t) = \sum_{j=0}^{\infty} G(j, \theta)\mathbf{e}(t-j),$$

where the $\mathbf{z}(t)$'s and the $\mathbf{e}(t)$'s have s components, the $\mathbf{e}(t)$'s are a fourth-order stationary process and $E(\mathbf{e}(t)) = 0$, $E(\mathbf{e}(t)\mathbf{e}(t)') = K(\mu)$, and the cumulant of $e_a(t_1)$, $e_b(t_2)$, $e_c(t_3)$, $e_d(t_4)$ is equal to K_{abcd} if $t_1 = t_2 = t_3 = t_4$ and is equal to 0 otherwise ($e_a(t)$, $e_b(t)$, $\cdots$, denote the components of $\mathbf{e}(t)$). The $G(j, \theta)$'s are $s \times s$ matrices such that $\sum_{j=0}^{\infty} \operatorname{tr} G(j, \theta) K(\mu) G(j, \theta)' < \infty$ $(G(0, \theta) = I_s)$ and hence the process has a spectral density matrix $f(\omega; \theta, \mu)$ such that

$$(1.5) \qquad f(\omega; \theta, \mu) = \left(\frac{1}{2\pi}\right) k(\omega, \theta) K(\mu) k(\omega, \theta)^*, \qquad -\pi \leqq \omega \leqq \pi,$$

where $k(\omega, \theta) = \sum_{j=0}^{\infty} G(j, \theta) e^{i\omega j}$. Let Ω_i, Ω_i^* be as defined above and let M be a region of R^r and let the hypothesis H_j^* be that $\theta \in \Omega_j^*$, $\theta \notin \Omega_{j-1}^*$ and $\mu \in M$ for $j = 1, 2, \cdots, p$ and H_0^* be that $(\theta, \mu) \in \Omega_0 \times M$. Assume the following conditions:

(i) If $(\theta^1, \mu^1) \neq (\theta^2, \mu^2)$, $f(\omega; \theta^1, \mu^1) \neq f(\omega; \theta^2, \mu^2)$ on a set of positive Lebesgue measure;

(ii) each component of $k(\omega, \theta)$ is twice-differentiable with respect to θ and the second derivative is continuous with respect to (ω, θ). $K(\mu)$ has continuous second derivatives with respect to μ;

(iii) $\sum_{j=0}^{\infty} j \operatorname{tr} \{G(j, \theta) G(j, \theta)^*\} < \infty$;

(iv) (1) for each β_1, β_2 and m

$$\operatorname{Var}\left[E\{e_{\beta_1}(t)e_{\beta_2}(t+m) \mid B(t-\tau)\} - \delta(m, 0)K_{\beta_1,\beta_2}(\mu)\right] = O(\tau^{-2-\varepsilon}),$$

$\varepsilon > 0$, uniformly in t as τ tends to ∞;

$$(2) \quad E|E\{e_{\beta_1}(t_1)e_{\beta_2}(t_2)e_{\beta_3}(t_3)e_{\beta_4}(t_4) \mid B(t_1-\tau)\} - E\{e_{\beta_1}(t_1)e_{\beta_2}(t_2)e_{\beta_3}(t_3)e_{\beta_4}(t_4)\}|$$
$$= O(\tau^{-1-\eta}), \qquad \eta > 0,$$

uniformly in t_1 where $t_1 \leqq t_2 \leqq t_3 \leqq t_4$, as $\tau \to \infty$;

(3) the spectral densities $f_{\beta\beta}$ ($\beta = 1, \cdots, s$) are square-integrable; where $B(t)$ is the σ-field generated from $\{e(s); s \leqq t\}$.

Denote by $I(\omega)$ the periodogram matrix constructed from the partial realization $\{z(1), \cdots, z(n)\}$; namely,

$$(1.6) \qquad I(\omega) = \left(\frac{1}{2\pi n}\right)\left\{\sum_{t=1}^{n} z(t)e^{i\omega t}\right\}\left\{\sum_{t=1}^{n} z(t)e^{i\omega t}\right\}^*, \qquad -\pi \leqq \omega \leqq \pi.$$

Suppose that for each l ($l = 0, 1, \cdots, p$) and for sufficiently large n there exists the unique value $(\hat{\theta}(l), \hat{\mu}(l))$ in $\Omega_l^* \times M$ which maximizes the quasi-Gaussian log-likelihood function $\bar{L}$ which is defined as

$$(1.7) \qquad \bar{L}(\theta, \mu) = -\left[\log \det K(\mu) + \left(\frac{1}{2\pi}\right)\int_{-\pi}^{\pi} \text{tr}\{f^{-1}(\omega; \theta, \mu)I(\omega)\}d\omega\right].$$

Define the quasi log-likelihood ratios $\bar{L}_{i,j}$ as

$$(1.8) \qquad \bar{L}_{i,j} = \bar{L}(\hat{\theta}(i), \hat{\mu}(i)) - \bar{L}(\hat{\theta}(j), \hat{\mu}(j)).$$

Now suppose that $(\theta^0, \mu^0) \in \Omega_0 \times M$ is the true value and suppose that $(\hat{\theta}(l), \hat{\mu}(l))$ tends to (θ^0, μ^0) in probability for $0 \leqq l \leqq p$; then it follows from the asymptotic theory with respect to $(\hat{\theta}(l), \hat{\mu}(l))$ (see Dunsmuir and Hannan (1976) and Hosoya and Taniguchi (1982)) that $\bar{L}_{0,l}$ ($l = 1, 2, \cdots, p$) has the same asymptotic distribution as

$$\begin{aligned}
S_{0l} = -(\tfrac{1}{2})[&\sqrt{n}\,(\hat{\theta}(l) - \theta^0(l))'\Psi(l)\sqrt{n}\,(\hat{\theta}(l) - \theta^0(l)) \\
&+ \sqrt{n}\,(\hat{\mu}(l) - \mu^0))'\Phi\sqrt{n}\,(\hat{\mu}(l) - \mu^0) \\
&- \sqrt{n}\,(\hat{\theta}(0) - \theta^0(0))'\Psi(0)\sqrt{n}\,(\hat{\theta}(0) - \theta^0(0)) \\
&- \sqrt{n}\,(\hat{\mu}(0) - \mu^0)'\Phi\sqrt{n}\,(\hat{\mu}(0) - \mu^0)],
\end{aligned}$$

(1.9)

where the $\Psi(j)$ ($j = 0, \cdots, p$) and Φ are the matrices defined as

$$\Psi(j) = \left\{\left(\frac{1}{4\pi}\right)\int_{-\pi}^{\pi} \text{tr}\,[f(\omega; \theta^0, \mu^0)^{-1}(\partial f(\omega; \theta^0, \mu^0)/\partial\theta_j)\right.$$
$$\left. \times f(\omega; \theta^0, \mu^0)^{-1}(\partial f(\omega; \theta^0, \mu^0)/\partial\theta_k)]d\omega;\, j, k = 1, \cdots, q_j\right\}$$

$$\Phi = \{\text{tr}\,(K^{-1}(\mu^0)(\partial K(\mu^0)/\partial\mu_j)K^{-1}(\mu^0)(\partial K(\mu^0)/\partial\mu_k);\, j, k = 1, \cdots, r\}$$

and $\hat{\theta}(j), \theta^0(j), j = 0, \cdots, p$, are regarded as q_j-vectors with the obvious identification. The cross-product terms of $\sqrt{n}\,(\hat{\theta}(j) - \theta^0)$ and $\sqrt{n}\,(\hat{\mu}(j) - \mu^0)$ are missing in (1.8) because as $n \to \infty$

$$\partial^2 \bar{L}/\partial\theta_j\partial\mu_k \to 2\,\mathrm{tr}\,(\partial K(\mu^0)/\partial\mu_k)K(\mu^0)\int_{-\pi}^{\pi} k^{*-1}(\omega, \theta^0)(\partial k^*(\omega, \theta^0)/\partial\theta_j)d\omega$$

in probability, whereas the integral in the right-hand side above vanishes. Since $\sqrt{n}(\hat{\theta}(j) - \theta^0(j))$ $(j = 0, \cdots, p)$ and $\sqrt{n}(\hat{\mu}(j) - \mu^0)$ are asymptotically independently distributed as $\Psi(j)^{-1}\sqrt{n}\partial\bar{L}(\theta^0, \mu^0)/\partial\theta(j)$ and $\Phi^{-1}\sqrt{n}\partial\bar{L}(\theta^0, \mu^0)/\partial\mu$ where $\partial\bar{L}(\theta^0, \mu^0)/\partial\theta(j)$ is the q_j-vector with components $\partial\bar{L}(\theta^0, \mu^0)/\partial\theta_k$ $(k = 1, \cdots, q_j)$ and $\partial\bar{L}(\theta^0, \mu^0)/\partial\mu$ is defined similarly, S_{0l} is asymptotically distributed as

$$-(\tfrac{1}{2})[(\sqrt{n}\partial\bar{L}(\theta^0, \mu^0)'/\partial\theta(l))\Psi(l)^{-1}(\sqrt{n}\partial\bar{L}(\theta^0, \mu^0)/\partial\theta(l))$$
$$-(\sqrt{n}\partial\bar{L}(\theta^0, \mu^0)/\partial\theta(0))\Psi(0)^{-1}(\sqrt{n}\partial\bar{L}(\theta^0, \mu^0)/\partial\theta(0))].$$

Since $\Psi(p)$ is symmetric and positive-definite, there exists a lower-triangular, non-singular real matrix such that $C(p)'\,C(p) = \Psi(p)^{-1}$. Denote by $C_{ij}(p)$ the (i, j)th element of $C(p)$. Set $M_{np} = \sqrt{n}\partial\bar{L}(\theta^0, \mu^0)/\partial\theta(p)$ and $w(n) = C(p)M_{np}$, and denote by $w_i(n)$ the ith element of $w(n)$ and by $w^{(i)}(n)$ the column q_i-vector with elements $w_j(n)$, $j = 1, \cdots, q_i$. Moreover denote by $C^{(i)}(p)$ the triangular matrix $\{C_{k,j}(p), k, j = 1, \cdots, q_i\}$. Then it follows from the triangularity of $C^{(i)}(p)$ that

$$C^{(i)}(p)^{-1}J^{(i)}(p)C^{(i)}(p)'^{-1} = I_{q_i}$$

where I_{q_i} is the identity. Consequently, the S_{0l}'s are expressed as

$$
\begin{aligned}
S_{0l} &= -(\tfrac{1}{2})\{w^{(l)}(n)'C^{(l)}(p)'^{-1}\Psi(l)^{-1}C^{(l)}(p)^{-1}w^{(l)}(n) \\
&\quad - w^{(0)}(n)'C^{(0)}(p)'^{-1}\Psi(0)^{-1}C^{(0)}(p)^{-1}w^{(0)}(n)\} \\
&= -(\tfrac{1}{2}) \sum_{j=q_0+1}^{q_i} w_j(n)^2, \qquad i = 1, \cdots, p.
\end{aligned}
$$

(1.10)

Since M_{np} has the asymptotic normal distribution with mean 0 and covariance matrix $\Psi(p)^{-1}$, $w_j(n)$, $j = 1, \cdots, q_p$ are asymptotically identically independently standard-normally distributed. Thus if χ_i, $i = 1, \cdots, p$, denote independent χ^2-random variables with d.f. $(q_i - q_{i-1})$, respectively, the joint asymptotic distribution of $\{\bar{L}_{01}, \cdots, \bar{L}_{0p}\}$ is the same as that of $-(\tfrac{1}{2})\sum_{j=1}^{i}\chi_j$, $i = 1, \cdots, p$. Consequently the GLR test whose rejection region is given as

(1.11)
$$\bar{L}_{0,l} < c_l' \quad \text{for some} \quad l \qquad (l = 1, \cdots, p)$$

is reduced for a large sample to the canonical test (1.3).

As for the determination of the critical values $\{d_j\}$ for a given significance level, there does not seem a uniformly optimal choice from the view of power.

Though such a criterion as minimax choice might be thought of, it will have only a very limited application because of the difficulty of the evaluation of the error rate under alternative hypotheses. Consequently certain conventions are called for. One way would be to determine the d_j's in such a way that the marginal probability of rejecting H_0 by $L_{0,j}$ when H_0 is true is the same for each $j = 1, \cdots, p$. When there are several alternative hypotheses which compete and when there is no *a priori* order of priority among them, it seems reasonable to compare the likelihood of the null hypothesis with that of every alternative instead of comparing it only with the most complicated one and at the same time to let each of the test criteria stand on an equal footing by letting it have the same chance of rejecting the null hypothesis. Namely, the d_j's in (1.3) are determined so as to satisfy

$$(1.12) \qquad \Pr\left\{\sum_{j=1}^{i} \chi_j > d_i\right\} = \Pr\{\chi_1 > d_1\}, \qquad i = 2, \cdots, p,$$

and

$$(1.13) \qquad \Pr\left\{\bigcup_{i=1}^{P}\left(\sum_{j=1}^{i} \chi_j > d_i\right)\right\} = \alpha$$

for a given significance level. Though admittedly conventional, as will be seen in the following two sections, this way of determining the d_j's has the advantage of enabling the systematic development of the whole test theory, making such procedures as actual evaluations of the critical values or the P-values (observed significance levels) routine ones. Besides, this way of determining the critical values is the only one which is consistent with the significance level which Stone (1969), p. 492, calls data-snooping. For an application of this kind of test to confidence set construction of a true model, see Hosoya (1984).

2. Distributions

Suppose that a significance level α $(0 < \alpha < 1)$ is set for the canonical test (1.3) and that the critical values d_j are to be determined according to the rule (1.12) of equal-marginal error rate; then the Newton–Raphson iteration method can be employed for the purpose of actual evaluation of the d_j's. As in the previous section, let χ_j $(j = 1, \cdots, p)$ be independent χ^2-random variables with $(q_i - q_{i-1})$ d.f. respectively. Set $l_i = q_i - q_{i-1}$ $(i = 1, \cdots, p)$, $r_i = q_i - q_0$ and set $\psi_j = \sum_{i=1}^{j} \chi_i$. Denote by $d_j(\beta)$, $j = 1, \cdots, p$, the critical value when the marginal significance level is β, namely

$$(2.1) \qquad \Pr\{\psi_j > d_j(\beta)\} = \beta$$

and thus $d_j(\beta) = F_{r_j}^{-1}(1 - \beta)$, where $F_k(\cdot)$ denotes throughout the distribution function of χ^2-distribution with k d.f. Since the problem is to numerically approximate the values of the $d_j(\beta)$ and β which satisfy the equations

(2.2)
$$\Pr\left\{\bigcap_{j=1}^{p}(\psi_j \le d_j(\beta))\right\} = 1 - \alpha$$

and

(2.3)
$$d_j(\beta) = F_{r_j}^{-1}(1 - \beta),$$

the Newton–Raphson iteration can be conducted as

(2.4)
$$\beta^{(i+1)} = \beta^{(i)} - (\partial \Pr\{\psi_j \le d_j(\beta^{(i)}), j = 1, \cdots, p\}/\partial\beta)^{-1}$$
$$\times [\Pr\{\psi_j \le d_j(\beta^{(i)}), j = 1, \cdots, p\} - (1 - \alpha)], \qquad i = 1, 2, \cdots$$

where $\beta^{(i)}$ denotes the *i*th-step value of the marginal significance level β. In order to compute the derivative term in (2.4), the fact that the sequence $\{\psi_j, j = 1, \cdots, p\}$ has the Markov property is employed. Denote by $f_k(\cdot)$ the density function of χ^2-distribution with k d.f. and by $\Pr(A \mid B)$ the conditional probability of the event A given B; then it holds that

(2.5)
$$\Pr\{\psi_m < d_m, m = 1, \cdots, p\} = \int_0^{d_k} \Pr\{\psi_h < d_h, h = 1, \cdots, k-1 \mid \psi_k = y\}$$
$$\times \Pr\{\phi_{n,k} < d_n - y, n = k+1, \cdots, p\} f_{r_k}(y)\,dy,$$

where $\phi_{n,k} = \sum_{k+1}^{n} \chi_j$, whence in view of the fact that $\partial d_k(\beta)/\partial\beta = -(1/f(d_k(\beta)))$ it follows that

$$\partial \Pr\{\psi_m < d_m(\beta), m = 1, \cdots, p\}/\partial\beta$$
$$= \sum_{k=1}^{p} [\partial \Pr\{\psi_m < d_m(\beta), m = 1, \cdots, p\}/\partial d_k](\partial d_k(\beta)/\partial\beta)$$

(2.6)
$$= -\sum_{k=2}^{p-2} \Pr\{\psi_h < d_h, h = 1, \cdots, k-1 \mid \psi_k = d_k\}$$
$$\times \Pr\{\phi_{n,k} < d_n - d_k, n = k+1, \cdots, p\}$$
$$- \Pr\{\phi_{n,1} < d_n - d_1, n = 2, \cdots, p\}$$
$$- \Pr\{\psi_h < d_h, h = 1, \cdots, p-1 \mid \psi_p = d_p\}.$$

The probabilities and the conditional probabilities which are involved in (2.4) and (2.6) can be determined by means of recursive formulas as shown below, if all l_j's are even numbers; but on the other hand if the number of odd l_j's is large, the numerical computation of those integrals is hardly tractable, since the multiple integrals cannot be reduced to a feasible form. The rest of this section, first of all, deals with the case where all $l_j = q_j - q_{j-1}$ are even, and then considers the general case.

Let the d_j's be constants such that $0 \le d_1 \le d_2 \le \cdots \le d_p$ and let $m_j, j =$

$1, \cdots, p$ be non-negative constants. Denote by $\{d_j\}$ and $\{m_j\}$ the ordered set of the d_j's and the m_j's. Set for a non-negative integer m, and for $k = 1, \cdots, p$,

$$(2.7) \quad A(k, m, \{d_j\}, \{m_j\}) = \int_0^{d_1} dz_1 \int_{z_1}^{d_2} dz_2 \cdots \int_{z_{k-1}}^{d_k} z_1^{m_1}(z_2 - z_1)^{m_2} \cdots$$

$$(z_{k-1} - z_{k-2})^{m_{k-1}}(z_k - z_{k-1})^m \exp(-z_k/2)dz_k;$$

$$(2.8) \quad B(k, m, \{d_j\}, \{m_j\}) = \int_0^{d_1} dz_1 \int_{z_1}^{d_2} dz_2 \cdots \int_{z_{k-2}}^{d_{k-1}} z_1^{m_1}(z_2 - z_1)^{m_2} \cdots$$

$$(z_{k-1} - z_{k-2})^{m_{k-1}}(d_k - z_{k-1})^m dz_{k-1};$$

then those integrals are computed recursively according to the following steps. Set $n_j = \sum_{i=j}^p (m_i + 1)$ and then

$$A(1, k, \{d_j\}, \{m_j\}) = -d_1^k \exp(-d_1/2) + 2kA(1, k-1), \qquad k \geq 1;$$

$$(2.9) \quad A(1, 0, \{d_j\}, \{m_j\} = -2\{\exp(-d_1/2) - 1\};$$

$$B(0, k, \{d_j\}, \{m_j\}) = B(1, k, \{d_j\}, \{m_j\}) = d_1^k, \qquad 1 \leq j \leq n_1,$$

and, for $k \geq 2$,

$$A(k, 0) = -2 \exp(-d_k/2)B(k-1, m_{k-1}+1)/(m_{k-1}+1)$$

$$(2.10) \quad + 2A(k-1, m_{k-1});$$

$$A(k, j) = -2 \exp(-d_k/2)B(k, j) + 2jA(k, j-1) \quad \text{for} \quad 0 < j \leq n_k;$$

$$B(k, j) = I(j \geq 1)(j!) \sum_{m=0}^{j-1} \frac{(d_k - d_{k-1})^{j-m}}{(j-m)! \, (m_{k-1}+1) \cdots (m_{k-1}+1+m)}$$

$$\times B(k-1, m_{k-1}+1+m)$$

$$(2.11) \quad + \frac{j!}{(m_{k-1}+1) \cdots (m_{k-1}+j+1)} B(k-2, m_{k-1}+j+1),$$

$$0 \leq j \leq n_k,$$

where $A(k, l, \{d_j\}, \{m_j\})$ and $B(k, l, \{d_j\}, \{m_j\})$ are abbreviated as $A(k, l)$ and $B(k, l)$, and $I(j \geq 1)$ is the indicator function such that $I = 1$ if $j \geq 1$ and $I = 0$ if $j = 0$.

Suppose $\chi_j^*, j = 1, 2, \cdots, k$, be independent χ^2-random variables with d.f. $2(n_j + 1)$ respectively and let $e_1 \leq e_2 \leq \cdots \leq e_k$ be non-negative constants. Set $\psi_l^* = \sum_{j=1}^l \chi_j^*$; then all the probabilities which are involved in (2.4) and (2.5) are either of the form $\Pr\{\psi_j^* < e_j, j = 1, \cdots, k\}$ or $\Pr\{\psi_j^* < e_j, j = 1, \cdots, k-1 \mid \psi_k^* = e_k\}$, and they are able to be expressed in terms of the functions A and B as

$$(2.12) \quad \Pr\{\psi_j^* < e_j, j = 1, \cdots, k\} = \left[\prod_{j=1}^k \{2^{(n_j+1)} n_j!\} \right]^{-1} A(k, n_k, \{e_j\}, \{n_j\});$$

and

$$\Pr\{\psi_j^* < e_j, j = 1, \cdots, k-1 \mid \psi_k^* = e_k\}$$

$$= \left\{ \int_0^{e_1} dx_1 \int_{x_2}^{e_2} dx_2 \cdots \int_{x_{k-2}}^{e_{k-1}} x_1^{n_1}(x_2 - x_1)^{n_2} \cdots \right.$$

(2.13)
$$\left. (x_{k-1} - x_{k-2})^{n_{k-1}}(e_k - x_{k-1})^{n_k} dx_{k-1} \right\}$$

$$\times \left\{ \int_0^{e_k} dx_1 \int_{x_2}^{e_k} dx_2 \cdots \int_{x_{k-2}}^{e_k} x_1^{n_1}(x_2 - x_1)^{n_2} \cdots (e_k - x_{k-1})^{n_k} dx_{k-1} \right\}^{-1}$$

$$= B(k, n_k, \{e_j\}, \{n_j\})/B(k, n_k, \{h_j\}, \{n_j\})$$

where $h_j = e_k, j = 1, 2, \cdots, k$.

When the $l_i = q_i - q_{i-1}$ are not necessarily even, the probability evaluations cannot be reduced to recursive formulas as in the above. When p is small, numerical integration methods can be employed as will be shown in the next section, but the applicability of such methods is limited in general. One device of general applicability would be to transform the χ_j's of odd d.f.'s into χ^2-random variables of even d.f.'s. Let the χ_j, $j = 1, \cdots, p$ be of l_j d.f. respectively and suppose that $l_{j_1}, \cdots, l_{j_r}$ $(j_1 < \cdots < j_r, r \leq p)$ are odd. Then define $\tilde{\chi}_j, j = 1, \cdots, p$ as

(2.14)
$$\tilde{\chi}_{l_{j_k}} = F_{l_{j_k}+1}^{-1}(F_{l_{j_k}}(\chi_j)) \quad \text{if} \quad j = j_k, \quad k = 1, \cdots, r$$
$$= \chi_j, \quad \text{otherwise,}$$

where $F_l(\cdot)$ denotes the χ^2-distribution function of l d.f.; then set

(2.15)
$$\tilde{\psi}_j = \sum_{i=1}^{j} \tilde{\chi}_i, \quad j = 1, \cdots, p,$$

and construct the rejection region of this modified GLR test as

(2.16)
$$\tilde{\psi}_j > \tilde{d}_j \quad \text{for some} \quad j = 1, \cdots, p.$$

Then the whole algorithm of the preceding arguments becomes applicable, making feasible the evaluation of the error rates and the critical values. In terms of the likelihood ratio (1.1), the modified GLR is given as follows. Set, for $j = 1, \cdots, p$,

(2.17)
$$M_j = F_{l_j+1}^{-1}\{F_{l_j}(-2L_{j-1,j})\} \quad \text{if} \quad j = j_k \quad (k = 1, \cdots, r)$$
$$= -2L_{j-1,j}, \quad \text{otherwise,}$$

and set $N_j = \sum_{i=1}^{j} M_i$; then the rejection region is given in the form $\{N_j > \tilde{d}_j$, for some $j = 1, \ldots, p\}$.

3. Some special cases

The cases where p is small and all l_j's are equal to 1 are special but important in practice; for such cases the direct application of numerical integration methods is feasible. The problem is again the computation of the probabilities in the form (2.12) or (2.13); however, the m_j's are not integers this time but are equal to $-\frac{1}{2}$.

First of all, define functions $B_1(x, a_1)$ and $B_2(x, a_1, a_2)$ for positive constants $a_1, a_2 (a_1 \leqq a_2)$ as follows:

$$
\begin{aligned}
B_1(x, a_1) &= 1 \quad \text{if} \quad x < a_1 \\
&= \{\pi - 2 \arctan \sqrt{(x - a_1)/a_1}\}/\pi \quad \text{if} \quad x \geqq a_1;
\end{aligned}
\tag{3.1}
$$

$$
\begin{aligned}
B_2(x, a_1, a_2) &= 1 \quad \text{if} \quad x \leqq a_1 \\
&= \sqrt{a_1}/\sqrt{x} \quad \text{if} \quad a_1 < x \leqq a_2 \\
&= H(a_1, a_2, x)/(2\pi \sqrt{x}) \quad \text{if} \quad a_2 < x
\end{aligned}
\tag{3.2}
$$

where

$$
\begin{aligned}
H(a_1, a_2, x) &= 2\pi(\sqrt{x} - \sqrt{x - a_2}) + 4\sqrt{x - a_2} \cdot \arctan \sqrt{(a_2 - a_1)/a_1} \\
&\quad - 2\sqrt{a_1}(2 \arctan \sqrt{(x - a_2)/(a_2 - a_1)} - \pi) \\
&\quad - 2\sqrt{x}[\arcsin(\{(x + a_1)a_2 - 2a_1 x\}\{a_2(x - a_1)\}^{-1}) + \pi/2].
\end{aligned}
$$

Let χ_j^*, $j = 1, \cdots, p$, be independent χ^2-random variables all with 1 d.f. and set $\psi_j^* = \sum_{i=1}^{j} \chi_i^*$. Set

$$
C_k(d_1, d_2, \cdots, d_k) = \Pr\{\psi_j^* < d_j, j = 1, \cdots, k\}, \quad \text{and}
\tag{3.3}
$$

$$
D_k(d_1, d_2, \cdots, d_k) = \Pr\{\psi_j^* < d_j, j = 1, \cdots, k-1 \mid \psi_k^* = d_k\};
\tag{3.4}
$$

then thanks to the Markov property of the sequence $\psi_1^*, \cdots, \psi_p^*$, the following relationships hold:

$$
C_2(d_1, d_2) = \int_0^{d_2} B_1(z, d_1)f_2(z)dz,
$$

$$
C_3(d_1, d_2, d_3) = \int_0^{d_3} B_2(z, d_1, d_2)f_3(z)dz,
$$

$$
C_4(d_1, \cdots, d_4) = \int_0^{d_3} B_2(z, d_1, d_2)F_1(d_4 - z)f_3(z)dz,
\tag{3.5}
$$

$$
C_5(d_1, \cdots, d_5) = \int_0^{d_3} dy \int_0^{d_5 - y} B_2(y, d_1, d_2)B_1(z, d_4 - y)f_2(z)f_3(y)dz,
$$

$$
C_6(d_1, \cdots, d_6) = \int_0^{d_3} dy \int_0^{d_6 - y} B_2(y, d_1, d_2)B_2(z, d_4 - y, d_5 - y)f_3(y)f_3(z)dz;
$$

and

$$D_2(d_1, d_2) = B_1(d_2, d_1)$$

$$D_3(d_1, d_2, d_3) = B_2(d_3, d_1, d_2),$$

$$D_4(d_1, \cdots, d_4) = \int_0^{d_2} B_1(z, d_1)B_1(d_4 - z, d_3 - z)f_2(z)dz,$$

(3.6)

$$D_5(d_1, \cdots, d_5) = \int_0^{d_2} B_1(z, d_1)B_2(d_5 - z, d_3 - z, d_4 - z)f_2(z)dz,$$

$$D_6(d_1, \cdots, d_6) = \int_0^{d_3} B_2(z, d_1, d_2)B_2(d_6 - z, d_4 - z, d_5 - z)f_3(z)dz,$$

where $f_l(\cdot)$ denotes the density of χ^2-distribution of l d.f. Actually C_7 and D_7 are expressed by double integration but that is not the case for $p \geq 8$. Table 1 exhibits the critical values $d_j(\beta)$ such that

$$\Pr\{\psi_j^* < d_j(\beta), j = 1, \cdots, p\} = 1 - \alpha \quad \text{and} \quad \Pr\{\psi_j^* > d_j(\beta)\} = \beta$$

for $2 \leq p \leq 7$ and for the significance level $\alpha = 0.01(0.01)0.1$ and $\alpha = 0.1(0.05)0.5$. The values were obtained by the Newton–Raphson method (2.4) and by the formulas (3.5) and (3.6); as for numerical integration, the Simpson $\frac{1}{3}$ method for single integration and the Gauss method for double integration were used, and the subdivision of the domain of integration was iterated until the absolute difference from the preceding estimate became less than $(0.1)^5$. The Newton–Raphson iterations were also continued until $|d_j(\beta^{(i+1)}) - d_j(\beta^{(i)})| < (0.1)^5$ (in effect, a few steps were enough for all cases).

TABLE 1

Critical values of the GLR test

$p = 2$									
α	0.01	0.02	0.03	0.04	0.05	0.06	0.07	0.08	0.09	0.10
β	0.0063	0.0128	0.0194	0.0261	0.0328	0.0396	0.0465	0.0534	0.0603	0.0673
d_1	7.4584	6.1950	5.4643	4.9505	4.5550	4.2340	3.9643	3.7320	3.5283	3.3471
d_2	10.1299	8.7149	7.8841	7.2929	6.8332	6.4567	6.1377	5.8607	5.6159	5.3965

α	0.15	0.20	0.25	0.30	0.35	0.40	0.45	0.50
β	0.1029	0.1395	0.1770	0.2155	0.2549	0.2953	0.3368	0.3795
d_1	2.6596	2.1833	1.8224	1.5342	1.2963	1.0953	0.9226	0.7724
d_2	4.5474	3.9392	3.4629	3.0698	2.7339	2.4395	2.1765	1.9379

$p = 3$									
α	0.01	0.02	0.03	0.04	0.05	0.06	0.07	0.08	0.09	0.10
β	0.0049	0.0101	0.0154	0.0207	0.0262	0.0317	0.0373	0.0430	0.0487	0.0545
d_1	7.9064	6.6204	5.8743	5.3483	4.9425	4.6124	4.3346	4.0949	3.8843	3.6996
d_2	10.6264	9.1940	8.3515	7.7511	7.2836	6.9003	6.5751	6.2926	6.0426	5.8183
d_3	12.8701	11.3272	10.4129	9.7576	9.2450	8.8229	8.4634	8.1500	7.8717	7.6213

TABLE 1 (*Continued*)

α	0.15	0.20	0.25	0.30	0.35	0.40	0.45	0.50		
β	0.0842	0.1152	0.1473	0.1807	0.2152	0.2511	0.2883	0.3271		
d_1	2.9812	2.4815	2.0997	1.7922	1.5359	1.3173	1.1275	0.9604		
d_2	4.9482	4.3223	3.8301	3.4222	3.0722	2.7640	2.4874	2.2351		
d_3	6.6417	5.9275	5.3588	4.8818	4.4673	4.0980	3.7622	3.4517		

$p = 4$

α	0.01	0.02	0.03	0.04	0.05	0.06	0.07	0.08	0.09	0.10
β	0.0042	0.0086	0.0131	0.0178	0.0225	0.0273	0.0322	0.0372	0.0422	0.0473
d_1	8.2105	6.9089	6.1523	5.6180	5.2053	4.8692	4.5859	4.3412	4.1260	3.9340
d_2	10.9621	9.5174	8.6666	8.0598	7.5868	7.1987	6.8693	6.5829	6.3293	6.1017
d_3	13.2299	11.6767	10.7555	10.0949	9.5777	9.1517	8.7887	8.4720	8.1908	7.9376
d_4	15.2742	13.6294	12.6491	11.9435	11.3895	10.9318	10.5410	10.1993	9.8953	9.6210

α	0.15	0.20	0.25	0.30	0.35	0.40	0.45	0.50		
β	0.0736	0.1013	0.1303	0.1605	0.1921	0.2252	0.2597	0.2960		
d_1	3.2000	2.6849	2.2893	1.9692	1.7011	1.4711	1.2701	1.0922		
d_2	5.2172	4.5793	4.0764	3.6586	3.2991	2.9818	2.6962	2.4349		
d_3	6.9460	6.2218	5.6442	5.1588	4.7365	4.3595	4.0160	3.6979		
d_4	8.5416	7.7469	7.1085	6.5684	6.0954	5.6702	5.2802	4.9164		

$p = 5$

α	0.01	0.02	0.03	0.04	0.05	0.06	0.07	0.08	0.09	0.10
β	0.0037	0.0076	0.0117	0.0158	0.0201	0.0245	0.0289	0.0334	0.0379	0.0426
d_1	8.4389	7.1256	6.3611	5.8206	5.4027	5.0621	4.7747	4.5263	4.3076	4.1124
d_2	11.2136	9.7594	8.9023	8.2905	7.8135	7.4218	7.0892	6.7997	6.5434	6.3132
d_3	13.4990	11.9378	11.0113	10.3465	9.8259	9.3968	9.0311	8.7119	8.4283	8.1729
d_4	15.5584	13.9066	12.9217	12.2125	11.6555	11.1952	10.8021	10.4582	10.1521	9.8760
d_5	17.4807	15.7488	14.7123	13.9641	13.3752	12.8876	12.4704	12.1049	11.7792	11.4849

α	0.15	0.20	0.25	0.30	0.35	0.40	0.45	0.50		
β	0.0666	0.0920	0.1188	0.1470	0.1765	0.2076	0.2403	0.2747		
d_1	3.3648	2.8382	2.4327	2.1033	1.8265	1.5881	1.3791	1.1932		
d_2	5.4180	4.7710	4.2601	3.8349	3.4684	3.1443	2.8521	2.5842		
d_3	7.1723	6.4404	5.8560	5.3643	4.9360	4.5532	4.2040	3.8802		
d_4	8.7887	7.9873	7.3431	6.7975	6.3192	5.8890	5.4940	5.1252		
d_5	10.3222	9.4606	8.7646	8.1724	7.6511	7.1800	6.7456	6.3381		

$p = 6$

α	0.01	0.02	0.03	0.04	0.05	0.06	0.07	0.08	0.09	0.10
β	0.0033	0.0069	0.0106	0.0145	0.0184	0.0224	0.0265	0.0306	0.0348	0.0391
d_1	8.6208	7.2983	6.5275	5.9821	5.5600	5.2158	4.9252	4.6739	4.4526	4.2548
d_2	11.4135	9.9519	9.0897	8.4739	7.9935	7.5989	7.2637	6.9720	6.7135	6.4813
d_3	13.7127	12.1452	11.2143	10.5461	10.0225	9.5910	9.2231	8.9019	8.6165	8.3594
d_4	15.7838	14.1266	13.1378	12.4256	11.8661	11.4037	11.0086	10.6631	10.3554	10.0777
d_5	17.7166	15.9799	14.9400	14.1891	13.5979	13.1085	12.6896	12.3227	11.9956	11.6999
d_6	19.5540	17.7451	16.6590	15.8731	15.2533	14.7395	14.2991	13.9130	13.5683	13.2565

α	0.15	0.20	0.25	0.30	0.35	0.40	0.45	0.50		
β	0.0615	0.0853	0.1105	0.1370	0.1651	0.1946	0.2258	0.2589		
d_1	3.4963	2.9609	2.5474	2.2108	1.9272	1.6823	1.4669	1.2748		
d_2	5.5772	4.9231	4.4059	3.9748	3.6028	3.2734	2.9759	2.7028		
d_3	7.3513	6.6132	6.0234	5.5267	5.0936	4.7062	4.3524	4.0241		
d_4	8.9838	8.1771	7.5281	6.9781	6.4956	6.0612	5.6622	5.2894		

Table 1 (*Continued*)

d_5	10.5313	9.6650	8.9648	8.3688	7.8436	7.3689	6.9308	6.5198		
d_6	12.0208	11.1011	10.3550	9.7178	9.1546	8.6438	8.1711	7.7260		

$p = 7$

α	0.01	0.02	0.03	0.04	0.05	0.06	0.07	0.08	0.09	0.10
β	0.0033	0.0068	0.0105	0.0143	0.0181	0.0221	0.0261	0.0302	0.0344	0.0386
d_1	8.6386	7.3178	6.5481	6.0036	5.5822	5.2385	4.9485	4.6975	4.4767	4.2793
d_2	11.4330	9.9736	9.1129	8.4982	8.0188	7.6250	7.2906	6.9994	6.7417	6.5100
d_3	13.7336	12.1686	11.2394	10.5725	10.0502	9.6196	9.2526	8.9322	8.6477	8.3913
d_4	15.8058	14.1513	13.1645	12.4538	11.8957	11.4344	11.0404	10.6956	10.3890	10.1122
d_5	17.7396	16.0058	14.9681	14.2189	13.6292	13.1410	12.7234	12.3573	12.0313	11.7366
d_6	19.5780	17.7722	16.6884	15.9043	15.2861	14.7736	14.3346	13.9494	13.6060	13.2952
d_7	21.3454	19.4731	18.3468	17.5306	16.8863	16.3515	15.8928	15.4901	15.1306	14.8051

α	0.15	0.20	0.25	0.30	0.35	0.40	0.45	0.50
β	0.0606	0.0839	0.1086	0.1346	0.1620	0.1909	0.2214	0.2537
d_1	3.5222	2.9878	2.5749	2.2387	1.9554	1.7106	1.4950	1.3026
d_2	5.6085	4.9564	4.4407	4.0110	3.6402	3.3119	3.0152	2.7428
d_3	7.3864	6.6510	6.0633	5.5686	5.1374	4.7516	4.3994	4.0724
d_4	9.0220	8.2185	7.5721	7.0247	6.5445	6.1123	5.7153	5.3444
d_5	10.5723	9.7096	9.0124	8.4193	7.8969	7.4248	6.9892	6.5806
d_6	12.0642	11.1485	10.4058	9.7719	9.2119	8.7041	8.2342	7.7919
d_7	13.5131	12.5489	11.7646	11.0935	10.4991	9.9589	9.4577	8.9848

4. Remarks

The nesting structure of the hypotheses given in the first paragraph of Section 1 implies that the test of this paper excludes the case where some of the alternative hypotheses are such that the null hypothesis consists of a set of boundary values of them (the case given by Chernoff (1954)) or some alternatives are separate from the null hypothesis (see Cox (1961)). For example, the present test does not apply to the ARMA $(1, 1)$ model with

$$(4.1) \qquad z(t) = \alpha z(t-1) + e(t) - \beta e(t-1)$$

where $H_0 : \alpha = \beta = 0$ and $H_1 : \alpha \neq \beta$. Also it should be noted that the present theory does not apply to the nesting model of the covariance-matrix parameters of the innovations, either, unless the generated process is Gaussian.

Acknowledgement

The author is very grateful to Professor Kei Takeuchi and the referee for helpful comments on this paper.

References

CHERNOFF, H. (1954) On the distribution of the likelihood ratio. *Ann. Math. Statist.* **25**, 573–578.

Cox, D. R. (1961) Tests of separate families of hypotheses. *Proc. 4th. Berkeley Symp. Math. Statist. Prob.* **1**, 105–123.

Dunsmuir, W. and Hannan, E. J. (1976) Vector linear time-series models. *Adv. Appl. Prob.* **8**. 334–364.

Hosoya, Y. (1984) Information criteria and tests for time-series models. In *Time Series Analysis: Theory and Practice* **5**, ed. O. D. Anderson, Elsevier, Amsterdam.

Hosoya, Y. and Taniguchi, M. (1982) A central limit theorem for stationary processes and the parameter estimation of linear processes. *Ann. Statist.* **10**, 132–153.

Stone, M. (1969) The role of significance testing: some data with a message. *Biometrika* **56**, 485–493.

Wald, A. (1943) Tests of statistical hypotheses concerning several parameters when the number of observations is large. *Trans. Amer. Math. Soc.* **54**, 426–482.

Testing for the Presence of Sinusoidal Components

B. G. QUINN

Abstract

Approximate and asymptotic distributional results are obtained for the likelihood ratio test of the hypothesis that a time series is composed from s sinusoidal components, at unknown frequencies, with additive Gaussian white noise, against the hypothesis that there are an additional r sinusoidal components at unknown frequencies. The work extends that of Fisher (1929), and contains a number of simulations illustrating the results.

PERIODOGRAM ORDINATES; LIKELIHOOD RATIO

1. Introduction

The search for periodic components in data has long been a matter of interest, dominating the time series literature during the early development of the field. The first test for the existence of a cyclic component at unknown frequency was derived by Walker (1914), but did not account for the sampling fluctuations of the variance of the data. An exact test, allowing for these fluctuations, was obtained by Fisher (1929), the test rejecting the null hypothesis that the data are derived from Gaussian white noise if the largest (Studentized) periodogram ordinate is too large. The procedure was modified so as to include the non-Gaussian, non-uniform-spectrum cases by Whittle (1952), Hannan (1961) and Walker (1965). A different kind of procedure, based on autocovariances, was introduced by Priestley (1962) and is discussed in Hannan (1970), p. 473 and Priestley (1981), pp. 626–642.

In this paper we develop the likelihood-ratio criterion for testing the null hypothesis that a time series is composed of s sinusoidal components with additive Gaussian white noise, against the hypothesis that, besides the original s sinusoidal components, there are r other sinusoidal components present, with the sinusoidal components under both hypotheses assumed to be at unknown frequencies of the form $2\pi j/T$, where $1 \leq j \leq [(T-1)/2]$ and T is the length of the sample. In Section 3 approximate distributional results are obtained which are exact when $s = 0$, while in Section 4 the asymptotic results are derived. Finally, in Section 5 the results are illustrated by a number of simulations.

2. The likelihood ratio criterion

We shall consider in this paper those sequences of random variables $\{Y_t\}$ satisfying equations of the form

$$(2.1) \qquad Y_t = \sum_{j=1}^{f} \{\alpha_j \cos (t\lambda_j) + \beta_j \sin (t\lambda_j)\} + \varepsilon_t; \qquad t = 1, \cdots, T$$

where

(i) $\{\alpha_j\}$ and $\{\beta_j\}$ are sets of real numbers with $\alpha_j^2 + \beta_j^2 \geqq \alpha_{j+1}^2 + \beta_{j+1}^2$, $j = 1, \cdots, (f-1)$.

(ii) Each λ_j is of the form $2\pi k/T$ where $k \in \Lambda_T$, a set with P_T elements, which is a subset of $\{n \in Z : 1 \leqq n \leqq [(T-1)/2]\}$.

(iii) $\{\varepsilon_t\}$ is a sequence of independent random variables, normally and identically distributed with means 0 and variances σ^2.

It is conceivable that condition (iii) could be relaxed along those lines taken by Whittle (1952) and Walker (1965) in their modifications of Fisher's procedure. It has, however, proved necessary to enforce condition (ii) in order that certain well-known properties of the periodogram ordinates may be used. The significance levels of the tests derived here could thus not be expected to be accurate when condition (ii) did not hold.

If (2.1) is fitted to a time series $\{y_t; t = 1, \cdots, T\}$ by least squares, when $\{\lambda_j; j = 1, \cdots, f\}$ is a *known* set of frequencies, then the residual sum of squares is

$$\sum_{t=1}^{T} y_t^2 - \sum_{j=1}^{f} I_y(\lambda_j),$$

where

$$I_X(\omega) = (2/T) \left| \sum_{t=1}^{T} X_t \exp (it\omega) \right|^2$$

for any set of numbers or random variables $\{X_t; t = 1, \cdots, T\}$. Consequently, when $\{\lambda_j; j = 1, \cdots, f\}$ is *unknown*, the residual sum of squares is $\sum_{t=1}^{T} y_t^2 - S_f$, where S_f is the sum of the f largest elements of $\{I_y(\omega_j); T\omega_j/(2\pi) \in \Lambda_T\}$.

The null hypothesis (H_0) which is to be tested is that only s frequencies are real (i.e. $\alpha_s^2 + \beta_s^2 > 0$ but $\alpha_{s+1}^2 + \beta_{s+1}^2 = 0$). The alternative hypothesis (H_A) is that all f frequencies are real. The likelihood ratio statistic l_T for testing H_0 against (H_A) is thus obtained from

$$(2.2) \qquad l_T^{-2/T} = \left(\sum_{t=1}^{T} Y_T^2 - S_s \right) \Big/ \left(\sum_{t=1}^{T} Y_t^2 - S_f \right)$$

the test rejecting H_0 if l_T is too small. In the following sections, approximate and asymptotic distributional results for $l_T^{2/T}$ will be obtained, the exact distribution when s is greater than 0 most likely being analytically intractable since S_s need not be composed from the 'true frequencies'.

3. Distributional theory

We first show that, under H_0, the statistic $l_T^{-2/T}$ is asymptotically equivalent to a random variable whose distributional properties will be determined later.

Lemma 1. Let $\{Y_t\}$ be a time series satisfying (2.1) and assumptions (i)–(iii), with $\alpha_s^2 + \beta_s^2 > 0$, $\alpha_{s+1}^2 + \beta_{s+1}^2 = 0$, and associated frequencies λ_j, $j = 1, \cdots, s$. Let $\Gamma_T = \{T\lambda_j/2\pi; j = 1, \cdots, s\}$, $\Delta_T = \Lambda_T \setminus \Gamma_T$, so that the set $(2\pi/T)\Gamma_T$ contains the 'true frequencies' and $(2\pi/T)\Delta_T$ the remaining frequencies considered. Let

$$(3.1) \qquad W_T = \left\{ \sum_{t=1}^{T} \varepsilon_t^2 - \sum_{j=1}^{s} I_\varepsilon(\lambda_j) \right\} \Big/ \left\{ \sum_{t=1}^{T} \varepsilon_t^2 - \sum_{j=1}^{s} I_\varepsilon(\lambda_j) - T_r \right\}$$

where T_r is the sum of the r greatest elements of $\{I_\varepsilon(2\pi k/T); k \in \Delta_T\}$ and $r = f - s$. Then $\Pr\{l_T^{-2/T} = W_T\}$ converges to 1 as T increases.

Proof. The lemma follows once it is proven that the event

$$B_T = \left\{ \min_{j=1,\cdots,s} I_Y(\lambda_j) \geq \max_{k \in \Delta_T} I_Y(2\pi k/T) \right\}$$

has probability converging to 1, since $\sum_{t=1}^{T} Y_t^2 - \sum_{j=1}^{s} I_Y(\lambda_j)$ is equal to $\sum_{t=1}^{T} \varepsilon_t^2 - \sum_{j=1}^{s} I_\varepsilon(\lambda_j)$, and $I_Y(2\pi k/T) = I_\varepsilon(2\pi k/T)$, for all $k \in \Delta_T$. Assuming without loss of generality that $\sigma^2 = 1$, the probability of the event B_T is given by

$$\Pr(B_T) = E\left\{ \left(1 - \exp\left(-\frac{1}{2} \min_{j=1,\cdots,s} I_Y(\lambda_j) \right) \right)^{P_T - s} \right\}.$$

Now,

$$I_Y(\lambda_j) = \left\{ U_j + \alpha_j \left(\frac{T}{2} \right)^{\frac{1}{2}} \right\}^2 + \left\{ V_j + \beta_j \left(\frac{T}{2} \right)^{\frac{1}{2}} \right\}^2,$$

where

$$U_j = \left(\frac{2}{T}\right)^{\frac{1}{2}} \sum_{t=1}^{T} \varepsilon_t \cos(t\lambda_j), \qquad V_j = \left(\frac{2}{T}\right)^{\frac{1}{2}} \sum_{t=1}^{T} \varepsilon_t \sin(t\lambda_j)$$

and $\{U_1, \cdots, U_s, V_1, \cdots, V_s\}$ is an independent sequence of standard normal random variables. Thus

$$\Pr(B_T) > \int_{A_{T,\delta}} \left\{1 - \exp\left(-\frac{1}{2} \min_{j=1,\cdots,s} I_y(\lambda_j)\right)\right\}^{P_T - s}$$
$$\times \exp\left(-\frac{1}{2} \sum_{j=1}^{s} (u_j^2 + v_j^2)\right) \prod_{j=1}^{s} (du_j dv_j)$$

where $A_{T,\delta}$ is the subset of R^{2s} defined by

$$A_{T,\delta} = \left\{(u_1, v_1, u_2, v_2, \cdots, u_s, v_s) : \left(u_j + \alpha_j\left(\frac{T}{2}\right)^{\frac{1}{2}}\right)^2 + \left(v_j + \beta_j\left(\frac{T}{2}\right)^{\frac{1}{2}}\right)^2\right.$$
$$\left. \geqq 2T\delta ; j = 1, \cdots, s\right\}.$$

Hence

$$(3.2) \qquad \Pr(B_T) > \{1 - \exp(-T\delta)\}^{P_T - s} \Pr\{(U_1, V_1, \cdots, U_s, V_s) \in A_{T,\delta}\}.$$

If δ is chosen so that $\alpha_j^2 + \beta_j^2 > 4\delta$, $j = 1, \cdots, s$, then the right-hand side of (3.2) converges to 1, since P_T is of order at most $O(T)$, and the $T^{-\frac{1}{2}}U_j$ and $T^{-\frac{1}{2}}V_j$ converge in probability to 0. Thus $\Pr(B_T)$ converges to 1.

The distribution of W_T under H_0 will not involve any of the parameters of the model (2.1), unlike the distribution of $l_T^{-2/T}$. Moreover, the asymptotic distributions of $l_T^{-2/T}$ and W_T under H_0 are the same, from Lemma 1, motivating the use, as an approximation to the distribution of $l_T^{-2/T}$, of that of W_T, the derivation of which is facilitated by the following two lemmas.

Lemma 2. Let $\{X_i; i = 1, \ldots, n\}$ be an i.i.d. sequence of random variables with probability density function (p.d.f.) $\exp(-x)$. Let $\{Y_i; i = 1, \cdots, n\}$ be $\{X_i; i = 1, \cdots, n\}$ ordered so that $Y_1 > Y_2 > \cdots > Y_n$, $U = \sum_{j=1}^{r} Y_j$ and $V = \sum_{j=r+1}^{n} Y_j$. Then the p.d.f. of $Z = U/(U + V)$ is given by

$$(3.3) \qquad f_Z(z) = \left((n-1)\binom{n}{r} \Big/ r^{n-r-1}\right) \sum_{k=1}^{n-r} (-1)^{n-r-k} \binom{n-r}{k}$$
$$\times \{(k+r)z - r\}_+^{n-2} H(1-z)/k^{r-1}$$

where $\{g(z)\}_+ = H(z)g(z)$ for any function $g(z)$, and

$$H(z) = \begin{cases} 1 & z \geqq 0 \\ 0 & z < 0. \end{cases}$$

Proof. The moment-generating function $\phi(s, t)$ of U and V is given by

$$\phi(s, t) = E\{\exp - (sU + tV)\}$$

$$= n\binom{n-1}{r-1} \int_0^\infty \exp\{-(s+1)y\}$$

$$\times \left\{\int_y^\infty \exp\{-(s+1)z\}dz\right\}^{r-1} \left\{\int_0^y \exp\{-(t+1)z\}dz\right\}^{n-r} dy$$

$$\text{(3.4)} \qquad = \left\{n\binom{n-1}{r-1}\middle/\{(s+1)^{r-1}(t+1)^{n-r}\}\right\}$$

$$\times \int_0^\infty \exp\{-r(s+1)y\}\{1 - \exp(-(t+1)y)\}^{n-r}dy$$

$$= n!\middle/\left\{r!(s+1)^r(t+1)^{n-r}\prod_{j=1}^{n-r}\{r(s+1)/(t+1)+j\}\right\}.$$

The p.d.f. of U and V is thus

$$f_{U,V}(u, v) = \left\{\binom{n}{r}\middle/(r^{n-r-1}(n-2)!)\right\}\exp\{-(u+v)\}$$

$$\times \sum_{k=1}^{n-r} (-1)^{n-r-k}\binom{n-r}{k}(ku - rv)_+^{n-2}/k^{r-1}.$$

The p.d.f. of $Z = U/(U+V)$ is obtained using the equation

$$f_Z(z) = (1-z)^{-2}\int_0^\infty xf_{U,V}(zx/(1-z), x)dx,$$

and is given by (3.3).

The distribution function $F_Z(z)$ of Z is given by

$$F_Z(z) = \Pr(Z \leq z)$$

$$\text{(3.5)} \qquad = \begin{cases} \left\{\binom{n}{r}\middle/r^{n-r-1}\right\}\displaystyle\sum_{k=1}^{n-r}(-1)^{n-r-k}\binom{n-r}{k} \\ \qquad\qquad\qquad \times \{(k+r)z - r\}_+^{n-2}/\{k^{r-1}(k+r)\}, \quad z \leq 1 \\ 1, \quad z > 1. \end{cases}$$

As will be seen in the next section, Z has most of its probability near $r \log n/n$ as n increases, so that typically we shall need of the order of $n - [n/\log n]$ terms to evaluate (3.5). For the case where $r = 1$, however, the usual expression for the distribution of Fisher's g-statistic contains only $O([n/\log n])$ terms. The analogous expression for the general case may be shown to be

$$
\begin{aligned}
F_Z(z) = 1 - \left\{ \binom{n}{r} \middle/ r^{n-r-1} \right\} \\
\times \left\{ \sum_{k=1}^{n-r} (-1)^{k+r+1} \binom{n-r}{k} \{r - (k+r)z\}_+^{n-1} H(z)/(k^{r-1}(k+r)) \right. \\
\left. + \sum_{j=0}^{r-1} (-1)^{r-j} r^{n-2-j} C_{n,r,j} \sum_{k=0}^{j} \binom{n-1}{k} z_+^k (1-z)_+^{n-1-k} \right\}
\end{aligned}
$$

(3.6)

where

$$
C_{n,r,j} = \sum_{k=1}^{n-r} (-1)^k \binom{n-r}{k} \middle/ k^{r-j-1}.
$$

Lemma 3. Let $\{X_i\}$, $\{Y_i\}$, U and V be defined as in Lemma 2, and let W be independent of $\{X_i\}$, $2W$ having a chi-square distribution with q degrees of freedom. Then the distribution function F_ξ of $\xi = U/(W+U+V)$ may be obtained from (3.7).

Proof. Firstly, $(U+V)$ is independent of $U/(U+V)$. For if the X_i's were to have as common p.d.f. $\alpha \exp(-\alpha x)$, then the redefined U and V would be such that $U/(U+V)$ had a distribution not depending on α, while $(U+V)$ would be complete sufficient for α. We may thus write $\xi = Z/(G+1)$, where $G = W/(U+V)$ and $G \cdot 2n/q$ is distributed as F with q and $2n$ degrees of freedom, independently of Z defined in Lemma 2. Hence,

$$
\Pr(\xi \geq z) = \int_0^\infty \Pr\{Z \geq z(x+1)\} \cdot x^{q/2-1}/\{B(q/2, n)(1+x)^{n+q/2}\} dx
$$

where $B(a, b) = \Gamma(a)\Gamma(b)/\Gamma(a+b)$. This expression reduces to

$$
\begin{aligned}
1 - F_\xi(z) = \left\{ \binom{n}{r} \middle/ r^{n-r-1+q/2} \right\} \left\{ \sum_{k=1}^{n-r} (-1)^{k+r+1} \binom{n-r}{k} (r - (k+r)z)_+^{n+q/2-1} \right. \\
\times H(z)/\{k^{r-1}(k+r)\} \\
+ \sum_{j=0}^{r-1} (-1)^{r-j} r^{n-2-j+q/2} C_{n,r,j} \sum_{k=0}^{j} \binom{n-1}{k} z_+^k (1-z)_+^{n-1-k+q/2} \\
\left. \times B(q/2, n-k)/B(q/2, n) \right\}
\end{aligned}
$$

(3.7)

where $C_{n,r,j}$ is defined below (3.6).

The distribution function of W_T is now easily obtained.

Theorem 1. The random variable W_T defined by (3.1) has the same distribution as $(1-\xi)^{-1}$, where ξ is defined in Lemmas 2 and 3, with $n = P_T - s$ and $q = T - 2P_T$.

Proof. By the Fisher–Cochran theorem,

$$\left\{ \sum_{t=1}^{T} \varepsilon_t^2 - \sum_{j=1}^{s} I_\varepsilon(\lambda_j) \right\} \Big/ \sigma^2 = A + B, \quad \text{where} \quad A = \sum_{k \in \Delta} I_\varepsilon(2\pi k/T)/\sigma^2$$

and B is distributed as chi-square with degrees of freedom $T - 2P_T$, independently of the elements of $\{I_\varepsilon(2\pi k/T)/\sigma^2; \ k \in \Delta_T\}$ which are distributed independently as chi-square with two degrees of freedom. Thus $W_T = \{1 - U/(W + U + V)\}^{-1}$ where U, V and W are identified with $T_r/(2\sigma^2)$, $\frac{1}{2}A - T_r/(2\sigma^2)$ and $\frac{1}{2}B$ respectively, and the associated degrees of freedom n and q with $(P_T - s)$ and $(T - 2P_T)$ respectively, T_r being defined in Lemma 1.

We note here that the results of this section may easily be modified to cover the case where (2.1) is extended to incorporate an unknown mean term μ. If the data are mean-corrected initially, the results follow if the parameter q is reduced by 1.

4. Asymptotic theory

We proceed to the main result via the following lemma.

Lemma 4. With the random variable U defined as in Lemma 2, $U - r \log n$ has a distribution which converges, as n increases, to that of a random variable ζ with the p.d.f.

(4.1)
$$f_\zeta(x) = (r^{r-1}/r!) \exp(-x) \left\{ \sum_{k=1}^{\infty} (-1)^{k-r-1} \exp(-kx/r)/(k^{r-1}k!) \right.$$
$$\left. + (1/(r-1)!) \sum_{j=0}^{r-1} (x/r)^{r-1-j} \binom{r-1}{j} \gamma_j \right\}$$

where $\gamma_j = d^j(\Gamma(x))/(dx)^j \big|_{x=1}$, $j \geq 1$ and $\gamma_0 = 1$.

Proof. From the proof of Lemma 2, the characteristic function $\psi_n(s)$ of $(U - r \log n)$ is given by

(4.2)
$$\psi_n(s) = E \exp\{-is(U - r \log n)\}$$
$$= \{\exp(isr \log n)n!/\Gamma(n+1+isr)\}$$
$$\times \Gamma\{1 + r(1+is)\}/\{r!(1+is)^r\}, \quad \text{Im}(s) < 1.$$

Since $\exp(isr \log n)n!/\Gamma(n+1+isr)$ converges to 1 for all $s \in C$, $\psi_n(s)$ converges to $\psi(s) = \Gamma\{1 + r(1+is)\}/\{r!(1+is)^r\}$, which is continuous at $s = 0$. Thus $U - r \log n$ has a distribution which converges to that of a random variable with characteristic function $\psi(s)$. It remains only to invert $\psi(s)$. To simplify matters, we invert instead $\tilde{\psi}(s) = \Gamma(1 + ris)/(is)^r$ and multiply the result by $\exp(-x)/r!$. The function $\tilde{\psi}$ has poles of order 1 at $s = ik/r$, $k = 1, 2, 3, \cdots$ and a pole of order r at $s = 0$. The pole at $s = ik/r$ contributes to $r!\exp(x)f_\zeta(x)$ the amount

$$(2\pi i/2\pi)\exp(-kx/r)/(-k/r)^r \lim_{s \to ik/r}(s - ik/r)\Gamma(1 + ris)$$

$$= (-1)^{k-r-1}r^{r-1}\exp(-kx/r)/\{k^{r-1}(k-1)!\}$$

while the pole at $s = 0$ contributes

$$\{2\pi i/(2\pi(r-1)!\, i^r)\}\frac{d^{r-1}}{(ds)^{r-1}}\{\exp(isx)\Gamma(1 + ris)\}\big|_{s=0}$$

$$= \sum_{j=0}^{r-1}\binom{r-1}{j}\gamma_j r^j x^{r-1-j}(r-1)!,$$

where $\gamma_j = \{d^j\Gamma(x)/(dx)^j\}\big|_{x=1}$, giving the expression (4.1) for the p.d.f. $f_\zeta(x)$. (*Note*: The contour of integration is the same for all $x \in R$ because of the nature of $\Gamma(1 + ris)$ in the region $\{s \in C : \mathrm{Im}(s) > 0\}$.)

An integral expression for $f_\zeta(x)$ is

$$f_\zeta(x) = \{r^{r-1}/(r!\,(r-1)!)\}\alpha_r(x)\exp(-x)\int_1^\infty (\ln v)^{r-1}\exp(-v\alpha_r(x))dv$$

where $\alpha_r(x) = \exp(-x/r)$, whence the distribution function $F_\zeta(x)$ may be written as

$$F_\zeta(x) = \Pr(\zeta \le x)$$

(4.3)
$$= \{r^r/(r-1)!\}\sum_{j=0}^r \{(\alpha_r(x))^j/j!\}\int_1^\infty v^{j-r-1}(\ln v)^{r-1}\exp(-v\alpha_r(x))dv.$$

For $r = 1$, $F_\zeta(x)$ reduces to the well-known expression used by Walker (1914): $\exp\{-\exp(-x)\}$, while for $r = 2$, it may be shown that

$$F_\zeta(x) = \{1 + \exp(-x/2)\}\exp\{-\exp(-x/2)\} - \exp(-x)E_1\{\exp(-x/2)\}$$

where $E_1(z) = \int_z^\infty \exp(-u)/u\,du$. For $r > 2$, $F_\zeta(x)$ is most likely best evaluated by truncating the infinite sum in the expression

$$1 - F_\zeta(x) = (r^r/r!)\exp(-x)\left\{\sum_{k=1}^\infty (-1)^{k-r-1}\exp(-kx/r)\bigg/\{k^{r-1}k!\,(k+r)\}\right.$$

(4.4)
$$\left. + \sum_{k=0}^{r-1}(x^k/k!)\sum_{j=0}^{r-1-k}\gamma_j r^{j-r}\bigg/j!\right\}$$

where γ_j is defined below (4.1) and may be calculated recursively from

$$\gamma_j = \frac{d^{j-1}}{(dx)^{j-1}}(\Gamma(x)\psi(x))\big|_{x=1}$$

$$= (j-1)!\sum_{k=0}^{j-1}(-1)^{j-k}\gamma_k C_{j-1-k}\bigg/k!$$

where $C_1 = -\gamma$ (Euler's constant), $C_k = \sum_{j=1}^{\infty} j^{-k}$, $k \geq 2$, and $\psi(x) = d/dx \ln \Gamma(x)$. The asymptotic distribution of $l_T^{-2/T}$, defined by (2.2), may now be obtained.

Theorem. Within the framework of Section 2, with $(T - 2P_T) = o(T)$ as T increases, $\Pr\{P_T(1 - l_T^{2/T}) - r \log P_T \geq x\}$ converges to $1 - F_\zeta(x)$, where $F_\zeta(x)$ is defined by (4.4).

Proof. From Lemma 1, we may replace $l_T^{-2/T}$ by the random variable W_T, so that the event $\{(P_T - s)(1 - l_T^{2/T}) - r \log (P_T - s) \geq x\}$ has the same asymptotic probability as $\{(P_T - s)\xi - r \log (P_T - s) \geq x\}$, where ξ is defined in Lemma 3 with the parameter set (n, r, q) identified with $(P_T - s, r, T - 2P_T)$. Using the notation of that lemma, we have

$$(4.5) \quad (P_T - s)\xi - r \log (P_T - s) = \{(P_T - s)/(U + V + W)\}\{U - r \log (P_T - s)\}$$
$$+ r \log (P_T - s)\{(P_T - s)/(U + V + W) - 1\}.$$

Now, since $2(U + V + W)$ is distributed as chi-square with degrees of freedom $T - 2s$, and $T - 2P_T = o(T)$, it follows that the right-hand side of (4.5) has the same asymptotic distribution as $U - r \log (P_T - s)$, whose distribution function is given by (4.4).

TABLE 1

Percentage points of S_T

N		$r = 1$ 99%	95%	90%	$r = 2$ 99%	95%	90%
64	$s = 0$:	3.749	2.455	1.870	3.871	2.505	1.845
	$s = 1$:	3.729	2.444	1.862	3.835	2.483	1.829
128	$s = 0$:	4.085	2.648	2.008	4.524	2.897	2.127
	$s = 1$:	4.079	2.645	2.006	4.511	2.890	2.121
256	$s = 0$:	4.295	2.774	2.100	4.952	3.161	2.319
	$s = 1$:	4.293	2.772	2.099	4.948	3.159	2.317
∞	$s = 0, 1$:	4.600	2.970	2.250	5.625	3.600	2.651

TABLE 2

Percentage of occurrences of S_T below percentage points in 1000 simulations

N		$r = 1$ 99%	95%	90%	$r = 2$ 99%	95%	90%
64	$s = 0$:	98.9	93.6	88.6	99.0	94.9	89.5
	$s = 1$:	98.9	93.4	88.0	99.1	95.4	89.5
128	$s = 0$:	98.7	94.8	90.2	99.1	94.9	90.5
	$s = 1$:	98.6	94.4	90.4	99.2	95.2	90.8
256	$s = 0$:	98.6	94.5	88.8	99.4	94.2	89.3
	$s = 1$:	98.6	94.5	89.0	98.9	94.3	89.2

5. Simulations

Equation (2.1) was used to generate 1000 time series for each combination of sample size (T) 64, 128 and 256, number of sinusoids (s) 0 and 1, and (for $s = 1$) values of α_1 1 and 3 and frequencies $5\pi/16$, $5\pi/8$ and $15\pi/16$, σ^2 and β_1 chosen respectively as 1 and 0 throughout. No *a priori* knowledge of the frequencies was assumed, so that $\Lambda_T = \{n \in Z : 1 \leqq n \leqq [(T-1)/2]\}$ and $P_T = [(T-1)/2]$. For each set of 1000 statistics $S_T = P_T\{1 - l_T^{2/T}\} - r \log P_T$, computed for $r = 1$ and $r = 2$, the number of occurrences above the 99%, 95% and 90% points of the theoretical distributions (as calculated from (3.7) and (4.3)) was recorded. Table 1 contains the theoretical percentage points, while Table 2 contains, for $s = 0$, the percentages of occurrences below the 99%, 95% and 90% points. For $s = 1$, only the worst results for each combination of T and r are reported. The results show that confidence in the theoretical results is justified.

References

FISHER, R. A. (1929) Tests of significance in harmonic analysis. *Proc. R. Soc. London* A **125**, 54–59.

HANNAN, E. J. (1961) Testing for a jump in the spectral function. *J. R. Statist. Soc.* B **23**, 394–404.

HANNAN, E. J. (1970) *Multiple Time Series*. Wiley, New York.

PRIESTLEY, M. B. (1962) Analysis of stationary processes with mixed spectra—II. *J. R. Statist. Soc.* B **24**, 511–529.

PRIESTLEY, M. B. (1981) *Spectral Analysis and Time Series*. Academic Press, London.

WALKER, A. M. (1965) Some asymptotic results for the periodogram of a stationary time series. *J. Austral. Math. Soc.* **5**, 107–128.

WALKER, G. T. (1914) Correlation in seasonal variation of weather. III. On the criterion for the reality of relationships or periodicities. *Mem. Indian Metereol. Dept.* **21** (9), 13–15.

WHITTLE, P. (1952) The simultaneous estimation of a time series' harmonic components and covariance structure. *Trab. Estadist.* **3**, 43–57.

Asymptotic Expansions for Time Series Statistics

KATSUTO TANAKA

Abstract

Asymptotic expansions for the distributions of estimators and test statistics are derived in connection with time series models. The expansions relate to marginal and joint distributions together with the percentiles of marginal distributions. We also consider transforming a statistic so that the transformed statistic has a distribution that coincides with its asymptotic distribution up to a higher order.

TEST STATISTICS; TIME SERIES MODEL; MAXIMUM LIKELIHOOD ESTIMATOR

1. Introduction

In this paper we are concerned with the derivation of asymptotic expansions for estimators and test statistics associated with time series models. The expansions relate to marginal and joint distributions together with the percentiles of marginal distributions. Though the idea of approximating sampling distributions by asymptotic expansions has a long history in statistics and there have been suggested some other types of expansions, we concentrate here on the Edgeworth and Cornish–Fisher expansions.

As for the Edgeworth expansions, Chambers (1967) gave conditions for validity and algorithms for computation dealing with quite general vector functions of multivariate statistics. The underlying assumption there is that the vector observations are independently and identically distributed with finite cumulants up to a certain order. Recently Sargan (1976) specialized the results by Chambers (1967) and suggested a general algorithm for obtaining Edgeworth expansions associated with econometric estimators. Sargan's algorithm is, however, restricted to cases of scalar statistics.

In Section 2, after giving an introductory discussion of Edgeworth expansions, we extend Sargan's algorithm to vector cases allowing for joint distributions. When we treat the maximum likelihood estimator (MLE), Edgeworth expansions are not directly applicable. The MLE is usually a solution to the likelihood equation that is highly non-linear in the parameters. We therefore need to express the MLE explicitly from the likelihood equation before applying the algorithm suggested in Section 2. Section 3 gives two methods for doing this together with some expansion results. It is an easy matter to obtain Cornish–Fisher expansions once Edgeworth expansions are derived. Cornish–Fisher expansions are useful for approximating the percentiles of a true, unknown distribution using those of a known distribution, and are extensively discussed in Hill and Davis (1968). In Section 4 we discuss the Cornish–Fisher expansions in connection with test statistics. Approximations to the power of some tests are also given.

Asymptotic expansions make it clear in an analytical way how different the exact distribution is from its asymptotic distribution in the usual sense. This leads us to transform a statistic so that the transformed statistic has a distribution that coincides with its asymptotic distribution up to a higher order. Section 5 gives two algorithms for obtaining such transformations, together with some examples. In Section 6, the last section, the usefulness of computerized algebra is emphasized for practical application of asymptotic expansions.

2. Edgeworth expansions

In this section we first present an algorithm for obtaining Edgeworth expansions developed by Sargan (1976) (see also Phillips (1977)) in cases where the scalar statistic is asymptotically normal, and then extend the algorithm to vector cases.

Let $e(q)$ be some scalar function of a $p \times 1$ random vector q with $e(0) = 0$. It is assumed that $\bar{q} = \sqrt{T} q$ has bounded moments of all orders as $T \to \infty$ where T is the sample size. Then it is straightforward following Sargan (1976) to obtain

$$P(\sqrt{T} e(q) < x) = \Phi\left(\frac{x}{\omega}\right) + \phi\left(\frac{x}{\omega}\right)\left\{c_0 + c_1\left(\frac{x}{\omega}\right) + c_2\left(\frac{x}{\omega}\right)^2\right.$$

(2.1)

$$\left. + c_3\left(\frac{x}{\omega}\right)^3 + c_5\left(\frac{x}{\omega}\right)^5\right\} + O(T^{-\frac{3}{2}}),$$

where $\Phi(x)$ and $\phi(x)$ are the standard normal distribution and its probability density respectively while

$$c_0 = -\frac{\alpha_4}{2\omega\sqrt{T}} + \frac{\alpha_1}{6\omega^3} + \frac{\alpha_3}{2\omega^3\sqrt{T}},$$

$$c_1 = -\frac{1}{\omega^2}\left(\frac{\alpha_5}{2\sqrt{T}} + \frac{\alpha_7}{2T} + \frac{\alpha_4^2}{8T} + \frac{\alpha_9}{4T}\right) + \frac{3}{\omega^4} A - \frac{15}{\omega^6} B,$$

$$c_2 = -\frac{1}{\omega^3}\left(\frac{\alpha_1}{6} + \frac{\alpha_3}{2\sqrt{T}}\right), \qquad c_3 = -\frac{1}{\omega^4} A + \frac{10}{\omega^6} B, \qquad c_5 = -\frac{1}{\omega^6} B,$$

and

$$A = \frac{\alpha_2}{24} + \frac{\alpha_{10}}{2\sqrt{T}} + \frac{\alpha_1\alpha_4}{12\sqrt{T}} + \frac{\alpha_6}{6T} + \frac{\alpha_3\alpha_4}{4T} + \frac{\alpha_8}{2T}, \qquad B = \frac{\alpha_1^2}{72} + \frac{\alpha_1\alpha_3}{12\sqrt{T}} + \frac{\alpha_3^2}{8T}.$$

To give a definition of ω^2 and the α_i's $(i = 1, \cdots, 10)$ called the Edgeworth expansion coefficients we put

$$e_j = \frac{\partial e(0)}{\partial q_j}, \qquad e_{jk} = \frac{\partial^2 e(0)}{\partial q_j \partial q_k}, \qquad e_{jkl} = \frac{\partial^3 e(0)}{\partial q_j \partial q_k \partial q_l},$$

and denote by $\psi(\theta)$ the cumulant generating function for $\bar{q}$. The second-, third- and fourth-order cumulants for $\bar{q}$ are therefore defined as

$$\psi_{jk} = \frac{\partial^2 \psi(0)}{\partial \theta_j \partial \theta_k}, \qquad \psi_{jkl} = \frac{\partial^3 \psi(0)}{\partial \theta_j \partial \theta_k \partial \theta_l}, \qquad \psi_{jklm} = \frac{\partial^4 \psi(0)}{\partial \theta_j \partial \theta_k \partial \theta_l \partial \theta_m}.$$

Then, using the tensor summation convention for indices j, k, l, $m = 1, \cdots, p$, we define ω^2 and α_i's as follows:

$$\omega^2 = \psi_{jk} e_j e_k, \qquad \alpha_1 = \psi_{jk} e_j e_k e_l, \qquad \alpha_2 = \psi_{jklm} e_j e_k e_l e_m, \qquad \alpha_3 = \gamma_j e_{jk} \gamma_k,$$

$$\alpha_4 = \psi_{jk} e_{jk}, \qquad \alpha_5 = \delta_{jk} e_{jk}, \qquad \alpha_6 = e_{jkl} \gamma_j \gamma_k \gamma_l, \qquad \alpha_7 = e_{jkl} \psi_{jk} \gamma_l,$$

$$\alpha_8 = \gamma_j e_{jk} \gamma_{kl} e_{lm} \gamma_m, \qquad \alpha_9 = \psi_{jm} e_{jk} \psi_{jl} e_{lm}, \qquad \alpha_{10} = \gamma_j e_{jk} \beta_k,$$

where $\gamma_j = \psi_{jk} e_k$, $\beta_j = \psi_{jkl} e_k e_l$ and $\delta_{jk} = \psi_{jkl} e_l$.

It is possible in principle to obtain the exact values of α_i's, but we need not do so as far as the asymptotic expansions considered here are concerned. We only have to retain the highest order of each α_i and thus the highest order of each cumulant noting that $\psi_{jk} = O(1)$, $\psi_{jkl} = O(T^{-\frac{1}{2}})$ and $\psi_{jklm} = O(T^{-1})$ in general. As for ω^2, however, we need to evaluate ψ_{jk} up to $O(T^{-1})$ if we do wish to obtain the expansion up to $O(T^{-1})$. On the other hand if we are satisfied with the expansion up to $O(T^{-\frac{1}{2}})$, we only have to compute ω^2 up to $O(1)$. In the expansion up to $O(T^{-\frac{1}{2}})$ we also only have to compute α_1, α_3 and α_4 implying that we need only c_0 and c_2 in (2.1).

The probability density of $\sqrt{T} e(q)$, if it exists, may be approximated from (2.1) as

$$\frac{1}{\omega} \phi\left(\frac{x}{\omega}\right)\left\{1 + c_1 + (2c_2 - c_0)\frac{x}{\omega} + (3c_3 - c_1)\left(\frac{x}{\omega}\right)^2 - c_2\left(\frac{x}{\omega}\right)^3 \right.$$

$$(2.2)$$

$$\left. + (5c_5 - c_3)\left(\frac{x}{\omega}\right)^4 - c_5\left(\frac{x}{\omega}\right)^6\right\}.$$

The algorithm for obtaining (2.1) or (2.2) is quite simple if ω^2 and the Edgeworth expansion coefficients α_i's are already available. To arrive at (2.1) or (2.2) we can use one of the general purpose algebraic manipulation languages, such as REDUCE (Hearn (1973)). The only complexity seems to be the computation of the cumulants ψ_{jk}, ψ_{jkl} and ψ_{jklm}. These are either traces of Toeplitz matrices or quadratic forms of a special type.

As an illustration let us consider the model

$$(2.3) \qquad y_t = m + \beta y_{t-1} + u_t \quad (t = \cdots, -1, 0, 1, \cdots; \ |\beta| < 1),$$

where $\{u_t\}$ is a sequence of $N(0, \sigma^2)$ independent random variables. Here we put $\mu = m/(1 - \beta)$, which is $E(y_t)$. Given the observations $y = (y_0, y_1, \cdots, y_T)'$ the least squares estimator (LSE) $\hat{\beta}$ of β is computed as

$$(2.4) \qquad \hat{\beta} = \frac{y' A_1 y - \dfrac{1}{T}(d_1' y)(d_2' y)}{y' A_0 y - \dfrac{1}{T}(d_1' y)^2} = \frac{q_2 + \mu_2 - (q_3 + \mu_3)(q_4 + \mu_4)}{q_1 + \mu_1 - (q_3 + \mu_3)^2},$$

where

$$A_0 = \begin{pmatrix} 1 & 0 & . & . & . & . & 0 \\ 0 & 1 & & & & & . \\ . & & . & & & & . \\ . & & & . & & 1 & 0 \\ 0 & . & . & . & . & 0 & 0 \end{pmatrix},$$

$$A_1 = \frac{1}{2}\begin{pmatrix} 0 & 1 & . & . & . & 0 \\ 1 & 0 & 1 & & & . \\ . & 1 & & . & & . \\ . & & . & & & 1 \\ 0 & . & . & . & 1 & 0 \end{pmatrix}; \quad (T+1) \times (T+1),$$

$$d_1 = (1, \cdots, 1, 0)', \quad d_2 = (0, 1, \cdots, 1)'; \quad (T+1) \times 1,$$

$$q_i = (y' A_{i-1} y - E(y' A_{i-1} y))/T \quad (i = 1, 2),$$

$$q_j = (d_{j-2}' y - E(d_{j-2}' y))/T \quad (j = 3, 4),$$

$$\mu_1 = E(y' A_0 y)/T = \frac{\sigma^2}{1 - \beta^2} + \mu^2, \qquad \mu_2 = E(y' A_1 y)/T = \frac{\beta \sigma^2}{1 - \beta^2} + \mu^2,$$

$$\mu_3 = E(d_1' y)/T = \mu, \qquad \mu_4 = E(d_2' y)/T = \mu.$$

Note here that $e(q) = \hat{\beta} - \beta$ with $e(0) = 0$ and that $\bar{q} = \sqrt{T} q$ has bounded moments of all orders. The cumulant generating function $\psi(\theta)$ for $\bar{q}_j$'s ($j = 1, 2, 3, 4$) is found to be

$$\psi(\theta) = -\tfrac{1}{2} \log |G(\theta_1/\sqrt{T})| + \tfrac{1}{2} a'(\theta_2/\sqrt{T}) G^{-1}(\theta_1/\sqrt{T}) \Sigma^{-1} a(\theta_2/\sqrt{T})$$

$$- \sqrt{T} \sum_{j=1}^{4} \theta_j \mu_j - \tfrac{1}{2} \mu^2 d' \Sigma^{-1} d,$$

where

$$\theta = (\theta_1, \cdots, \theta_4)', \quad \boldsymbol{\theta}_1 = (\theta_1, \theta_2)', \quad \boldsymbol{\theta}_2 = (\theta_3, \theta_4)',$$

$$d = (1, \cdots, 1); (T+1) \times 1,$$

$$a(\boldsymbol{\theta}_2) = \mu d + \Sigma(\theta_3 d_1 + \theta_4 d_2), \quad G(\boldsymbol{\theta}_1) = I - 2(\theta_1 A_0 + \theta_2 A_1)\Sigma.$$

Here I is the identity matrix of $(T+1) \times (T+1)$ while Σ is the $(T+1) \times (T+1)$ covariance matrix of y with $\sigma^2 \beta^{|t-s|}/(1-\beta^2)$ in the (t, s)th place. We now obtain the cumulants ψ_{jk}, ψ_{jkl} and ψ_{jklm} by taking partial derivatives of $\psi(\theta)$ evaluated at $\theta = 0$. As for the second-order cumulants we have

$$\psi_{jk} = \frac{2}{T} \operatorname{tr}(A_{j-1}\Sigma A_{k-1}\Sigma) + \frac{4}{T} \mu^2 d' A_{j-1}\Sigma A_{k-1} d \quad (j, k = 1, 2),$$

$$\psi_{jr} = \frac{2}{T} \mu d' A_{j-1}\Sigma d_{r-2} \qquad\qquad (j = 1, 2; r = 3, 4),$$

$$\psi_{rs} = \frac{1}{T} d'_{r-2}\Sigma d_{s-2} \qquad\qquad (r, s = 3, 4).$$

These cumulants are either traces of Toeplitz matrices or quadratic forms of a special type, as described before. The same is true of the higher-order cumulants. To evaluate these quantities up to the highest order we have the following theorems.

Theorem 1 (Hannan (1970), pp. 353–355).

$$(2.5) \qquad \frac{1}{T} \operatorname{tr}(A_{j-1}\Sigma A_{k-1}\Sigma) = \frac{1}{2\pi} \int_{-\pi}^{\pi} f^2(\omega)(\cos \omega)^{j+k-2} d\omega + O(T^{-1}),$$

where $f(\omega) = \sigma^2/|1 - \beta e^{i\omega}|^2$.

The above integration in (2.5) can be evaluated using Cauchy's residue theorem.

Theorem 2.

$$(2.6) \qquad \frac{1}{T} d' A_{j-1}\Sigma A_{k-1} d = \frac{1}{2\pi} \int_{-\pi}^{\pi} f(\omega)(\cos \omega)^{j+k-2} g(\omega) d\omega + O(T^{-1})$$

$$= \frac{1}{2\pi} \int_{-\pi}^{\pi} f(\omega) g(\omega) d\omega + O(T^{-1}) = f(0) + O(T^{-1}),$$

where $g(\omega) = 1 + 2(\cos \omega + \cos 2\omega + \cos 3\omega + \cdots)$. The other quantities such as $d' A_{j-1}\Sigma d_{r-2}/T$ and $d'_{r-2}\Sigma d_{s-2}/T$ have the same limiting value as $d' A_{j-1}\Sigma A_{k-1} d/T$.

Proof. We only outline the proof here. We note that

$$\frac{1}{T}\, d'A_{j-1}\Sigma A_{k-1}d = \frac{1}{T}\,\mathrm{tr}\,(A_{j-1}\Sigma A_{k-1}dd')$$

$$= \frac{1}{T}\,\mathrm{tr}\,\{A_{j-1}\Sigma A_{k-1}(I+2A_1+2A_2+\cdots+2A_T)\},$$

where $A_r\ (r\geq 1)$ is a $(T+1)\times(T+1)$ symmetric matrix which appears in expressing $\Sigma y_t y_{t-r}$ as $y'A_r y$. This observation justifies the first equality in the theorem using Theorem 1. The second equality comes from the property of $g(\omega)$ while the last equality is obtained by noting that $g(\omega)/(2\pi)$ corresponds to the spectrum which has the jump of 1 only at $\omega = 0$.

The asymptotic expansions for marginal distributions of the LSE of β and m in (2.3) are given in Tanaka (1983) while the results for $m = 0$ were initially obtained in Phillips (1977).

We now extend the Edgeworth expansions to those for vector statistics. Let $e(q)$ be some $n\times 1$ vector function of a $p\times 1$ random vector q with $e(0)=0$. We also define by $\xi(s)$ the moment-generating function for $\bar{q}=\sqrt{T}\,q$. Then the moment-generating function $\lambda(t)$ for $\tilde{\theta}=\sqrt{T}\,e(q)$ may be approximated, up to $O(T^{-\frac{1}{2}})$, as

$$\lambda(t) = E(\exp t'\tilde{\theta}))$$

$$(2.7)\qquad \sim \xi((e_1't,\cdots,e_p't)') + \frac{1}{2\sqrt{T}}\sum_{a=1}^{n}\sum_{j,k=1}^{p} t_a e_{jk}^{(a)}\frac{\partial^2\xi((e_1't,\cdots,e_p't)')}{\partial s_j\partial s_k},$$

where $e_j\ (j=1,\cdots,p)$ now denotes the $n\times 1$ vector with $e_j^{(a)}$ as the ath component which is $\partial e^{(a)}(0)/\partial q_j\ (a=1,\cdots,n)$ while $e_{jk}^{(a)}=\partial^2 e^{(a)}(0)/(\partial q_j\partial q_k)$. We then put $\psi(s)=\log\xi(s)$, and, expanding $\xi(s)$ and $\partial^2\xi(s)/(\partial s_j\partial s_k)$ around $s=0$, we have

$$(2.8)\qquad \lambda(t)\sim \exp\left(\tfrac{1}{2}\omega_{ab}t_a t_b\right)\left\{1+\frac{1}{\sqrt{T}}\left(\eta_1^{(abc)}t_a t_b t_c + \eta_2^{(a)}t_a\right)\right\},$$

where

$$\omega_{ab}=\psi_{jk}e_j^{(a)}e_k^{(b)}, \qquad \eta_1^{(abc)}=\frac{\sqrt{T}}{6}\psi_{jkl}e_j^{(a)}e_k^{(b)}e_l^{(c)}+\tfrac{1}{2}\psi_{jl}e_l^{(a)}e_{jk}^{(b)}\psi_{km}e_m^{(c)},$$

$$\eta_2^{(a)}=\tfrac{1}{2}\psi_{jk}e_{jk}^{(a)}.$$

Here the tensor summation convention has been used for indices $a,\ b,\ c = 1,\cdots,n$ and $j,k,l,m = 1,\cdots,p$. Finally the probability density $p(x)$ of $\tilde{\theta}=\sqrt{T}\,e(q)$, if it exists, may be approximated by inverting $\lambda(t)$ as

$$(2.9)\quad p(x)\sim\frac{1}{|\Omega|^{\frac{1}{2}}}\phi(\Omega^{-\frac{1}{2}}x)-\frac{1}{\sqrt{T}}\frac{1}{|\Omega|^{\frac{1}{2}}}\left\{\eta_1^{(abc)}\frac{\partial^3\phi(\Omega^{-\frac{1}{2}}x)}{\partial x_a\partial x_b\partial x_c}+\eta_2^{(a)}\frac{\partial\phi(\Omega^{-\frac{1}{2}}x)}{\partial x_a}\right\},$$

where Ω is an $n \times n$ matrix having ω_{ab} in the (a, b)th place while $\phi(z)$ is now the probability density of an n-dimensional standard normal distribution. It is seen that (2.9) for $n = 1$ is equivalent to (2.2) though the former is an expansion up to $O(T^{-\frac{1}{2}})$.

The approximate marginal density of any subvector of $\tilde{\theta}$ is obtained similarly. The Edgeworth expansion for the distribution function $P(\tilde{\theta} < x)$ may be obtained by integrating (2.9).

So far we have dealt with the cases where the asymptotic distribution is normal. The other cases such as χ^2 are also important in connection with test statistics. The derivation of the Edgeworth expansions can be done almost in the same way as above. Some examples will be shown later.

3. Expansions for the MLE

The MLE $\hat{\theta}$ of an $n \times 1$ vector θ is usually a solution to the likelihood equation $\partial L(\theta)/(\partial \theta) = 0$, where $L(\theta)$ is the log-likelihood function. The equation is highly non-linear in θ and thus we need to derive an explicit expression for $\tilde{\theta} = \sqrt{T}(\hat{\theta} - \theta)$ before applying the algorithms obtained in Section 2. In this section we give two methods for doing this. For this purpose we define the following quantities:

$$g = \frac{1}{\sqrt{T}} \frac{\partial L(\theta)}{\partial \theta}; \quad n \times 1, \qquad W = \frac{1}{\sqrt{T}} \left(\frac{\partial^2 L(\theta)}{\partial \theta \partial \theta'} - E\left(\frac{\partial^2 L(\theta)}{\partial \theta \partial \theta'} \right) \right); \quad n \times n,$$

$$A = \lim_{T \to \infty} \frac{1}{T} E\left(\frac{\partial^2 L(\theta)}{\partial \theta \partial \theta'} \right); \quad n \times n, \qquad K_{abc} = \lim_{T \to \infty} \frac{1}{T} E\left(\frac{\partial^3 L(\theta)}{\partial \theta_a \partial \theta_b \partial \theta_c} \right); \quad 1 \times 1.$$

Here we assume that the matrix A, the negative of which is Fisher's information matrix, is non-singular.

The first method starts by expanding $\partial L(\hat{\theta})/(\sqrt{T} \, \partial \theta) = 0$ around the true value θ in suffix notation. We have

(3.1)
$$0 = g_a + \left(A_{ab} + \frac{1}{\sqrt{T}} W_{ab} \right) \tilde{\theta}_b + \frac{1}{2\sqrt{T}} K_{abc} \tilde{\theta}_b \tilde{\theta}_c + o_p(T^{-\frac{1}{2}})$$
$$\equiv f_a(\tilde{\theta}, \bar{q}) + o_p(T^{-\frac{1}{2}}) \quad (a = 1, \cdots, n),$$

where $\bar{q}$ is the vector composed of g_a and distinct W_{ab}. Let the dimension of $\bar{q}$ be p. We then solve the equation $f_a(\tilde{\theta}, \bar{q}) = 0$ $(a = 1, \cdots, n)$ in (3.1) for $\tilde{\theta}$ as a continuously differentiable function $\tilde{\theta}(\bar{q})$ of $\bar{q}$. The explicit form of $\tilde{\theta}(\bar{q})$ is expressed in suffix notation as

(3.2)
$$\tilde{\theta}_a(\bar{q}) = e_j^{(a)} \bar{q}_j + \frac{1}{2\sqrt{T}} e_{jk}^{(a)} \bar{q}_j \bar{q}_k + o_p(T^{-\frac{1}{2}}) \quad (a = 1, \cdots, n),$$

where $e_j^{(a)} = \partial \tilde{\theta}_a(0)/(\partial \bar{q}_j)$ and $e_{jk}^{(a)} = \sqrt{T} \partial^2 \tilde{\theta}_a(0)/(\partial \bar{q}_j \partial \bar{q}_k)$ for $a = 1, \cdots, n$ and $j, k = 1, \cdots, p$. Note that $f(0, 0) = 0$ and thus $\tilde{\theta}_a(0) = 0$. The $e_j^{(a)}$ and $e_{jk}^{(a)}$ can be found by the implicit function theorem as a solution to the system of equations:

$$(3.3) \qquad \frac{\partial f_a(0, 0)}{\partial \bar{q}_j} + A_{ab} e_j^{(b)} = 0 \quad (a = 1, \cdots, n; j = 1, \cdots, p),$$

$$(3.4) \qquad \frac{\partial^2 f_a(0, 0)}{\partial \bar{q}_j \partial \tilde{\theta}_b} e_k^{(b)} + \left(\frac{\partial^2 f_a(0, 0)}{\partial \bar{q}_k \partial \tilde{\theta}_b} + \frac{1}{\sqrt{T}} K_{abc} e_k^{(c)} \right) e_j^{(b)} + \frac{1}{\sqrt{T}} A_{ab} e_{jk}^{(b)} = 0$$

$$(a = 1, \cdots, n; j, k = 1, \cdots, p).$$

The system of equations (3.3) and (3.4) can be solved recursively solving first for $e_j^{(b)}$ and then (3.4) for $e_{jk}^{(b)}$ given $e_j^{(b)}$. Once the solution to (3.3) and (3.4) is obtained, the Edgeworth expansions developed in Section 2 are readily applicable. The above method and some expansion results are found in Tanaka (1984).

The second method starts by expanding an $n \times 1$ vector $\partial L(\hat{\theta})/(\sqrt{T} \partial \theta) = 0$ itself. We have

$$(3.5) \qquad 0 = g + \left(A + \frac{1}{\sqrt{T}} W \right) \tilde{\theta} + \frac{1}{2\sqrt{T}} K \cdots \circ \tilde{\theta} \circ \tilde{\theta} + o_p(T^{-\frac{1}{2}}),$$

where $K \cdots \circ \tilde{\theta} \circ \tilde{\theta}$ is an $n \times 1$ vector with $K_{abc} \tilde{\theta}_b \tilde{\theta}_c$ as the ath component. The operator $\circ$ is found in the literature of multivariate analysis (see, for example, Hayakawa (1977)). To solve (3.5) for $\tilde{\theta}$ we put $\tilde{\theta} = x + y/\sqrt{T}$ and, substituting this for (3.5) and using the distributive property of $\circ$, we find x and y to obtain

$$(3.6) \quad \tilde{\theta} = -A^{-1}g + \frac{1}{\sqrt{T}} \{ A^{-1}WA^{-1}g - \tfrac{1}{2}A^{-1}(K \cdots \circ A^{-1}g \circ A^{-1}g) \} + o_p(T^{-\frac{1}{2}}).$$

The solution $\tilde{\theta}$ in (3.6) should be compared with $\tilde{\theta}_a(\bar{q})$ in (3.2) to identify $e_j^{(a)}$ and $e_{jk}^{(a)}$. It is evident that $e_j^{(a)}$ for $a, j = 1, \cdots, n$ is given as the elements of $-A^{-1}$ and that $e_j^{(a)} = 0$ for $j > a$. Anyway, to obtain $\tilde{\theta}$ as in (3.6), use of computerized algebra is strongly recommended.

As an illustration let us consider the model

$$(3.7) \qquad y_t = \mu + u_t - \alpha u_{t-1} \quad (t = \cdots, -1, 0, 1, \cdots; |\alpha| < 1),$$

where $\{u_t\}$ is a sequence of $N(0, \sigma^2)$ independent random variables. Given the observations $y = (y_1, \cdots, y_T)'$ the MLE $\hat{\theta}$ of $\theta = (\sigma, \sigma^2, \mu)'$ is the one that maximizes

$$(3.8) \qquad L(\theta) = -\frac{T}{2} \log \sigma^2 - \frac{1}{2} \log |\Sigma| - \frac{1}{2\sigma^2} (y - \mu d)' \Sigma^{-1} (y - \mu d),$$

where d is a $T \times 1$ vector whose components are all 1's while Σ is a $T \times T$

matrix with $1+\alpha^2$ in the main diagonal and $-\alpha$ in the first diagonal above and below the main diagonal and zeroes elsewhere. The non-zero A_{ab} and K_{abc} in (3.6) are obtained as

$$A_{11} = -\frac{1}{1-\alpha^2}, \qquad A_{22} = -\frac{1}{2\sigma^4}, \qquad A_{33} = -\frac{1}{(1-\alpha)^2\sigma^2},$$

$$K_{111} = \frac{-6\alpha}{(1-\alpha^2)^2}, \qquad K_{112} = \frac{1}{(1-\alpha^2)\sigma^2}, \qquad K_{133} = -\frac{2}{(1-\alpha)^3\sigma^2},$$

$$K_{222} = \frac{2}{\sigma^6}, \qquad K_{233} = \frac{1}{(1-\alpha)^2\sigma^4}.$$

The vector $\bar{q}$ composed of g and distinct W_{ab} is defined as

$$\bar{q}_j = \frac{1}{\sqrt{T}} \frac{\partial L(\theta)}{\partial \theta_j} \quad (j = 1, 2, 3)$$

$$\bar{q}_4 = \frac{1}{\sqrt{T}} \left(\frac{\partial^2 L(\theta)}{\partial \theta_1^2} - E\left(\frac{\partial^2 L(\theta)}{\partial \theta_1^2} \right) \right),$$

$$\bar{q}_5 = \frac{1}{\sqrt{T}} \left(\frac{\partial^2 L(\theta)}{\partial \theta_1 \partial \theta_3} - E\left(\frac{\partial^2 L(\theta)}{\partial \theta_1 \partial \theta_3} \right) \right),$$

and is related to g and W as $g_j = \bar{q}_j$ $(j = 1, 2, 3)$, $W_{11} = \bar{q}_4$, $W_{12} = -\bar{q}_1/\sigma^2$, $W_{13} = \bar{q}_5$, $W_{22} = -2\bar{q}_2/\sigma^2$, $W_{23} = -\bar{q}_3/\sigma^2$ and $W_{33} = 0$. Solving for $\tilde{\theta} = \sqrt{T}(\hat{\theta} - \theta)$ from (3.3), (3.4) or (3.6) we have $\tilde{\theta}_a(\bar{q})$ in (3.2) as follows.

$$\tilde{\theta}_1(\bar{q}) \sim (1-\alpha^2)\bar{q}_1 + \frac{(1-\alpha^2)}{2\sqrt{T}}$$

$$\times \{-6\alpha\bar{q}_1^2 + 2(1-\alpha^2)\bar{q}_1\bar{q}_4 - 2(1-\alpha)\sigma^2\bar{q}_3^2 + 2(1-\alpha)^2\sigma^2\bar{q}_3\bar{q}_5\},$$

$$\tilde{\theta}_2(\bar{q}) \sim 2\sigma^4\bar{q}_2 + \frac{\sigma^2}{2\sqrt{T}}\{-2(1-\alpha^2)\bar{q}_1^2 - 2\sigma^2(1-\alpha)^2\bar{q}_3^2\},$$

$$\tilde{\theta}_3(\bar{q}) \sim (1-\alpha)^2\sigma^2\bar{q}_3 + \frac{(1-\alpha^2)(1-\alpha)\sigma^2}{2\sqrt{T}}\{-4\bar{q}_1\bar{q}_3 + 2(1-\alpha)\bar{q}_1\bar{q}_5\}.$$

We then identify $e_j^{(a)}$ and $e_{jk}^{(a)}$ from these expansions and apply the algorithm developed in Section 2 to obtain the approximate probability density of $\tilde{\theta}$ as in (2.9). The probability density $p(x)$ of $\tilde{\theta}$ is

$$(3.9) \qquad p(x) \sim \frac{1}{|\Omega|^{\frac{1}{2}}} \phi(\Omega^{-\frac{1}{2}}x)\left\{1 + \frac{1}{\sqrt{T}}\left(\frac{2\alpha+1}{1-\alpha^2}x_1 - \frac{3}{2\sigma^2}x_2 - \frac{x_1^2 x_2}{2(1-\alpha^2)\sigma^2} + \frac{x_2^3}{6\sigma^6}\right)\right\},$$

where Ω is a 3×3 diagonal matrix defined as $\Omega = -A^{-1}$. It is seen from (3.9) that $\hat{\alpha}$ and $\hat{\sigma}^2$ are dependent in the sense of higher order. The marginal densities are easily obtained from (3.9) by integration. For example we have

$$p_1(x_1) \sim \frac{1}{\sqrt{2\pi}\,\sqrt{1-\alpha^2}} \exp\left(-\frac{x_1^2}{2(1-\alpha^2)}\right)\left(1+\frac{1}{\sqrt{T}}\frac{2\alpha+1}{1-\alpha^2}x_1\right),$$

$$p_2(x_2) \sim \frac{1}{\sqrt{2\pi}\,\sqrt{2\sigma^2}} \exp\left(-\frac{x_2^2}{4\sigma^4}\right)\left(1+\frac{1}{\sqrt{T}}\left(-\frac{2x_2}{\sigma^2}+\frac{x_2^3}{6\sigma^6}\right)\right).$$

Therefore $E(\hat{\alpha}-\alpha)=(2\alpha+1)/T$ and $E(\hat{\sigma}^2-\sigma^2)=-2\sigma^2/T$ up to $O(T^{-1})$. When $\mu=0$ in (3.7), it can be shown that $E(\hat{\alpha}-\alpha)=\alpha/T$ and $E(\hat{\sigma}^2-\sigma^2)=-\sigma^2/T$ up to $O(T^{-1})$.

4. Cornish–Fisher expansions and approximations to power

This section considers approximating percentiles of a distribution which is asymptotically normal or χ^2 as the sample size T goes to ∞. The expansions for percentiles are called Cornish–Fisher expansions and are extensively discussed in Hill and Davis (1968). Here we obtain the Cornish–Fisher expansions associated with time series test statistics. For this purpose we denote by $\tilde{\theta}_T$ a scalar test statistic based on a sample of size T with a distribution function $P(\tilde{\theta}_T < x)$.

Let us consider first the asymptotically normal case where $\tilde{\theta}_T$ tends to normality as $T \to \infty$. Then $P(\tilde{\theta}_T < x)$ is formally expanded in Edgeworth series as

$$(4.1) \qquad P(\tilde{\theta}_T < x) = \Phi(x-\delta) + \phi(x-\delta)\left\{\frac{1}{\sqrt{T}}c_\delta(x)+\frac{1}{T}d_\delta(x)\right\} + o(T^{-1}),$$

where δ is a location parameter while $c_\delta(x)$ and $d_\delta(x)$ are polynomials in x, the coefficients of which depend on δ and the other parameters associated with $P(\tilde{\theta}_T < x)$. Suppose $\delta=0$ and put $P(\tilde{\theta}_T < x_0)=\Phi(u)$ under $\delta=0$. Then we express the percentile x_0 as a function of u. It is an easy matter to obtain

$$(4.2) \qquad x_0 \sim u - \frac{1}{\sqrt{T}}c_0(u) + \frac{1}{T}\{c_0(u)c_0'(u)-\tfrac{1}{2}uc_0^2(u)-d_0(u)\}.$$

The percentile x_0 usually depends on the true parameters. When $\tilde{\theta}_T$ is a test statistic, the percentile x_0 in (4.2) can be used to compute the higher-order power. The expansion (4.1) with x_0 given in (4.2) is

$$P(\tilde{\theta}_T < x_0) \sim \Phi(u-\delta) + \phi(u-\delta)$$

$$(4.3) \qquad \times\left[\frac{1}{\sqrt{T}}\{c_\delta(u)-c_0(u)\}+\frac{1}{T}\{c_0(u)(c_0'(u)-c_\delta'(u)-uc_0(u)\right.$$

$$\left. +\tfrac{1}{2}\delta c_0(u)+(u-\delta)c_\delta(u))+d_\delta(u)-d_0(u)\}\right].$$

Note that $P(\tilde{\theta}_T < x_0)$ in (4.3) is regarded as the power of a left-hand-sided test of size $\Phi(u)$.

We next consider the asymptotically χ^2 case where $\tilde{\theta}_T$ converges to $K_n(x; \lambda^2)$, a non-central χ^2 distribution with n degrees of freedom and non-centrality parameter λ^2. Then $P(\tilde{\theta}_T < x)$ is expanded in Edgeworth series as

$$(4.4) \quad P(\tilde{\theta}_T < x) = K_n(x; \lambda^2) + \sum_{j=0}^{m} \left(\frac{1}{\sqrt{T}} c_j(\lambda) + \frac{1}{T} d_j(\lambda) \right) K_{n+2j}(x; \lambda^2) + o(T^{-1}),$$

where the $c_j(\lambda)$ and $d_j(\lambda)$ are constants depending on λ and the other parameters associated with $P(\tilde{\theta}_T < x)$ and satisfy

$$(4.5) \qquad \sum_{j=0}^{m} c_j(\lambda) = \sum_{j=0}^{m} d_j(\lambda) = 0.$$

When $\lambda = 0$, it is assumed that $c_j(0) = 0$ for all j. Suppose $\lambda = 0$ and put $P(\tilde{\theta}_T < x_0) = K_n(v)$, where $K_n(v) = K_n(v; 0)$. We then express the percentile x_0 as a function of v under $\lambda = 0$. We have

$$(4.6) \qquad x_0 \sim v + \frac{1}{T} A(v; d),$$

where

$$A(v; d) = -2 \left\{ \frac{d_0 v}{n} + \frac{(d_0 + d_1) v^2}{n(n+2)} + \cdots + \frac{(d_0 + d_1 + \cdots + d_{m-1}) v^m}{n(n+2) \cdots (n+2m-2)} \right\},$$

and $d_j = d_j(0)$. The function $A(v; d)$ usually depends on the true parameters and so does x_0. The corresponding expansion (4.4) with x_0 given in (4.6) is

$$
\begin{aligned}
(4.7) \quad P(\tilde{\theta}_T < x_0) &\sim K_n(v; \lambda^2) + \frac{1}{T} A(v; d) k_n(v; \lambda^2) \\
&\quad + \sum_{j=0}^{m} \left(\frac{1}{\sqrt{T}} c_j(\lambda) + \frac{1}{T} d_j(\lambda) \right) K_{n+2j}(v; \lambda^2).
\end{aligned}
$$

Unlike the expansion (4.4), the above expansion (4.7) contains the probability density $k_n(v; \lambda^2) = \partial K_n(v; \lambda^2)/(\partial v)$, but this can be expressed in terms of distribution functions by noting that

$$v k_n(v; \lambda^2) = n k_{n+2}(v; \lambda^2) + \lambda^2 k_{n+4}(v; \lambda^2)$$

and

$$k_n(v; \lambda^2) = \tfrac{1}{2}(K_{n-2}(v; \lambda^2) - K_n(v; \lambda^2)).$$

To illustrate the above methodology let us consider the model

$$(4.8) \qquad y_t = \beta y_{t-1} + u_t \quad (t = \cdots, -1, 0, 1, \cdots; |\beta| < 1),$$

where $\{u_t\}$ is a sequence of $N(0, \sigma^2)$ independent random variables. Given the observations $y = (y_1, \cdots, y_T)'$ suppose we conduct a test

(4.9)
$$H_0 : \beta = \beta_0 \quad \text{against} \quad H_1 : \beta \neq \beta_0.$$

Here we consider two test statistics; one is the t-ratio statistic defined as

(4.10)
$$t_{\hat{\beta}} = \sqrt{T} \, (\hat{\beta} - \beta_0)/(1 - \hat{\beta}^2)^{\frac{1}{2}},$$

where $\hat{\beta} = \sum_{t=2}^{T} y_t y_{t-1}/\sum_{t=2}^{T} y_{t-1}^2$; the other is the Wald statistic defined as

(4.11)
$$W = T(\hat{\beta} - \beta_0)^2/(1 - \hat{\beta}^2).$$

To deal with not only H_0, but also H_1, we make H_1 more specific by considering the following local composite hypothesis:

(4.12)
$$H_1 : \beta = \beta_0 + \frac{\gamma}{\sqrt{T}},$$

where γ is any unknown constant satisfying $|\beta| < 1$.

Under the local alternative H_1 in (4.12) it is not hard to obtain the Edgeworth expansion for the distribution of $\hat{\beta}$. Especially we have

$$P\left(\frac{\sqrt{T}\,(\hat{\beta} - \beta_0)}{\omega} < x\right) \sim \Phi(x - \delta) + \phi(x - \delta)\left\{\frac{1}{\sqrt{T}} \frac{\beta(1 + (x - \delta)^2)}{\omega} \right.$$
$$\left. + \frac{1}{4T}\left(\delta - x + \frac{(\beta^2 + 1)(x - \delta)^2}{\omega^2} - \frac{2\beta^2(x - \delta)^5}{\omega^2}\right)\right\},$$

where $\omega = (1 - \beta^2)^{\frac{1}{2}}$ and $\delta = \gamma/\omega$. The distributions of $t_{\hat{\beta}}$ in (4.10) and W in (4.11) can be obtained on the basis of this expansion. It can be shown that, under H_1 in (4.12),

$$P(t_{\hat{\beta}} < x) \sim \Phi(x - \delta) + \phi(x - \delta)$$

(4.13)
$$\times \left\{\frac{1}{\sqrt{T}} \frac{\beta(1 - \delta(x - \delta))}{\omega} + \frac{1}{4T}\left(-x - x^3 + \frac{(\beta^2 + 1)x^2 - \omega^2}{\omega^2}\delta \right.\right.$$
$$\left.\left. + \frac{-2\beta^2 x^3 + \omega^2 x}{\omega^2}\delta^2 + \frac{6\beta^2 x^2 - \beta^2 - 1}{\omega^2}\delta^3 - \frac{6\beta^2 x \delta^4}{\omega^2} + \frac{2\beta^2 \delta^5}{\omega^2}\right)\right\}.$$

Therefore, using (4.2), the percentile x_0 of $P(t_{\hat{\beta}} < x_0)$ under the null hypothesis $H_0 : \beta = \beta_0$ is

(4.14)
$$x_0 \sim u - \frac{\beta_0}{\sqrt{T}\,\omega_0} + \frac{1}{4T}\left\{\frac{(1 - 3\beta_0^2)u}{\omega_0^2} + u^3\right\},$$

where $\omega_0 = (1 - \beta_0^2)^{\frac{1}{2}}$ and u satisfies $P(t_{\hat{\beta}} < x_0) = \Phi(u)$ under H_0. Using (4.3) the power of the left-hand-sided t-ratio test of size $\Phi(u)$ is approximated as

$$\Phi(u-\delta)+\phi(u-\delta)\left[\frac{1}{\sqrt{T}}\frac{\beta\delta(\delta-u)}{\omega}+\frac{1}{4T\omega^2}\right.$$
$$\times\{(1-3\beta^2)u^2+3\beta^2+3)\delta-(2\beta^2u^3-(7\beta^2+1)u)\delta^2$$
$$\left.+(6\beta^2u^2-5\beta^2-1)\delta^3-6\beta^2u\delta^4+2\beta^2\delta^5\}\right].$$

On the other hand, the distribution of W in (4.11) is approximated, under H_1 in (4.12), as

$$P(W<x)\sim K_1+\frac{\beta\delta^3}{\sqrt{T}\omega}(K_1-2K_3+K_5)$$

$$+\frac{1}{T}\left\{\left(-\frac{1}{4}-\frac{(7\beta^2+1)\delta^4}{4\omega^2}+\frac{\beta^2\delta^6}{2\omega^2}\right)K_1\right.$$

(4.15)
$$+\left(-\frac{1}{2}+\frac{\delta^2}{2}+\frac{(12\beta^2+1)\delta^4}{2\omega^2}-\frac{2\beta^2\delta^6}{\omega^2}\right)K_3$$

$$+\left(\frac{3}{4}-2\delta^2-\frac{7\beta^2\delta^4}{\omega^2}+\frac{3\beta^2\delta^6}{\omega^2}\right)K_5$$

$$\left.+\left(\frac{3\delta^2}{2}+\frac{(6\beta^2-1)\delta^4}{2\omega^2}-\frac{2\beta^2\delta^6}{\omega^2}\right)K_7+\left(\frac{\delta^4}{4}+\frac{\beta^2\delta^6}{2\omega^2}\right)K_9\right\},$$

where $K_j(x;\delta^2)$ is written as K_j for short. Therefore the percentile x_0 of $P(W<x_0)$ is, under $H_0: \beta=\beta_0$, just given by

(4.16)
$$x_0\sim v+\frac{v(v+1)}{2T},$$

where v satisfies $P(W<x_0)=K_1(v;0)$ under H_0. The percentile x_0 in this special case does not depend on β_0. Using (4.7) the power of the Wald test of size $1-K_1(v)$ is approximated as

$$1-K_1-\frac{\beta\delta^3}{\sqrt{T}\omega}(K_1-2K_3+K_5)-\frac{1}{T}\left\{\left(-\frac{(7\beta^2+1)\delta^4}{4\omega^2}+\frac{\beta^2\delta^6}{2\omega^2}\right)K_1\right.$$

(4.17)
$$+\left(\frac{3\delta^2}{4}+\frac{(12\beta^2+1)\delta^4}{2\omega^2}-\frac{2\beta^2\delta^6}{\omega^2}\right)K_3+\left(-\frac{3\delta^2}{4}-\frac{7\beta^2\delta^4}{\omega^2}+\frac{3\beta^2\delta^6}{\omega^2}\right)K_5$$

$$\left.+\left(\frac{(11\beta^2-1)\delta^4}{4\omega^2}-\frac{2\beta^2\delta^6}{\omega^2}\right)K_7+\frac{\beta^2\delta^6}{2\omega^2}K_9\right\},$$

where K_j here is defined as $K_j(v;\delta^2)$.

In the above example we considered the tests based on the LSE of β. We can also conduct tests based on the MLE. It is of interest to compare the power between the tests based on different estimation criteria. It can be shown that the difference of the power between the left-hand-sided t-ratio tests of size $\Phi(u)$ is

$$(4.18) \qquad \text{Power(LSE)} - \text{Power(MLE)} \sim \frac{\phi(u-\delta)(3\beta^2-1)\delta}{2T\omega^2} \quad (\delta < 0).$$

It is seen from (4.18) that the test based on the LSE is not necessarily worse than the MLE-based test if $|\beta|$ is not large. The same is true of the right-hand-sided and two-sided tests. We also naturally obtain that the difference of the power between the LSE- and MLE-based Wald tests of size $1 - K(v)$ is $k_3(v; \delta^2)(1 - 3\beta^2)\delta^2/(T\omega^2)$.

5. Transformation of statistics

As we noted in Section 4, the percentiles usually depend on true parameters associated with $P(\tilde{\theta}_T < x)$. This is quite irrelevant if $\tilde{\theta}_T$ is a test statistic for conducting a test of correct size. In this section we consider transforming $\tilde{\theta}_T$ so that the percentiles of the distribution of a transformed $\tilde{\theta}_T$ coincide, up to $O(T^{-1})$, with those of the usual, asymptotic distribution.

We treat first the asymptotically normal case and consider a transformation $z_T = g(\tilde{\theta}_T)$ with inverse

$$(5.1) \qquad \tilde{\theta}_T = r(z_T) = z_T + \frac{1}{\sqrt{T}} a(z_T) + \frac{1}{T} b(z_T) + o_p(T^{-1}),$$

where $a(z_T)$ and $b(z_T)$ are to be determined below. We have

$$\begin{aligned} P(z_T < x) &= P(g(\tilde{\theta}_T) < x) \\ &= P(\tilde{\theta}_T < r(x)) \\ &\sim \Phi(r(x)) + \phi(r(x))\left\{\frac{1}{\sqrt{T}} c(r(x)) + \frac{1}{T} d(r(x))\right\} \\ &\sim \Phi(x) + \phi(x)\left\{\frac{1}{\sqrt{T}}(a(x) + c(x)) + \frac{1}{T}(b(x) \right. \\ &\qquad \left. -\tfrac{1}{2}xa^2(x) - xa(x)c(x) + a(x)c'(x) + d(x))\right\}. \end{aligned}$$

Then we determine $a(x)$ and $b(x)$ so that the terms of $O(T^{-\frac{1}{2}})$ and $O(T^{-1})$ vanish in the above expansion, which yields

$$(5.2) \qquad a(x) = -c(x), \qquad b(x) = -\frac{x}{2} c^2(x) + c(x)c'(x) - d(x).$$

Therefore $r(x)$ can be determined approximately by using $a(x)$ and $b(x)$ in

(5.2). This, in turn, determines approximately $g(\tilde{\theta}_T)$ noting that $g = r^{-1}$. The transformation $g(\tilde{\theta}_T)$ of $\tilde{\theta}_T$ we finally find is

$$(5.3) \qquad g(\tilde{\theta}_T) = \tilde{\theta}_T + \frac{1}{\sqrt{T}}\, c(\tilde{\theta}_T) + \frac{1}{T}\{\tfrac{1}{2}\tilde{\theta}_T c^2(\tilde{\theta}_T) + d(\tilde{\theta}_T)\}.$$

The transformation $g(\tilde{\theta}_T)$ in (5.3) is, however, not what we want because $c(\tilde{\theta}_T)$ and $d(\tilde{\theta}_T)$ may depend on the true parameters. We therefore first replace $c(\tilde{\theta}_T)$ by its estimate $\hat{c}(\tilde{\theta}_T)$, which yields the first-stage transformation $g_1(\tilde{\theta}_T) = \tilde{\theta}_T + \hat{c}(\tilde{\theta}_T)/\sqrt{T}$. The distribution of $g_1(\tilde{\theta}_T)$ really does not contain the $O(T^{-\frac{1}{2}})$ term, but may still depend on the $O(T^{-1})$ term which we denote by $d_1(\tilde{\theta}_T)$. The second-stage transformation $g_2(\tilde{\theta}_T)$ which is computable and whose distribution $P(g_2(\tilde{\theta}_T) < x)$ coincides with $\Phi(x)$ up to $O(T^{-1})$ is thus given by

$$(5.4) \qquad g_2(\tilde{\theta}_T) = g_1(\tilde{\theta}_T) + \frac{1}{T}\, \hat{d}_1(g_1(\tilde{\theta}_T)),$$

where $g_1(x) = x + \hat{c}(x)/\sqrt{T}$ and $d_1(x)$ is the $O(T^{-1})$ term in the distribution of $g_1(\tilde{\theta}_T)$.

The asymptotically χ^2 case is much easier to deal with. When $\lambda = 0$, the distribution function $P(\tilde{\theta}_T < x)$ in (4.4) reduces to

$$P(\tilde{\theta}_T < x) = K_n(x) + \frac{1}{T}\sum_{j=0}^{m} d_j K_{n+2j}(x) + o(T^{-1})$$

$$= K_n(x) - \frac{1}{T}\, k_n(x) A(x;d) + o(T^{-1}),$$

where $K_n(x) = K_n(x; 0)$ and $k_n(x) = K_n'(x)$ while $A(x; d)$ is given in (4.6) and use is made of the assumption (4.5) and the facts that

$$K_{n+2j}(x) = K_{n+2m}(x) + 2\sum_{i=j}^{m-1} k_{n+2i+2}(x) \quad (j < m),$$

$$k_{n+2j}(x) = \frac{x^j}{n(n+2)\cdots(n+2j-2)}\, k_n(x) \quad (j > 0).$$

Then it is fairly easy to obtain a transformation $h_1(\tilde{\theta}_T)$ which is computable and whose distribution $P(h_1(\tilde{\theta}_T) < x)$ coincides with $K_n(x)$ up to $O(T^{-1})$ as

$$(5.5) \qquad h_1(\tilde{\theta}_T) = \tilde{\theta}_T - \frac{1}{T}\, A(\tilde{\theta}_T; \hat{d}).$$

As an example let us take up the AR(1) model given in (4.8) and consider the t-ratio statistic $t_{\hat{\beta}}$ in (4.10) and the Wald statistic W in (4.11). The null distributions of these statistics are given in (4.13) and (4.15) respectively by putting $\delta = 0$. Then the second-stage transformation $g_2(t_{\hat\beta})$ of $t_{\hat\beta}$ whose distribution $P(g_2(t_{\hat\beta}) < x)$ coincides with $\Phi(x)$ up to $O(T^{-1})$ is given by

$$(5.6) \qquad g_2(t_{\hat\beta}) = g_1(t_{\hat\beta}) + \frac{(3\hat\beta^2 - 5)g_1(t_{\hat\beta}) - \hat\omega^2 g_1^3(t_{\hat\beta})}{4T\hat\omega^2},$$

where $g_1(t_{\hat\beta}) = t_{\hat\beta} + \hat\beta/(\sqrt{T}\,\hat\omega)$ and $\hat\omega = (1 - \hat\beta^2)^{\frac{1}{2}}$. On the other hand the transformation $h_1(W)$ of W whose distribution $P(h_1(W) < x)$ coincides with $K_1(x)$ up to $O(T^{-1})$ is just given by

$$(5.7) \qquad h_1(W) = W - \frac{W(W+1)}{2T}.$$

6. Concluding remarks

In this paper we discussed how to obtain the Edgeworth and Cornish–Fisher expansions associated with time series models. These expansions are useful in studying the higher-order performance of estimators and test statistics and can be used to make statistical inference more rigorous.

Some complexities, however, may arise for practical application of these expansions. As the model becomes complicated, the computational burden necessarily increases. The computation of cumulants is most troublesome. The second- and third-order cumulants are required for the expansion up to $O(T^{-\frac{1}{2}})$ while the cumulants up to the fourth order for the expansion up to $O(T^{-1})$. Though these cumulants may be evaluated by the complex integral of a function of the spectral density of the underlying process, computations really become serious as the model complexity increases. Now that computerized algebra is readily available and is useful for integration and other algebraic manipulations, it is hoped that the method of asymptotic expansions will be incorporated into statistical practice in the future.

References

CHAMBERS, J. M. (1967) On methods of asymptotic approximation for multivariate distributions. *Biometrika* **54**, 367–383.

HANNAN, E. J. (1970) *Multiple Time Series*. Wiley, New York.

HAYAKAWA, T. (1977) The likelihood ratio criterion and the asymptotic expansion of its distribution. *Ann. Inst. Statist. Math.* **29**, 359–378.

HEARN, A. C. (1973) REDUCE-2 User's Manual. UCP-19, University of Utah.

HILL, G. W. AND DAVIS, A. W. (1968) Generalized asymptotic expansions of Cornish–Fisher type. *Ann. Math. Statist.* **39**, 1264–1273.

PHILLIPS, P. C. B. (1977) Approximations to some finite sample distributions associated with a first order stochastic difference equation. *Econometrica* **45**, 463–486.

SARGAN, J. D. (1976) Econometric estimators and the Edgeworth approximation. *Econometrica* **44**, 421–448.

TANAKA, K. (1983) Asymptotic expansions associated with the AR(1) model with unknown mean. *Econometrica* **51**, 1221–1231.

TANAKA, K. (1984) An asymptotic expansion associated with the maximum likelihood estimators in ARMA models. *J. R. Statist. Soc.* B **46**, 58–67.

PART 4

NON-LINEAR AND NON-STATIONARY SYSTEMS IN TIME SERIES

The Box–Jenkins Approach to Random Coefficient Autoregressive Modelling

D. F. NICHOLLS

Abstract

Recent time series research has been directed towards the relaxation of the assumption that time series models have constant coefficients. One class of models to emerge as a result of this has been that of random coefficient autoregressive models. This paper demonstrates how the Box–Jenkins three-step approach of model specification, estimation and diagnostic checking may be applied to this class of models.

IDENTIFICATION; ESTIMATION; DIAGNOSTIC CHECKS; SECOND-ORDER STATIONARITY; SCORE TEST; TWO-STEP ESTIMATION PROCEDURE

1. Introduction

When one examines the literature relating to the analysis of time series it soon becomes obvious that the subject has developed, at least from a practical point of view, as a direct result of the continued increase in computing capabilities and the rapid increase in the availability of related software.

The earlier research in time series concentrated on the analysis of sets of time-dependent data using constant coefficient models. One of the major contributions in this area was that by Box and Jenkins (1976), who proposed an integrated approach to time series modelling (in the time domain) which involved three stages: namely, the specification (or identification) of an appropriate model to fit to the data, the estimation of the specified model and diagnostic checks to confirm that the model fitted the data satisfactorily.

More recently attention has been directed towards the relaxation of the requirement that time series models have constant coefficients. Indeed, the amount of research which has been directed towards stochastic parameter models can be gauged by examining the annotated bibliographies compiled by Johnson (1977), (1980).

One class of stochastic parameter model which has been shown to be statistically quite tractable is the class of random coefficient autoregressive models, discussed in Nicholls and Quinn (1982) (to be referred to as N–Q in what follows). Such models can be analysed with little difficulty given the capacity and speed of computers now available.

The Box–Jenkins method applied to constant-coefficient linear time series models has achieved popularity through its logical three-step approach to the analysis of time series data and its suitability for computer analysis. It seems appealing therefore to extend this approach to include the class of random coefficient autoregressions. While many of the statistical results for such models have already been derived by Andel (1976) and N–Q for example, what we shall do here is to organize and extend these results in such a way as to fit into the specification–estimation–diagnostic checking cycle as proposed by Box and Jenkins.

For a scalar time series $\{y_t, t = 1, 2, 3, \cdots\}$ of equally spaced data points the random coefficient autoregression can be written in the form

$$y_t = \sum_{j=1}^{p} \beta_{t,j} y_{t-j} + \varepsilon_t$$

(1)

$$= \sum_{j=1}^{p} (\bar{\beta}_j + \eta_{t,j}) y_{t-j} + \varepsilon_t$$

where the $p \times 1$ vector of coefficients $\beta_t = \bar{\beta} + \eta_t$, with $\bar{\beta}$ constant and η_t stochastic.

In order to successfully apply the Box–Jenkins approach to this random coefficient autoregressive (RCA) model a number of fairly general assumptions must be made. Indeed for the model (1) we assume;

(i) $\{\varepsilon_t, t = 0, 1, 2, \cdots\}$ is a sequence of identically and independently distributed (i.i.d.) random variables with mean 0 and variance σ^2.

(ii) $\bar{\beta}' = (\bar{\beta}_1, \cdots, \bar{\beta}_p)$ is a vector of constants.

(iii) $\{\eta_t, t = 1, \cdots, N\}$ is a sequence of i.i.d. random vectors with zero mean and $E(\eta_t \eta_t') = \Sigma$.

(iv) $\{\eta_t\}$ and $\{\varepsilon_t\}$ are mutually independent.

In order to describe an easily checkable condition for second-order stationarity of the model (1) let

$$M = \begin{bmatrix} 0 & \vdots & I \\ \hline \bar{\beta}_p & \cdots & \bar{\beta}_1 \end{bmatrix}$$

with the $(1, 1)$ block being the $(p-1) \times 1$ null matrix and the $(1, 2)$ block the $(p-1) \times (p-1)$ identity matrix. If $\mathscr{F}_t$ is the σ-field generated by $\{\varepsilon_s, \eta_s; s \leq t\}$ and $\Sigma > 0$ then there exists a unique $\mathscr{F}_t$-measurable second-order stationary solution to (1) if and only if M has all its eigenvalues within the unit circle and (vec Σ)' vec $A < 1$, where vec A is the last column of the matrix $(I - M \otimes M)^{-1}$ (see Andel (1976) or N–Q).

2. Model specification

In the application of the Box–Jenkins approach to models of the form (1) we shall assume that the data set is second-order stationary. In practice we should of course check this out by examining the sample autocorrelation coefficient to determine whether or not it failed to decay to 0 as the number of lags increased. If it were not stationary it would then be necessary to achieve stationarity by means of some suitable transformation, such as differencing, or by taking logarithms.

Assuming the set of data $\{y_t, t = 1, 2, \cdots, N\}$ to be modelled is stationary, the first step in the modelling process is to determine the order p of the model (1). If $Y'_t = (y_t, \cdots, y_{t+1-p})$, rewriting (1) in the form

$$(2) \qquad y_t = \sum_{j=1}^{p} \bar{\beta}_j y_{t-j} + u_t = Y'_{t-1}\bar{\beta} + u_t,$$

with

$$(3) \qquad u_t = \sum_{j=1}^{p} \eta_{t,j} y_{t-j} + \varepsilon_t = Y'_{t-1}\eta_t + \varepsilon_t,$$

the (ordinary) least squares estimates of $\bar{\beta}_p, p = 1, 2, \cdots$ are just the partial autocorrelation coefficients. As we shall see these estimators are strongly consistent and asymptotically normally distributed. Consequently the order determination process can be carried out in a manner identical to that proposed by Box and Jenkins in the case of constant coefficient autoregressions. If the first coefficient which is not significantly different from 0 occurs at lag $(p+1)$, and all higher-order coefficients are not significantly different from 0, then the model is of order p.

There are numerous different procedures for order determination, a number of these being discussed in Priestley (1981). Most recent model specification procedures have centred around Akaike's AIC, BIC or related criteria. The AIC criterion is now being used as one of the more reliable methods for order determination.

Shibata (1976) considered the properties of the estimated order derived from the AIC criterion and showed, in the case of univariate autoregressive models, that the order estimate determined from the minimization of AIC is not a consistent estimate of the true order of the model but tends to overestimate the true value for large samples. Hannan (1980) considered a modified form of AIC for fitting multivariate ARMA(p, q) models and showed, under fairly general conditions, that the estimated order converges almost surely to the true values. A further refinement, in the case of univariate autoregressive models, has been given in Hannan and Quinn (1979) where these results, based on the law of the iterated logarithm, lead to a strongly consistent estimate of the true order. This procedure is shown, asymptotically, to underestimate the true order to a lesser degree than Hannan's earlier procedure.

Having determined the order of the model, the second step in the specification stage is to determine whether or not a constant coefficient autoregressive model would fit the data just as well.

The obvious way to test for coefficient constancy is to test the hypothesis that the variance of the coefficients is 0, i.e. $\Sigma = 0$. The usual theory associated with tests based on maximum likelihood estimates will not hold in this context however since the vector of unknown coefficient (vech Σ in this case) under the null hypothesis (of a constant coefficient model) lies on the boundary of the parameter space. As we shall later see, to derive the asymptotic distribution of maximum likelihood estimates of models of the form (1), it is necessary to restrict the parameter space Θ in such a way that parameters do not lie on the boundary of the parameter space.

If a boundary value was to be permitted, e.g. $\Sigma = 0$, the maximum likelihood estimate of Σ would be obtained from finding the maximum of the likelihood function $l(\Sigma)$ subject to $\Sigma \geqq 0$. As Moran (1971) and Chant (1974) have shown, the restricted estimate $\hat{\Sigma}$ of Σ obtained from this restricted maximization procedure has a very complex distribution. Consequently this approach is unattractive from a computational point of view.

One way to overcome this problem is to base a test for $\Sigma = 0$ upon the vector of scores $\partial l / \partial \theta$, where $\theta = \text{vech}\,\Sigma$, which ignores the constraint $\Sigma \geqq 0$. This approach, which is essentially Neyman's (1959) $C(\alpha)$ test, was suggested by Moran as the test to use. The test which we state below can be regarded as a score test (or Lagrange multiplier test). This approach loses power compared to that based on the scores of the restricted likelihood but it does at least have a tractable solution.

To test the null hypothesis that $\Sigma = 0$ or, equivalently $\gamma = \text{vech}\,\Sigma = 0$, let $\hat{\beta}$ and $\hat{\sigma}^2$ be the maximum likelihood estimates of the parameters of (1) under the null hypothesis. That is to say, $\hat{\bar{\beta}}$ and $\hat{\sigma}^2$ are just the constant coefficient maximum likelihood estimates for the constant coefficient autoregressive model.

If K_p is a $\{p(p+1)/2\} \times p^2$ matrix such that

$$\text{vec}\,A = K_p'\,\text{vech}\,A$$

(see Henderson and Searle (1979)), then for a sample of size N, if

$$z_t = K_p \operatorname{vec}(Y_{t-1}Y'_{t-1}),$$

$$\bar{z} = N^{-1}\sum_{t=1}^{N} z_t, \qquad \hat{\varepsilon}_t = y_t - \hat{\bar{\beta}}'Y_{t-1},$$

$$g_N = N^{-1}\sum_{t=1}^{N} \hat{\varepsilon}_t^2 z_t - \hat{\sigma}^2\bar{z}, \qquad W_n = N^{-1}\sum_{t=1}^{N}(z_t-\bar{z})(z_t-\bar{z})'$$

$$\nu_N = N^{-1}\sum_{t=1}^{N}\left[\frac{\hat{\varepsilon}_t^2}{\hat{\sigma}^2}-1\right]^2,$$

and

$$(4) \qquad \hat{\psi} = (\nu_N\hat{\sigma}^4)^{-1}Ng_N'W_N^{-1}g_N$$

then under fairly general conditions and assuming that $\gamma = 0$ it can be shown that $\hat{\psi}$ given by (4) is asymptotically distributed as χ^2 with $p(p+1)/2$ degrees of freedom. The proof of this result, along with the required conditions for the result to hold is given in Chapter 6 of N–Q.

This result allows us to construct a test to determine whether or not the coefficients of the model are fixed or random. It is worth noting that a test which is more powerful than that based on (4) is derived in N–Q, pp. 105–108; because of its complexity we shall not discuss it here but refer the interested reader to that book.

3. Model estimation

For the model (1) the parameters to be estimated are the coefficients $\bar{\beta}' = (\beta_1, \cdots, \beta_p)$, the variance σ^2 and the covariance matrix Σ. Since Σ is symmetric it will only be necessary to estimate $\gamma = \operatorname{vech}\Sigma$.

Given a sample of size $(N+p)$ say $y_{1-p}, \cdots, y_0, y_1, \cdots, y_N$, the least squares estimates of the parameters may be obtained by means of a two-step estimation procedure, the first step being to estimate $\bar{\beta}$ and then using this estimate in the second step to estimate the remaining parameters. As we shall see, the least squares estimates are strongly consistent and prove useful in obtaining the maximum likelihood estimates. Indeed, the maximum likelihood method is an iterative process requiring initial estimates of the parameters to commence the iterations. The consistent least squares estimates are extremely suitable for this purpose.

Now $\mathscr{F}_t$ is the σ-field generated by $\{(\varepsilon_s, \eta_s); s \leq t\}$, ε_t and η_t are independent of $\{(\varepsilon_{t-j}, \eta_{t-k}); j, k = 1, 2, \cdots\}$ and Y_{t-1} is a measurable function of $\mathscr{F}_{t-1}$. From (3),

$$E(u_t \mid \mathscr{F}_{t-1}) = Y'_{t-1}E(\eta_t) + E(\varepsilon_t) = 0$$

and

$$E(u_t^2 \mid \mathscr{F}_{t-1}) = h_t = \sigma^2 + Y'_{t-1}\Sigma Y_{t-1}.$$

Since

$$\{\operatorname{vec}(Y_{t-1}Y'_{t-1})\}' = Y'_{t-1} \otimes Y'_{t-1}$$

it follows that

(5)
$$E(u_t^2 \mid \mathscr{F}_{t-1}) = \sigma^2 + \{\operatorname{vec}(Y_{t-1}Y'_{t-1})\}' K'_p \gamma$$
$$= \sigma^2 + z'_t \gamma.$$

The two-step least squares estimation procedure now operates on (2) and (5). The first step is to apply least squares to (2) to give the estimate $\tilde{\tilde{\beta}}$ of β as

(6)
$$\tilde{\tilde{\beta}} = \left(\sum_{t=1}^{N} Y_{t-1}Y'_{t-1} \right)^{-1} \sum_{t=1}^{N} Y_{t-1}y_t.$$

The second step involves using $\tilde{\tilde{\beta}}$ to form

$$\tilde{u}_t = y_t - Y'_{t-1}\tilde{\tilde{\beta}}$$

and if, from (5), we let

$$\tilde{u}_t^2 = \sigma^2 + z'_t \gamma + \xi_t,$$

then the estimates $\tilde{\sigma}^2$ and $\tilde{\gamma}$ of σ^2 and γ respectively are obtained by regressing $\tilde{u}_t^2$ on 1 and z_t. Doing this gives

(7)
$$\tilde{\gamma} = \left\{ \sum_{t=1}^{N} (z_t - \bar{z})(z_t - \bar{z})' \right\}^{-1} \sum_{t=1}^{N} \tilde{u}_t^2 (z_t - \bar{z})$$

and

(8)
$$\tilde{\sigma}^2 = N^{-1} \sum_{t=1}^{N} \tilde{u}_t^2 - \bar{z}'\tilde{\gamma}.$$

Expressions (6), (7) and (8) provide the least squares estimates of the parameters of the model (1).

To obtain the asymptotic properties of the least squares estimates, conditions (i)–(iv) are required. The boundedness of the second moment of $\{y_t\}$ is also required. If the two criteria required for the second-order stationarity of (1) are bounded away from unity this moment condition will be satisfied. Consequently to achieve this we assume:

(v) The largest eigenvalue of m is less than or equal to $(1-\delta_1)$ and $(\operatorname{vec}\Sigma)'$ vec $W \leq (1-\delta_2)$ where $\delta_1 > 0$ and $\delta_2 > 0$ are both arbitrarily small.

The parameters $\bar{\beta}$ and Σ must be such that the solution $\{y_t\}$ to (1) is strictly stationary and ergodic. A sufficient condition for these conditions to hold is that a second-order stationary solution to (1) exists (which is guaranteed by (v)) together with the fact that $\{\varepsilon_t\}$ and $\{\eta_t\}$ are strictly stationary, which follows immediately from (i) and (iii).

The proofs of the asymptotic results also require that $E\{(z_t - E(z_t)) \times$

$(z_t - E(z_t))'\}$ be positive definite. This will be so if we assume:

(vi) There is no non-zero constant vector α such that $\alpha'(z_t - E(z_t)) = 0$ almost everywhere.

We now state the following theorem relating to the asymptotic properties of the least squares estimates. A proof of this theorem is given in N–Q, pp. 46–50. Hannan and Kavalieris (1983) have also shown under very general conditions that the least squares estimates of the parameters, as well as the estimate of the order of the model using BIC, converge almost surely to their theoretical values.

Theorem 1. For the process $\{y_t\}$ satisfying (1) under Conditions (i)–(vi) and for $\tilde{\bar{\beta}}, \tilde{\gamma}$ and $\tilde{\sigma}^2$ defined by (6), (7) and (8) respectively, if $E(y_t^4) < \infty$ the vector $\tilde{\theta}' = (\tilde{\bar{\beta}}', \tilde{\gamma}', \tilde{\sigma}^2)$ converges almost surely to $\theta' = (\beta', \gamma', \sigma^2)$. If in addition $E(y_t^8) < \infty$, $N^{\frac{1}{2}}(\tilde{\theta} - \theta)$ has a distribution which converges to that of a normally distributed random vector with mean 0 and covariance matrix Ω.

The eighth moment condition is only required for the existence of the covariance matrix Ω (whose form is derived in N–Q, pp. 57–58). Such a condition may not of course be easy to check; even the fourth moment condition may be complex though for most of the simpler models one would consider in practice this would be quite straightforward.

The maximum likelihood estimates (MLE) of the parameters of (1) are obtained by setting up the likelihood function as though $\{\varepsilon_t\}$ and $\{\eta_t\}$ are jointly normal. The estimates obtained using this procedure will be shown to be strongly consistent and to satisfy a central limit theorem. The moment conditions in Theorem 1 will not be required in order to obtain the asymptotic properties of the MLE, what will be required in this case is the finiteness of the fourth moment of $\{\varepsilon_t\}$ and $\{\eta_t\}$, conditions (i)–(vi) and

(vii) The variance σ^2 of ε_t is bounded below by δ_3 while the smallest eigenvalue of Σ is bounded below by δ_4, with $\delta_3 > 0$ and $\delta_4 > 0$ both arbitrarily small.

This latter condition eliminates the possibility of the vector of parameters of (1) lying on the boundary of the parameter space. The implications of boundary-value problems have been discussed earlier and, in the interest of simplicity, are best avoided.

As is shown in Chapter 4 of N–Q, if

$$\bar{l}_N(\bar{\beta}, \sigma^2, \lambda) = \ln \sigma^2 + N^{-1} \sum_{t=1}^{N} \ln(1 + \lambda'z_t) + \sigma^{-2}N^{-1} \sum_{t=1}^{N} \frac{(y_t - \bar{\beta}'y_{t-1})^2}{1 + \lambda'z_t}$$

where $\lambda = \sigma^{-2}\gamma$,

$$l_N(\bar{\beta}, \lambda) = \inf_{\sigma^2} \bar{l}_N(\bar{\beta}, \sigma^2, \lambda) - 1$$

then the MLE $\hat{\bar{\beta}}$, $\hat{\sigma}^2$ and $\hat{\gamma}$ of $\bar{\beta}$, σ^2 and $\gamma = \text{vech } \Sigma$ are given by

$$(9) \qquad l_N(\hat{\bar{\beta}}, \hat{\lambda}) = \inf_{(\bar{\beta}',\lambda')' \in \theta} l_N(\bar{\beta}, \lambda)$$

$$(10) \qquad \hat{\sigma}^2 = N^{-1} \sum_{t=1}^{N} \frac{(y_t - \hat{\bar{\beta}}' Y_{t-1})^2}{1 + \hat{\lambda}' z_t}$$

and

$$(11) \qquad \hat{\gamma} = \text{vech } (\hat{\Sigma}) = \hat{\sigma}^2 \hat{\lambda}.$$

The set θ in (9) is a compact subset of $R^{p(p+3)/2}$ and is such that condition (v) is satisfied along with:

(viii) $\sigma^{-2}\Sigma$ has strictly positive eigenvalues, all of which are greater than or equal to $\delta_5 > 0$.

This form of θ is required in order to prove the strong consistency of, and obtain a central limit theorem for, the MLE, which are contained in the following theorem.

Theorem 2. For y_t satisfying (1) under conditions (i)–(viii) the maximum likelihood estimates $\hat{\bar{\beta}}$, $\hat{\gamma}$ and $\hat{\sigma}^2$ defined by (9), (10) and (11) are strongly consistent. Furthermore, assuming finiteness of the fourth moments of $\{\varepsilon_t\}$ and $\{\eta_t\}$, if $\hat{\theta}' = (\hat{\bar{\beta}}', \hat{\gamma}', \hat{\sigma}^2)$ then $N^{\frac{1}{2}}(\hat{\theta} - \theta)$ has a limiting normal distribution with mean 0 and covariance matrix $I^{-1}JI^{-1}$.

For a proof of this theorem, as well as expressions for I and J, we refer the reader to Chapter 4 of N–Q. It should also be pointed out that when $\{\varepsilon_t\}$ and $\{\eta_t\}$ are jointly normal, the covariance matrix of the estimates reduces to $2J^{-1}$.

Thus under fairly general conditions the maximum likelihood estimates for the model (1) can be obtained in a straightforward fashion. As the likelihood function to be optimized is non-linear in the parameters to be estimated however, it is necessary to use an iterative procedure to obtain the estimator. In practice it is desirable to commence the iterative procedure as close to the global maximum as possible in order to avoid the possibility of convergence to a local optimum. Consequently, as we indicated earlier, it is appealing to use the least squares estimates to commence the iterations since they are strongly consistent. This approach is recommended in practice.

4. Diagnostic checks

After specifying and estimating an appropriate model, the final step in the Box–Jenkins modelling cycle is to check that the estimated model is acceptable. In the case of the model (1), from (3) it is not hard to see that

$$E(u_t u_{t-k} \mid \mathscr{F}_{t-1}) = \delta_{0,k}(Y'_{t-1}\Sigma Y_{t-k-1} + \sigma^2)$$

where $\delta_{0,k} = 1$ when $k = 0$ and 0 otherwise. Consequently

$$E(u_t u_{t-k} \mid \mathscr{F}_{t-1}) = 0, \qquad k \neq 0$$

and

$$E(u_t^2 \mid \mathscr{F}_{t-1}) = h_t$$

say, where $h_t = \sigma^2 + z_t'\gamma$ in agreement with (5).

If $e_t = h_t^{-\frac{1}{2}} u_t$, then $E(e_t e_{t-k}) = 0$, $k \neq 0$. This implies that diagnostic checks may be based on the autocorrelation function of the $\hat{e}_t = \hat{h}_t^{-\frac{1}{2}} \hat{u}_t$, where

$$\hat{u}_t = y_t - \sum_{j=1}^{p} \hat{\beta}_j y_{t-j} \quad \text{and} \quad \hat{h}_t = \hat{\sigma}^2 + z_t'\hat{\gamma}.$$

The autocorrelation coefficients $r(k)$ $k = 0, \pm 1, \pm 2, \cdots$ are estimated by

$$\hat{r}(k) = \frac{\displaystyle\sum_{t=k+1}^{p} \hat{e}_t \hat{e}_{t-k}}{\displaystyle\sum_{t=1}^{N} \hat{e}_t^2}.$$

A direct application of Theorems 2 and 3 of Heyde and Hannan (1972) shows that $\hat{r}(k) \xrightarrow{\text{a.s.}} r(k)$ and $N^{\frac{1}{2}}(\hat{r}(k) - r(k))$ converges to a normal distribution. Consequently the $\hat{r}(k)$ can be used to form an analogue of the Box–Pierce (1970) test (in the case of constant coefficient linear time series models):

$$Q = N \sum_{k=1}^{K} \hat{r}^2(k) \sim \chi_{K-p}^2.$$

Durbin (1970) pointed out that $N^{\frac{1}{2}} r(k)$ was not distributed as a standard normal variable under the null hypothesis that $r(k) = 0$ when the $\hat{e}_t$ were residuals from an autoregression, because the asymptotic variance of $N^{\frac{1}{2}} \hat{r}(k)$ was less than unity. He gave a number of procedures for obtaining a test statistic of the correct size. One of these, which has become known as Durbin's second method, involved regressing $\hat{e}_t$ against $\hat{e}_{t-k}$ and testing if the coefficient of $\hat{e}_{t-k}$ in this regression was in fact 0. This would be an appropriate method for testing model adequacy in the current context particularly in the case of small sample sizes when the analogue of the Box–Pierce test may lead to a test statistic for which the true significance level may be much lower than that given by the χ^2 distribution.

5. Conclusion

We have here demonstrated how the recently developed theory relating to random coefficient autoregressive models can be used to develop a Box–Jenkins modelling approach for this class of model. It has been shown that the three stages of specification, estimation and diagnostic checking follow in a similar manner to that for constant coefficient autoregression.

There is one extra step included in the specification stage, however, where the score test is used to determine whether or not a random coefficient model should be selected in preference to a constant coefficient model. Once this decision has been made one can then determine the order of the model to be fitted. Both least squares and maximum likelihood procedures are applicable to the models with the estimator obtained having desirable (asymptotic) properties. Finally, a diagnostic test based on estimates of the residuals can be carried out to determine the goodness-of-fit of the model.

References

ANDEL, J. (1976) Autoregressive series with random parameters. *Math. Operationsforsch. u. Statist.* **7**, 735–741.

BOX, G. E. P. AND JENKINS, G. M. (1976) *Time Series Analysis: Forecasting and Control* (revised edition). Holden-Day, San Francisco.

BOX, G. E. P. AND PIERCE, D. A. (1970) Distribution of residual autocorrelations in autoregressive integrated moving average time series models. *J. Amer. Statist. Assoc.* **65**, 1509–1526.

CHANT, D. (1974) On asymptotic tests of composite hypotheses in non-standard conditions. *Biometrika* **61**, 291–298.

DURBIN, J. (1970) Testing for serial correlation in least squares regression when some of the regressors are lagged dependent variables. *Econometrica* **38**, 410–421.

HANNAN, E. J. (1980) The estimation of the order of an ARMA process. *Ann. Statist.* **8**, 1071–1081.

HANNAN, E. J. AND KAVALIERIS, L. (1983) The convergence of autocorrelations and autoregression. *Austral. J. Statist.* **25**, 287–297.

HANNAN, E. J. AND QUINN, B. G. (1979) The determination of the order of an autoregression. *J. R. Statist. Soc.* B **41**, 190–195.

HENDERSON, H. V. AND SEARLE, S. R. (1979) Vec and vech operators for matrices with some uses in Jacobian and multivariate statistics. *Canad. J. Statist.* **7**, 65–81.

HEYDE, C. C. AND HANNAN, E. J. (1972) On limit theorems for quadratic functions of discrete time series. *Ann. Math. Statist.* **43**, 2058–2066.

JOHNSON, L. W. (1977) Stochastic parameter regression; an annotated bibliography. *Internat. Statist. Rev.* **45**, 257–272.

JOHNSON, L. W. (1980) Stochastic parameter regression: an additional annotated bibliography. *Internat. Statist. Rev.* **48**, 95–102.

MORAN, P. A. P. (1971) Maximum likelihood estimation in non standard conditions. *Proc. Camb. Phil. Soc.* **70**, 441–445.

NEYMAN, J. (1959) Optimal asymptotic tests for composite statistical hypotheses. In *Probability and Statistics*, ed. U. Grenander, Wiley, New York, 213–234.

NICHOLLS, D. F. AND QUINN, B. G. (1982) *Random Coefficient Autoregressive Models: An Introduction.* Springer-Verlag, New York.

PRIESTLEY, M. B. (1981) *Spectral Analysis and Time Series*, Vols 1 and 2. Academic Press, New York.

SHIBATA, R. (1976) Selection of the order of an autoregressive model by Akaike's information criterion. *Biometrika* **63**, 117–126.

Local Gaussian Modelling of Stochastic Dynamical Systems in the Analysis of Non-Linear Random Vibrations

TOHRU OZAKI

Abstract

Stochastic dynamical system models for non-linear random vibrations are considered and their discrete-time version, non-linear time series models are introduced using the local Gaussian modelling method. Some computational problems and implications of the present method in non-linear time series analysis are discussed.

DIFFUSION PROCESS; DYNAMICS OF MOORED VESSEL; LOCAL LINEARIZATION; NON-LINEAR MARKOVIAN REPRESENTATION; AUTOREGRESSIVE-MOVING AVERAGE MODELS: AMPLITUDE-DEPENDENT AUTORE-GRESSIVE MODELS

1. Introduction

Dynamical system models have played a very important role as models which describe dynamic phenomena in many scientific fields such as astronomy, physics, mechanical engineering, electrical engineering, biology and economics. The introduction of stochastic concepts has also stimulated development of stochastic dynamical system models based on Brownian motion theory, and has made the dynamical system model more acceptable in many other scientific and engineering fields. On the other hand, recent development of measurement technology and the introduction of high-speed computers have stimulated the development of numerical analysis by which means discrete-time deterministic dynamical system models may be studied in order to simulate and to analyse physical phenomena with such models. However, such methods are not useful for the analysis of stochastic dynamical systems driven by random noise. What is desired in this situation is a discrete-time stochastic dynamical system model whose structure and parameters can be estimated from observational data. Such a model is called a time series model in statistics. Recently it was pointed out (Ozaki (1983)) that continuous-time stochastic dynamical system models and non-linear time series models are closely related by a time discretization scheme for deterministic dynamical systems, called local linearization. The purpose of the present paper is to show that it is possible to extend the idea of Ozaki (1983) to the multidimensional situation using an example of non-linear random vibrations in nautical engineering. For this purpose, in Section 2 stochastic dynamical systems of non-linear random vibrations and their relation to diffusion processes are discussed. In Section 3 the local Gaussian modelling method of multivariate stochastic dynamical system is given, and a non-linear Markovian representation model is introduced by the modelling method. Some computational problems are discussed in Section 4.

2. Stochastic dynamical systems of non-linear random vibrations

One of the best-known and most interesting non-Gaussian processes is the dynamic behaviour of a vessel moored at sea. In this case, the vessel is excited by the random force exerted by the sea waves and, because it is restricted by the mooring rope, its motion is asymmetric in type. One particular problem, in practice, is that when the sea becomes rough, it is possible for the vessel to capsize or for the mooring rope to break. In order to prevent the occurrence of this kind of disaster, a thorough knowledge of the dynamic and stochastic character of the mooring system is required, so that the system may be satisfactorily controlled. The dynamics of the vessel's motion, represented in the form of its rolling angle, have been studied in nautical engineering, where the motion is usually approximated by the stochastic differential equation

$$(2.1) \qquad \ddot{x} + a(\dot{x}) + b(x) = n(t) - T(x)$$

where $x(t)$ is the rolling angle of the ship, $n(t)$ is the random force exerted by the waves, $a(\dot{x})$ and $b(x)$ are the damping and restoring forces of the ship, respectively, and $T(x)$ is the force on the mooring rope which is non-zero only when x exceeds some angle x_0 in one direction.

The stochastic process $\{x(t)\}$ defined by (2.1) is asymmetric and non-Gaussian even when $n(t)$ is assumed to be a Gaussian white noise process and $a(\dot{x})$ and $b(x)$ are approximated by the linear functions

$$(2.2) \qquad a(\dot{x}) \doteqdot ax$$

$$(2.3) \qquad b(x) \doteqdot bx.$$

This is obvious if the asymmetric effect of $T(x)$ is considered. It is usual to approximate $T(x)$ by

$$(2.4) \qquad T(x) \doteqdot c(x - x_0)_+^3$$

where

$$c(x - x_0)^3_+ = \begin{cases} c(x - x_0)^3 & \text{for} \quad x \geqq x_0 \\ 0 & \text{for} \quad x < x_0. \end{cases}$$

In this case the total restoring force of the ship including the effect of the rope is $bx + c(x - x_0)^3_+$, and we obtain the second-order stochastic differential equation

$$(2.5) \qquad \ddot{x} + a\dot{x} + bx + c(x - x_0)^3_+ = n(t).$$

This equation is equivalent to the following bivariate stochastic dynamical system:

$$(2.6) \qquad \dot{\boldsymbol{v}} = \boldsymbol{f}(\boldsymbol{v}) + \boldsymbol{n}(t)$$

where

$$\boldsymbol{v} = (v_1, v_2)' = (\dot{x}, x)',$$
$$\boldsymbol{f}(\boldsymbol{v}) = (f_1(\boldsymbol{v}), f_2(\boldsymbol{v}))',$$
$$f_1(\boldsymbol{v}) = -a\dot{x} - bx - c(x - x_0)^3_+,$$
$$f_2(\boldsymbol{v}) = \dot{x},$$

and
$$\boldsymbol{n}(t) = (n(t), 0)'.$$

When we assume that $n(t)$ is a Gaussian white noise process with variance σ^2, (2.6) defines a bivariate diffusion process whose Fokker–Planck equation is

$$(2.7) \qquad \frac{\partial p}{\partial t} = \sum_{i=1}^{2} \frac{\partial}{\partial v_i} [f_i(v)p] + \tfrac{1}{2} \sum_{i=1}^{2} \sum_{j=1}^{2} \frac{\partial^2}{\partial v_i \, \partial v_j} [b_{ij}p]$$

where p is the transition probability of the diffusion process and $\boldsymbol{f}(\boldsymbol{v}) = (f_1(\boldsymbol{v}), f_2(\boldsymbol{v}))'$ and $\{b_{i,j}\}$ are given by

$$\boldsymbol{f}(\boldsymbol{v}) = \begin{pmatrix} -a\dot{x} - bx - c(x - x_0)^3_+ \\ \dot{x} \end{pmatrix}$$

$$B = (b_{ij}) = \begin{pmatrix} \sigma^2 & 0 \\ 0 & 0 \end{pmatrix}.$$

If we assume $\partial p / \partial t = 0$ in Equation (2.7) we obtain the equilbrium distribution $p(\boldsymbol{v} \mid \infty)$ of this bivariate diffusion process (Caughey (1963)) which is given by

$$(2.8) \qquad p(\boldsymbol{v} \mid \infty) = p(\dot{x}, x) = c \exp\left(-\frac{a\dot{x}^2}{\sigma^2}\right) \exp\left\{-\frac{bx^2}{\sigma^2} - \frac{c(x - x_0)^4_+}{2\sigma^2}\right\}.$$

Hence, it may be seen from (2.8) that the distribution of $\dot{x}$ is Gaussian, although the distribution of x is not Gaussian because of the presence of the term $c(x - x_0)^3_+$ which makes the tail of the distribution $p(x)$ die out more quickly, for $x > x_0$, than the Gaussian distribution.

3. Local Gaussian modelling

As we saw in the previous section, multivariate stochastic dynamical systems and multivariate diffusion processes are useful in the theoretical analysis of certain dynamic and stochastic phenomena. However, they are not so useful in practical applications, where we are concerned with the inference problem for such phenomena, based on a finite set of observation data on the process. In such situations, prediction and control have been accomplished by using a multivariate time series model, which may be identified from a finite set of multivariate time series data (Otomo et al. (1972), Ohtsu et al. (1979)). The multivariate time series models used in these applications were linear Markovian models or autoregressive moving average (ARMA) models (Hannan (1969), Akaike (1974), Hannan (1981)) which are essentially suitable only for Gaussian or quasi-Gaussian time series. Such models are not really appropriate for the statistical modelling of such non-Gaussian time series data as may arise from the dynamics of the moored vessel discussed in the previous section.

A method for extending the time series modelling approach to the case of non-Gaussian processes is discussed, for the univariate case, in Ozaki (1983). This method, the local linearization method, is, as we shall see in this section, also applicable to the multivariate case where the phenomenon is expressible by a multivariate stochastic dynamical system or by a multivariate diffusion equation.

Suppose we have a bivariate stochastic dynamical system,

$$(3.1) \qquad \dot{v} = f(v) + n(t)$$

where $f(v) = (f_1(v), f_2(v))'$ and $n(t) = (n_1(t), n_2(t))'$ is a bivariate Gaussian white noise with variance–covariance matrix

$$\Sigma = \begin{pmatrix} \sigma_1^2 & 0 \\ 0 & \sigma_2^2 \end{pmatrix}.$$

Using the local linearization method, the discrete-time version of the deterministic dynamical system

$$(3.2) \qquad \dot{v} = f(v)$$

may be obtained, where it is assumed that the Jacobian

$$J(v) = \left(\frac{\partial f_i(v)}{\partial v_j} \right)$$

is constant for a sufficiently small time interval Δt. In this way, we obtain the expression (Ozaki (1983))

$$(3.3) \qquad v_{t+\Delta t} = v_t + J_t^{-1}\{\mathrm{Exp}\,(J_t \Delta t) - I\}f(v_t)$$

which is obtained by integrating, with respect to τ, over $[0, \Delta t)$

$$\dot{\boldsymbol{v}}(t+\tau) = \mathrm{Exp}\,(J_t\tau)\dot{\boldsymbol{v}}(t)$$

which, in turn, is obtained by integrating, over $[t, t+\tau)$,

$$\ddot{\boldsymbol{v}}(t) = J_t\dot{\boldsymbol{v}}(t)$$

where

$$\boldsymbol{v}_{t+\Delta t} = \boldsymbol{v}(t+\Delta t), \qquad J_t = J(v_t)$$

and

$$\mathrm{Exp}\,(J_t\Delta t) = I + \Delta t J_t + \frac{\Delta t^2}{2!}\,J_t^2 + \frac{\Delta t^3}{3!}\,J_t^3 + \cdots .$$

(3.3) can be expressed in the following non-linear Markovian form:

$$\boldsymbol{v}_{t+\Delta t} = A_t\boldsymbol{v}_t$$

where

(3.4)
$$A_t = I + J_t^{-1}\{\mathrm{Exp}\,(\Delta t J_t) - I\}F_t$$
$$F_t \cdot \boldsymbol{v}_t = \boldsymbol{f}(\boldsymbol{v}_t).$$

For the special case where the dynamical system (3.2) is linear, i.e.,

(3.5)
$$\dot{\boldsymbol{v}} = A\boldsymbol{v}$$

where A is some non-singular constant matrix, the solution is

(3.6)
$$\boldsymbol{v}(t) = \mathrm{Exp}\,(tA)\boldsymbol{v}_0$$

where

$$\mathrm{Exp}\,(tA) = I + tA + \frac{t^2}{2!}\,A^2 + \frac{t^3}{3!}\,A^3 + \cdots$$

and $\boldsymbol{v}_0$ is the initial value. The corresponding discrete-time dynamical system (3.4) is given by

(3.7)
$$\boldsymbol{v}_k = \mathrm{Exp}\,(\Delta t A)\boldsymbol{v}_{k-1}$$
$$= \mathrm{Exp}\,(k\Delta t A)\boldsymbol{v}_0$$

which coincides with the continuous-time solution $\boldsymbol{v}(t)$ at each time point $k\Delta t$ $(k = 0, 1, 2, \cdots)$, where

$$\boldsymbol{v}_k = \boldsymbol{v}(k\Delta t).$$

It is easy to check that, in general, the discrete-time dynamical system (3.4) is consistent, i.e. it converges to the original dynamical system for $\Delta t \rightarrow 0$.

A natural idea of extending the local linearization of the dynamical system to the stochastic situation is to consider the process to be Gaussian on each time interval $[t, t + \Delta t)$ as

$$(3.8) \qquad\qquad \dot{\boldsymbol{v}} = K_t \boldsymbol{v} + \boldsymbol{n}(t)$$

where K_t satisfies the equation

$$\mathrm{Exp}\,(K_t \Delta t) = A_t$$

where A_t is given by (3.4). This idea is reasonable because any Markov diffusion process is locally Gaussian. Then a discrete-time dynamical system model for the Gaussian process $\boldsymbol{v}(t)$ of (3.8) is given as follows:

$$(3.9) \qquad\qquad \boldsymbol{v}_{t+\Delta t} = \mathrm{Exp}\,(K_t \Delta t)\boldsymbol{v}_t + \boldsymbol{n}_{t+\Delta t}$$

where

$$K_t = \frac{1}{\Delta t}\,\mathrm{Log}\,A_t$$

$$(3.10) \qquad\qquad A_t = I + J_t^{-1}\{\exp\,(J_t \Delta t) - I\}F_t$$

$$\mathrm{Log}\,A_t = \sum_{k=1}^{\infty} \frac{(-1)^k}{k!}(A_t - I)^k$$

and

$$(3.11) \qquad\qquad \boldsymbol{n}_{t+\Delta t} = \int_t^{t+\Delta t} \mathrm{Exp}\,\{K_t(t + \Delta t - \tau)\}\boldsymbol{n}(\tau)d\tau.$$

$\mathrm{Exp}\,(J_t \Delta t)$ is given as follows using the eigenvalues μ_1 and μ_2 of J_t:

$$\mathrm{Exp}\,(J_t \Delta t) = \frac{1}{(\mu_1 - \mu_2)^2}$$
$$\times \begin{pmatrix} \mu_1 \exp\,(\mu_1 \Delta t) - \mu_2 \exp\,(\mu_2 \Delta t) & \mu_1 \mu_2(\exp\,(\mu_1 \Delta t) - \exp\,(\mu_2 \Delta t)) \\ \exp\,(\mu_1 \Delta t) - \exp\,(\mu_2 \Delta t) & \mu_1 \exp\,(\mu_2 \Delta t) - \mu_2 \exp\,(\mu_1 \Delta t) \end{pmatrix}.$$

We note that

$$\mathrm{Exp}\,(K_t \Delta t) = \mathrm{Exp}\,(\mathrm{Log}\,A_t)$$
$$= A_t$$

when A_t satisfies the condition

$$(3.12) \qquad\qquad \|A_t - I\| < 1,$$

which is satisfied by A_t of (3.10) when Δt is sufficiently small. The norm $\|A\|$ of an $n \times n$ matrix A is defined by

$$\|A\| = \sum_{i,j=1}^{n} a_{ij}^2$$

where $a_{i,j}$ is the (i, j)th element of A. The variance–covariance matrix Σ_0 of $\boldsymbol{n}_{t+\Delta t}$ of (3.9) is given by

$$\Sigma_0 = E\left[\int_t^{t+\Delta t} \mathrm{Exp}\,\{K_t(t-\tau)\boldsymbol{n}(\tau)\boldsymbol{n}(\tau')\,\mathrm{Exp}\,\{K_t(t-\tau)\}'d\tau\right]$$

(3.13)
$$= V_t\begin{pmatrix} (\exp\,(2\mu_1\Delta t)-1)s_{11}/2\mu_1 & \\ \{\exp\,((\mu_1+\mu_2)\Delta t)-1\}s_{12}/(\mu_1+\mu_2) & \end{pmatrix}$$
$$\begin{pmatrix} & \{\exp\,((\mu_1+\mu_2)\Delta t)-1\}s_{12}/(\mu_1+\mu_2) \\ & (\exp\,(2\mu_2\Delta t)-1)s_{22}/2\mu_2 \end{pmatrix}V_t'$$

where

$$\Sigma = \begin{pmatrix} \sigma_1^2 & 0 \\ 0 & \sigma_2^2 \end{pmatrix}$$

$$s_{11} = (\sigma_1^2 + \sigma_2^2\mu_2^2)/(\mu_1-\mu_2)^2$$

$$s_{22} = (\sigma_1^2 + \sigma_2^2\mu_1^2)/(\mu_1-\mu_2)^2$$

$$s_{12} = (\sigma_1^2 + \sigma_2^2\mu_1\mu_2)/(\mu_1-\mu_2)^2$$

$$= s_{21}$$

and V_t is non-singular matrix

$$V_t = \begin{pmatrix} \mu_1 & \mu_2 \\ 1 & 1 \end{pmatrix}$$

which transforms the matrix K_t into a diagonal matrix thus:

$$V_t^{-1}K_tV_t = \begin{pmatrix} \mu_1 & 0 \\ 0 & \mu_2 \end{pmatrix}.$$

It is obvious from the above definition that the autocovariance matrix function of the discrete model (3.9) coincides with the autocovariance matrix function of the continuous model (3.8) at each discrete time value $t = 0, \Delta t, 2\Delta t, 3\Delta t, \cdots$.

If we approximate the exponential functions in (3.13) by the first-order Taylor expansions, we have the following discrete-time model:

(3.14)
$$\boldsymbol{v}_{t+\Delta t} = A_t\boldsymbol{v}_t + \boldsymbol{n}_{t+\Delta t}$$

where

$$\Sigma_1 = E[\boldsymbol{n}_{t+\Delta t} \cdot \boldsymbol{n}'_{t+\Delta t}]$$

$$= \Delta t\Sigma.$$

The model (3.14) is known to have a sample path $\boldsymbol{v}_t, \boldsymbol{v}_{t+\Delta t}, \boldsymbol{v}_{t+2\Delta t}, \cdots$ whose distribution converges to the distribution of $\boldsymbol{v}(t)$ of the continuous Markov diffusion process (3.1) on a finite interval $[0, T]$ for $\Delta t \to 0$ (Gikhman and Skorohod (1965), Maruyama (1983)).

4. Discussion

The local Gaussian modelling method in the previous section is valid not only for the bivariate case but also for any finite-dimensional case. However the approximation of Σ_0, up to order Δt, described at the end of the last section, is not appropriate for the stochastic dynamical system of random vibrations where the variance–covariance matrix of the continuous-time white noise is not full rank but is given by

$$\Sigma = \begin{pmatrix} \sigma_1^2 & 0 \\ 0 & 0 \end{pmatrix}.$$

This is because the noise variance of the univariate process $x(t)$ of the vibration process is of order Δt^3 although the noise variance of the process $\dot{x}(t)$ is of order Δt. To see this, let us consider the linear random vibration, defined by the second stochastic differential equation

$$(4.1) \qquad \ddot{x} + a\dot{x} + bx = n(t),$$

for which time series models have been considered by many statisticians (Yule (1927), Kendall (1944), Bartlett (1946), Durbin (1961), Hannan (1970) and Pandit and Wu (1975)). By regarding the random vibration (4.1) as a stochastic dynamical system, we have, from (4.1) the following representation,

$$(4.2) \qquad \dot{v} = J \cdot v + n(t)$$

where

$$v = (\dot{x}, x)',$$

$$J = \begin{pmatrix} -a & -b \\ 1 & 0 \end{pmatrix},$$

and

$$n(t) = (n(t), 0)'.$$

From (4.2) we have the following discrete-time model:

$$(4.3) \qquad v_{t+\Delta t} = \mathrm{Exp}\,(J \cdot \Delta t)v_t + n_{t+\Delta t}$$

where

$$n_{t+\Delta t} = \int_t^{t+\Delta t} \mathrm{Exp}\,\{J(t+\Delta t - \tau)\}n(\tau)d\tau.$$

The variance–covariance matrix Σ_0 of $\boldsymbol{n}_t$ in (4.3) is given by

$$(4.4) \qquad \begin{aligned} \Sigma_0 &= E[\boldsymbol{n}_t \cdot \boldsymbol{n}'_t] \\ &= \sigma_1^2 \Sigma_1 \end{aligned}$$

where the elements σ_{11}, σ_{12}, σ_{21} and σ_{22} of Σ_1 are given by

$$\sigma_{11} = \frac{1}{(\mu_1 - \mu_2)^2} \left[\frac{\mu_1}{2} (\exp(2\mu_1 \Delta t) - 1) \right.$$
$$\left. - \frac{2\mu_1 \mu_2}{(\mu_1 + \mu_2)} \{\exp((\mu_1 + \mu_2)\Delta t) - 1\} + \frac{\mu_2}{2} (\exp(2\mu_2 \Delta t) - 1) \right]$$

$$\sigma_{22} = \frac{1}{(\mu_1 - \mu_2)^2} \left[\frac{1}{2\mu_1} (\exp(2\mu_1 \Delta t) - 1) \right.$$
$$(4.5) \qquad \left. - \frac{2}{\mu_1 + \mu_2} \{\exp((\mu_1 + \mu_2)\Delta t) - 1\} + \frac{1}{2\mu_2} (\exp(2\mu_2 \Delta t) - 1) \right]$$

$$\sigma_{12} = \frac{1}{(\mu_1 - \mu_2)^2} \left[\tfrac{1}{2}(\exp(2\mu_1 \Delta t) - 1) \right.$$
$$\left. - \{\exp((\mu_1 + \mu_2)\Delta t) - 1\} + \tfrac{1}{2}(\exp(2\mu_2 \Delta t) - 1) \right]$$

$$= \sigma_{21}.$$

μ_1 and μ_2 are eigenvalues of J. We note that, although the variance–covariance matrix of continuous-time white noise $\boldsymbol{n}(t)$ has rank 1, the variance–covariance matrix of discrete-time white noise $\boldsymbol{n}_t$ has full rank 2 as the result of the integration over $[t, t + \Delta t]$. Σ_1 is decomposed as follows:

$$(4.6) \qquad \Sigma_1 = U \begin{pmatrix} \lambda_1 & 0 \\ 0 & \lambda_2 \end{pmatrix} U'$$

using a unitary matrix U whose elements u_{11}, u_{12}, u_{21} and u_{22} are obtained by

$$u_{11} = \sigma_{12}/\sqrt{\{\sigma_{12}^2 + (\lambda_1 - \sigma_{11})^2\}}$$
$$(4.7) \qquad u_{22} = \sigma_{12}/\sqrt{\{\sigma_{12}^2 + (\lambda_2 - \sigma_{22})^2\}}$$
$$u_{12} = (\lambda_1 - \sigma_{11})/\sqrt{\{\sigma_{12}^2 + (\lambda_1 - \sigma_1)^2\}}$$
$$u_{21} = -u_{12}.$$

Eigenvalues λ_1 and λ_2 are given as the roots of

$$\lambda^2 - (\sigma_{11} + \sigma_{22})\lambda + \sigma_{11}\sigma_{22} - \sigma_{12}^2 = 0.$$

From (4.3), (4.4) and (4.6) we have the following Markovian state space representation for the process x_t:

$$(4.8) \qquad \begin{aligned} \boldsymbol{v}_{t+\Delta t} &= A\boldsymbol{v}_t + B\boldsymbol{z}_{t+\Delta t} \\ x_t &= C\boldsymbol{v}_t \end{aligned}$$

where

$$A = \mathrm{Exp}\,(J\Delta t),$$

$$B = U\begin{pmatrix} \sqrt{\lambda_1} & 0 \\ 0 & \sqrt{\lambda_2} \end{pmatrix}$$

$$C = (0, 1)$$

and z_t is a discrete-time bivariate Gaussian white noise of variance–covariance $\sigma_1^2 I$. Model (4.8) is known (Akaike (1974)) to be equivalent to the following ARMA model:

$$(4.9) \qquad x_t - \phi_1 x_{t-\Delta t} - \phi_2 x_{t-2\Delta t} = C \cdot Bz_t + C(A - \phi_1 I)Bz_{t-\Delta t}$$

where ϕ_1 and ϕ_2 are found from the characteristic polynomial of A, i.e.

$$A^2 - \phi_1 A - \phi_2 I = 0.$$

Model (4.9) is rewritten, using the elements a_{ij} of A and u_{ij} of U, in the following way:

$$(4.10) \qquad x_t - \phi_1 x_{t-\Delta t} - \phi_2 x_{t-2\Delta t} = u_{21}\sqrt{\lambda_1}\varepsilon_t^{(1)} + u_{22}\sqrt{\lambda_2}\varepsilon_t^{(2)}$$
$$+ \pi_1\sqrt{\lambda_1}\varepsilon_{t-\Delta t}^{(1)} + \pi_2\sqrt{\lambda_2}\varepsilon_{t-\Delta t}^{(2)}$$

where

$$(4.11) \qquad \pi_1 = \{a_{11}u_{11} + (a_{22} - \phi_1)u_{21}\}/u_{21},$$

$$(4.12) \qquad \pi_2 = \{a_{21}u_{12} + (a_{22} - \phi_1)u_{22}\}/u_{22},$$

and both $\varepsilon_t^{(1)}$ and $\varepsilon_t^{(2)}$ are Gaussian white noise processes of variance σ_1^2. If we consider that $\varepsilon_t^{(1)}$ and $\varepsilon_t^{(2)}$ are mutually independent, we can see that (4.10) is equivalent to the following ARMA model:

$$(4.13) \qquad x_t - \phi_1 x_{t-\Delta t} - \phi_2 x_{t-2\Delta t} = e_t - \theta e_{t-\Delta t},$$

where

$$(4.14) \qquad \sigma_e^2 = E[e_t^2] = (u_{21}^2\lambda_1 + u_{22}^2\lambda_2)\sigma_1^2$$

$$(4.15) \qquad \theta = -(\pi_1 u_{21}^2\lambda_1 + \pi_2 u_{22}^2\lambda_2)/(u_{21}^2\lambda_1 + u_{22}^2\lambda_2).$$

We note that tedious calculations show that θ of (4.15) and σ_e^2 of (4.14) are equivalent to the expressions (4.16) and (4.17) given by Pandit and Wu (1975). (There is a misprint in the expression for θ in Pandit and Wu (1975).)

$$(4.16) \qquad \begin{aligned} \sigma_e^2 = &\frac{\sigma_1^2}{2\mu_1\mu_2(\mu_1^2 - \mu_2^2)(\mu_1 - \mu_2)} \\ &\times \{(\mu_2 \exp\,(\mu_1\Delta t) - \mu_1 \exp\,(\mu_2\Delta t))^2 \\ &+ \mu_1\mu_2(\exp\,(\mu_1\Delta t) - \exp\,(\mu_2\Delta t))^2 - (\mu_1 - \mu_2)^2\} \end{aligned}$$

TABLE 4.1.

The values of parameters σ_e^2, θ, ϕ_1, ϕ_2, λ_1 and λ_2 of the discrete model (4.13) for
$$\ddot{x} + 0.25\dot{x} + 14.8x = n(t)$$

Δt	1.0	1.0×10^{-1}	1.0×10^{-2}	1.0×10^{-3}	1.0×10^{-4}
σ_e^2	0.264×10^{-1}	0.318×10^{-3}	0.333×10^{-6}	0.333×10^{-9}	0.333×10^{-12}
θ	-0.593	-0.504	-0.500	-0.500	-0.500
ϕ_1	-0.135×10^1	0.183×10^1	0.200×10^1	0.200×10^1	0.200×10^1
ϕ_2	-0.779	-0.975	-0.998	-0.100×10^1	-0.100×10^1
λ_1	0.491	0.931×10^{-1}	0.997×10^{-2}	0.100×10^{-2}	0.100×10^{-3}
λ_2	0.262×10^{-1}	0.856×10^{-4}	0.834×10^{-7}	0.833×10^{-10}	0.833×10^{-13}

$$(4.17) \qquad \theta = \frac{(\mu_1 - \mu_2)\{\mu_2 \exp(\mu_2 \Delta t) - \mu_1 \exp(\mu_1 \Delta t) - \exp((\mu_1 + \mu_2)\Delta t)(\mu_2 \exp(\mu_1 \Delta t) - \mu_1 \exp(\mu_2 \Delta t))\}}{\{(\mu_2 \exp(\mu_1 \Delta t) - \mu_1 \exp(\mu_2 \Delta t))^2 + \mu_1 \mu_2 (\exp(\mu_1 \Delta t) - \exp(\mu_2 \Delta t))^2 - (\mu_1 - \mu_2)^2\}}.$$

Table 4.1 shows the values of σ_e^2, θ, ϕ_1, ϕ_2, λ_1, λ_2 for several values of Δt where we used the following linear random vibrations:

$$\ddot{x} + 0.25\dot{x} + 14.8x = n(t).$$

We see from the table that the noise variance σ_e^2 of the ARMA model (4.11) is of order Δt^3, θ, ϕ_1 and ϕ_2 are of order 1, λ_1 is of order Δt and λ_2 is order Δt^3. On the other hand, if we employ the approximation of order Δt for Σ_0 of (4.4) we have a model with a white noise of variance of order Δt. This means that, for the stochastic dynamical system of non-linear random vibrations of a moored vessel, the following model is more appropriate:

$$(4.18) \qquad v_{t+\Delta t} = \mathrm{Exp}(K_t \Delta t)v_t + B_t \cdot z_{t+\Delta t}$$

$$x_t = Cv_t$$

$$T_t = c(x_t - x_0)_+^3$$

where

$$K_t = \frac{1}{\Delta t} \mathrm{Log}\, A_t$$

$$A_t = I + J_t^{-1}\{\mathrm{Exp}(J_t \Delta t) - I\}F_t$$

$$B_t = U \begin{pmatrix} \sqrt{\lambda_1} & 0 \\ 0 & \sqrt{\lambda_2} \end{pmatrix}, \qquad C = (0, 1)$$

$$J_t = \begin{pmatrix} -a & -b - 3c(x_t - x_0)_+^2 \\ 1 & 0 \end{pmatrix} \quad \text{and} \quad F_t = \begin{pmatrix} -a & -b - c(x_t - x_0)_+^3/x_t \\ 1 & 0 \end{pmatrix}.$$

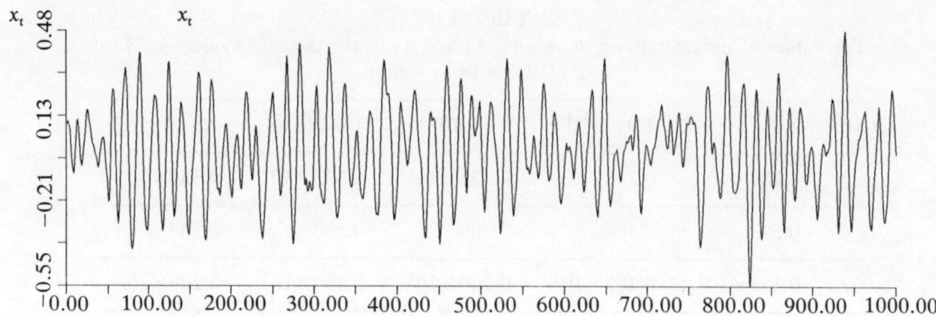

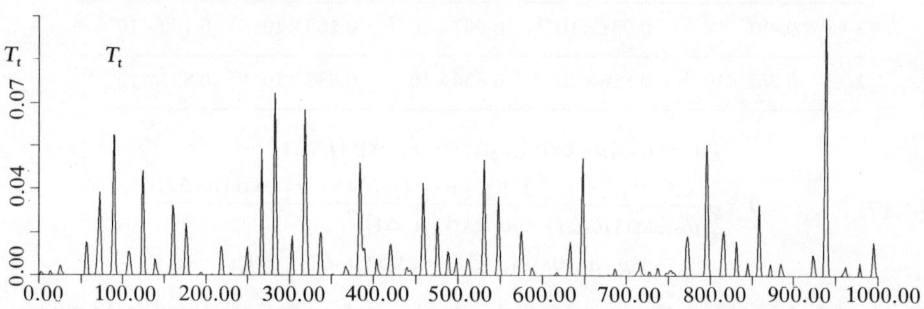

Figure 4.1. Simulations of x_t and T_t by the model (4.18) for $\ddot{x} + 0.25\dot{x} + 14.8x = n(t) - x_+^3$.

z_t is a Gaussian white noise of variance–covariance matrix $\sigma_1^2 I$, and U is given by (4.7), where each element of U is given as a function of eigenvalues μ_1 and μ_2 of K_t. Simulation of x_t and T_t obtained by using model (4.18) is shown in Figure 4.1.

In the non-linear random vibration case, K_t is not always equal to $K_{t+\Delta t}$. Therefore in the non-linear case, we cannot have an ARMA representation of the process x_t such as (4.9). However, when Δt is sufficiently small, the difference of the eigenvalues of K_t and those of $K_{t+\Delta t}$ will also be small. Figure 4.2 shows the figures of $\phi_1(x_t)$ and $\phi_2(x_t)$, which are

(4.19) $$\phi_1(x_t) = \exp\{\mu_1(x_t)\Delta t\} + \exp\{\mu_2(x_t)\Delta t\}$$

(4.20) $$\phi_2(x_t) = -\exp[\{\mu_1(x_t) + \mu_2(x_t)\}\Delta t]$$

for the non-linear random vibrations of a moored vessel, where $\mu_1(x_t)$ and $\mu_2(x_t)$ are eigenvalues of

$$K_t = \frac{1}{\Delta t}\,\mathrm{Log}\,A_t$$

$$= \frac{1}{\Delta t}\,\mathrm{Log}\,[I + J_t^{-1}\{\mathrm{Exp}\,(J_t\Delta t) - I\}F_t]$$

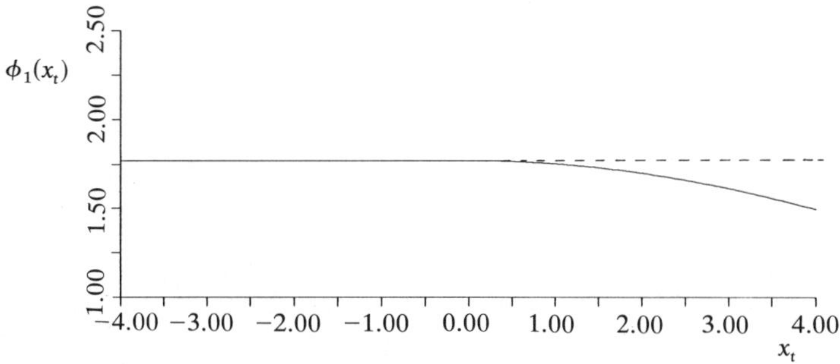

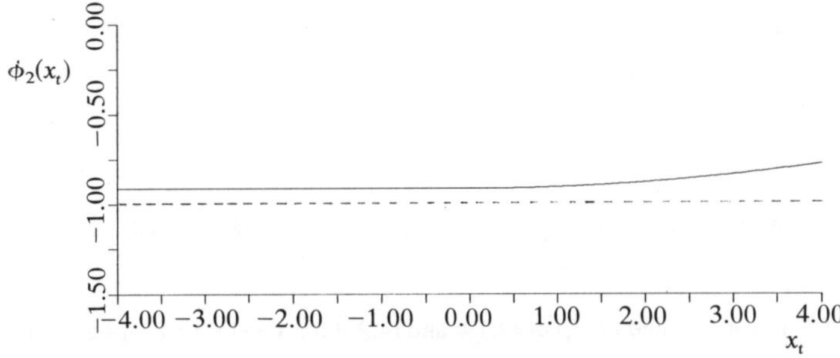

Figure 4.2. $\phi_1(x_t)$ and $\phi_2(x_t)$ of (4.19) and (4.20) for $\ddot{x} + 0.25\dot{x} + 14.8x + x_+^3 = n(t)$

and are functions of x_t. Figure 4.3 shows the figures of $\phi_1(x_t)$ and $\phi_2(x_t)$ of non-linear random vibrations of the van der Pol type specified by

$$\ddot{x} - a(1 - x^2)\dot{x} + bx = n(t)$$

where J_t and F_t are given as follows:

$$J_t = \begin{pmatrix} a(1 - x_t^2) & -b - 2a\dot{x}_t x_t \\ 1 & 0 \end{pmatrix}$$

$$F_t = \begin{pmatrix} a(1 - x_t^2) & -b \\ 1 & 0 \end{pmatrix}.$$

The figures suggest that when we use one of the conventional amplitude-dependent non-linear AR models (Haggan and Ozaki (1981), Ozaki (1982), Tong and Lim (1980)) for the modelling of non-linear random vibrations, we should choose the one whose AR coefficients are at least continuous functions of x_t.

Model estimations and applications of the models presented in the paper to real ship-rolling data will be discussed in a forthcoming paper.

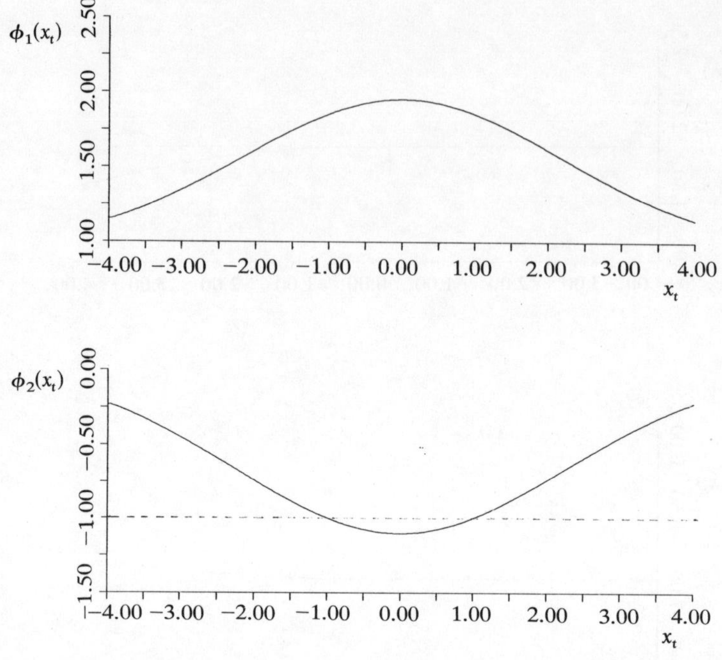

Figure 4.3. $\phi_1(x_t)$ and $\phi_2(x_t)$ of (4.19) and (4.20) for $\ddot{x} - (1 - x^2)\dot{x} + 14.8x = n$

Acknowledgement

The author is grateful to Professor Yamanouchi and Mr Oda for directing his interest to non-linear systems in nautical engineering. Thanks are also due to Professor Hannan, Professor Akaike and Dr Haggan for their valuable discussions. Part of the work was done while the author was visiting the Department of Statistics, Institute of Advanced Studies, of The Australian National University.

References

AKAIKE, H. (1974) Markovian representation of stochastic processes and its application to the analysis of autoregressive moving average processes. *Ann. Inst. Statist. Math.* **26**, 363–387.

BARTLETT, M. S. (1946) On the theoretical specification and sampling properties of autocorrelated time series. *J. R. Statist. Soc.* B**8**, 27–41.

CAUGHEY, T. K. (1963) Derivation and application of the Fokker–Planck equation to discrete nonlinear dynamic systems subjected to white random excitation. *J. Acoust. Soc. Amer.* **35**, 1683–1692.

DURBIN, J. (1961) Efficient fitting of linear models for continuous stationary time series from discrete data. *Bull. Inst. Internat. Statist.* **38**, 273–282.

GIKHMAN, I. I. AND SKOROHOD, A. V. (1965) *Introduction to the Theory of Random Processes* (translated by Scripta Technica, Inc.). W. B. Saunders Company, Philadelphia.

HAGGAN, V. AND OZAKI, T. (1981) Modelling nonlinear random vibrations using an amplitude-dependent autoregressive time series model. *Biometrika* **68**, 189–196.

HANNAN, E. J. (1969) The identification of vector-mixed autoregressive-moving average systems. *Biometrika* **57**, 223–225.

HANNAN, E. J. (1970) *Multiple Time Series.* Wiley, New York.

HANNAN, E. J. (1981) System identification. In *Stochastic Systems: The Mathematics of Filtering and Identification and Applications.* Proceeding of NATO Advanced Study Institute. D. Reidel, Dordrecht, 221–246.

KENDALL, M. G. (1944) On autoregressive time series. *Biometrika* **33**, 105–122.

MARUYAMA, T. (1983) Stochastic theory of population genetics. *Bull. Math. Biol.* **45**, 521–554.

OHTSU, K., HORIGOME, M. AND KITAGAWA, G. (1979) A new ship's auto pilot design through a stochastic model. *Automatica* **15**, 255–268.

OTOMO, T., NAKAGAWA, T. AND AKAIKE, H. (1972) Statistical approach to computer control of cement rotary kilns. *Automatica* **8**, 35–48.

OZAKI, T. (1982) Non-linear threshold autoregressive models for non-linear random vibrations. *J. Appl. Prob.* **18**, 443–451.

OZAKI, T. (1983) Non-linear time series models and dynamical systems. In *Handbook of Statistics* **5**, ed. E. J. Hannan et al., North-Holland, Amsterdam.

PANDIT, S. M. AND WU, S. M. (1975) Unique estimates of the parameters of a continuous stationary stochastic process. *Biometrika* **62**, 497–501.

TONG, H. AND LIM, K. S. (1980) Threshold autoregression, limit cycles and cyclical data. *J. R. Statist. Soc.* **B42**, 245–292.

YULE, G. U. (1927) On a method of investing periodicities in disturbed series, with special reference to Wolfer's sunspot numbers. *Phil. Trans. R. Soc. London* **A226**, 267–298.

Identification of Non-Linear Systems Using General State-Dependent Models

M. B. PRIESTLEY

S. M. HERAVI

Abstract

This paper describes an extension of the SDM scheme introduced by Priestley (1980) to the problem of the identification of non-linear systems using only input/output records. It is shown that, under quite general assumptions on the structure of the system, it is possible to identify a non-linear relationship between the input and output processes via a modified version of the Kalman-type algorithm previously applied to the study of non-linear time series models. The paper includes a numerical study of data generated from various types of non-linear systems.

NON-LINEAR MODELS; KALMAN FILTERING; NON-PARAMETRIC FUNCTION FITTING

1. Introduction

Among the many outstanding contributions which Ted Hannan has made to time series analysis, his work on the fundamental structure and identification of linear systems is among the most notable. (See, e.g. Hannan (1973), (1976), (1979), and more recently, Hannan and Kavalieris (1984).) This work has provided a deep study of both structural and inferential problems relating to linear systems. By contrast, the study of non-linear models is still in a very early stage of development, and previous work in this area has been largely empirical in nature, dealing mainly with the fitting of special types of non-linear models to various types of data. The present paper is written in the same spirit, but describes a fairly general and flexible class of non-linear schemes called *state-dependent models* (SDM). These form a natural extension of the class of state-dependent models previously developed for describing time series data, and by way of background information we describe briefly in the next section the essential ideas underlying the construction of the SDM time series models. (A more extensive discussion of these models is given in Priestley (1980) and Haggan et al. (1984).)

2. sdm schemes for time series data

Given a discrete-parameter time series $\{X_t; t = 0, \pm1, \pm2, \cdots\}$, a general non-linear 'autoregressive' model of order k takes the form

$$(2.1) \qquad X_t = h(X_{t-1}, X_{t-2}, \cdots, X_{t-k}) + \varepsilon_t,$$

where h is some arbitrary function and $\{\varepsilon_t\}$ is a sequence of independent zero-mean random variables. Assuming that h is an analytic function we may expand the right-hand side of (2.1) in a Taylor series about some fixed time point t_0. Using only a first-order expansion we obtain

$$(2.2) \qquad X_t = h(X_{t_0-1}, X_{t_0-2}, \cdots, X_{t_0-k}) + \sum_{u=1}^{k} f_u(x_{t-1})(X_{t-u} - X_{t_0-u}) + \varepsilon_t,$$

where x_t denotes the vector

$$(2.3) \qquad x_t = (X_{t-k+1}, \cdots, X_t)'$$

and the $\{f_u\}$ depend on the first partial derivatives of h. We note that at time $(t-1)$ the evolution of the process described by (2.1) is completely determined by the vector x_{t-1} (together with the future values of ε_t), and thus x_{t-1} may be interpreted as the 'state-vector' at time $(t-1)$. We can now rewrite (2.2) in the form

$$(2.4) \qquad X_t + \phi_1(x_{t-1})X_{t-1} + \cdots + \phi_k(x_{t-1})X_{t-k} = \mu(x_{t-1}) + \varepsilon_t$$

say, and we call (2.4) an '(autoregressive) state-dependent model of order k'. It has an appealing interpretation as a locally linear AR(k) model in which the evolution of the process at time $(t-1)$ is governed by a set of 'AR coefficients' $\{\phi_u\}$, and a 'local mean' μ, all of which depend on the current 'state' of the process at time $(t-1)$.

These ideas may be readily extended to an 'ARMA' type of non-linear model. Thus if we consider a relationship of the form

$$(2.5) \qquad X_t = h(X_{t-1}, X_{t-2}, \cdots, X_{t-k}, \varepsilon_{t-1}, \cdots, \varepsilon_{t-l}) + \varepsilon_t,$$

then we may construct an SDM representation of (2.5) in the form

$$(2.6) \qquad X_t + \sum_{u=1}^{k} \phi_u(x_{t-1})X_{t-u} = \mu(x_{t-1}) + \varepsilon_t + \sum_{u=1}^{l} \psi_u(x_{t-1})\varepsilon_{t-u}$$

where now the state-vector x_t is given by

$$x_t = (\varepsilon_{t-l+1}, \cdots, \varepsilon_t, X_{t-k+1}, \cdots, X_t)'.$$

A formal state-space description of (2.6) is easily derived in the form

$$x_{t+1} = \mu(x_t) + \{F(x_t)\}x_t + \varepsilon_{t+1},$$

$$X_t = Hx_t,$$

where $\boldsymbol{F}$, $\boldsymbol{H}$ are suitably defined matrices, the elements of $\boldsymbol{F}$ being functions of $\boldsymbol{x}_t$ (Priestley (1980)). Regarded in this way, the SDM model becomes a special case of the 'linear analytic' systems considered by Brockett (1976). The model (2.6) possesses a considerable degree of generality, and includes as special cases the main types of specific non-linear time series model previously studied, such as the *bilinear model* (Granger and Andersen (1978), Subba Rao (1981)), the *threshold autoregressive model* (Tong and Lim (1980)), and the *exponential autoregressive model* (Haggan and Ozaki (1981)).

3. Fitting an SDM

The problem of fitting an SDM to a given set of data lies in the determination of the 'coefficients' μ, $\psi_1, \cdots, \psi_l$, $\phi_1, \cdots, \phi_k$. However, these coefficients depend on the state vector $\boldsymbol{x}_t$, and the problem thus becomes the estimation of the functional forms of this dependency. The simplest non-trivial assumption that can be made is that these coefficients are linear functions of the components of $\boldsymbol{x}_t$, so that we may write

$$\mu(\boldsymbol{x}_t) = \mu^{(0)} + \boldsymbol{x}_t' \boldsymbol{\alpha},$$

and for each u,

$$\psi_u(\boldsymbol{x}_t) = \psi_u^{(0)} + \boldsymbol{x}_t' \boldsymbol{\beta}_u,$$

$$\phi_u(\boldsymbol{x}_t) = \phi_u^{(0)} + \boldsymbol{x}_t' \boldsymbol{\gamma}_u.$$

Although this assumption is too restrictive for many types of non-linear model (with the exception of bilinear models), it seems reasonable to assume that these coefficients may be represented *locally* as linear functions of $\boldsymbol{x}_t$, i.e. we write

$$\mu(\boldsymbol{x}_{t-1}) = \mu(\boldsymbol{x}_t) + \Delta \boldsymbol{x}_{t+1}' \boldsymbol{\alpha}^{(t+1)},$$

(3.1) $$\psi_u(\boldsymbol{x}_{t+1}) = \psi(\boldsymbol{x}_t) + \Delta \boldsymbol{x}_{t+1}' \boldsymbol{\beta}_u^{(t+1)},$$

$$\phi_u(\boldsymbol{x}_{t+1}) = \phi_u(\boldsymbol{x}_t) + \Delta \boldsymbol{x}_{t+1}' \boldsymbol{\gamma}_u^{(t+1)},$$

where $\Delta \boldsymbol{x}_{t+1} = \boldsymbol{x}_{t+1} - \boldsymbol{x}_t$, and allow the 'gradient parameters' $\boldsymbol{\alpha}^{(t)}$, $\{\boldsymbol{\beta}_u^{(t)}\}$, $\{\boldsymbol{\gamma}_u^{(t)}\}$ to wander in the form of 'random walks'. This model may be represented in matrix form by

$$\boldsymbol{B}_{t+1} = \boldsymbol{B}_t + \boldsymbol{V}_{t+1},$$

where

(3.2) $$\boldsymbol{B}_t = (\boldsymbol{\alpha}^{(t)}, \boldsymbol{\beta}_1^{(t)}, \boldsymbol{\beta}_2^{(t)}, \cdots, \boldsymbol{\beta}_l^{(t)}, \boldsymbol{\gamma}_1^{(t)}, \cdots, \boldsymbol{\gamma}_k^{(t)})'$$

and $\boldsymbol{V}_t$ is a sequence of independent matrix-valued random variables, with $\boldsymbol{V}_t = N(\boldsymbol{0}, \boldsymbol{\Sigma}_v)$. The estimation procedure then determines, for each t, those values of $\boldsymbol{B}_t$ which, roughly speaking, minimise the discrepancy between the observed value of X_{t+1} and its predicted value computed from the model. The algorithm is thus sequential in nature, and of a similar form to the procedures used in the Kalman filter (Kalman (1963)). In fact, the Kalman algorithm can be applied directly if we first rewrite the SDM scheme in a state-space form in which the state-vector is no longer x_t, but is replaced by the vector

$$\theta_t = (\mu_{t-1}, \psi_1^{(t-1)}, \cdots, \psi_l^{(t-1)}, \phi_1^{(t-1)}, \cdots, \phi_k^{(t-1)},$$
$$\alpha^{(t)'}, \beta_1^{(t)'}, \cdots, \beta_l^{(t)'}, \gamma_1^{(t)'}, \cdots, \gamma_k^{(t)'})',$$

(where, for brevity, we have written μ_{t-1} in place of $\mu(x_{t-1})$, etc.). Thus, θ_t is the vector of all the current parameters of the model. The reformulated state-space equations become (Priestley (1980))

$$(3.3) \qquad X_t = H_t^* \theta_t + \varepsilon_t,$$

$$(3.4) \qquad \theta_t = F_{t-1}^* \theta_{t-1} + W_t$$

where

$$H_t^* = (1, \hat{\varepsilon}_{t-1}, \cdots, \hat{\varepsilon}_{t-l}, -X_{t-1}, \cdots, -X_{t-k}, 0, 0, \cdots, 0)$$

(with $\hat{\varepsilon}_{t-u} = X_{t-u} - H_{t-u}^* \hat{\theta}_{t-u}$),

$$F_t^* = \left[\begin{array}{c|cc} I_{k+l+1} & \Delta x'_{t-1} & 0 \\ & 0 & \Delta x'_{t-1} \\ \hline 0 & & I_{(k+1)(k+l+1)} \end{array} \right],$$

and

$$W_t = (0, 0, \cdots, 0, v'_{1,t}, \cdots, v'_{k+l+1,t})',$$

$v_{1,t}, \cdots, v_{k+l+1,t}$ being the columns of V_t. Application of the Kalman algorithm to (3.3), (3.4) gives the recursion (Priestley (1980)),

$$(3.5) \qquad \hat{\theta}_t = F_{t-1}^* \hat{\theta}_{t-1} + K_t^* \{ X_t - H_t^* F_{t-1}^* \theta_{t-1} \},$$

where K_t^*, the Kalman gain matrix, is given by

$$K_t^* = \Phi_t (H_t^*)' \sigma_e^{-2},$$

Φ_t being the variance–covariance matrix of the one-step prediction error of θ_t, i.e.

$$\Phi_t = E[\{\theta_t - F_{t-1}^* \hat{\theta}_{t-1}\}\{\theta_t - F_{t-1}^* \hat{\theta}_{t-1}\}']$$

and σ_e^2 is the variance of the one-step prediction error of X_t, i.e. σ_e^2 is the variance of $e_t = \{X_t - H_t^* F_{t-1}^* \hat{\theta}_{t-1}\}$. Denoting the variance–covariance matrix of $(\theta_t - \hat{\theta}_t)$ by C_t, successive values of $\hat{\theta}_t$ may be computed using the standard recursive equations for the Kalman filter, namely,

$$(3.6) \qquad \boldsymbol{K}_t^* = \boldsymbol{\Phi}_t(\boldsymbol{H}_t^*)'[\boldsymbol{H}_t^*\boldsymbol{\Phi}_t(\boldsymbol{H}_t^*)' + \sigma_\varepsilon^2]^{-1},$$

$$(3.7) \qquad \boldsymbol{\Phi}_t = \boldsymbol{F}_{t-1}^*\boldsymbol{C}_{t-1}(\boldsymbol{F}_{t-1}^*)' + \boldsymbol{\Sigma}_w,$$

$$(3.8) \qquad \boldsymbol{C}_t = \boldsymbol{\Phi}_t - \boldsymbol{K}_t^*[\boldsymbol{H}_t^*\boldsymbol{\Phi}_t(\boldsymbol{H}_t^*)' + \sigma_\varepsilon^2]\boldsymbol{K}_t^{*'},$$

where

$$\boldsymbol{\Sigma}_w = \begin{pmatrix} \boldsymbol{0} & \boldsymbol{0} \\ \boldsymbol{0} & \boldsymbol{\Sigma}_v \end{pmatrix}.$$

This recursive procedure must be started at some time point t_0, say, and hence initial values are required for $\hat{\theta}_{t_0-1}$ and $\boldsymbol{C}_{t_0-1}$. These initial values may be obtained by fitting a linear ARMA model to an initial stretch of the data, and the details of this technique are given in Haggan et al. (1984). The choice of $\boldsymbol{\Sigma}_v$, the variance–covariance matrix of $\boldsymbol{V}_t$, depends on the assumed 'smoothness' of the coefficients as functions of $\boldsymbol{x}_t$. In practice, one usually sets the diagonal elements of $\boldsymbol{\Sigma}_v$ reasonably large relative to σ_ε^2, thereby giving the updating algorithm the freedom to make rapid changes in the model coefficients. The diagonal elements of $\boldsymbol{\Sigma}_v$ are set equal to $\hat{\sigma}_\varepsilon^2$ multiplied by some constant α called the *smoothing factor*, and the off-diagonal elements are set equal to 0. If the elements of $\boldsymbol{\Sigma}_v$ are set too large the estimated coefficients may become unstable and 'explode' to very large values. On the other hand, if the elements of $\boldsymbol{\Sigma}_v$ are made too small, the algorithm has very little freedom to make changes in the coefficients and it is then difficult to detect non-linearity in the data, the procedure in this case being essentially equivalent to the recursive fitting of a linear model. In practice, a useful procedure is to reduce the magnitude of the smoothing factor until the coefficients exhibit non-explosive behaviour. If the coefficients still appear to be erratic the smoothing factor may be reduced further.

At time $(t+1)$ we obtain estimates $\hat{\mu}(\boldsymbol{x}_t), \hat{\psi}_1(\boldsymbol{x}_t), \cdots, \hat{\psi}_l(\boldsymbol{x}_t), \hat{\phi}_1(\boldsymbol{x}_t), \cdots, \hat{\phi}_k(\boldsymbol{x}_t)$, and we may then construct a $(k+l)$-dimensional plot of each of these coefficients, plotted against the corresponding value of $\boldsymbol{x}_t$. Thus, as each new observation becomes available we obtain a new ordinate in this graph, and as the observations continue we gradually build up a picture of each of the 'parameter surfaces', $\hat{\mu}(\boldsymbol{x}_t), \hat{\psi}_1(\boldsymbol{x}_t), \cdots, \hat{\phi}_k(\boldsymbol{x}_t)$. In addition, the ordinates of the estimated coefficients may be smoothed by a multidimensional form of the *non-parametric function fitting* technique of Priestley and Chao (1972). Using this procedure, the resulting parameter surfaces give a clearer idea of the type of non-linearity present in the data, and thus provide indications of special types of non-linear models which may be appropriate for particular data sets. Haggan et al. (1984) report the results obtained in applying the SDM algorithm to a large variety of data sets (including both simulated and real time series), and show that, for simulated data, the algorithm enjoyed considerable success in correctly identifying the particular type of non-linear structure from which the data were generated. These authors also discuss various modifications to the SDM algorithm as described above, including the use of stationary AR(1) type models for the gradient parameters (in place of the random walk model), and the use of an updating algorithm in which, e.g. $\hat{\mu}(\mathbf{x}_t)$ is related to $\hat{\mu}(\mathbf{x}_s)$, $\mathbf{x}_s$ being the 'nearest neighbour' vector to $\mathbf{x}_t$ (with similar modifications for the other model coefficients).

4. SDM schemes for non-linear systems

It is well known that a conventional linear time series model may be interpreted as a 'dynamical systems' model in which the observed series is regarded as the output of a linear system 'driven' by an (unobservable) white-noise input, $\{\varepsilon_t\}$. Thus, if we are dealing with a physical system in which the input $\{U_t\}$ and the output $\{X_t\}$ are both observable, we may adapt a linear time series model to this situation by replacing the white-noise process $\{\varepsilon_t\}$ by the observable input process $\{U_t\}$ (see, e.g. Priestley (1981), Chapter 10). The ·same approach may be applied to the non-linear SDM schemes. Consider the single-input–single-output system shown schematically in Figure 4.1. Here, U_t denotes the input, X_t the 'uncorrupted' output, N_t is an additive noise disturbance, and Y_t is the observed output. We assume that the system is operating in 'open-loop' form, so that it is reasonable to assume that $\{N_t\}$ and $\{U_t\}$ are uncorrelated processes. The 'systems' model corresponding to (2.5) takes the form

$$
\begin{aligned}
(4.1) \quad X_t = {}& h(X_{t-1}, X_{t-2}, \cdots, X_{t-k}, U_{t-1}, U_{t-2}, \cdots, U_{t-l}) \\
& + g(X_{t-1}, X_{t-2}, \cdots, X_{t-k}, U_{t-1}, U_{t-2}, \cdots, U_{t-l}) U_t
\end{aligned}
$$

together with the observation equation

$$
(4.2) \qquad\qquad Y_t = X_t + N_t.
$$

Equation (4.1) may now be rewritten in SDM form as

$$
\begin{aligned}
(4.3) \quad & X_t + \phi_1(\mathbf{x}_{t-1}) X_{t-1} + \cdots + \phi_k(\mathbf{x}_{t-1}) X_{t-k} \\
& = \mu(\mathbf{x}_{t-1}) + \psi_0(\mathbf{x}_{t-1}) U_t + \psi_1(\mathbf{x}_{t-1}) U_{t-1} + \cdots + \psi_l(\mathbf{x}_{t-1}) U_{t-l}
\end{aligned}
$$

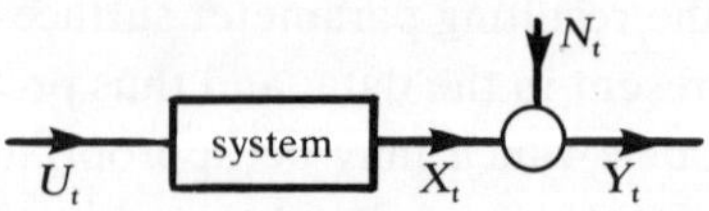

Figure 4.1

where the state vector at time $(t-1)$ is here given by

$$(4.4) \qquad \boldsymbol{x}_{t-1} = (U_{t-l}, \cdots, U_{t-1}, X_{t-k}, \cdots, X_{t-1})'.$$

(Note that since U_t is an observable process, the coefficient of U_t, ψ_0, is no longer unity but is now an additional parameter of the model.) If we augment the state vector by including the constant unity as an additional element, i.e. if we now write

$$(4.5) \qquad \boldsymbol{x}_{t-1} = (1, U_{t-l}, \cdots, U_{t-1}, X_{t-k}, \cdots, X_{t-1})',$$

then we can write (4.2), (4.3) in the following state-space form:

$$(4.6) \qquad \boldsymbol{x}_{t+1} = \{\boldsymbol{F}(\boldsymbol{x}_t)\}\boldsymbol{x}_t + \{\boldsymbol{G}(\boldsymbol{x}_t)\}U_{t+1}$$

$$(4.7) \qquad Y_t = \boldsymbol{H}\boldsymbol{x}_t + N_t$$

where the system matrix $\boldsymbol{F}$ is given by

$$
\begin{bmatrix}
1 & 0 & 0 & \cdots & 0 & 0 & 0 & & \cdots & 0 \\
0 & 0 & 1 & \cdots & 0 & 0 & 0 & & \cdots & 0 \\
0 & 0 & 0 & \cdots & 1 & 0 & 0 & & \cdots & 0 \\
0 & 0 & 0 & \cdots & 0 & 0 & 0 & & \cdots & 0 \\
0 & 0 & 0 & \cdots & 0 & 0 & 1 & & \cdots & 0 \\
0 & 0 & 0 & \cdots & 0 & 0 & 0 & 1 & \cdots & 0 \\
0 & 0 & 0 & \cdots & 0 & 0 & 0 & 0 & \cdots & 1 \\
\mu & \psi_l & \psi_l & \cdots & \psi_1 & -\phi_k & -\phi_{k-1} & & \cdots & -\phi_1
\end{bmatrix}
$$

the input matrix $\boldsymbol{G}$ by

$$\boldsymbol{G} = (0, 0, \cdots, 1 \mid 0 \cdots \psi_0)',$$

and the observation matrix $\boldsymbol{H}$ by

$$\boldsymbol{H} = (0, 0, \cdots, 0 \mid 0 \cdots 1).$$

Assuming, as previously, that the coefficients μ, $\{\phi_u\}$, $\{\psi_u\}$ may be represented as locally linear functions of $\boldsymbol{x}_t$, the 'up-dating' equations then take the same form as (3.1). (Note, however, that due to the presence of the additional coefficient, ψ_0, there is an additional gradient vector, $\boldsymbol{\beta}_0^{(t)}$.) The gradient parameters $(\boldsymbol{\alpha}^{(t+1)}, \boldsymbol{\beta}_u^{(t+1)}, \boldsymbol{\gamma}_u^{(t+1)})$ can similarly be allowed to wander in the form of random walks, or as stationary AR(1) models. These may be written respectively as

$$\boldsymbol{B}_{t+1} = \boldsymbol{B}_t + \boldsymbol{V}_{t+1},$$

or

$$\boldsymbol{B}_{t+1} = \lambda \boldsymbol{B}_t + \boldsymbol{V}_{t+1}$$

where $\boldsymbol{B}_t$ is the matrix defined in (3.2), and $\{V_t\}$ is again a sequence of independent matrix-valued random variables, with $\boldsymbol{V}_t = N(\boldsymbol{0}, \boldsymbol{\Sigma}_v)$. Our estimation procedure then determines, for each t, those values of $\boldsymbol{B}_t$ which minimise the discrepancy between the observed value of Y_{t+1} and its predictor $\hat{Y}_{t+1}$ computed from the model fitted at time t. As in the case of the pure time series models, we implement this recursive fitting by rewriting the state-space representation of the SDM model in terms of a new state vector which represents the current values of all the model parameters. Specifically, we define the $(k+l+1)(k+l+2) \times 1$ column vector $\boldsymbol{\theta}_t$ by

$$\boldsymbol{\theta}_t = (\mu_{t-1}, \psi_{0,t-1}, \cdots, \psi_{l,t-1}, \phi_{1,t-1}, \cdots, \phi_{k,t-1},$$

$$\boldsymbol{\alpha}^{(t)'}, \boldsymbol{\beta}_0^{(t)'}, \cdots, \boldsymbol{\beta}_l^{(t)'}, \boldsymbol{\gamma}_1^{(t)'}, \cdots, \boldsymbol{\gamma}_k^{(t)'})',$$

and the $1 \times (k+l+2)(k+l+1)$ row vector $\boldsymbol{H}_t^*$ by $\boldsymbol{H}_t^* = (1, U_t, U_{t-1}, \cdots, U_{t-l}, -X_{t-1}, \cdots, -X_{t-k}, 0, \cdots, 0)$, so that we may write

(4.8) $$Y_t = \boldsymbol{H}_t^* \boldsymbol{\theta}_t + N_t.$$

The evolution of $\boldsymbol{\theta}_t$ over time is given by

(4.9) $$\boldsymbol{\theta}_t = \boldsymbol{F}_{t-1}^* \boldsymbol{\theta}_{t-1} + \boldsymbol{W}_t,$$

where the $(k+l+1)(k+l+2) \times (k+l+1)(k+l+2)$ matrix $\boldsymbol{F}_{t-1}^*$ is given by

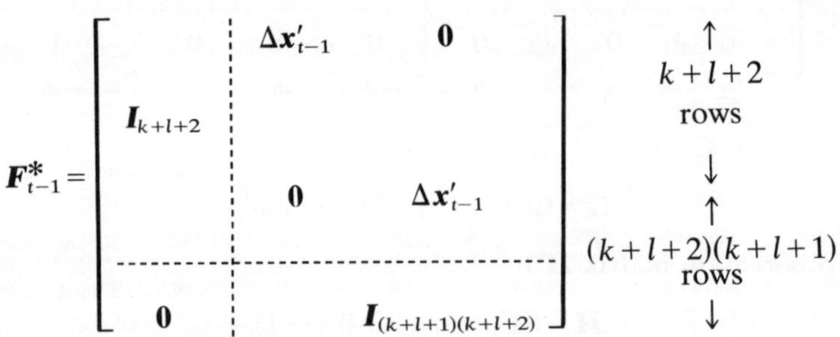

and

$$\boldsymbol{W}_t = (0, 0, \cdots, 0, \boldsymbol{v}_{1,t}', \cdots, \boldsymbol{v}_{k+l+2,t}')',$$

$\boldsymbol{v}_{1,t}, \cdots, \boldsymbol{v}_{k+l+2,t}$ being the columns of $\boldsymbol{V}_t$. (Note that $\boldsymbol{W}_t$ contains $(k+l+2)$ zero elements.) The vector $\Delta \boldsymbol{x}_{t-1}'$ is given by

$$\Delta \boldsymbol{x}_{t-1}' = (U_{t-1} - U_{t-2}, \cdots, U_{t-l} - U_{t-l-1}, X_{t-1} - X_{t-2}, \cdots, X_{t-k} - X_{t-k-1}).$$

Applying the Kalman algorithm to (4.7), (4.8) yields the recursion,

$$\hat{\boldsymbol{\theta}}_t = \boldsymbol{F}_{t-1}^* \hat{\boldsymbol{\theta}}_{t-1} + \boldsymbol{K}_t^* \{Y_t - \boldsymbol{H}_t^* \boldsymbol{F}_{t-1}^* \hat{\boldsymbol{\theta}}_{t-1}\},$$

where $\boldsymbol{K}_t^*$, the Kalman gain matrix, is given by (as previously)

$$\boldsymbol{K}_t^* = \boldsymbol{\Phi}_t (\boldsymbol{H}_t^*)' \sigma_e^{-2},$$

$\boldsymbol{\Phi}_t$ being the variance–covariance matrix of the one-step of $\boldsymbol{\theta}_t$, i.e.

$$\boldsymbol{\Phi}_t = E[(\boldsymbol{\theta}_t - \boldsymbol{F}_{t-1}^* \hat{\boldsymbol{\theta}}_{t-1})(\boldsymbol{\theta}_t - \boldsymbol{F}_{t-1}^* \hat{\boldsymbol{\theta}}_{t-1})'],$$

and σ_e^2 is the variance of $e_t = \{Y_t - H_t F_{t-1}^* \hat{\boldsymbol{\theta}}_{t-1}\}$. Again, we may compute successive values of $\hat{\boldsymbol{\theta}}_t$ using the standard recursive equations for the Kalman filter. Thus if we denote the variance–covariance matrix of $(\boldsymbol{\theta}_t - \hat{\boldsymbol{\theta}}_t)$ by $\boldsymbol{C}_t$, then the recursion is given by

$$\boldsymbol{K}_t^* = \boldsymbol{\Phi}_t (\boldsymbol{H}_t^*)' [\boldsymbol{H}_t^* \boldsymbol{\Phi}_t (\boldsymbol{H}_t^*)' + \sigma_N^2]^{-1},$$
$$\boldsymbol{\Phi}_t = \boldsymbol{F}_{t-1}^* \boldsymbol{C}_{t-1} (\boldsymbol{F}_{t-1}^*)' + \boldsymbol{\Sigma}_w,$$
$$\boldsymbol{C}_t = \boldsymbol{\Phi}_t - \boldsymbol{K}_t^* [\boldsymbol{H}_t^* \boldsymbol{\Phi}_t (\boldsymbol{H}_t^*)' + \sigma_N^2](\boldsymbol{K}_t^*)',$$

where σ_N^2 here denotes the variance of the noise disturbance, N_t, and $\boldsymbol{\Sigma}_w$ denotes, as previously, the variance–covariance matrix of $\boldsymbol{W}_t$.

In implementing this recursion we need to know both the matrices $\boldsymbol{F}_{t-1}^*$ and $\boldsymbol{H}_t^*$ at time t. This means that we need to know $X_{t-1}, \cdots, X_{t-k}$ in addition to $U_t, \cdots, U_{t-l}$. Given a sufficiently long sequence of observations on (U_s, Y_s), the relevant X_t's can be estimated recursively by setting $\hat{X}_{t-u} = \boldsymbol{H}_{t-u}^* \hat{\boldsymbol{\theta}}_{t-u}$, and X_{t-u} then replaced by $\hat{X}_{t-u}$ in the expressions for $\boldsymbol{H}_t^*$ and $\boldsymbol{F}_t^*$. Starting the recursion at $t = t_0$, say, the algorithm requires, as starting values, $\hat{\boldsymbol{\theta}}_{t_0-1}$ and $\boldsymbol{C}_{t_0-1}$. These values may be obtained by following a similar starting procedure to that used in fitting SDMS to pure time series data (Priestley (1980)). We take an initial stretch of data, say the first $2m$ observations on (U_t, Y_t), and fit a linear 'ARMA'-type model. This gives us initial parameter estimates, together with an estimate of the noise variance. We may then start the recursion midway along this initial stretch of data, i.e. we may take $t_0 = m$. Treating the coefficients in (4.3) now as constants, we may write the initially fitted model as

$$(4.10) \qquad Y_t + \phi_1 Y_{t-1} + \cdots + \phi_k Y_{t-k} = \mu + \psi_0 U_t + \cdots + \psi_l U_{t-l} + \zeta_t$$

where $\zeta_t = N_t + \phi_1 N_{t-1} + \cdots + \phi_k N_{t-k}$.

Given initial observations on (U_t, Y_t), (4.10) may be fitted by the standard Box–Jenkins procedure (Box and Jenkins (1970), Chapter 11). In essence, this procedure involves (a) computing preliminary estimates of $\mu, \phi_1, \cdots, \phi_k, \psi_0, \cdots, \psi_l$ by applying least-squares to (4.10), (b) using these estimated parameters to construct an estimated realisation of $\{\zeta_t\}$ from which we estimate the covariance structure of $\{\zeta_t\}$, (c) re-fitting (4.10) by weighted least-squares using the estimated covariances of $\{\zeta_t\}$. This procedure thus provides us with initial parameter estimates, $\hat{\mu}, \hat{\phi}_1, \cdots, \hat{\phi}_k, \hat{\psi}_0, \cdots, \hat{\psi}_l$, together with an estimate of the variance–covariance matrix of these parameters, $\hat{\boldsymbol{R}}_{\mu,\phi,\psi}$, say.

Finally, we may construct an estimated realisation of the noise process, using $\hat{N}_t = Y_t - \hat{X}_t$, from which we obtain an estimate $\hat{\sigma}_N^2$ of the noise variance.

5. Numerical examples

We illustrate the SDM model-fitting procedure described in the preceding section by applying this approach to simulated input/output data generated from various types of model. In all the simulation studies 500 observations were generated for the input and output processes. In each case the input process $\{U_t\}$ was generated from the stationary AR(1) model,

$$U_t + 0.5 U_{t-1} = \varepsilon_t,$$

$\{\varepsilon_t\}$ being a sequence of independent N(0, 1) variables, and the noise disturbance $\{N_t\}$ was also generated from a Gaussian white-noise process, with zero mean and unit variance.

The estimated parameter 'surfaces' were constructed using the non-parametric function fitting technique of Priestley and Chao (1972) based on a Gaussian smoothing kernel. The range of smoothing was determined by choosing a Gaussian kernel with 'standard deviation' equal to 10% of the total range of the abscissa variable. However, spurious end-effects often arise in this context since here the ordinates are not equally spaced, and if there are one or two outlying states excessive importance is given to the parameter values at these states by the Gaussian smoothing kernel. (See the discussion in Haggan et al. (1984).) For each model, a linear ARMA-type model was fitted to the first 50 observations and the recursion started at time $t = 25$. Graphs of the $\{\hat{\phi}_u\}$ and $\{\hat{\psi}_u\}$ were then plotted against the corresponding values of X_{t-1} or U_{t-1}, which are then compared with the theoretical functional forms corresponding to each simulated model. It should be noted that, in common with the 'time series' case, the SDM algorithm will not necessarily produce the 'natural' parameter forms, except for the parameter ψ_0.

Consider the general non-linear system (4.1). Since the SDM approach is based on a local linearization, the coefficients $\phi_1, \cdots, \phi_k, \psi_1, \cdots, \psi_l$ will follow the pattern of the first partial derivatives of h with respect to $X_{t-1}, \cdots, X_{t-k}, U_{t-1}, \cdots, U_{t-l}$, giving

$$\hat{\phi}_u = -\frac{\partial h}{\partial X_{t-u}} \qquad U = 1, \cdots, K,$$

$$\hat{\psi}_u = \frac{\partial h}{\partial U_{t-u}} \qquad U = 1, \cdots, l.$$

Thus, only $\hat{\psi}_0$ will follow its natural functional form, i.e.

$$\hat{\psi}_0 = g(X_{t-1}, \cdots, X_{t-k}, U_{t-1}, \cdots, U_{t-l}).$$

A. *Linear systems simulation.* The particular system chosen was given by

$$X_t + 0.8X_{t-1} = U_t - 0.6U_{t-1},$$

$$Y_t = X_t + N_t.$$

Here, the functional form of ϕ_1 should be equal to 0.8, and ψ_0 and ψ_1 should be equal to 1 and -0.6 respectively.

The preliminary estimates were:

$$\hat{\hat{\mu}} = -0.2242$$
$$\hat{\hat{\phi}}_1 = 0.5568$$
$$\hat{\hat{\psi}}_0 = 1.034$$
$$\hat{\hat{\psi}}_1 = -0.9373$$
$$\hat{\sigma}_N^2 = 1.142$$

$$\hat{\hat{R}} = \begin{bmatrix} 0.0256 & -0.0041 & -0.0040 & -0.0028 \\ -0.0041 & 0.0136 & 0.0040 & 0.0027 \\ -0.0040 & 0.0040 & 0.0232 & -0.0107 \\ -0.0028 & 0.0027 & -0.0104 & 0.0224 \end{bmatrix}.$$

In this case the smoothing factor α was taken to be 10^{-4}, and the resulting graphs of $\hat{\phi}_1$, $\hat{\psi}_0$, $\hat{\psi}_1$ are given in Figures 5.1, 5.2 and 5.3. We note that, except for the usual end-effects, they are consistent with their theoretical values.

The constancy in the estimated parameters $\hat{\phi}_1$, $\hat{\psi}_0$, $\hat{\psi}_1$ gives a clear indication that these series were generated by a linear system.

B. *Bilinear system simulation.* The system chosen was,

$$X_t - 0.4X_{t-1} = 0.8 + 0.4X_{t-1}U_t$$

$$Y_t = X_t + N_t.$$

If X_t is written in the form (4.1) it can be seen that the functional form of ϕ_1 is given by

$$\phi_1 = -\frac{\partial h}{\partial X_{t-1}} = -0.4$$

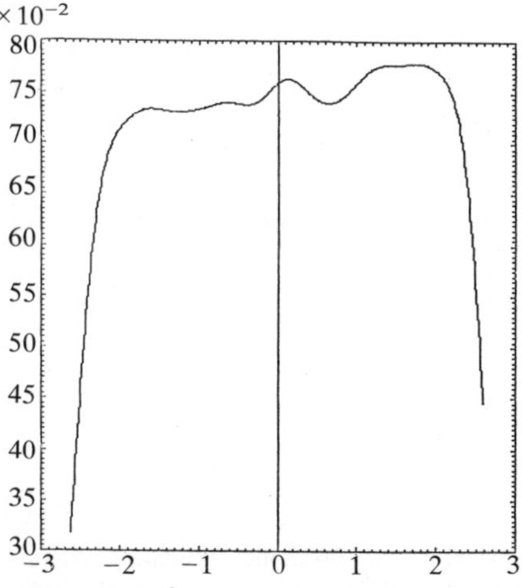

Figure 5.1. $\hat{\phi}_1$ for model A (linear system).

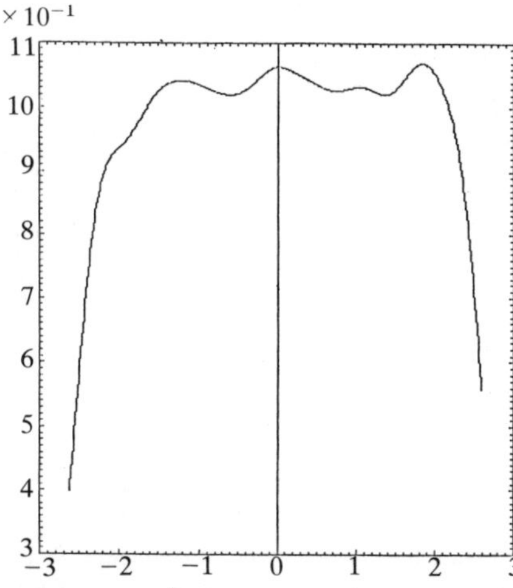

Figure 5.2. $\hat{\psi}_0$ for model A (linear system).

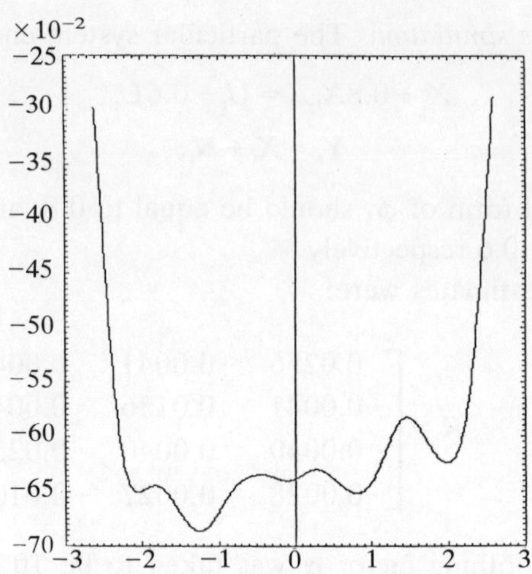

Figure 5.3. $\hat{\psi}_1$ for model A (linear system)

and the functional form of ψ_0 is

$$\psi_0 = g(X_{t-1}) = 0.4X_{t-1}.$$

The preliminary estimates were:

$$\hat{\hat{\mu}} = 1.301 \qquad \hat{\hat{R}} = \begin{bmatrix} 0.1276 & 0.0496 & 0.0200 \\ 0.0496 & 0.0273 & 0.0143 \\ 0.0200 & 0.0143 & 0.0326 \end{bmatrix}.$$

$$\hat{\hat{\phi}}_1 = -0.3436$$
$$\hat{\hat{\psi}}_0 = 0.8391$$
$$\hat{\hat{\sigma}}_N^2 = 1.768$$

Figures 5.4 and 5.5 show the resulting graphs of ϕ_1 and ψ_0 (as functions of

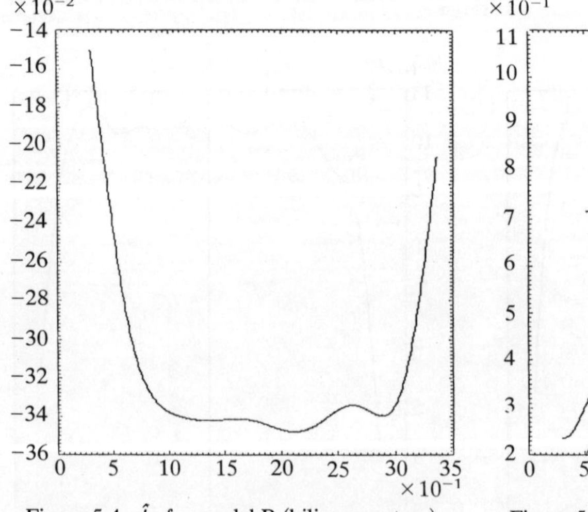

Figure 5.4. $\hat{\phi}_1$ for model B (bilinear system)

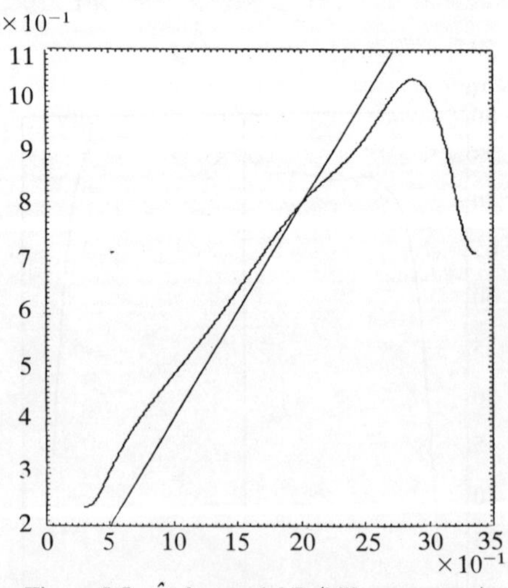

Figure 5.5. $\hat{\psi}_0$ for model B (bilinear system)

X_{t-1}) respectively, and the straight line in Figure 5.5 represents the true functional form of ψ_0. Again, apart from the end-effects, the estimated values of $\hat{\phi}_1$ are constant, and very close to the value of -0.4 expected for this parameter. The graph of $\hat{\psi}_0$ shows a clear linear trend, with the slope very close to that of the expected functional form.

C. *Exponential system simulation.* Here the system is given by the exponential AR model

$$X_t + (-0.9 + 0.1 \exp(-X_{t-1}^2))X_{t-1} = U_t,$$

$$Y_t = X_t + N_t.$$

The functional forms of ϕ_1 and ψ_0 are thus given by

$$\phi_1 = -\frac{\partial h}{\partial X_{t-1}} = -0.9 + 0.1 \exp(-X_{t-1}^2) - 0.2X_{t-1}^2 \exp(-X_{t-1}^2),$$

with

$$\psi_0 = g(X_{t-1}) = 1.$$

The functional form of ϕ_1 is shown in Figure 5.6. The preliminary estimates were:

$$\hat{\hat{\mu}} = 0.2516 \qquad \hat{\hat{R}} = \begin{bmatrix} 0.0504 & 0.0064 & -0.0026 \\ 0.0064 & 0.0039 & 0.0025 \\ -0.0020 & 0.0025 & 0.0284 \end{bmatrix}.$$
$$\hat{\hat{\phi}}_1 = -0.8742$$
$$\hat{\hat{\psi}}_0 = 0.8578$$
$$\hat{\sigma}_N^2 = 1.8806$$

The graph of $\hat{\phi}_1$ is shown in Figure 5.7. The shape of the graph is again very similar to the true form (except at each end), and the range of the numerical values of $\hat{\phi}_1$ is in good agreement with the corresponding range for the

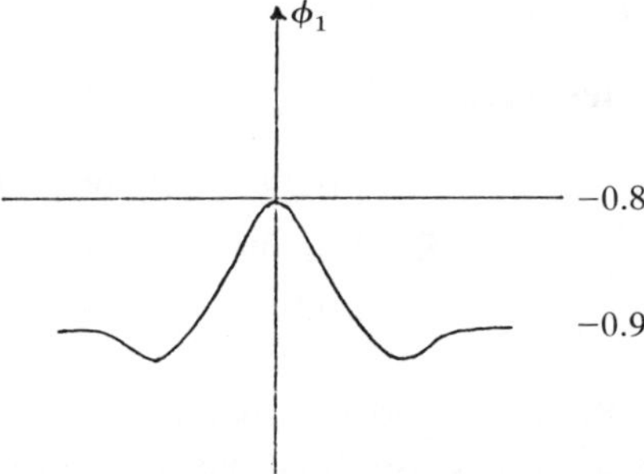

Figure 5.6. Graph of ϕ_1 for the exponential system C

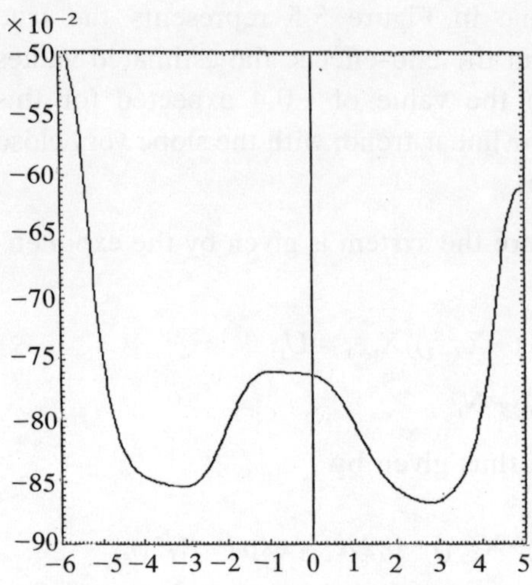

Figure 5.7. $\hat{\phi}_1$ for the exponential system C   Figure 5.8. $\hat{\psi}_0$ for the exponential system C

theoretical form of this parameter. The graph of the estimated parameter $\hat{\psi}_0$, Figure 5.8, is also quite consistent with the theoretical value of 1.

D. *Threshold* AR *system simulation.* In this example we consider the threshold AR model,

$$X_t + 0.9X_{t-1} = U_t, \quad \text{if} \quad X_{t-1} \leqq 2,$$
$$X_t - 0.9X_{t-1} = U_t, \quad \text{if} \quad X_{t-1} > 2,$$
$$Y_t = X_t + N_t.$$

The functional forms of ϕ_1 and ψ_0 are here given by

$$\phi_1 = -\frac{\partial h}{\partial X_{t-1}} = \begin{cases} 0.9, & \text{if} \quad X_{t-1} \leqq 2 \\ -0.9, & \text{if} \quad X_{t-1} > 2, \end{cases}$$

with

$$\psi_0 = g(X_{t-1}) = 1.$$

The preliminary estimates were:

$$\hat{\hat{\mu}} = 1.173$$
$$\hat{\hat{\phi}}_1 = -0.6077$$
$$\hat{\hat{\psi}}_0 = 0.8992 \qquad \hat{\hat{R}} = \begin{bmatrix} 0.4252 & 0.1091 & 0.0539 \\ 0.1091 & 0.0402 & 0.0273 \\ 0.0534 & 0.0273 & 0.1051 \end{bmatrix}.$$
$$\hat{\sigma}_N^2 = 6.101$$

The resulting graph of the estimated parameter $\hat{\phi}_1$ is given in Figure 5.9. The shape of the parameter $\hat{\phi}_1$ is essentially of the correct form, and clearly differs from the shapes derived for the other systems. The 'threshold' effect is fairly

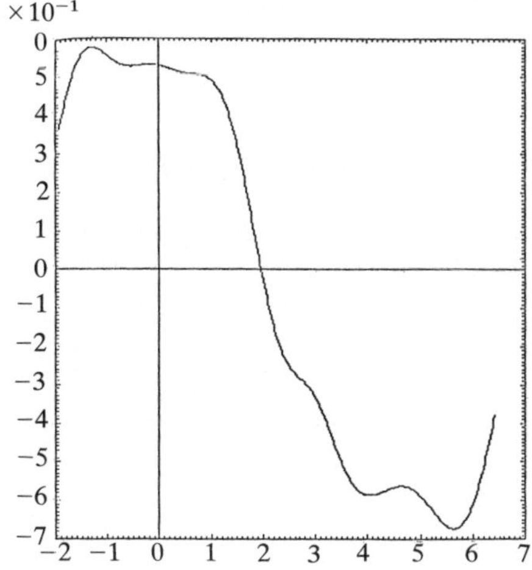

Figure 5.9. $\hat{\phi}_1$ for the threshold system D

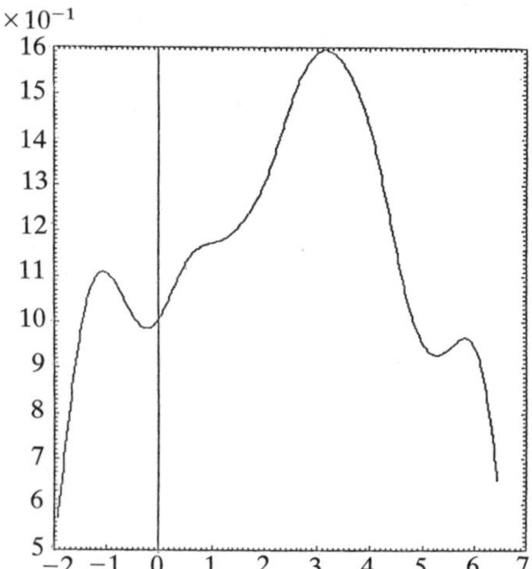

Figure 5.10. $\hat{\phi}_0$ for the threshold system D

obvious at the value 2, but the values taken by the parameter are smaller than the theoretical ones.

However, the graph of $\hat{\psi}_0$ (Figure 5.10) is not consistent with the value of 1 since, in order to make the 'threshold' effect in $\hat{\phi}_1$ obvious, we have used a large smoothing factor ($\alpha = 0.05$), which, in turn, produces an erratic behaviour in the parameter $\hat{\psi}_0$.

References

Box, G. E. P. and Jenkins, G. M. (1970) *Time Series, Forecasting, and Control.* Holden-Day, San Francisco.

Brockett, R. W. (1976) Volterra series and geometric control theory. *Automatica* **12**, 167–172.

Granger, C. W. J. and Andersen, A. P. (1978) *An Introduction to Bilinear Time Series Models.* Vandenhoeck and Ruprecht, Gottingen.

Haggan, V., Heravi, S. M., and Priestley, M. B. (1984) A study of the application of state-dependent models in non-linear time series analysis. *J. Time Series Anal.* **5**, 69–102.

Haggan, V. and Ozaki, T. (1981) Modelling non-linear random vibrations using an amplitude-dependent autoregressive time series model. *Biometrika* **68**, 189–196.

Hannan, E. J. (1973) The asymptotic theory of linear time series models. *J. Appl. Prob.* **10**, 130–145.

Hannan, E. J. (1976) The identification and parametrisation of ARMAX and state space forms. *Econometrica* **44**, 713–723.

Hannan, E. J. (1979) Estimating the dimension of a linear system. Paper presented at the International Time Series Meeting, University of Nottingham, England, March 1979.

Hannan, E. J. and Kavalieris, L. (1984) Multivariate linear time series models. *Adv. Appl. Prob.* **16**, 492–561.

Kalman, R. E. (1963) New methods of Wiener filtering theory. In *Proc. 1st Symp. Eng. Appns. of Random Functions Theory and Prob.*, ed. J. L. Bogdanoff and F. Kozin. Wiley, New York.

KUO, B. C. (1974) *Discrete Data Control Systems*. Science Tech., Illinois.

PRIESTLEY, M. B. (1980) State-dependent models: a general approach to non-linear time series analysis. *J. Time Series Anal.* **1**, 47–71.

PRIESTLEY, M. B. (1981) *Spectral Analysis and Time Series*, Vols. I and II, Academic Press, London.

PRIESTLEY, M. B. AND CHAO, M. T. (1972) Non-parametric function fitting. *J. R. Statist. Soc.* B **34**, 385–392.

SUBBA RAO, T. (1981) On the theory of bilinear models. *J. R. Statist. Soc.* B **43**, 244–255.

TONG, H. AND LIM, K. S. (1980) Threshold autoregression, limit cycles, and cyclic data. *J. R. Statist. Soc.* B **42**, 245–292.

PART 5

RANDOM FIELDS AND POINT PROCESSES

Regression for Randomly Sampled Spatial Series: The Trigonometric Case

DAVID R. BRILLINGER

Abstract

The model $Y(t) = s(t \mid \theta) + \varepsilon(t)$ is studied in the case that observations are made at scattered points τ_j in a subset of R^p and θ is a finite-dimensional parameter. The particular cases of

$$s(t \mid \theta) = \alpha \cos(\omega, t) + \beta \sin(\omega, t)$$

$\theta = (\alpha, \beta)$ and (α, β, ω) are considered in detail. Consistency and asymptotic normality results are developed assuming that the spatial series $\varepsilon(\cdot)$ and the point process $\{\tau_j\}$ are independent, stationary and mixing. The estimates considered are equivalent to least squares asymptotically and are not generally asymptotically efficient.

Contributions of the paper include: study of the R^p case, management of irregularly placed observations, allowance for abnormal domains of observation and the discovery that aliasing complications do not arise when the point process $\{\tau_j\}$ is mixing. There is a brief discussion of the construction and properties of maximum likelihood estimates for the spatial–temporal case.

POINT PROCESS; PROCESS WITH STATIONARY INCREMENTS; NON-LINEAR REGRESSION; TRIGONOMETRIC REGRESSION; PERIODOGRAM; RANDOM MEASURE; HIDDEN PERIODICITIES; ARRAY DATA; BLUE

1. Introduction

The problems of linear and non-linear regression have been much investigated for the case of discrete time series. Basic references for linear regression include Grenander (1954), Grenander and Rosenblatt (1957), Hannan (1970). One fundamental finding is that for trigonometric regression with known frequencies, the least squares estimates are asymptotically efficient. General non-linear regression has been studied in Hannan (1971) and Robinson (1972). Trigonometric regression with unknown frequencies ('hidden periodicities') was considered in Whittle (1952), Hannan (1971), (1973) and Walker (1971), (1973). The case of linear regression for continuous time has been examined in Chiang (1959), Kholevo (1969), (1971) and Hannan (1975). Again, least squares estimates are found to be asymptotically efficient for trigonometric

Paper prepared with the partial support of NSF Grant DMS-8316634 and NIH Grant BRSG-2-S07-RRO 7006.

regression with known frequencies. The case of continuous-time trigonometric regression with unknown frequencies was studied in Ivanov (1980). Estimation of the mean level of a stationary time series from sampled values was studied in Brillinger (1973). (After the paper was submitted the author found that Isokawa (1983) had considered the case of trigonometric regression for time series and randomly sampled values.) (An extensive literature exists on the problem of estimating the power spectrum of a stationary series from sampled values—see Brillinger (1972), Masry (1980), Roberts and Gaster (1980), Dunsmuir (1984) for example.) There do not appear to be many papers devoted to the study of properties of regression estimates for spatial series. For lattices Leonenko (1981) investigates linear least squares estimates while Hinich and Shaman (1972) estimate unknown frequencies.

The concern of this paper is trigonometric regression in the spatial series case when observations are available on a piece of a realization of a stationary spatial point process. Amongst others, the reasons motivating the research were: the increasing availability of array data, the desires to pre-whiten and detrend such data, the occurrence of missing values in the discrete case and the question of whether the linear least squares trigonometric estimates remained asymptotically efficient in the sampled case. (Concerning this last, it was found that the Ornstein–Uhlenbeck process with Poisson sampling showed the answer to be no. The referee later pointed out that this result was previously developed in McDunnough and Wolfson (1979) and so it will not be repeated here.)

The approach to the problems is via the calculus of processes with stationary increments, developed in Brillinger (1972), and of spatial series, developed in Brillinger (1970). The formulation of the problems, the style of argument, the particular results obtained owe so much, however, to the work of E. J. Hannan. He wrote the seminal papers. He recognized the dominant role of central limit theorems. He showed how to structure proofs of consistency. He worked through the particular cases.

2. Notation and assumptions

In this paper R^p will denote Euclidean p-space, $t = (t_1, \cdots, t_p)$ and $\lambda = (\lambda_1, \cdots, \lambda_p)$ are in R^p with $(\lambda, t) = \lambda_1 t_1 + \cdots + \lambda_p t_p$, $|t| = (t, t)^{1/2}$ and $t \cdot \lambda = (t_1 \lambda_1, \cdots, t_p \lambda_p)$. For $\mathcal{T}$ a measurable set $\subseteq R^p$, $|\mathcal{T}|$ will denote its Lebesgue measure. When it promotes clarity, vectors and matrices will be written in bold-face type. The transpose of a vector or matrix $\mathbf{A}$ will be $\mathbf{A}^\tau$.

For jointly distributed random variates $Y_1, \cdots, Y_K$, their joint cumulant will be denoted cum$\{Y_1, \cdots, Y_K\}$. The properties of this functional are described in Brillinger (1975). In the present context, the essential property will be that it measures the level of statistical dependence of the variates involved.

For a stationary spatial series $\varepsilon(t)$, t in R^p, the cumulant functions will be denoted by

$$(2.1) \qquad c_{\varepsilon \cdots \varepsilon}(u_1, \cdots, u_K) = \operatorname{cum}\{\varepsilon(t+u_1), \cdots, \varepsilon(t+u_K), \varepsilon(t)\}.$$

The cumulant function of order $K+1$ is a function of K variables. When it exists the power spectrum of $\varepsilon(\cdot)$ will be given by

$$(2.2) \qquad f_{\varepsilon\varepsilon}(\lambda) = (2\pi)^{-p} \int \exp\{-i(\lambda, u)\} c_{\varepsilon\varepsilon}(u)\,du.$$

For $dV(t) = V(dt)$ a stationary random measure, the cumulant measures will be denoted by

$$(2.3) \qquad C_{V \cdots V}(du_1, \cdots, du_K)\,dt = \operatorname{cum}\{dV(t+u_1), \cdots, dV(t+u_K), dV(t)\}.$$

These last are discussed in Brillinger (1972) in the instance of $p=1$. The particular case of a point process corresponds to $V(dt) = N(dt)$ giving the number of points in the p-dimensional interval $(t, t+dt)$. The power spectrum of the process $V(\cdot)$ is given by

$$(2.4) \qquad f_{VV}(\lambda) = (2\pi)^{-p} \int \exp\{-i(\lambda, u)\}\,dC_{VV}(u).$$

The model to be studied in this paper is

$$(2.5) \qquad Y(t) = s(t \mid \theta) + \varepsilon(t)$$

with $s(\cdot)$ a regression function of interest and $\varepsilon(t)$ a disturbance. The data available will be $\{\tau_j, Y(\tau_j)\}$ with the τ_j in $\mathcal{T}$ a (measurable) subset of R^p. Asymptotic results will be developed for 'large' $\mathcal{T}$.

A variety of assumptions will be required. These include the following.

Assumption I. $\varepsilon(t)$, t in R^p, is a stationary series whose cumulant functions exist and satisfy

$$(2.6) \qquad \int \cdots \int |c_{\varepsilon \cdots \varepsilon}(u_1, \cdots, u_K)|\,du_1 \cdots du_K < \infty$$

for $K = 1, 2, \cdots$. The mean, c_ε, of $\varepsilon(\cdot)$ is 0.

Assumption II. $V(dt)$, t in R^p, is a stationary random measure whose cumulant measures exist and satisfy

$$(2.7) \qquad \int \cdots \int |C_{V \cdots V}(du_1, \cdots, du_K)| < \infty$$

for $K = 1, 2, \cdots$.

Assumption III. $N(dt)$ is a point process satisfying Assumption II and statistically independent of the series $\varepsilon(\cdot)$.

In the work to be presented the sampled processes $Y(\tau_j)$ and $\varepsilon(\tau_j)$, with the τ_j points of a point process $N(\cdot)$ will occur often. They correspond to random measures via the relationships

$$(2.8) \qquad X(dt) = Y(t)N(dt) : V(dt) = \varepsilon(t)N(dt).$$

In connection with the latter one has the following result.

Theorem 1. Let the process $\varepsilon(\cdot)$ satisfy Assumption I and the process $N(\cdot)$ satisfy Assumption III. Then the process $V(\cdot)$ given by (2.8) satisfies Assumption II.

(The proofs of the theorems are presented in the appendix of the paper.) The parameters of the processes $X(\cdot)$ and $V(\cdot)$ may be given in terms of those of the processes $Y(\cdot)$ and $\varepsilon(\cdot)$. In particular for V of (2.8), $C_V = 0$ and

$$(2.9) \qquad f_{VV}(\lambda) = C_N^2 f_{\varepsilon\varepsilon}(\lambda) + \int f_{NN}(\lambda - \alpha) f_{\varepsilon\varepsilon}(\alpha) d\alpha.$$

This last parameter will play a fundamental role in describing the variability of the estimates constructed.

Concerning $\mathcal{T}$, the domain of observation, we make the following assumption.

Assumption IV. $\mathcal{T}$, a subset of R^p, has the form $\mathcal{T} = T \cdot U$ with U a bounded set of measure 1 of R^p and T in R^p.

The limiting process to be adopted later will that of $T = (T_1, \cdots, T_p) \to \infty$. The finite Fourier transform, given by

$$(2.10) \qquad d_X^T(\lambda) = \int_{\mathcal{T}} \exp\{-i(\lambda, t)\} Y(t)N(dt),$$

will be the principal statistic employed in the construction of estimates. For example the frequency of a hidden periodicity will be estimated as the location of its maximum absolute value in a restricted domain. Concerning that domain and $s(t \mid \theta)$ we require the following.

Assumption V. The regression function is given by

$$(2.11) \qquad s(t \mid \theta) = \alpha \cos(\omega, t) + \beta \sin(\omega, t)$$

with $|\rho|^2 = \alpha^2 + \beta^2 \neq 0$, $\omega \neq 0$, and ω in Λ a bounded region that does not also contain $-\omega$.

3. Estimation

Suppose that $Y(t) = \alpha \cos(\omega, t) + \beta \sin(\omega, t) + \varepsilon(t)$ and that measurements are available at $t = \tau_j$ within the region $\mathcal{T}$. In the case that ω is unknown, it will be estimated by $\hat{\omega}$ the value of λ maximizing the periodogram-type statistic

$$(3.1) \qquad |d_X^T(\lambda)|^2 = \left| \sum_j \exp\{-i(\lambda, \tau_j)\} Y(\tau_j) \right|^2$$

for λ in Λ. The coefficients α, β will be estimated by $\hat{\alpha}$, $\hat{\beta}$ given by

$$(3.2) \qquad \begin{bmatrix} \sum_j \cos(\hat{\omega}, \tau_j)^2 & \sum_j \cos(\hat{\omega}, \tau_j) \sin(\hat{\omega}, \tau_j) \\ \sum_j \sin(\hat{\omega}, \tau_j) \cos(\hat{\omega}, \tau_j) & \sum_j \sin(\hat{\omega}, \tau_j)^2 \end{bmatrix} \begin{bmatrix} \hat{\alpha} \\ \hat{\beta} \end{bmatrix} = \begin{bmatrix} \sum_j \cos(\hat{\omega}, \tau_j) Y(\tau_j) \\ \sum_j \sin(\hat{\omega}, \tau_j) Y(\tau_j) \end{bmatrix}$$

with $\hat{\omega}$ replaced by the known value when the latter is available. One has the following result.

Theorem 2. Suppose that Assumptions I, III, IV, V are satisfied. Then

$$(3.3) \qquad \hat{\alpha} - \alpha, \hat{\beta} - \beta \to 0 \quad \text{in probability}$$

$$(3.4) \qquad T \cdot (\hat{\omega} - \omega) \to 0 \quad \text{in probability}$$

as $T \to \infty$.

It may be noted that almost sure results may be developed here, but the present concern is the asymptotic distribution of the estimates and almost sure results are not needed to derive this.

In the course of the proof of this theorem, use will be made of the following result, which is of some independent interest.

Theorem 3. Let the stationary random measure $V(dt)$ satisfy Assumption II. Let the region $\mathcal{T}$ satisfy Assumption IV. Then for any $\eta > 0$, and Λ bounded

$$(3.5) \qquad \sup_{\lambda \in \Lambda} \left| \int_{\mathcal{T}} t^l \exp\{-i(\lambda, t)\}[V(dt) - C_V dt] \right| = o_p(|\mathcal{T}|^{\eta + 1/2} T^l)$$

for $l_j = 0, 1, 2, \cdots$ with t^l a monomial of degree l in the coordinates of t.

Almost sure results of this type may also be set down, see for example Theorem 3.3 in Brillinger (1970). When $p = 1$, results are given in Hannan (1973), Ivanov (1980), Vere-Jones (1982) for the discrete-time, continuous-time and point-process cases respectively. In the latter two cases Λ is allowed to be unbounded and to be bounded but growing respectively.

The asymptotic distribution of the estimates will now be studied. In the case of known frequency $\hat{\alpha}$, $\hat{\beta}$ are linear in the $Y(\tau_j)$ and it is a matter of studying the asymptotic distribution of a linear functional of the process $X(\cdot)$. The required distribution will be a particular case of a general theorem, about to be presented, that will also be made use of in deriving the asymptotic distribution in the case of unknown frequency. Of concern will be L-vectors of the form

$$(3.6) \qquad \int_{\mathcal{T}} \boldsymbol{\phi}^T(t) V(dt)$$

with the $\boldsymbol{\phi}^T(\cdot)$ satisfying the following condition.

Assumption VI. $\boldsymbol{\phi}^T(t)$, $t \in R^p$ is L-vector-valued and satisfying

$$(3.7) \qquad \int_{\mathcal{T}} \|\boldsymbol{\phi}^T(t)\|^2 dt = O(1),$$

satisfying

$$(3.8) \qquad \boldsymbol{d}_\phi^T(\lambda)\overline{\boldsymbol{d}_\phi^T(\lambda)}^\tau d\lambda \to \boldsymbol{v}(d\lambda) \quad \text{weakly}$$

with $\boldsymbol{d}_\phi^T(\lambda) = \int_{\mathcal{T}} \boldsymbol{\phi}^T(t) \exp\{-i(\lambda, t)\} dt$ and $\boldsymbol{v}(d\lambda)$ a matrix-valued measure, and finally satisfying

$$(3.9) \qquad \sup_{u_1,\cdots,u_K} \int_{\mathcal{T}} |\phi_{l_1}^T(t+u_1) \cdots \phi_{l_K}^T(t+u_K)\phi_{l_{K+1}}^T(t)| dt = o(1)$$

for $l_1, \cdots, l_{K+1} = 1, \cdots, L$ and $K = 2, 3, \cdots$. All the limiting results are to hold as $T \to \infty$, with $\mathcal{T}$ satisfying Assumption IV.

One may now state the following theorem.

Theorem 4. Let the random measure $V(dt)$ satisfy Assumption II and have mean 0. Let the $\boldsymbol{\phi}^T$ satisfy Assumption VI. Then as $T \to \infty$, the variate (3.6) is asymptotically normal with mean 0 and covariance matrix

$$(3.10) \qquad \int f_{VV}(\lambda)\boldsymbol{v}(d\lambda).$$

As a particular case take $\phi_1^T(t) = \cos(\omega, t)/|\mathcal{T}|^{\frac{1}{2}}$, $\phi_2(t) = \sin(\omega, t)/|\mathcal{T}|^{\frac{1}{2}}$. It is quickly checked that Assumption VI is satisfied with $\boldsymbol{v}(d\lambda)$ given by

$$(3.11) \qquad \frac{(2\pi)^p}{4} \begin{bmatrix} \delta(\lambda-\omega)+\delta(\lambda+\omega) & i[\delta(\lambda-\omega)-\delta(\lambda+\omega)] \\ -i[\delta(\lambda-\omega)-\delta(\lambda+\omega)] & \delta(\lambda-\omega)+\delta(\lambda+\omega) \end{bmatrix} d\lambda$$

with $\delta(\cdot)$ the Dirac delta function. This particular case leads to the following result.

Theorem 5. Let Assumptions I, II, IV, V hold with ω known and $\neq 0$. Let $\hat{\alpha}$, $\hat{\beta}$ be given by (3·2) with $\hat{\omega}$ the known value. Then $|\mathcal{T}|^{\frac{1}{2}}(\hat{\alpha}-\alpha)$ and $|\mathcal{T}|^{\frac{1}{2}}(\hat{\beta}-\beta)$

are asymptotically independent normals with mean 0 and variance

$$2(2\pi)^p C_N^{-2} f_{VV}(\omega) = 2(2\pi)^p [f_{\varepsilon\varepsilon}(\omega) + C_N^{-2} \int f_{NN}(\omega - \mu) f_{\varepsilon\varepsilon}(\mu) d\mu].$$

In the usual case, the asymptotic variance is proportional to the error spectrum $f_{\varepsilon\varepsilon}(\omega)$. Here it is found to be proportional to $f_{VV}(\omega)$.

The final theorem concerns the asymptotic distribution of the estimates $\hat{\alpha}$, $\hat{\beta}$, $\hat{\omega}$. Its statement involves the $(p+2)\times(p+2)$ matrix

(3.12)
$$\Sigma_T = \begin{bmatrix} |\mathcal{T}| & 0 & \beta \int t\,dt \\[2ex] 0 & |\mathcal{T}| & -\alpha \int t\,dt \\[2ex] \beta \int t^\tau dt & -\alpha \int t^\tau dt & |\rho|^2 \int t^\tau t\,dt \end{bmatrix}$$

with the integrals appearing being over the region $\mathcal{T}$.

Theorem 6. Let Assumptions I, II, IV, V hold. Let $\hat{\alpha}$, $\hat{\beta}$ be given by (3.2) and $\hat{\omega}$ by maximizing $|d_X^T(\lambda)|^2$ over Λ. Then the variate $(\hat{\alpha}, \hat{\beta}, \hat{\omega})$ is asymptotically normal with mean (α, β, ω) and covariance matrix

(3.13)
$$C_N^{-2} 2(2\pi)^p f_{VV}(\omega) \Sigma_T^{-1}.$$

It is notable that once again the asymptotic variances only depend on the power present at frequency ω. The matrix Σ_T may be inverted as a partitioned matrix leading to the following corollary.

Corollary. Under the conditions of the theorem $\hat{\omega}$ is asymptotically normal with mean ω and covariance matrix

(3.14)
$$C_N^{-2} 2(2\pi)^p f_{VV}(\omega) |\rho|^{-2} \left[\int t^\tau t\,dt - \left(\int t^\tau dt \right) \left(\int t\,dt \right) \Big/ |\mathcal{T}| \right]^{-1}.$$

As was to be expected the variance of the estimate decreases as the signal amplitude, $|\rho|$, increases. For sufficiently regular regions $\mathcal{T}$, the matrix inverse appearing in (3.14) may be determined explicitly.

4. Further Remarks

In order to avoid messy statements, the case of $\omega = 0$ was excluded from the theorems. This case corresponds to estimating a constant level and corresponding results exist. (The case of $p = 1$ is studied in Brillinger (1973).) Likewise, in order to avoid messy proofs, the case of several unknown frequencies was not studied. As is done in Walker (1971), Hinich and Shaman (1972), Ivanov (1980), results for the several-frequency case may be obtained under an assumption like the following.

Assumption VI. The regression function is given by

$$(4.1) \qquad s(t \mid \theta) = \sum_{i=1}^{I} [\alpha_i \cos(\omega_i, t) + \beta_i \sin(\omega_i, t)]$$

with $|\rho_i|^2 = \alpha_i^2 + \beta_i^2 \neq 0$, $\omega_i \neq 0$. The ω_i are in Λ a bounded region that does not include any $-\omega_i$. Λ_T is a subset of Λ^I with properties

$$(4.2) \qquad \lim \inf |T \cdot \lambda_i| = \infty$$

$$(4.3) \qquad \lim_{i \neq i'} \inf |T \cdot (\lambda_i - \lambda_{i'})| = \infty$$

for $(\lambda_1, \cdots, \lambda_I)$ in Λ_T.

Here the estimation of $(\omega_1, \cdots, \omega_I)$ is carried out by seeking the I largest relative maxima of $|d_X^T(\lambda)|^2$ in Λ_T.

It is to be noted that no difficulties have arisen due to aliasing beyond being unable to distinguish ω and $-\omega$. This also was the situation for spectrum estimation as studied in Brillinger (1972). It results from the process $N(dt)$ being mixing.

From Theorems V and VI it is apparent that if confidence regions are to be constructed for α, β, ω then an estimate of $f_{VV}(\omega)$ is needed. In fact such an estimate may be constructed directly. Form the values

$$(4.4) \qquad \hat{\varepsilon}(\tau_j) = Y(\tau_j) - \hat{\alpha} \cos(\hat{\omega}, \tau_j) - \hat{\beta} \sin(\hat{\omega}, \tau_j)$$

$$(4.5) \qquad d_{\hat{\varepsilon}}^T(\lambda) = \sum_j \exp\{-i(\lambda, \tau_j)\}\hat{\varepsilon}(\tau_j)$$

$$(4.6) \qquad I_{\hat{\varepsilon}\hat{\varepsilon}}^T(\lambda) = (2\pi)^{-p} |d_{\hat{\varepsilon}}^T(\lambda)|^2 |\mathcal{T}|^{-1}$$

and smooth the latter in the neighborhood of λ to estimate $f_{VV}(\lambda)$ generally.

Condition (4.3) has the effect of restricting the acceptable degree of closeness of frequencies when several components are present. For the discrete time series case, Pisarenko (1973) announced some results concerning the estimation of a pair of close frequencies.

It was mentioned in the introduction that the Ornstein–Uhlenbeck process with Poisson sampling shows that the least squares estimates are not always asymptotically efficient. The paper concludes with a presentation of some preliminary results concerning the maximum likelihood estimation of the parameters of a sinusoidal model when the errors are intercorrelated and array data is available. (In the time series case with continuous data some asymptotic efficiency results are announced in Kutoyants (1977).)

It is convenient to alter the notation somewhat from that employed so far in the paper. Suppose

$$(4.7) \qquad Y(x, y, t) = \rho \cos(\alpha x + \beta y + \gamma t + \delta) + \varepsilon(x, y, t)$$

with (x, y) planar coordinates and t time. Suppose data is available for $t = 0, \ldots, T-1$ and J sensors located at positions (x_j, y_j), $j = 1, \ldots, J$. The limiting conditions to be considered are now $T \to \infty$, but J fixed. (Previously the domain of observation had been large in all directions.) Discussion will concentrate on the problem of whether or not a signal is present ($\rho \neq 0$) and on the estimation of (α, β). Since T is 'large', estimation of γ is not difficult. The approach is that of Fourier inference.

Set

$$(4.8) \qquad d_{Y,j}^T(\lambda) = \sum_{t=0}^{T-1} Y(x_j, y_j, t) \exp\{-i\lambda t\}$$

with a similar definition for $d_{\varepsilon,j}^T(\lambda)$. Further, set

$$(4.9) \qquad \Delta^T(\lambda) = \sum_{t=0}^{T-1} \exp\{-i\lambda t\}.$$

Then, from (4.7), supposing λ to be near $\gamma \neq 0$

$$(4.10) \qquad d_{Y,j}^T(\lambda) = \frac{\rho}{2} e^{i\delta} \exp\{i(\alpha x_j + \beta y_j)\}\Delta^T(\lambda - \gamma) + d_{\varepsilon,j}^T(\lambda).$$

Consider these Fourier values at n frequencies of the form $2\pi s/T$, s an integer, near γ. It is then convenient to rewrite (4.10) as

$$(4.11) \qquad \mathbf{Y} = \mathbf{B}\xi\mathbf{A} + \boldsymbol{\epsilon}$$

with

$$(4.12) \qquad \mathbf{Y} = \left[\frac{1}{T} d_{Y,j}^T\left(\frac{2\pi s}{T}\right)\right] \qquad \boldsymbol{\epsilon} = \left[\frac{1}{T} d_{\varepsilon,j}^T\left(\frac{2\pi s}{T}\right)\right]$$

$$\mathbf{B} = [\exp\{i(\alpha x_j + \beta y_j)\}] \qquad \xi = \frac{\rho}{2} e^{i\delta} \qquad \mathbf{A} = \left[\frac{1}{T}\Delta^T\left(\frac{2\pi s}{T} - \gamma\right)\right]$$

with $\mathbf{Y}$ $J \times n$, with $\mathbf{B}$ $J \times 1$, with ξ 1×1, and with $\mathbf{A}$ $1 \times n$. Supposing the J vector-valued noise series $\boldsymbol{\varepsilon}(t) = [\varepsilon(x_j, y_j, t)]$ to be stationary and mixing, the Fourier values $d_{\varepsilon}^T(\lambda)$ satisfy a central limit theorem (see Brillinger (1983) for example.) Specifically the columns of $\boldsymbol{\epsilon}$ may be viewed as independent complex normal variates with mean $\mathbf{0}$ and covariance matrix $\boldsymbol{\Sigma} = (2\pi/T)\mathbf{f}(\gamma)$, $\mathbf{f}(\gamma)$ being the spectral density matrix of the series $\boldsymbol{\varepsilon}(t)$ at frequency γ.

The reason for rewriting the model in the form (4.11) is that the latter is a classical growth curve model that has been much studied (see Khatri (1966), Geiser (1980) for example). Here (4.11) involves complex-valued variates, but that is of no great consequence. Set

$$(4.13) \qquad \mathbf{S} = \mathbf{Y}\bar{\mathbf{Y}}^\tau - \mathbf{Y}\bar{\mathbf{A}}^\tau(\mathbf{A}\bar{\mathbf{A}}^\tau)^{-1}\mathbf{A}\bar{\mathbf{Y}}^\tau$$

then the maximum likelihood estimates of ξ and Σ are given by

$$(4.14) \qquad \hat{\xi} = (\bar{B}^\tau S^{-1} B)^{-1} \bar{B}^\tau S^{-1} Y \bar{A}^\tau (A\bar{A}^\tau)^{-1}$$

$$(4.15) \qquad n\hat{\Sigma} = (Y - B\hat{\xi}A)(\overline{Y - B\hat{\xi}A})^\tau$$

respectively, and the maximum of the likelihood is proportional to $|n\hat{\Sigma}|^{-n}$. In the case at hand, of array data, B is a function of further parameters namely (α, β). To estimate them, in maximum likelihood fashion, one will want to minimize the determinant $|\hat{\Sigma}|$. It is convenient to suppose that γ has the form $2\pi S/T$, although this is no essential restriction. Now $A\bar{A}^\tau = 1$, for example. By algebraic manipulations it may be shown that minimizing $|\hat{\Sigma}|$ as a function of (α, β) now comes down to maximizing

$$(4.16) \qquad |\bar{B}^\tau S^{-1} Y \bar{A}^\tau|^2 / \bar{B}^\tau S^{-1} B$$

as a function of (α, β). It would seem worthwhile to plot (4.16) as of function of (α, β). The significance of a particular value of ξ, e.g. $\xi = 0$ may be examined via the result that

$$(4.17) \qquad |\hat{\xi} - \xi|^2 (\bar{B}^\tau S^{-1} B)(n - J)/(1 + A\bar{Y}^\tau S^{-1} Y \bar{A}^\tau - |\hat{\xi}|^2 (\bar{B}^\tau S^{-1} B))$$

is distributed as $F(2, 2(n - J))$ in the growth curve case. (In the case that $\Sigma = I$, this type of approach is propounded in Shumway (1983).) As a matter of interest it may be remarked that, in the present notation, the Capon (1969) high-resolution spectrum estimate has the form

$$(4.18) \qquad (\bar{B}^\tau Y \bar{Y}^\tau B)^{-1}$$

while the least squares estimate, with which the bulk of this paper has been concerned, has the form

$$(4.19) \qquad |\bar{B}^\tau Y \bar{A}^\tau|^2.$$

The practical value of this maximum likelihood estimate is being examined at present. The theoretical validity is clear.

Appendix

Proof of Theorem 1. This follows via the same argument as that of the proof of Lemma 2 in Brillinger (1973).

Proof of Theorem 2. This proof makes use of two lemmas and Theorem 3.

Lemma 1. Suppose $a < 1$ and set

$$(A.1) \qquad \delta^T(\lambda) = \int_{\mathcal{T}} \exp\{-i(\lambda, t)\}dt$$

with $\mathcal{T}$ satisfying Assumption IV. Then for given $b, c > 0$ there exists d in R^p such that for $T > d$

$$(A.2) \qquad \sup_{|T^a \cdot \lambda| \geq b} |\delta^T(\lambda)| < c\,|\mathcal{T}|.$$

Further there exists $e > 0$ such that

$$(A.3) \qquad \sup_{|T \cdot \lambda| \geq b} |\delta^T(\lambda)| < (1 - e)\,|\mathcal{T}|.$$

Proof. Taking note of the assumed form of $\mathcal{T}$ one has

$$(A.4) \qquad \delta^T(\lambda) = |\mathcal{T}|\,\delta^U(T \cdot \lambda)$$

with U bounded. Now by the Riemann–Lebesgue lemma (e.g. Edwards (1965), p. 723) given $c > 0$, there exists Q a compact set such that $|\delta^U(\gamma)| < c$ for $\gamma \in \bar{Q}$. Next note that for t sufficiently large $\{\gamma : |T^{1-a} \cdot \gamma| \geq b\} \subset \bar{Q}$. It follows that $|\delta^T(\lambda)| < c\,|\mathcal{T}|$ for $|T^a \cdot \lambda| \geq b$ and T sufficiently large. Turning to (A.3), from (A.4) one sees that the concern is with

$$\sup_{|\gamma| \geq b} |\delta^U(\gamma)|.$$

Let $I_U(u)$, $u \in R^p$, denote the indicator function of the set U. Then

$$1 - |\delta^U(\gamma)|^2 = \int [1 - \cos(\gamma, u)] I_U * I_U(u)\,du.$$

The right-hand side here is strictly positive for, using the Riemann–Lebesgue lemma consideration of the integral may be restricted to a bounded region including $u = 0$, and there one has a continuous non-negative function (of γ).

Lemma 2. Under the conditions of the theorem,

$$(A.5) \qquad d_X^T(\lambda) = C_N \frac{\rho}{2} \delta^T(\lambda - \omega) + o_p(|\mathcal{T}|)$$

uniformly in $\lambda \in \Lambda$, with $\rho = \alpha - i\beta$.

Proof. One has

$$(A.6) \qquad d_X^T(\lambda) = \frac{\rho}{2} d_N^T(\lambda - \omega) + \frac{\bar{\rho}}{2} d_N^T(\lambda + \omega) + d_V^T(\lambda)$$

with X and V given by (2.8). Now

$$d_N^T(\lambda) = C_N \delta^T(\lambda) + \int \exp(-i(\lambda, t))[N(dt) - C_N dt] = C_N \delta^T(\lambda) + o_p(|\mathcal{T}|^{n+1/2})$$

from (3.5). The second term is $o(|\mathcal{T}|)$ from (A.2) and the assumption that $-\omega \notin \Lambda$. The final term is $o_p(|\mathcal{T}|^{n+\frac{1}{2}})$ from (3.5), giving the result.

Returning to the proof of the theorem, to begin consider (3.4). Given $b > 0$, let $P = \{\lambda : |T \cdot (\lambda - \omega)| \geq b\}$. Now

$$\Pr\{|T \cdot (\hat{\omega} - \omega)| \geq b\} \leq \Pr\left\{\sup_{\lambda \in P} |d_X^T(\lambda)|^2 \geq |d_X^T(\omega)|^2\right\}$$

$$= \Pr\left\{\sup_{\lambda \in P} |\delta^T(\lambda - \omega)|^2 \geq |\mathcal{T}|^2 + o_p(|\mathcal{T}|^2)\right\}$$

from (A.5). In view of (A.3) this last probability tends to 0, and one has (3.4). In the case of (3.3), first note that

$$(A.7) \quad |\delta^T(\lambda) - |\mathcal{T}|| = \left|\int_{\mathcal{T}} [\exp(-i(\lambda, t)) - 1]dt\right| \leq |\mathcal{T}||T \cdot \lambda| \int_U |u|\, du.$$

Next note that, for example,

$$\sum_j \exp\{i(\hat{\omega}, \tau_j)\} = \int_T \exp\{i(\hat{\omega}, t)\} N(dt)$$

$$(A.8) \qquad\qquad = C_N \delta^T(\hat{\omega}) + o_p(|\mathcal{T}|^{n+\frac{1}{2}})$$

$$\qquad\qquad = C_N \delta^T(\omega) + o_p(|\mathcal{T}|) + o_p(|\mathcal{T}|^{n+\frac{1}{2}})$$

using (3.5) at the second last step and (A.7) at the last. Now if $\boldsymbol{\Psi}_T$ denotes the 2×2 matrix of (3.2), then this last relationship may be used to see that $\boldsymbol{\Psi}_T \sim \frac{1}{2} C_N |\mathcal{T}| \boldsymbol{I}$, whether ω is known or not. Since

$$(A.9) \qquad\qquad \begin{bmatrix} \hat{\alpha} - \alpha \\ \hat{\beta} - \beta \end{bmatrix} = \boldsymbol{\Psi}_T^{-1} \begin{bmatrix} \displaystyle\int_{\mathcal{T}} \cos(\hat{\omega}, t) V(dt) \\ \displaystyle\int_{\mathcal{T}} \sin(\hat{\omega}, t) V(dt) \end{bmatrix}$$

one obtains (3.3) by another application of (3.5) and the argument of (A.8).

Proof of Theorem 3. Since (3.5) concerns $V(dt) - C_V dt$, it may be assumed that $C_V = 0$. Let $\sigma = \sup |t_j|$ for $t = (t_1, \cdots, t_p)$ in $\mathcal{T}$. Let $B_{\sigma, \ldots, \sigma}$ denote the space of entire functions of several variables of degree σ in each variable and bounded on the real axes. (This space is studied in Timan (1963), e.g. pp. 235, 253.) With probability 1 $d_V^T(\cdot)$ lies in this space. Let ψ be a number such that $0 < \psi |\lambda_j| < \pi$ for $\lambda = (\lambda_1, \cdots, \lambda_p)$ in Λ. Now

$$g(\lambda) = d_V^T(\lambda)[\sin \psi \lambda_1 \cdots \sin \psi \lambda_p]/[(\psi \lambda_1) \cdots (\psi \lambda_p)]$$

is in $B_{\sigma+\psi, \ldots, \sigma+\psi}$ and, as may be seen by taking Fourier transforms,

$$g(\lambda) = \int g(\lambda - \mu) \prod_j \frac{\sin(\sigma + \psi)\mu_j/2}{\mu_j/2}\, d\mu.$$

Now, arguing as on p. 409 of Brillinger (1975), by Holder's inequality

$$|g(\lambda)|^{2k} \leqq \int |g(\mu)|^{2k} d\mu \left[\int \prod_j \left| \frac{\sin (\sigma + \psi)\mu_j/2}{\mu_j/2} \right|^{2k/(2k-1)} d\mu \right]^{2k-1}.$$

The last term on the right here is $O(|\mathcal{T}|)$. Next, because of Assumption II, the cumulants of $d_V^T(\mu)$ are each $O(|\mathcal{T}|)$ uniformly and so $E\,|d_V^T(\mu)|^{2k} = O(|\mathcal{T}|^{2k})$ uniformly. Continuing, it follows from the integrability of $|[\sin \psi\mu_1]/[\psi\mu_1]|^{2k}$ therefore that

$$E \left[\sup_\Lambda |g(\lambda)| \right]^{2k} = O(|\mathcal{T}|^{k+1})$$

giving (3.5) for $l = 0$. The result for general l follows from this last via expression (27) on p. 253 of Timan (1963). (Because of the constraint on ψ, the bound on d_V^T is deducible from the one on g.)

Proof of Theorem 4. The covariance matrix of the variate of concern is given by

$$\int d_\phi^T(\lambda) \, \overline{d_\phi^T(\lambda)}^\tau f_{VV}(\lambda) d\lambda.$$

This tends to (3.10) because $f_{VV}(\cdot)$ is bounded and continuous. Next consider a cumulant of order $K + 1$. This is bounded by

$$\int \cdots \int |\phi_{l_1}^T(t + u_1) \cdots \phi_{l_K}^T(t + u_K)\phi_{l_{K+1}}^T(t)| \, |C_{V\ldots V}(du_1, \cdots, du_K)| \, dt$$

and tends to 0 as $T \to \infty$ following (3.9) and (2.7) for $K > 2$. This gives the desired result because the normal is determined by its moments.

Proof of Theorem 5. This follows immediately from the representation (A.9), Theorem 4, (3.11) and (2.9).

Before beginning to prove Theorem 6, it is first noted that the estimates $\hat\alpha$, $\hat\beta$, $\hat\omega$ being considered are equivalent to the least squares estimates. For a particular λ, the estimates $\hat\alpha$, $\hat\beta$ were defined to be least squares values so that much is clear. Next, suppose that for a given value of $\hat\omega$ least squares estimates of α, β have been determined. In traditional statistical notation the residual sum of squares is the $Y'Y - Y'X(X'X)^{-1}X'Y$. The full least squares values are then found by minimizing this as a function of λ. Now $X'X$ is here the 2×2 matrix (3.2) with $\hat\omega$ replaced by λ. The argument commencing at (A.8) may be used to show that $X'X \sim \frac{1}{2}C_N|\mathcal{T}|\,I$, so in seeking the least squares values one is seeking to maximize the equivalent $Y'XX'Y$. This last is expression (3.1).

Proof of Theorem 6. The asymptotic distribution of the least squares estimates will be derived. With $\theta = (\hat\alpha, \hat\beta, \hat\omega)$ let

$$S_T(\theta) = \frac{1}{2} \sum_j [Y(\tau_j) - \alpha \cos (\omega, \tau_j) - \beta \sin (\omega, \tau_j)]^2.$$

Let $S'_T(\cdot)$ denote the $1 \times (p+2)$ vector of first derivatives with respect to θ and $S''_T(\cdot)$ the $(p+2) \times (p+2)$ matrix of second derivatives. Then the least squares estimate $\hat{\theta}$ satisfies

$$S'_T(\hat{\theta}) = 0 = S'_T(\theta) + (\hat{\theta} - \theta) \cdot S''_T(\bar{\theta})$$

with $\bar{\theta}$ lying between $\hat{\theta}$ and θ. One has

$$(\hat{\theta} - \theta) = S'_T(\theta) \cdot S''_T(\bar{\theta})^{-1}$$

and the desired result follows once one shows that $S''_T(\bar{\theta}) \sim \frac{1}{2} C_N \Sigma_T$ in probability and that $S'_T(\theta)$ is asymptotically normal with mean 0 and covariance matrix $\frac{1}{2}(2\pi)^p f_{VV}(\omega) \Sigma_T$.

The entries of $S'_T(\theta)$ when evaluated are found to be linear in the $\varepsilon(\tau_j)$. The asymptotic distribution is found by an application of Theorem 4. In the case of $S''_T(\bar{\theta})$, Theorem 3 allows the replacement of $\bar{\theta}$ by θ. The limiting forms of its entries are found making use of Lemma 1.

References

BRILLINGER, D. R. (1970) The frequency analysis of relations between stationary spatial series. In *Proc. Twelfth Bien. Sem. Canadian Math. Congr.*, ed. R. Pyke. Canadian Mathematical Congress, Montreal, 39–81.

BRILLINGER, D. R. (1972) The spectral analysis of stationary interval functions. *Proc. 6th Berkeley Symp. Math. Statist. Prob.* **1**, 483–513.

BRILLINGER, D. R. (1973) Estimation of the mean of a stationary time series by sampling. *J. Appl. Prob.* **10**, 419–431.

BRILLINGER, D. R. (1975) *Time Series: Data Analysis and Theory*. Holt, Rinehart and Winston, New York.

BRILLINGER, D. R. (1983) The finite Fourier transform of a stationary process. In *Handbook of Statistics* **3**, ed. D. R. Brillinger and P. R. Krishnaiah, North-Holland, Amsterdam, 21–38.

CAPON, J. (1969) High resolution frequency-wavenumber spectrum analysis. *Proc. IEEE* **57**, 1408–1418.

CHIANG, T-P. (1959) On the estimation of regression coefficients of a continuous parameter time series with a stationary residual. *Theory Prob. Appl.* **4**, 373–390.

DUNSMUIR, W. (1984) Large sample properties of estimation in time series observed at unequally spaced times. In *Proc. Symp. Time Series Analysis of Irregularly Observed Data*, ed. E. Parzen. Springer-Verlag, New York.

EDWARDS, R. E. (1965) *Functional Analysis*. Holt, Rinehart and Winston, New York.

GEISSER, S. (1980) Growth curve analysis. In *Handbook of Statistics*, **1**, ed. P. R. Krishnaiah, North-Holland, Amsterdam, 89–116.

GRENANDER, U. (1954) On the estimation of regression coefficients in the case of autocorrelated disturbance. *Ann. Math. Statist.* **25**, 252–272.

GRENANDER, U. AND ROSENBLATT, M. (1957) *Statistical Analysis of Stationary Time Series*. Wiley, New York.

HANNAN, E. J. (1970) *Multiple Time Series*. Wiley, New York.

HANNAN, E. J. (1971) Non-linear time series regression. *J. Appl. Prob.* **8**, 767–780.

HANNAN, E. J. (1973) The estimation of frequency. *J. Appl. Prob.* **10**, 510–519.

HANNAN, E. J. (1975) Linear regression in continuous time. *J. Austral. Math. Soc.* **19**, 146–159.

HINICH, M. J. AND SHAMAN, P. (1972) Parameter estimation for an R-dimensional plane wave observed with additive independent Gaussian errors. *Ann. Math. Statist.* **43**, 153–169.

ISOKAWA, Y. (1983) Estimation of frequency by random sampling. *Ann. Inst. Statist. Math.* **35**, 201–213.

IVANOV, A. V. (1980) A solution of the problem of detecting hidden periodicities. *Theory Prob. Math. Statist.* **20**, 51–68.

KHATRI, C. G. (1966) A note on a MANOVA model applied to problems in growth curve. *Ann. Inst. Statist. Math.* **18**, 75–86.

KHOLEVO, A. S. (1969) On estimates of regression coefficients. *Theory Prob. Appl.* **14**, 79–104.

KHOLEVO, A. S. (1971) Almost-periodic signal estimation. *Theory Prob. Appl.* **16**, 249–263.

KUTOYANTS, YU. A. (1977) Estimation of a parameter of a signal in Guassian noise. *Probl. Pered. Inform.* **13**, 29–36.

LEONENKO, N. N. (1981) On the invariance principle for estimators of linear regression coefficients of a random field. *Theory Prob. Math. Statist.* **23**, 87–96.

MASRY, E. (1980) Discrete-time spectral estimation of continuous-time processes—the orthogonal series method. *Ann. Statist.* **5**, 1100–1109.

McDUNNOUGH, P. AND WOLFSON, D. B. (1979) On some sampling schemes for estimating the parameters of a continuous time series. *Ann. Inst. Statist. Math.* **31**, 487–497.

PISARENKO, V. F. (1973) Estimation of parameters of two harmonics of close frequencies on a background of noise. *Theory Prob. Appl.* **18**, 826.

ROBERTS, J. B. AND GASTER, M. (1980) On the estimation of spectra from randomly sampled signals; a method of reducing variability. *Proc. R. Soc. London* **371**, 235–258.

ROBINSON, P. M. (1972) Non-linear regression for multiple time-series. *J. Appl. Prob.* **9**, 758–768.

SHUMWAY, R. H. (1983) Replicated time-series regression: an approach to signal estimation and detection. In *Handbook of Statistics* **3**, ed. D. R. Brillinger and P. R. Krishnaiah, North-Holland, Amsterdam, 383–408.

TIMAN, A. F. (1963) *Theory of Approximation of Functions of a Real Variable.* Pergamon, Oxford.

VERE-JONES, D. (1982) On the estimation of frequency in point process data. *Essays in Statistical Science*, ed. J. Gani and E. J. Hannan, Applied Probability Trust, Sheffield, 383–394.

WALKER, A. M. (1971) On the estimation of a harmonic component in a times series with stationary independent residuals. *Biometrika* **58**, 21–36.

WALKER, A. M. (1973) On the estimation of a harmonic component in a time series with stationary dependent residuals. *Adv. Appl. Prob.* **5**, 217–241.

WHITTLE, P. (1952). The simultaneous estimation of a time series' harmonic components and covariance structure. *Trab. Estadist.* **3**, 43–57.

Point-Process Models with Linearly Parametrized Intensity for Application to Earthquake Data

YOSIHIKO OGATA
KOICHI KATSURA

Abstract

It is demonstrated that linear parametrization of the conditional intensity provides systematic classes of flexible models which are reasonably useful for calculating maximum likelihoods. To exemplify the modelling, seismic activity around Canberra is decomposed into components of evolutionary trend, clustering and periodicity. The causal relationship between earthquake sequences from two seismic regions is also analysed for a certain Japanese earthquake data set.

Some technical aspects of the modelling and calculations are described.

CONDITIONAL INTENSITY FUNCTION; MAXIMUM LIKELIHOOD ESTIMATES; TREND, CYCLIC AND CLUSTERING DECOMPOSITION; CAUSAL RELATION OF SHOCK SEQUENCES; RESPONSE FUNCTIONS; AKAIKE INFORMATION CRITERION; EARTHQUAKE MIGRATION

1. Introduction

The published lists of earthquakes usually consist of occurrence times, locations, depths and magnitudes. In our present analysis the origin times only are considered, other elements being largely ignored. Such records from a self-contained seismic region reveal time series of extremely complex structure. Large fluctuations in the numbers of shocks per month, complicated sequences of related shocks, dependence on activity from other seismic regions, fluctuation of seismicity on a larger time scale, and changes in the detection level of shocks, all appear to be characteristic features of such records. As a preliminary analysis, one may be interested in knowing the approximate shapes of trend, periodic and clustering effects, and the relation to the activity in other seismic areas, if such exist. A great number of papers in seismology as well as in mathematical statistics refer to such analyses (see for example Vere-Jones (1970), Hawkes and Adamopoulos (1973), and Vere-Jones and Ozaki (1982)). This paper is concerned with systematic classes of flexible parametric models (like ARIMA models in time series) of point processes to investigate such effects simultaneously by using likelihood analysis.

A key to the likelihood theory of point processes is the conditional intensity function (CIF) defined by

$$(1) \qquad \lambda(t \mid F_t) = \lim_{\Delta \downarrow 0} \Pr\{\text{an event in } [t, t+\Delta) \mid F_t\}/\Delta,$$

where F_t is a family of informations (σ-algebra) based on the observations up to time t, including the history of events at time t. A CIF characterizes a point process completely (see for example Chapter 18 in Liptzer and Shiryaev (1978)). This means that samples of a point process can be obtained by a CIF. Ogata (1981) developed a direct simple simulation algorithm for point processes including multivariate processes, extending the 'thinning method' of Lewis and Shedler (1979).

Also, given a CIF we can write the likelihood for the realization of events $\{t_i\}$ on the time interval $[0, T]$ as

$$(2) \qquad L(t_1, t_2, \cdots, t_I; T) = \left\{\prod_{i=1}^{I} \lambda(t_i \mid F_{t_i})\right\} \exp\left\{-\int_0^T \lambda(t \mid F_t)dt\right\}.$$

Parametrizing a CIF $\lambda_\theta(t)$ by a K-dimensional vector $\theta = (\theta_1, \cdots, \theta_K)$, the parameter estimates are then formed by maximizing L, or rather its logarithm

$$(3) \qquad \log L = \sum_{i=1}^{I} \log \lambda_\theta(t_i \mid F_{t_i}) - \int_0^T \lambda_\theta(t \mid F_t)dt,$$

with respect to the parameters. This is carried out by using the gradients of (3) with respect to θ and the standard non-linear optimization technique such as that in Fletcher and Powell (1963).

2. Linearly parametrized intensity

Ogata (1983) recommended a systematic use of the following parametrization for the CIF

$$(4) \qquad \lambda_\theta(t \mid F_t) = \sum_{k=1}^{K} \theta_k \cdot Q_k(t \mid F_t), \qquad \theta = (\theta_1, \cdots, \theta_K),$$

where each $Q_k(t \mid F_t)$ is independent of the parameter θ to be adjusted. A main advantage of such parametrization is that the log likelihood function (3) has at most one maximum no matter how the dimension of the parameter increases. This is because the Hessian matrix is everywhere negative definite in θ; i.e. for any real u_k, $k = 1, 2, \cdots, K$,

$$\sum_{j,k} u_j u_k \partial^2 \log L/\partial\theta_j\partial\theta_k = -\sum_{i=1}^{I} \left\{\sum_{k=1}^{K} u_k Q_k(t_i \mid F_t)/\lambda_\theta(t_i \mid F_{t_i})\right\}^2.$$

Thus if the analytic calculation of the integral $\int Q_k(t \mid F_t)dt$ is feasible, then the maximization algorithm for the log likelihood function (3) is implemented efficiently.

However the major disadvantage of (4) is that the CIF happens to be negative on some part of t outside the neighbourhood of $\{t_i\}$. This occurs more easily the larger the number of the parameters. This causes some difficulty in obtaining the maximum likelihood estimates, because the negativity of the CIF contributes to the apparent increase of log likelihood due to the second term of (3). In order to avoid negative CIF we impose restrictions on the parameters. We use a certain concave smooth penalty function $R(\theta)$ which takes the value 0 in the restricted region but takes large values on the outside, and then minimize the concave function

$$(5) \qquad G(\theta) = -\log L(\theta) + R(\theta),$$

the details of which are described later.

2.1. *Trend/cyclic/clustering decomposition of earthquake risk.* The model suggested in Ogata (1983) takes the form

$$(6) \qquad \lambda_\theta(t \mid F_t) = a_0 + P_J(t) + C_K(t) + \sum_{t_i < t} g_M(t - t_i).$$

The first and second components represent the evolutionary trend where

$$(7) \qquad P_J(t) = \sum_{j=1}^{J} a_j \phi_j(t), \qquad 0 < t < T,$$

T is the total length of the observed interval and $\phi_j(t)$ is a polynomial of order j. We introduce these components to stand for the seismic trend or the evolutionary change of the detection rate of shocks. The third component is the Fourier expansion

$$(8) \qquad C_K(t) = \sum_{k=1}^{K} \{b_{2k-1} \cos(2k\pi t/T_0) + b_{2k} \sin(2k\pi t/T_0)\},$$

for cyclic effects with a given fixed cycle length T_0, such as seasonal, monthly, diurnal and semidiurnal. The last component in (6) stands for the clustering effects such as aftershocks, foreshocks and earthquake swarms. The function $g_M(x)$ measures the increase in risk due to a shock. We may call this function a response function of a shock, and parametrize it as

$$(9) \qquad g_M(x) = \sum_{m=1}^{M} c_m x^{m-1} e^{-\alpha x}.$$

If the scaling parameter α is fixed in (9), then the model (6) is linearly parametrized. This model is used for establishing the existence of each component and estimating its shape.

2.2. *Detection of the causal relations between two series of events.* We extend the model (6) to

$$(10) \qquad \lambda_\theta(t \mid F_t) = a_0 + P_J(t) + C_K(t) + \sum_{t_i < t} g_M(t - t_i) + \sum_{u_j < t} h_N(t - u_j),$$

where $\{u_j\}$ is another series of events considered as inputs of the CIF system. The response function $h_N(x)$ is parametrized by

$$(11) \qquad h_N(x) = \sum_{n=1}^{N} d_n x^{n-1} e^{-\beta x}.$$

If there is no causal relation of $\{u_j\}$ to the CIF $\lambda_\theta(t \mid F_t)$, or to $\{t_j\}$, then $h_N(x) = 0$ is expected. Otherwise we are interested in knowing the influential time scale and the approximate shape of the response function.

A parallel model to (10) for $\{u_j\}$ versus $\{t_i\}$ as the inputs is also considered. Since the joint density distribution of $\{t_i, u_j\}$ with bivariate intensity $(\lambda_\theta(t \mid F_t), \mu_\zeta(s \mid F_s))$ is given by

$$L_t = \prod_{t_i < t} \lambda_\theta(t_i) \prod_{u_j < t} \mu_\zeta(u_j) \exp\left\{-\int_0^t (\lambda_\theta(s) + \mu_\zeta(s) - 1) ds\right\},$$

it is noted that the maximization of the log likelihood for the mutually exciting bivariate point process $\{t_i, u_j\}$ is equivalent to the maximization of both of the two partial log likelihoods of the type in (3), as far as the sets of parameters between θ and ζ of the two CIF are disjoint. Therefore the detection of mutual or one-way stimulation is reduced to the independent analysis of model (10) using the partial likelihood. It is also noted that the existence of the second-order correlation between the processes $\{t_i\}$ and $\{u_j\}$, or between their CIF, does not necessarily imply the existence of a stimulation. For example a similar shape or periodicity of corresponding $P_J(t)$ or $C_K(t)$ can make some correlation even if $h_N(x) = 0$ holds. There is also the case where some other factors which are not contained in the above models stimulate both $\{t_i\}$ and $\{u_j\}$.

2.3. *Akaike information criterion for the model selection.* To examine the existence of each component and to see the shapes of their response functions, we find the best model among different feasible combinations and different orders. For this purpose we adopt the Akaike information criterion, AIC (Akaike (1977)) as a measure to see which combination and orders of the model (6) (or (10) in the case where the data $\{u_j\}$ are given) most frequently reproduce similar features to the given data set of events $\{t_i\}$. AIC is defined by

$$(12) \qquad \text{AIC} = (-2) \max (\log \text{likelihood}) + 2(\text{number of parameters used}),$$

where log denotes the natural logarithm. The model with the smaller AIC shows the better fit to the data. It is sometimes useful to note that the log likelihood ratio statistic takes the form

$$(-2) \log \Lambda(H_0; H_1) = \text{AIC}\,(H_0) - \text{AIC}\,(H_1) + 2k,$$

where the model H_1 contains the model H_0 as a restricted family, and k denotes the difference of dimensions of parameters in H_0 and H_1. Under the null hypothesis H_0, the statistic is expected to have the chi-squared distribution with degree of freedom k. The relation of the minimum AIC procedure to the conventional likelihood ratio test is discussed in Akaike (1977), (1983).

It is reasonably feasible for us to calculate AIC values for all orders of a pair of the components up to a moderately high order, but it is time-consuming to do so for all these components. Instead we first fit only the clustering component

$$(13) \qquad \lambda_\theta(t \mid F_t) = a_0 + \sum_{t_i < t} g_M(t - t_i)$$

up to a suitable order M of the Laguerre-type polynomial (9), since the range of influence of $g_M(x)$ is usually significantly shorter than any other components. Thus we obtain the optimal order M_0 with minimum AIC value and estimate of corresponding exponential coefficient α of (9), before fitting (6) for all pairs of J and K up to suitable orders. In the case where the data u_j are available, fixing the coefficient α and the order M_0 above, we fit (Ogata et al. (1982a, b))

$$(14) \qquad \lambda_\theta(t \mid F_t) = a_0 + \sum_{t_i < t} g_M(t - t_i) + \sum_{u_j < t} h_N(t - u_j)$$

for all orders N to find the optimal N_0 and to estimate the corresponding exponential coefficient β in (11), before examining (10) for the pairs such as (J, N) with fixed α and β.

Some of the procedure above is not strict maximum likelihood estimation. It is possible for us to get the strict maximum likelihood using the very good initial estimates obtained above. Usually, according to our experiments, the improvement of the log likelihood values is not sufficiently significant to alter the preference of model according to AIC. Also, the re-estimated intensity rate is nearly the same as the previous estimate. The reason may be that the scaling parameters α and β are not similar to one another. Nevertheless, the exact log likelihood should be calculated for at least a few essential models, if the AIC values are used in a hypothesis-testing context.

2.4. *Some technical remarks.* The model (10) (including (6)) is linearly parameterized if the exponential coefficients α and β are fixed. It is useful for the accurate and efficient calculation of the log likelihood (3) with many parameters to reparametrize (6) and (10) as follows:

(i) The polynomial $\phi_j(t)$ is set by $\psi_j(2t/T - 1)$ where T is the total length of the observation interval and ψ_j is a suitable orthogonal polynomial such as Legendre functions defined on $[-1, 1]$, in which case we can use a recursive relation

$$\psi_0(x) = 1, \qquad \psi_1(x) = x,$$
$$\psi_j(x) = 2x\psi_{j-1}(x) - \psi_{j-2}(x) - (x\psi_{j-1}(x) - \psi_{j-2}(x))/j, \quad \text{and}$$
$$\int \psi_j(x)dx = \{\psi_{j+1}(x) - \psi_{j-1}(x)\}/(2j+1).$$

(ii) For the cyclic components, we put $c_{2k-1}(t) = \cos(2\pi kt/T_0)$ and $c_{2k}(t) = \sin(2\pi kt/T_0)$.

(iii) For each term in (9) put $p_m(x) = (\alpha x)^{m-1}e^{-\alpha x}$, and then set

$$P_m(t) = \sum_{t_i < t} p_m(t - t_i) = \int_0^t p_m(t - s)dN_s \qquad m = 1, 2, \cdots, M.$$

Then we can use a recursive relation for

$$P_m(t) = p_m(t - t_i) + \sum_{\nu=1}^{m} \binom{m-1}{k-1} P_\nu(t_i)p_{m-\nu+1}(t - t_i), \qquad t_i < t \leq t_{i+1}.$$

Moreover by the change of integral signs with respect to dt and dN_s

$$(15) \qquad R_m = \int_0^T P_m(t)dt = \int_0^T \int_s^T p_m(t - s)dtdN_s = \sum_{i=1}^{I} \int_0^{T-t_i} p_m(t)dt.$$

(iv) For each term in (11) put $q_n(x) = (\beta x)^{n-1}e^{-\beta x}$, and then set

$$Q_n(t) = \sum_{u_j < t} q_n(t - u_j) = \int_0^t q_n(t - s)dU_s, \qquad n = 1, 2, \cdots, N.$$

Thus we can use another recursive relation

$$Q_n(t) = \sum_{t_i < u_j < t} q_n(t - u_j) + \sum_{\nu=1}^{n} \binom{n-1}{\nu-1} Q(t_i)q_{n-\nu+1}(t - t_i), \qquad t_i < t \leq t_{i+1}.$$

Again by the change of integral signs with respect to dt and dU_s

$$(16) \qquad S_n = \int_0^T Q_n(t)dt = \int_0^T \int_s^T q_n(t - s)dtdU_s = \sum_{j=1}^{J} \int_0^{T-u_j} q_n(t)dt.$$

Furthermore it is useful for the accurate and efficient maximization of the log likelihood (3) with many parameters to make the sizes of the diagonals of the Hessian matrix similar. Thus we reparametrize (6) and (10) to get a favourable $\{Q_k(t \mid F_t)\}$ in (4) as follows:

(v) Replace $\phi_j(t)$ in (7) by $\phi_j(t)/\{\sum_{i=1}^{I} \phi_j^2(t_i)\}^{\frac{1}{2}}$, $j = 1, 2, \cdots, J$.

(vi) Replace $c_{2k-1}(t)$ and $c_{2k}(t)$ by $c_{2k-1}(t)/I^{\frac{1}{2}}$ and $c_{2k}(t)/I^{\frac{1}{2}}$, respectively, $k = 1, 2, \cdots, K$.

(vii) Replace $p_m(x)$ by $p_m(x)/\{\sum_{i=1}^{I} P_m(t_i)^2\}^{\frac{1}{2}}$, and $q_n(x)$ by $q_n(x)/\{\sum_{i=1}^{I} Q_n(t_i)^2\}^{\frac{1}{2}}$, $m = 1, 2, \cdots, M$ and $n = 1, 2, \cdots, N$, respectively.

By such replacement the maximum log likelihood value does not change, while the estimates of parameters thus obtained are transformed into those for the original parametrization (6) and (10).

2.5. *Constraints of the parameters and penalty functions.* It is not easy to define an explicit constraint on the parameters that will ensure the CIF remains non-negative throughout the time interval. Even an approximation needs much computing time. Instead, we consider the following natural constraints for each J, K, M and N in (10):

$$g_M(x) \geq 0 \quad \text{for} \quad x \geq 0,$$

(17)
$$h_N(x) \geq 0 \quad \text{for} \quad x \geq 0, \text{ and}$$

$$a_0 + P_J(t) + C_K(t) \geq 0 \quad \text{for} \quad 0 \leq t \leq T.$$

In the rest of the subsection we incorporate these constraints (17) into the gradient for use with the optimizing subroutine.

For the Mth- and Nth-order response functions $g_M(x)$ and $h_N(x)$ we choose a partition of $[0, \infty)$ such as $\{\delta_\nu = 4(\nu - 1)/3; \nu = 1, 2, \cdots, 12 \max(M, N)\}$, and put

$$g_{m\nu} = \int_{\delta_\nu/\alpha}^{\delta_{\nu+1}/\alpha} p_m(x)dx \quad \text{and} \quad h_{n\nu} = \int_{\delta_\nu/\beta}^{\delta_{\nu+1}/\beta} q_n(x)dx,$$

$$m, n = 1, 2, \cdots, \max(M, N),$$

where $p_m(x)$ and $q_n(x)$ are given in (vii) of Section 2.4. Then the first inequality in (17) is reduced to the linear constraints for the reparametrized c_m

(18)
$$G_\nu = \sum_{m=1}^{M} c_m g_{m\nu} \geq 0, \quad \nu = 1, 2, \cdots, 12M.$$

Similarly the second inequality for $h_N(x)$ is approximately reduced to the constraints for the reparametrized d_n

(19)
$$H_\nu = \sum_{n=1}^{N} d_n h_{n\nu} \geq 0, \quad \nu = 1, 2, \cdots, 12N.$$

For the polynomial $P_J(t)$, the interval $[0, T]$ is divided into equal subintervals by the points, $w_\nu = \nu T/120$, $\nu = 1, 2, \cdots, 120$, say. And for the $C_K(t)$ the interval $[0, T_0]$, where T_0 is periodicity, is divided by $w'_\nu = \nu T_0/60$, $\nu = 1, 2, \cdots, 60$, say. Then the last inequality in (17) is reduced to the following linear restrictions;

$$a_0 \geqq 0, \qquad a_0 = y_1 + y_2,$$

$$(20) \qquad P_\nu = y_1 + \sum_{j=1}^{J} a_j \Phi_{j\nu} \geqq 0, \qquad \nu = 1, 2, \cdots, 120, \quad \text{and}$$

$$D_\nu = y_2 + \sum_{k=1}^{K} (b_{2k-1} C_{2k-1,\nu} + b_{2k} C_{2k,\nu}) \geqq 0, \qquad \nu = 1, 2, \cdots, 60,$$

where

$$\Phi_{j\nu} = \int_{w_{\nu-1}}^{w_\nu} \phi_j(t)\,dt, \quad \text{and} \quad C_{k\nu} = \int_{w'_{\nu-1}}^{w'_\nu} c_k(t)\,dt.$$

Thus the penalty function $G(\theta)$ in (5) may be given by

$$(21) \qquad G(\theta) = W \left\{ (a_0^-)^2 + \sum_{\nu=1}^{120} (P_\nu^-)^2 + \sum_{\nu=1}^{60} (C^-)^2 + \sum_{\nu=1}^{12N} (G^-)^2 + \sum_{\nu=1}^{12N} (H_\nu^-)^2 \right\},$$

where W is a large constant, $a_0^- = \min(a_0, 0)$, $P^- = \min(P_\nu, 0)$, and so on.

3. Analysis of earthquake data

3.1. *Seismic activity around Canberra.* The source of the seismicity data examined here was the south-eastern Australia catalogue compiled by the Research School of Earth Sciences, Australian National University. Here the data set for 21 years from 1960 to 1980 was considered, and the region used was a rectangle from 148°E to 150°E and from 34°S to 37°S (Figure 1(a)) including the Australian Capital Territory (ACT). The graphs in Figure 1(b) are plots of log cumulative number of shocks and log frequency of shocks versus magnitude (M) of the region, respectively, which should be approximately linear if no shocks are missing. It is noted for the subsequent analysis that the magnitudes were reclassified as in Figure 1(b), i.e. M 2.5 stands for M 2.3–2.7, to keep the homogeneity of the magnitude data set throughout the period.

The minimum magnitude of completeness ($M \min_c$) is the magnitude above which all events which occurred are reported. The plots in Figure 1(b) are fitted with a straight line and $M \min_c$ (the level at which the data fall below the lines) for the data set of the considered area is magnitude 2.5; 583 shocks were obtained by cutting off the shocks with smaller magnitudes. This is about 23 per cent of all events detected in the area. The selected data set is reproduced by the transformation into i-day, i.e., the time scale unit is one-day with the origin of the time being equated to 1 January 1960. Figure 1(c) includes the cumulative number of such shocks versus time, as well as those of other magnitude cutoffs.

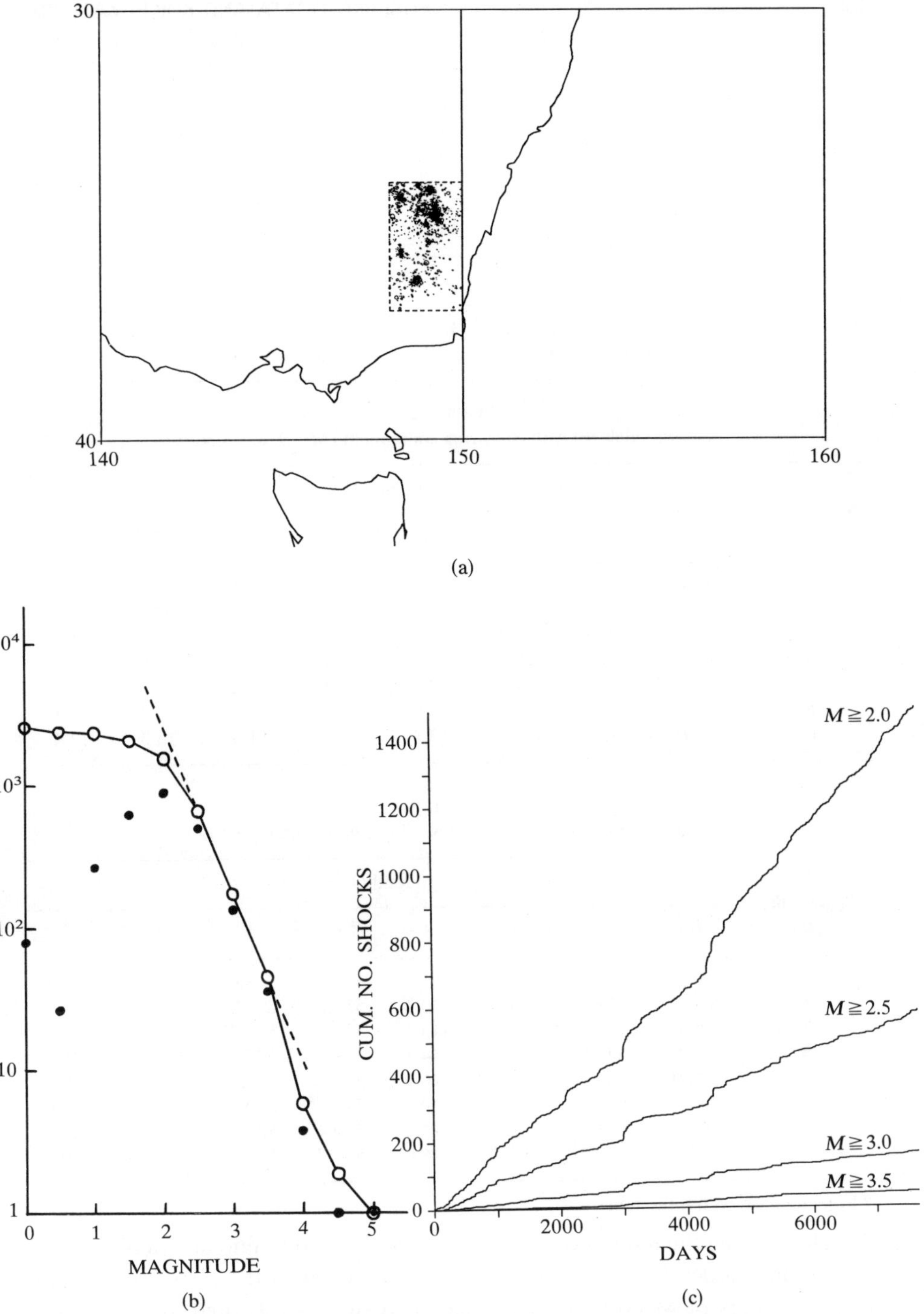

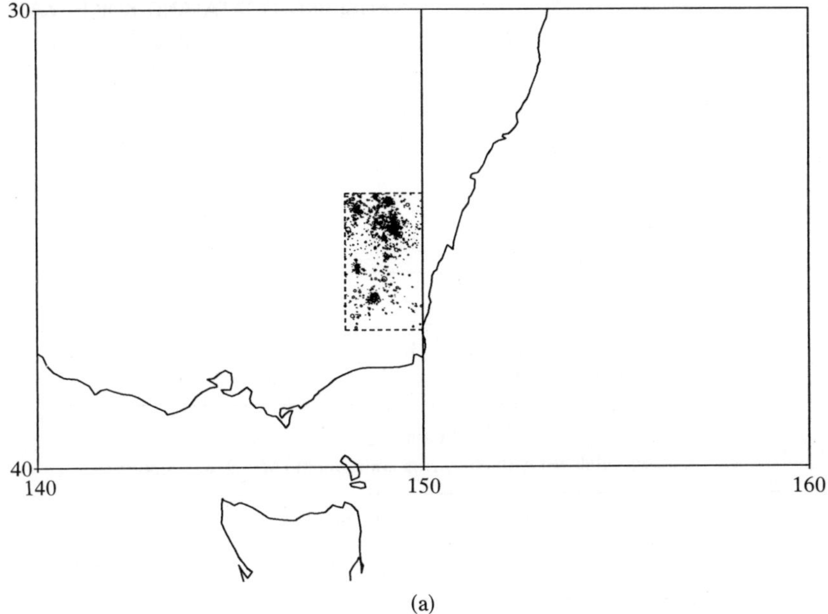

Figure 1. (a) Considered area in south-eastern Australia. (b) Cumulative number of shocks (○) and frequency of shocks (●) versus magnitude. (c) Cumulative number of shocks versus time for the corresponding magnitude thresholds

TABLE 1
AIC and estimated exponential coefficient of (13)

M	0	1	2	3	4	5	6	7
AIC	557.5	68.0	70.0	27.4	29.4	9.6	11.6	1.5
α	—	1.668	1.668	3.685	3.685	2.860	2.860	3.912

M	8	9	10	11	12	13	14
AIC	3.5	0.0*	2.1	3.5	5.6	7.6	9.6
α	3.912	3.368	3.334	3.021	4.373	4.386	4.473

TABLE 2
AIC of models (6) with clustering component ($M = 9$, $\alpha = 3.368$)

Seasonality (K)	Trend (J)							
	0	1	2	3	4	5	6	7
0	0.0	2.0	3.6	5.6	7.6	4.2	6.1	6.4
1	−0.4	1.6	3.3	5.2	7.2	3.4	5.4	6.0
2	−2.3*	−0.6	1.3	3.2	5.2	1.5	3.5	3.2
3	−1.0	0.7	2.6	4.4	6.4	3.0	5.0	4.2
4	1.5	3.2	5.2	7.0	9.0	4.8	6.8	6.3
5	5.4	7.1	9.1	10.9	12.9	8.7	10.7	10.1
6	7.9	9.7	11.6	13.4	15.4	11.3	13.3	12.4

TABLE 3
AIC of models (6) without clustering component ($M = 0$)

Seasonality (K)	Trend (J)							
	0	1	2	3	4	5	6	7
0	557.5	559.5	559.8	561.2	559.9	548.0	548.9	548.5
1	556.5	558.5	558.8	560.4	557.8	543.2	544.3	544.6
2	539.6	541.4	541.5	543.4	539.5	525.4*	526.5	526.6
3	540.9	542.7	542.2	544.1	539.5	527.5	528.4	528.5
4	543.9	545.7	545.0	546.9	542.3	529.6	530.9	529.2
5	546.3	548.3	547.9	549.7	545.1	532.5	533.7	531.6
6	540.2	542.1	542.0	544.0	540.4	527.1	528.5	527.4
7	542.5	544.2	544.1	546.0	541.7	529.3	530.7	529.6

* Minimum AIC within the table

Before fitting the model (6), we used the model (13) to find an initial guess of the optimal order M and the exponential coefficient in the response function (9). All orders of M up to 14 were examined successively, and AIC values and estimated exponential coefficients are listed in Table 1, which suggested $M = 9$ and $\alpha = 3.368$ (1/day). Thus the model (6) was fitted to calculate AIC for all pairs (K, J) up to 6 and 7, respectively, with fixed M, α obtained above and $T_0 = 365.25$. Also assuming $M = 0$ (i.e. no clustering), AIC were calculated for each (K, J). These are listed in Tables 2 and 3.

Comparing the AIC values in Tables 2 and 3, we see that existence of the clustering is very clear. Moreover it appears that a seasonal effect exists, but that evolutionary trend does not, since the overall minimum AIC is attained at $(M, K, J) = (9, 2, 0)$. Figures 2(a)–(c) display the estimated shapes of each component. The constant trend means no change of the seismic activity around

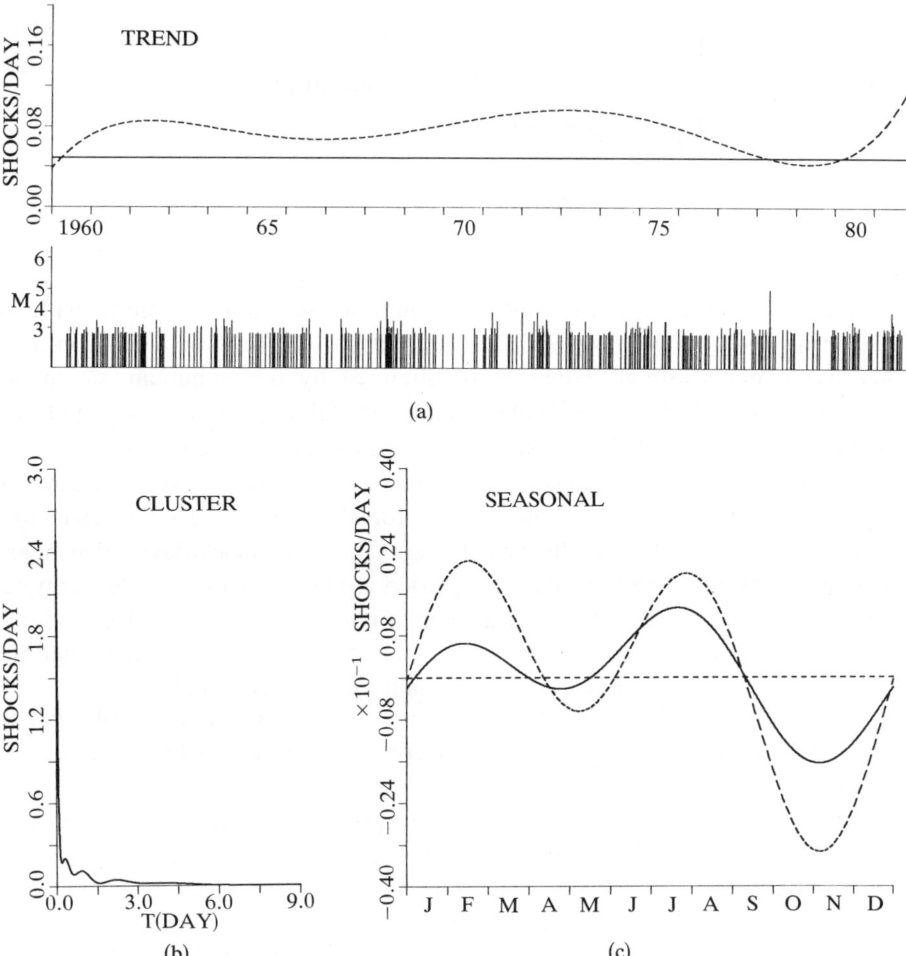

Figure 2. (a) Estimated trends for shocks with $M \geqq 2.5$. Solid and dotted lines are obtained by the minimum AIC models in Tables 2 and 3, respectively. The graph below shows the occurrences and magnitude of shocks. (b) Estimated response function (9) for the clustering component. (c) Estimated shapes of the seasonal effect. Solid and dotted lines are obtained by the minimum AIC models in Tables 2 and 3, respectively. (d) Seasonal rainfall patterns

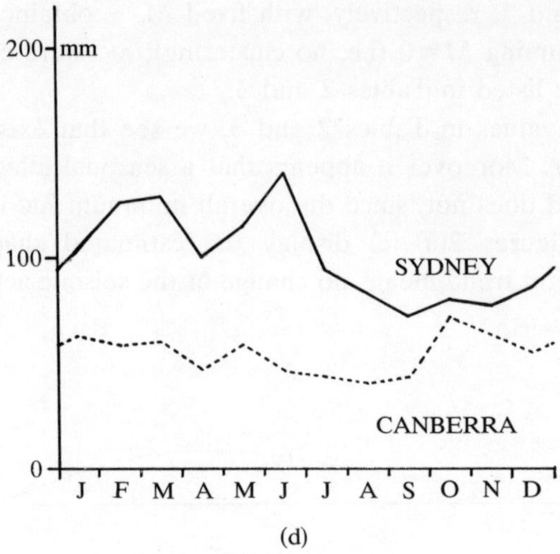

Figure 2 (*Continued*)

Canberra for these 21 years, which we could not see from the model without clustering components $(M = 9)$; the solid curves for the estimated evolutionary trend and seasonal change were obtained by the minimum AIC model $(K, J) = (2, 5)$ with $M = 0$ in Table 3. Also the difference between dotted and real curves in Figure 2(c) suggests that the seasonal component can be affected by the numbers of clusterings at each time of the year, unless the clustering component is considered in the models for the analysis. By the estimated response function in Figure 2(b) we get $\int_0^\infty g(s)ds = 0.16$ (shock/day) which means that about 93 out of 583 events are regarded as cluster members. Furthermore, since the level $a_0 = 0.049$ (shock/day) in (6) in the time interval considered $(T = 7669$ days), it would appear according to the present model that the number of shocks affected by the seasonality is not very small.

As discussed in Ogata (1983), we looked at the seasonal distribution of rainfall data. Monthly data only at Canberra and Sydney, nearby the area, were available in Rika Nenpyo (Scientific Tables, 1982) averaged over 21 and 31 years from 1931 and 1941 through 1960, respectively. Figure 2(d) displays such data. The seasonality in Figure 2(c) looks similar to the Sydney rainfall pattern rather than the Canberra one. This might suggest some correlation between earthquake occurrence around ACT and the migration of underground water across the south-eastern highlands. It is widely reported in seismological journals that drastic change of underground water migration can be a trigger for earthquakes. For example Muirhead (1981) discusses the incidence of seismic activity associated with the filling of a reservoir in south-eastern New South Wales, within the rectangular region in Figure 1(a).

3.2. *Examining a causal relation between shallow and deep seismicity.* Utsu (1975) discussed the correlation between the intermediate earthquake in Hida (the region around Takayama City) and the shallower earthquakes in the central Kwanto (the region around Tokyo). He tested the independence of the Kwanto earthquakes from those of the Hida area, assuming the former shocks to form a stationary Poisson process, and concluded that there was a significant dependence which could be attributed to the mechanical connection between the two seismic regions which belong to the same segment of the Pacific plate underthrusting at the north-eastern Japan arc. The data compiled by Utsu is composed of 61 earthquakes with magnitude $M \geqq 5.5$ in central Kwanto and 16 earthquakes of $M \geqq 5.0$ in Hida during the 51 years from 1924 through 1974.

Since the Kwanto shocks seem to have some slight clustering, we (Ogata et al. (1982a, b)) fitted the model (13) and then (14) against the Hida shocks as inputs, to estimate the scale and shape of the response functions. Similar analysis was made for Hida shocks versus Kwanto shocks as inputs. The results of our analysis showed that the earthquakes in central Kwanto region are not only self-exciting but also significantly receive one-way stimulation from earthquake occurrences in Hida region, and that Hida shocks are identified as a

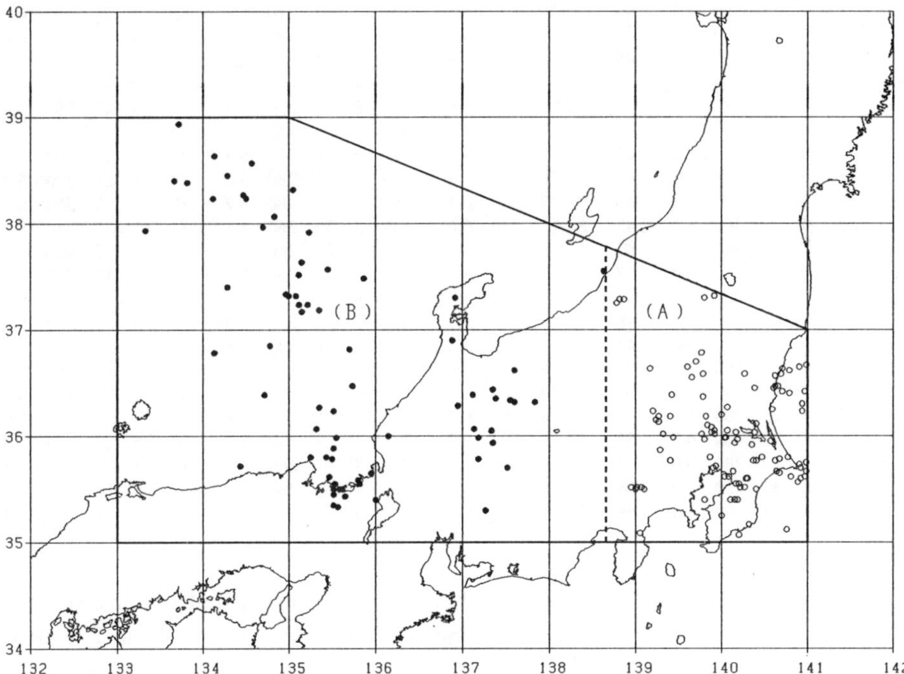

Figure 3. Spatial distributions of the shocks given in Table 6. Open circles and solid circles stand for the shallower and deeper shocks, respectively

TABLE 4

JMA data set in Figure 3

	Date				M	Date				M	Date				M	Date				M
	y	m	d	km		y	m	d	km		y	m	d	km		y	m	d	km	
Region (A)	26	8	3	20	6.3	33	2	13	0	5.6	46	1	30	60	5.6	60	1	14	60	5.7
	26	12	13	40	5.7	33	10	4	0	6.1	46	9	14	60	5.5	60	4	25	70	5.7
	27	9	5	40	5.5	34	5	31	10	5.8	48	10	29	50	6.3	62	2	6	120	5.8
	27	12	31	140	5.6	34	6	21	50	5.6	49	12	26	0	6.4	65	1	27	80	5.6
	28	1	1	20	5.6	35	2	20	60	6.0	49	12	27	0	5.8	65	4	6	60	5.5
	28	2	12	30	5.9	35	9	17	80	5.5	50	4	17	30	5.6	65	5	31	120	5.5
	28	5	21	70	6.2	35	9	30	50	5.9	50	9	10	0	6.3	67	3	19	80	5.5
	28	10	5	30	5.5	37	5	5	50	5.7	51	1	9	40	6.2	67	9	15	40	5.6
	29	4	18	40	5.9	38	2	7	100	6.2	52	5	8	50	6.0	68	7	1	50	6.1
	29	7	27	0	6.3	38	6	6	50	6.1	52	8	16	50	5.5	69	4	9	100	5.6
	30	5	1	10	5.7	40	7	15	40	5.8	52	11	2	60	5.5	71	9	21	180	5.7
	30	6	1	30	6.5	40	8	25	60	5.5	53	5	11	40	5.6	73	9	30	50	5.9
	30	8	17	0	6.1	40	11	27	130	5.5	54	1	17	60	5.6	73	10	1	60	5.8
	30	8	20	0	5.5	41	6	16	50	5.5	54	2	25	60	5.7	74	3	3	60	6.1
	30	11	26	0	7.3	42	2	19	0	5.8	54	6	5	60	5.5	74	8	4	50	5.8
	31	6	9	0	6.0	42	11	7	20	5.7	54	7	18	70	6.4	76	6	16	20	5.5
	31	6	11	0	5.9	43	1	20	60	5.5	55	1	17	70	5.7	76	12	29	130	5.8
	31	6	17	0	6.3	43	7	1	50	6.0	55	7	24	50	6.0	78	3	20	60	5.5
	31	7	10	0	5.5	43	8	12	0	6.2	56	2	14	50	6.0	79	10	28	90	5.5
	31	9	16	0	6.3	43	10	19	0	5.6	56	4	26	40	5.7	80	9	25	80	6.1
	31	9	21	0	6.9	44	6	16	50	6.1	56	9	30	60	6.3	82	3	7	60	5.5
	31	9	28	0	5.5	44	12	29	60	5.5	56	11	4	70	5.8	83	2	27	72	6.0
	31	10	3	40	5.6	45	1	5	50	5.5	59	3	18	80	5.6	83	8	8	22	6.0
	32	12	2	50	5.5	45	10	24	60	5.8	59	9	8	50	5.5					
Region (B)	27	1	15	380	6.3	39	9	22	360	5.5	60	8	20	220	5.1	73	8	15	280	5.0
	27	4	8	340	5.1	40	3	12	360	5.8	60	10	4	240	5.1	73	8	22	400	5.3
	27	6	12	320	5.2	42	3	28	380	5.3	62	4	11	320	5.3	73	12	25	400	5.5
	27	8	13	400	5.1	42	5	16	320	5.8	62	5	6	280	5.0	74	11	22	360	5.2
	27	11	11	220	5.3	43	9	27	360	6.1	65	3	21	280	5.2	76	9	9	380	5.1
	29	7	5	460	5.5	45	8	15	280	5.6	65	3	30	360	5.6	77	11	30	360	5.0
	31	4	21	380	6.1	47	4	7	360	5.3	65	6	24	360	5.5	78	3	7	440	5.7
	31	6	2	240	6.4	47	4	11	360	5.5	66	2	9	360	5.2	78	6	14	360	5.3
	31	6	29	400	5.1	47	7	30	420	5.7	67	8	14	360	6.6	78	10	9	260	5.7
	31	8	9	420	5.9	48	1	19	340	5.3	69	1	20	360	5.3	79	2	28	380	5.1
	32	7	25	360	6.4	48	7	13	340	5.2	69	4	1	420	6.8	79	5	28	300	5.4
	35	3	18	280	5.1	50	3	8	340	5.2	69	4	10	340	5.1	79	7	31	280	5.6
	35	4	15	240	6.1	50	9	24	240	5.0	69	10	15	360	5.1	80	1	19	440	6.3
	35	5	9	400	5.4	52	5	28	360	6.7	69	11	1	340	5.1	80	3	31	360	5.9
	35	5	31	480	6.7	53	6	28	260	5.7	70	7	7	380	5.2	80	6	22	380	5.2
	35	10	15	280	6.0	54	5	15	240	7.0	70	8	29	280	5.6	82	2	4	340	5.8
	36	10	20	420	6.0	58	9	2	400	6.1	71	5	12	380	5.0	82	6	16	460	5.8
	37	11	22	360	5.4	59	3	5	200	5.5	72	9	27	340	5.1	82	9	7	420	5.4

stationary Poisson process. However, there was a criticism that the size of the Hida data set may not be enough to conclude the independency from Kwanto shocks.

Therefore a data set was taken from the Seismological Bulletin of Japan Meteorological Agency (JMA) and its supplementary volume (1982) with the pentagonal area with vertices at the points (141°E, 35°N), (133°E, 35°N), (133°E, 39°N), (135°E, 39°N) and (141°E, 37°N) (Figure 3) to discuss the relation between the shallow and deep shocks in this area. That is, the region (A) with the longitude $\geqq 138.3°$ (including the whole Kwanto area) has shallower shocks with depth down to 200 km, and the other region (B) contains the deeper shocks. Hereafter we remove all shallow shocks (down to 50 km) from the region (B). The region (B) has three spatial clusters in the depths of 200–300 km (Hida area) and 300–400 km (Wakasa Bay and Sea of Japan). The occurrence origin times of shocks with $M \geqq 5.5$ in the region (A) and deep shocks with $M \geqq 5.0$ in the region (B) from 1926 through September 1983 are listed in Table 4. Firstly the shallow shocks with $M \geqq 5.5$ in (A) are fitted to the model (13). AIC values and estimated exponential coefficients are listed in Table 5. Then fixing the order $K = 1$ and $\alpha = 0.2519$ of (9) with the minimum AIC, the data set is fitted to the model (14) versus the deep shocks with $M \geqq 5.5$ in (B) (including 35 events), and the AIC values and estimated exponential coefficients are obtained in Table 6. Thus the model (10) for each pair of (J, N) is used by fixing $K = 0$, $M = 1$, α and β estimated above. Table 7 gives the AIC value for each pair, which suggest no stimulating effect. This is confirmed by the AIC values of Table 8 with simultaneously adjusted α, β and the trend component. The overall minimum AIC model among Tables 5–8 suggests a significant stimulating effect from (B) and a linearly decreasing trend of seismicity for the area (A). The graphs of each estimated response function are shown in Figures 4(a) and 4(b), and estimated intensity is given in Figure 4(c). According to these estimated response functions it might be said that respective average of 4

TABLE 5

AIC and estimated exponential coefficient of (13)

Order	0	1	2	3	4	5	6
AIC	0.0	−3.6*	−1.6	−1.3	0.7	2.5	4.5
α	—	0.2519	0.2519	0.2611	0.2611	0.2185	0.2185

TABLE 6

AIC and estimated exponential coefficient of (14) with $M = 1$ and $\alpha = 0.2519$ fixed

Order (N)	0	1	2	3	4	5
AIC	−3.6	−6.0*	−4.0	−2.3	−0.3	1.7
β	—	0.02117	0.02117	0.01096	0.01096	0.01070

TABLE 7
AIC of model (10) with $M = 1$, $\alpha = 0.2519$ and $\beta = 0.02117$ fixed

Input (N)	Trend (J)					
	0	1	2	3	4	5
0	−3.6	−11.6	−10.5	−8.8	−7.8	−5.8
1	−6.0	−15.1*	−13.9	−12.7	−11.7	−9.7
2	−4.0	−13.1	−12.0	−10.8	−9.7	−7.8
3	−2.0	−11.2	−10.0	−8.8	−7.7	−5.8
4	−0.8	−10.9	−9.4	−8.0	−6.6	−4.7
5	0.5	−8.9	−7.5	−6.2	−4.9	−3.0

TABLE 8
AIC of model (10) with $M = 1$ but α, β adjusted

Models	$N = 0, J = 0$	$N = 0, J = 1$	$N = 1, J = 0$	$N = 1, J = 1$
AIC	−3.6	−11.6	−6.1	−15.4*
α	—	0.3008	0.3185	0.4032
β	—	—	0.01995	0.02190

and 12 shocks out of 95 shocks with $M \geqq 5.5$ in the region (A) are likely to have taken place as the effects of clustering and stimulation from the region (B). A simulation of events for (A) region was made by the estimated intensity, given the same 35 occurrence times in region (B), and using the method in Ogata (1981). Figure 4(d) shows the result.

To see if a similar effect exists in the opposite direction, we again used the models (13), (14) and (10) for the shocks with $M \geqq 5.0$ in (B) against those $M \geqq 5.5$ in (A) as inputs. Corresponding AIC values are listed in Tables 9, 10 and 11, which suggest no effect of stimulation. Then the shocks in (A) for the input were restricted to those with larger magnitudes, but still no stimulations were found; for example AIC values are listed in Tables 12 and 13 for the restricted inputs with $M \geqq 6.0$. The minimum AIC model among Tables 9–13 suggest the non-stationary Poisson whose trend is shown in Figure 5. This trend can be regarded as seismic activity in deep region (B), since the minimum magnitude of completeness of the present JMA catalogues is not greater than $M \geqq 5.0$.

In fact, the present analyses could be another confirmation of the suggestion of Mogi (1973), i.e. deep to shallow, or west to east, migration of earthquakes within the descending lithosphere across the main island of Honshu in Japan.

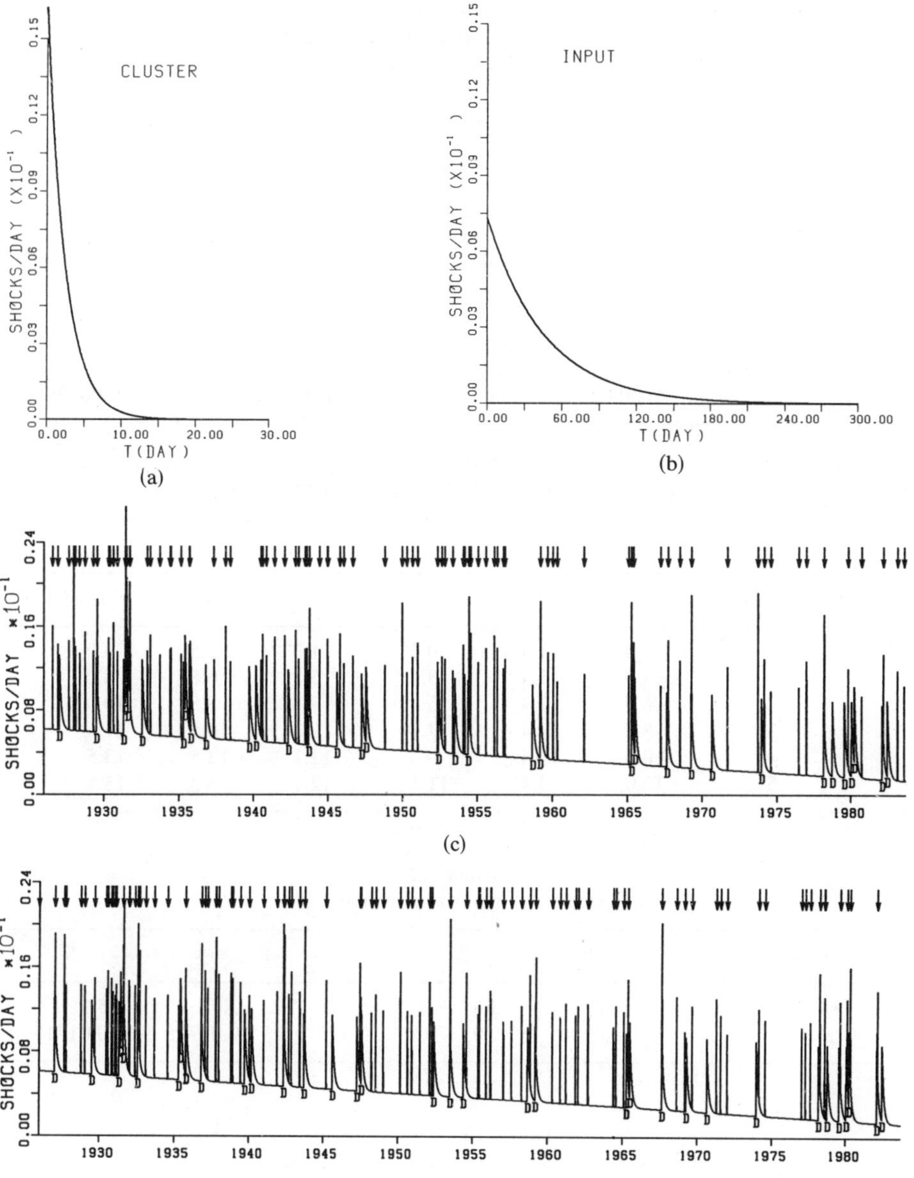

Figure 4. (a) Estimated response function (9) for clustering effect in the region (A). (b) Estimated response function (11) for stimulating effect from shocks ($M \geqq 6.0$) in region (B). (c) Estimated intensity of (A) shocks. The intensity has form

$$\lambda(t) = a_0 + a_1 t + \sum_{t_i < t} c \exp\left(-\alpha(t - t_i)\right) + \sum_{u_j < t} d \exp\left(-\beta(t - u_j)\right)$$

(shocks/day) with $a_0 = 0.006187$, $a_1 = -0.2292 \times 10^{-6}$, $c = 0.01623$, $d = 0.007306$, $\alpha = 0.4032$ and $\beta = 0.02190$. The arrows and symbol D indicate the occurrence times of (A) shocks and (B) shocks, respectively. (d) A simulated example of shocks by the intensity given in Figure 5(c). The 35 events D are same as the data

TABLE 9
AIC and exponential coefficient in (13)

Order (M)	0	1	2	3
AIC	0.0*	4.0	3.2	5.2
α	—	0.0244	0.0244	0.0244

TABLE 10
AIC for the model (10) with $M = 0$

Order (N)	0	1	2	3	4	5
AIC	0.0*	4.0	4.1	6.1	7.5	9.5
α	—	0.003165	0.003165	0.003165	0.003286	0.003286

TABLE 11
AIC of model (10) with $M = 0$ and $\beta = 0.003165$ fixed

	Trend (J)					
Input (N)	0	1	2	3	4	5
0	0.0	1.4	−0.6*	1.3	2.5	4.0
1	4.0	5.4	3.4	5.3	6.5	8.0
2	6.0	7.4	5.4	7.3	8.5	10.0
3	8.0	9.4	7.4	9.3	10.5	11.9
4	10.0	11.4	9.3	11.1	12.5	13.5
5	12.0	13.4	11.1	12.9	14.3	15.1

TABLE 12
AIC for the model (10) with $M = 0$

Order (N)	0	1	2	3	4	5
AIC	0.0*	4.0	3.8	5.8	3.9	5.9
β	—	0.0100	0.01270	0.01270	0.01715	0.01715

TABLE 13
AIC of model (10) with $M = 0$ and $\beta = 0.01270$ fixed

	Trend (J)					
Input (N)	0	1	2	3	4	5
0	0.0	1.4	−0.6*	1.3	2.5	4.0
1	4.0	5.3	3.3	5.2	6.4	7.8
2	5.7	6.9	4.7	6.5	7.7	9.1
3	7.7	8.8	6.6	8.5	9.6	11.0
4	9.6	10.5	8.5	10.4	11.6	12.9
5	10.9	11.2	9.7	11.5	12.9	14.1

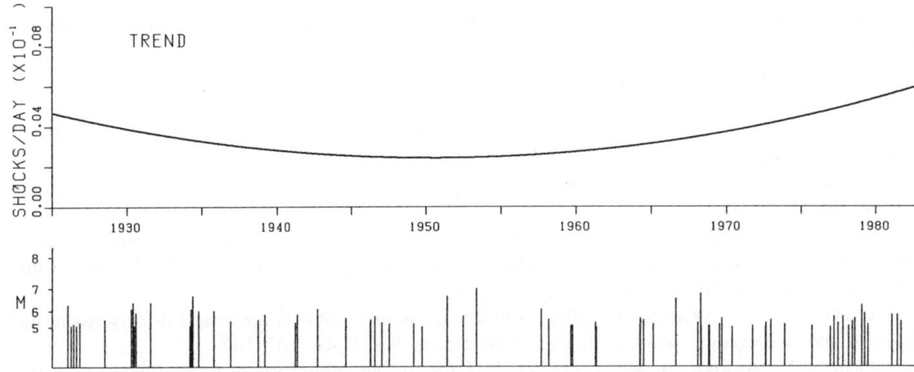

Figure 5. Estimated intensity of (B) shocks. The intensity has form $\lambda(t) = a_0 + a_1 t + a_2 t^2$ (shocks/day) with $a_0 = 0.46786 \times 10^{-2}$, $a_1 = -0.4792 \times 10^{-6}$ and $a_2 = 0.2557 \times 10^{-10}$. The graph below shows the occurrences and magnitudes of shocks with $M \geq 5.0$

Acknowledgements

We should like to express our appreciation of this opportunity to pay tribute to Professor Hannan's outstanding contributions to time series analysis and inference on stochastic processes. We are grateful to Dr K. J. Muirhead of the Australian National University, Professor K. Shimazaki of the Earthquake Research Institute, Tokyo University, and to the Japan Meteorological Agency for discussions and for supplying the data for the present analyses. We also thank the referee for his critical and helpful comments. The present research is partly supported by Grant-in-Aid for Scientific Research 58530014 from the Ministry of Education, Science and Culture of Japan. This work was also done as a part of the special research project on the statistical information processing of massive time series data in the Institute of Statistical Mathematics, and some of the computer programs used in the present paper will be released as a part of TIMSAC (Time Series Analysis and Control Program Package) series.

References

AKAIKE, H. (1977) On entropy maximization principle. In *Application of Statistics* ed. Krishnaiah, P. R., North-Holland, Amsterdam, 27–41.

AKAIKE, H. (1983) Information measure and model selection. *Proc. 44th Session Internat. Statistical Inst., Madrid, Spain, September 12–22, 1983. Bull. Internat. Statist. Inst.* **50**(1), 277–291.

FLETCHER, R. AND POWELL, M. J. D. (1963) A rapidly convergent descent method for minimization. *Computer J.* **6**, 163–168.

HAWKES, A. G. AND ADAMOPOULOS, L. (1973) Cluster models for earthquakes—regional comparison. *Bull. Internat. Statist. Inst.* **45**, 454–461.

LEWIS, P. A. W. AND SHEDLER, G. S. (1979) Simulation of non-homogeneous Poisson processes by thinning. *Naval Res. Logist. Quart.* **26**, 403–413.

LIPTZER, R. S. AND SHIRYAEV, A. N. (1978) *Statistics of Random Processes II, Applications.* Springer-Verlag, New York.

MOGI, K. (1973) Relationship between deep and shallow seismicity in the Western Pacific region. *Tectonophysics* **17**, 1–22.

MUIRHEAD, K. J. (1981) Seismicity induced by the filling of the Talbingo Reservoir. *J. Geol. Soc. Austral.* **28**, 291–298.

OGATA, Y. (1981) On Lewis' simulation method for point processes. *IEEE Trans. Inform. Theory* **IT-27**, 23–31.

OGATA, Y. (1983) Likelihood analysis of point process and its application to seismological data. *Proc. 44th Session Internat. Statistical Inst., Madrid, Spain, September 12–22, 1983. Bull. Internat. Statist. Inst.* **50**(2), 943–961.

OGATA, Y. AND AKAIKE, H. (1982a) On linear intensity models for mixed doubly stochastic Poisson and self-exciting point processes. *J. R. Statist. Soc.* B **44**, 102–107.

OGATA, Y., AKAIKE, H. AND KATSURA, K. (1982b) The application of linear intensity models to the investigation of causal relations between a point process and another stochastic process. *Ann. Inst. Statist. Math.* B **34**, 373–387.

UTSU, T. (1975) Correlation between shallow earthquakes in Kwanto region and intermediate earthquakes in Hida region, central Japan. *Zisin (J. Seism. Soc., Japan)* 2nd Ser. **28**, 303–311 (in Japanese).

VERE-JONES, D. (1970) Stochastic models for earthquake occurrence (with discussion). *J. R. Statist. Soc.* B **32**, 1–62.

VERE-JONES, D. AND OZAKI, T. (1982) Some examples of statistical inference applied to earthquake data. *Ann. Inst. Stat. Math.* **34**, 189–207.

Parameter Estimation for Finite-Parameter Stationary Random Fields

M. ROSENBLATT

Abstract

The concept of strong mixing is used to obtain a generalization of results on the asymptotic distribution of finite-parameter estimates of linear processes and extend them for stationary sequences and random fields.

STRONG MIXING; ASYMPTOTIC DISTRIBUTION; SMOOTHED PERIODOGRAM

Introduction

A notion of asymptotic independence for stationary sequences was introduced in Rosenblatt (1956). An extension of this concept for stationary random fields was discussed in Rosenblatt (1972). The object of this paper is to show how this idea can be used to get an extension of some results on asymptotic distribution of finite-parameter estimates for stationary sequences and random fields.

We briefly discuss strong mixing for random fields. Let S and S' be two sets of indices. Let $\mathscr{B}(S) = \mathscr{B}(X_\tau, \tau \in S)$ and $\mathscr{B}(S') = \mathscr{B}(X_\tau, \tau \in S')$ be the σ-fields generated by the random variables with index τ belonging to S and S' respectively. Consider $d(S, S')$, the Euclidean distance between the sets of indices S and S'. The random field X is said to be strongly mixing if

$$\sup_{A \in \mathscr{B}(S), B \in \mathscr{B}(S')} |P(AB) - P(A)P(B)| \leq \varphi(d(S, S'))$$

with φ a function such that $\varphi(d) \to 0$ as $d \to \infty$.

Discussion

The following generalization of a central limit theorem for stationary sequences satisfying a strong mixing condition (see Rosenblatt (1956)) holds for

Research supported in part by Office of Naval Research Contract N00014-81-K-0003 and National Science Foundation Grant DMS-8312106.

stationary random fields. Let $\{Y_{n_1,\cdots,n_k}\}$, $n_i = \cdots, -1, 0, 1, \cdots$, $i = 1, \cdots, k$ be a stationary random field with $EY_{n_1,\cdots,n_k} \equiv 0$, $EY^2_{n_1,\cdots,n_k} < \infty$. Assume that

$$(1) \qquad E \left| \sum_{\substack{n_i = a_i \\ i=1,\cdots,k}}^{b_i} Y_{n_1,\cdots,n_k} \right|^2 = h(b_1 - a_1, \cdots, b_k - a_k) \to \infty$$

as $b_1 - a_1, \cdots, b_k - a_k \to \infty$. Further let

$$(2) \qquad h(\alpha_1, \cdots, \alpha_k) = o(h(\beta_1, \cdots, \beta_k))$$

if $\alpha_1, \cdots, \alpha_k \to \infty$, $\alpha_i = O(\beta_i)$, $i = 1, \cdots, k$ but for some $j, \alpha_j = o(\beta_j)$. The derivation of the theorem parallels that given in Rosenblatt (1956).

Theorem 1. Let $\{Y_{n_1,\cdots,n_k}\}$ be a strictly stationary field with $EY_{n_1,\cdots,n_k} \equiv 0$ and $E|Y_{n_1,\cdots,n_k}|^{2+\delta} < \infty$ for some $\delta > 0$. Assume that Conditions (1) and (2) are satisfied and that the process is strongly mixing. Further, if

$$(3) \qquad E \left| \sum_{\substack{n_i = a_i \\ i=1,\cdots,k}}^{b_i} Y_{n_1,\cdots,n_k} \right|^{2+\delta} = O(h(b_1 - a_1, \cdots, b_k - a_k))^{1+\delta/2}$$

as $b_i - a_i \to \infty$, $i = 1, \cdots, k$, then

$$\sum_{\substack{n_i = 1 \\ i=1,\cdots,k}}^{\alpha_i m} Y_{n_1,\cdots,n_k}$$

when appropriately normalized is asymptotically $N(0, 1)$ as $m \to \infty$ for each fixed $\alpha_1, \cdots, \alpha_k > 0$. In particular, there are sequences $q_1(m), \cdots, q_k(m), p(m) \to \infty$ with $m/q_j(m) - p(m) = o(p(m))$, $q_j(m)p(m) \cong \alpha_j m, j = 1, \cdots, k$ such that the appropriate normalization is

$$\left\{ \left(\prod_{j=1}^{k} q_j(m) \right) h(p(m), \cdots, p(m)) \right\}^{\frac{1}{2}}.$$

If $q_1(m) \cdots q_k(m) h(p(m), \cdots, p(m)) \cong h(\alpha_1 m, \cdots, \alpha_k m)$ the normalization can be replaced by $\{h(\alpha_1 m, \cdots, \alpha_k m)\}^{\frac{1}{2}}$.

This central limit theorem can be directly applied to obtain results on the asymptotic distribution of quadratic functions of discrete time series or of random fields. Notice that no condition is made on the rate of decay of the mixing coefficient as is often the case for central limit theorems making use of mixing or mixingale conditions. The condition (3) can be replaced by a Lindeberg-like condition. Let us first make some remarks on the one-dimensional case. In a paper of Hannan and Heyde (1972) they deal with covariance estimates and assume that the process they deal with is of the form

$$(4) \qquad X_n = \sum_{j=0}^{\infty} \alpha_j \xi_{n-j}, \qquad \sum_{j=0}^{\infty} \alpha_j^2 < \infty$$

where the random variables ξ_j not only are a white-noise process, that is,

$$E\xi_j \equiv 0$$
$$E\xi_j \xi_k = \delta_{jk}\sigma^2$$

but also satisfy

$$(5) \qquad \begin{aligned} E[\xi_n \mid \mathcal{F}_{n-1}] &\equiv 0 \\ E[\xi_n^2 \mid \mathcal{F}_{n-1}] &= \sigma^2 \end{aligned}$$

with $\mathcal{F}_n$ the σ-field generated by the random variables $\xi_j, j \leq n$. If $\mathcal{F}_n$ is the σ-field generated by $X_j, j \leq n$, this implies that the process $\{X_n\}$ is such that the best predictor in mean square is a linear predictor. Of course, there is a very large class of processes having a representation of the form (4) in terms of a white-noise process for which the best predictor in mean square is non-linear. The theorem given below is valid for a reasonably large class of these processes. We shall give a version of the theorem applicable for random fields. Notice that in the case of random fields with $k \geq 2$, a martingale difference formulation like (5) does not appear to be a natural one.

Assume that one observes a stationary random field $X_{n_1,\cdots,n_k}$, $EX_{n_1,\cdots,n_k} \equiv 0$, over the cube of lattice points

$$n_i = 1, \cdots, m, \qquad i = 1, \cdots, k.$$

Consider the covariance estimates

$$r_{u_1,\cdots,u_k}(m) = \frac{1}{m^k} \sideset{}{'}\sum_{\substack{j_i=1 \\ i=1,\cdots,k}}^{m} X_{j_1,\cdots,j_k} X_{j_1+u_1,\cdots,j_k+u_k}$$

where the primed sum indicates that one only sums over indices $j_1, \cdots, j_k$ such that $j_1 + u_1, \cdots, j_k + u_k$ fall in the range $1, \cdots, m$. Let

$$(6) \qquad r_{u_1,\cdots,u_k} = \mathrm{Cov}\,(X_{j_1,\cdots,j_k}, X_{j_1+u_1,\cdots,j_k+u_k}).$$

We also shall assume that moments up to eighth order exist. This, of course, means that equivalently cumulants up to eighth order exist. In the following result absolute summability of (6) and of cumulants up to the eighth order will be assumed. From this point on we shall simply use boldface letters to represent k-vectors. This will avoid the tedious notation carried up to this point. Let $c_{a,b,d}$ be the fourth cumulant

$$c_{a,b,d} = \operatorname{cum}(X_n, X_{n+a}, X_{n+b}, X_{n+d}).$$

Theorem 2. Let $\{X_n\}$ be a strongly mixing k-dimensional strictly stationary random field with $EX_n \equiv 0$ and $EX_n^8 < \infty$. Assume that the cumulants of $\{X_n\}$ up to eighth order are absolutely summable. It then follows that $m^{k/2}(r_u(m) - r_u)$ for a fixed finite number of lags u are jointly asymptotically normal as $m \to \infty$ with mean 0 and covariances

$$R_{u,v} = \sum_a \{r_a r_{a+v-u} + r_{a+v} r_{a-u} + c_{u,a,a+v}\}.$$

This result is obtained by applying Theorem 1 with $\delta = 2$ to any linear combination of the terms $r_u(m) - r_u$. Thus, up to a negligible error, one is dealing with a partial sum

$$\sum_{j_1,\cdots,j_k=1}^{m} \alpha_s X_j X_{j+u(s)}$$

with weights α_s. Thus the random variables

$$Y_j = \sum_s \alpha_s X_j X_{j+u(s)}.$$

Estimates of the function $h(\alpha)$ as well as (3) are obtained from the summability conditions on the cumulants. They are good enough to ensure that one normalizes $r_u(m) - r_u$ by $m^{k/2}$.

The conditions we assume consist of strong mixing plus a limited number of moment conditions. Now it will be seen that conditions of this type will imply asymptotic normality of a class of smoothed periodograms. Let

$$I_m(\lambda) = I_m(\lambda_1, \cdots, \lambda_k)$$

$$= \frac{1}{(2\pi m)^k} \left| \sum_{j_1,\cdots,j_k=1}^{m} X_j \exp(-ij \cdot \lambda) \right|^2$$

$$= \frac{1}{(2\pi)^k} \sum_{\substack{|s_j| \leq m-1 \\ j=1,\cdots,k}} r_s(m) \exp(-is \cdot \lambda)$$

be the k-dimensional periodogram. We are interested in smoothed periodograms

$$\underbrace{\int_{-\pi}^{\pi} \cdots \int_{-\pi}^{\pi}}_{k} I_m(\lambda) A(\lambda) d\lambda$$

with a smoothing function $A(\lambda)$. The fourth-order cumulant spectral density

$$f_4(\lambda, \mu, \eta) = (2\pi)^{-3k} \sum_{a,b,d} c_{a,b,d} \exp\left(-i(a \cdot \lambda + b \cdot \mu + d \cdot \eta)\right)$$

is introduced because it arises naturally in the following theorem.

Theorem 3. *Let $X = \{X_n\}$ be an ergodic strictly stationary field that satisfies the assumptions of Theorem 2. Let*

$$\int I_m(\lambda)A_j(\lambda)d\lambda, \qquad j = 1, \cdots, s$$

be quadratic forms in X with real-valued weight functions $A_j(\lambda)$ symmetric about zero and square integrable. The quadratic forms then are asymptotically normal with means

$$\int f(\lambda)A_j(\lambda)d\lambda, \qquad j = 1, \cdots, s$$

and covariances

$$\cong \left(\frac{2\pi}{m}\right)^k \left\{2\int A_j(\lambda)A_k(\lambda)f^2(\lambda)d\lambda + \int\int f_4(\lambda, -\mu, \mu)A_j(\lambda)A_k(\mu)d\lambda d\mu\right\},$$

$j, k = 1, \cdots, s.$

This result is obtained by approximating the integrals $\int I_m(\lambda)A_j(\lambda)d\lambda$ by expansions in terms of a finite number of Fourier coefficients of the weight functions $A_j(\lambda)$. Theorem 2 is then directly applied to these approximations.

There is a class of finite parameter estimation problems for which there are now standard results under rather stringent conditions. These results are to a great extent based on insights of Whittle (1954). We shall qualitatively give results of this type under broader conditions of the type noted in Theorems 2 and 3. Further, the natural extension of these results to the case of finite-parameter stationary fields will be given. However, before giving these results it is best to discuss their character in the context of simple schemes, in particular that of linear moving averages in terms of independent, identically distributed random variables. Let

$$(7) \qquad X_n = \sum_{k=0}^{s} a_k \xi_{n-k}$$

with the ξ_k's independent, identically distributed and with $E\xi_k \equiv 0, a_0 = 1, E\xi_k^2 = 1$. A natural problem is that of estimating the coefficients a_k and σ^2 (the linear prediction error variance) from a sample $X_1, \cdots, X_m$ as $m \to \infty$. The ideas of Whittle are based on a Gaussian model. In effect, the parameters are estimated by maximizing an approximation to the likelihood function of the Gaussian process with the same covariance structure as that of the process $X = \{X_n\}$. In fact, if X were Gaussian, the a_k's would not be uniquely determined by full knowledge of the probability structure of the process. The problem is normalized in the following manner. Consider the polynomial

$$a(z) = \sum_{k=0}^{s} a_k z^k.$$

The spectral density of the process (7) is given by

$$f(\lambda) = \frac{1}{2\pi} |a(e^{-i\lambda})|^2.$$

In the Gaussian case only, the modulus of $a(e^{-i\lambda})$, $|a(e^{-i\lambda})|$ is specified by knowledge of the probability structure. It is conventional to specify the a_k's by requiring that they correspond to the polynomial $a(z)$ with all its zeros outside the unit disc in the complex plane, $|z| \leq 1$. This condition which is sometimes referred to as a *minimum phase condition* is often appealing because it corresponds to the linear prediction problem. However, it should be mentioned that non-Gaussian models (7) with the a_k's not satisfying the minimum phase condition arise in a number of geophysical applications (see Wiggins (1978) and Donoho (1981)). In fact, if $E\xi_k^3 \neq 0$, the a_k's can essentially be estimated whether they are minimum phase or not by quite different techniques. A detailed discussion of these alternative procedures can be found in Rosenblatt (1980) and Lii and Rosenblatt (1982).

The result on estimation of parameters will be phrased in the context of a k-dimensional random field. Assume that $X_{n_1,\cdots,n_k}$, $n_i = 1, \cdots, m$, $i = 1, \cdots, k$, are observed and that $I_m(\boldsymbol{\lambda})$ is the periodogram computed in terms of this data. If the field were Gaussian it is plausible that the m^{-k} times the log likelihood of the observations could be approximated by

$$(8) \qquad -\tfrac{1}{2}\log 2\pi\sigma^2 - (2\sigma^2)^{-1} \int I_m(\boldsymbol{\lambda}) g(\boldsymbol{\lambda}, \theta)^{-1} d\boldsymbol{\lambda}$$

with

$$\sigma^2 = (2\pi)^k \exp\left\{ \frac{1}{(2\pi)^k} \int \log f(\boldsymbol{\lambda}) d\boldsymbol{\lambda} \right\}$$

under appropriate conditions. Here

$$g(\boldsymbol{\lambda}, \theta) = \left\{ \frac{\sigma^2}{(2\pi)^k} \right\}^{-1} f(\boldsymbol{\lambda}, \theta; \sigma^2)$$

with f the spectral density of the field and $\theta = (\theta_1, \cdots, \theta_p)$ the p unknown parameters other than the variance σ^2. Estimates of the unknown parameters are obtained by computing the values $\hat{\theta}_m$ and $\hat{\sigma}_m^2$ maximizing this approximate log likelihood (in the Gaussian case). Nonetheless, it will be shown that these estimates are asymptotically consistent and Gaussian even in a range of non-Gaussian and non-linear models.

The assumptions we make will parallel those made by Walker (1963).

Assumptions. (1) *An absolute maximum of* (8) *is assumed for* $\theta = \hat{\theta}_n$, $\sigma^2 = \hat{\sigma}^2_m$.

(2) *The admissible values lie in a region* $0 < \sigma^2 < \infty$, $\theta \in \Theta$ *with* Θ *a bounded closed set in an open set S of p-dimensional Euclidean space.*

(3) $g(\lambda, \theta)$ *and* $g(\lambda, \theta)^{-1}$ *are continuous functions of* λ.

(4) *Given two distinct points* θ_1, θ_2, $g(\lambda, \theta_1)$ *and* $g(\lambda, \theta_2)$ *are not identically the same.* (Notice that it is at this point that in the one-dimensional case a condition like minimum phase is required. It should also be remarked that in the case of random fields ($k \geq 2$), there is no uniquely persuasive generalization of the minimum phase notion. There are many.)

(5) *The process* X_n *satisfies the conditions of Theorem* 3.

(6) *If* $h(\lambda, \theta) = \{g(\lambda, \theta)\}^{-1}$ *then h is continuously differentiable in* θ *up to third order as a function of* (λ, θ). *Further, let* $f(\lambda, \theta; \sigma^2)$ *have bounded first derivatives in* λ.

It should be clear that the assumptions cited are not intended as minimal conditions.

Theorem 4. *Let the assumptions cited above hold. Further assume that the matrix* $W = \{w_{ij}\}$ *with*

$$w_{ij} = \frac{\sigma_0^2}{(2\pi)^k} \int \frac{\partial}{\partial \theta_i} \log g(\lambda, \theta) \frac{\partial}{\partial \theta_j} \log g(\lambda, \theta) d\lambda \,|_{\theta = \theta_0}$$

is non-singular. Then $m^{k/2}\{\hat{\theta}_m - \theta_0\}$ (θ_0 *and* σ_0^2 *the true values) is asymptotically normal (as* $m \to \infty$) *with mean zero and covariance matrix* $W^{-1}QW^{-1}$ *where* $Q = (q_{ij})$ *is given by*

$$q_{ij} = \left\{ 2\sigma_0^2 w_{ij} + \int\int f_4(\lambda, -\mu, \mu) \frac{\partial}{\partial \theta_i} g^{-1} \frac{\partial}{\partial \theta_j} g^{-1} d\lambda d\mu \right\}_{\theta = \theta_0}.$$

The proof of this result can be obtained by using the scheme of Walker's derivation and employing our Theorem 3. Without pursuing it any further we note that one can show that $m^{k/2}\{\hat{\theta}_m - \theta_0\}$ and $m^{k/2}\{\hat{\sigma}^2_m - \sigma_0^2\}$ are jointly asymptotically normal.

The results have been derived for a cubic layout of indices of data points. However, the same arguments can be used for a rectangular parallelepiped as the layout of indices of data points.

A discussion of related estimation procedures for Gaussian random fields can be found in Larimore (1977). A more extended treatment of topics is given in Rosenblatt (1985).

References

DONOHO, D. (1981) On minimum entropy deconvolution. In *Applied Time Series Analysis II*, ed. D. F. Findley, Academic Press, New York, 565–608.

HANNAN, E. J. AND HEYDE, C. C. (1972) On limit theorems for quadratic functions of discrete time series. *Ann. Math. Statist.* **43**, 2058–2066.

LARIMORE, W. E. (1977) Statistical inference on stationary random fields. *Proc. IEEE* **65**, 961–970.

LII, K. S. AND ROSENBLATT, M. (1982) Deconvolution and estimation of transfer function phase and coefficients for non-Gaussian linear processes. *Ann. Statist.* **10**, 1195–1208.

ROSENBLATT, M. (1956) A central limit theorem and a strong mixing condition. *Proc. Nat. Acad. Sci. U.S.A.* **42**, 43–47.

ROSENBLATT, M. (1972) Central limit theorem for stationary processes. *Proc. 6th Berkeley Symp. Math. Statist. Prob.* **2**, 551–561.

ROSENBLATT, M. (1980) Linear processes and bispectra. *J. Appl. Prob.* **17**, 265–270.

ROSENBLATT, M. (1985) *Stationary Sequences and Random Fields*. Birkhäuser-Verlag, Basel.

WALKER, A. M. (1963) Asymptotic properties of least squares estimates of parameters of the spectrum of a stationary non-deterministic time-series. *J. Austral. Math. Soc.* **4**, 363–384.

WHITTLE, P. (1954) *A Study in the Analysis of Stationary Time Series*. Almquist and Wiksell, Uppsala.

WIGGINS, R. A. (1978) Minimum entropy deconvolution. *Geoexploration* **16**, 21–35.

PART 6

ALLIED STOCHASTIC PROCESSES

Mixed Cox Processes, with an Application to Accident Statistics

M. S. BARTLETT

Abstract

A doubly stochastic switching model previously used by the author to check against doubly stochastic 'white-noise' models is generalized in the case of Poisson processes to any distribution of fluctuating intensity rates. The resulting Cox process is used in place of a simple Poisson process as the basis of a mixed model, which is then fitted to some data on automobile accidents previously analysed by Seal (1980) in terms of a mixed Poisson process. It is shown that there is just significant evidence in favour of some additional heterogeneity 'within' each year for individual drivers; the nature of the data analysed (yearly time intervals) does not, however, allow discrimination between heterogeneity which is definitely attributable to a common fluctuating environment, and that which is in effect independent for each driver.

DOUBLY STOCHASTIC; SWITCHING MODELS; MIXED MODELS; AUTOMOBILE ACCIDENTS; NEGATIVE BINOMIAL

1. Introduction

In earlier papers (e.g. Bartlett (1982), (1983)) I have made considerable use of doubly stochastic switching models largely as a precautionary check against the convenient but sometimes dangerous use of doubly stochastic 'white-noise' models. In particular I have noted the more precise asymptotic results available for a doubly stochastic Poisson switching model (an example of a Cox process). For simplicity the switch in the intensity rate λ was confined to two possible values (with equal rates of switching from one value to the other). However, if the model is to be used in its own right an extension to a distribution of possible λ values permits greater flexibility (and is not difficult to simulate). Even if the relevance of the model in real situations may still seem sometimes a little artificial, it represents an exact approach enabling a study of the conditions under which various asymptotic results common with related Cox processes, such as Gauss–Poisson processes, are adequate. The model is then used to develop and apply the properties of such mixed Cox processes, in contrast with mixed Poisson processes, being perhaps rather more realistic and more general models for certain types of accident data.

2. The general Poisson switching model

To consider first the general Poisson switching model, the intensity rate λ_t at time t is assumed to switch at rate ε from its existing value to some new value taken at random from an arbitrary distribution specified by its differential element $dG(\lambda)$. (An even more general Markovian model would suppose that the transition rate ε varies with the existing value of λ, but this seems a rather unnecessary further complication.) It should be noted that, for convenience, a switch to the *same* value of λ is not ruled out; this convention is of no consequence for distributions $dG(\lambda) = g(\lambda)d\lambda$, but its effect should be remembered for discrete distributions. For example, in the simple case of $dG(\lambda) = \frac{1}{2}$ for $\lambda = \lambda_1$ or λ_2, the effective rate of switching to the alternative value of λ is $\frac{1}{2}\varepsilon$.

For the total moment-generating function $M_t(\theta) \equiv E\{\exp(\theta N_t)\}$, where N_t is the number of events in the time interval $(0, t)$, or equivalent probability-generating function $\Pi_t(z) \equiv E\{z^{N_t}\}$, define

$$(1) \qquad M_t(\theta) = \int_0^\infty d_\lambda M_t(\theta \mid \lambda),$$

where $d_\lambda M_t(\theta \mid \lambda)$ is the component of $M_t(\theta)$ with the current intensity rate λ_t for the process in $(\lambda, \lambda + d\lambda)$, and the forward equation for $d_\lambda M_t(\theta \mid \lambda)$ is

$$(2) \qquad \frac{\partial d_\lambda M_t(\theta \mid \lambda)}{\partial t} = \lambda(e^\theta - 1)d_\lambda M_t(\theta \mid \lambda) - \varepsilon d_\lambda M_t(\theta \mid \lambda) + \varepsilon M_t(\theta)dG(\lambda).$$

Write

$$A_t(s) = \int_0^\infty e^{-s\lambda} d_\lambda M_t(\theta \mid \lambda),$$

so that $M_t(\theta) = A_t(0)$, then from (2) we obtain

$$(3) \qquad \frac{\partial A_t(s)}{\partial t} = -u\frac{\partial A_t(s)}{\partial s} + \varepsilon[A_t(0)L(s) - A_t(s)],$$

where $L(s) = \int_0^\infty e^{-s\lambda}dG(\lambda)$ and $u = e^\theta - 1$. In particular,

$$(4) \qquad \frac{\partial A_t(0)}{\partial t} = -u\left[\frac{\partial A_t(s)}{\partial s}\right]_{s=0},$$

but of course Equation (4) is not sufficient to obtain $A_t(0)$ until we have solved (3) for $A_t(s)$.

For many purposes it seems adequate to solve (3) in terms of successive moments, or, a little more conveniently in view of the way θ appears in (3), in terms of $u = e^\theta - 1$. We write

$$A_t(s) = a_0(s, t) + ua_1(s, t) + u^2 a_2(s, t)/2! + \cdots$$

where $a_0(0, t) = 1$. Then

$$(5) \qquad \frac{\partial a_0(s, t)}{\partial t} = \varepsilon[L(s) - a_0(s, t)]$$

so that

$$(6) \qquad a_0(s, t) = (1 - e^{-\varepsilon t})L(s) + e^{-\varepsilon t}a_0(s, 0).$$

As t increases, or if $a_0(s, 0) = L(s)$ already,

$$(7) \qquad a_0(s, t) = L(s).$$

Alternatively, if $\lambda = \Lambda$ at $t = 0$, so that $A_0(s) = e^{-s\Lambda}$ then $a_0(s, 0) = e^{-s\Lambda}$. For $a_i(s, 0) = 0$ for $i > 0$, we then have in place of (7)

$$(8) \qquad a_0(s, t) = L(s) + e^{-\varepsilon t}(e^{-s\Lambda} - L(s)).$$

Unless otherwise stated, we use the result (7) rather than (8). Then, as from (3)

$$(9) \qquad \frac{\partial a_i(s, t)}{\partial t} = -i\frac{\partial a_{i-1}(s, t)}{\partial s} + \varepsilon[a_i(0, t)L(s) - a_i(s, t)]$$

or, for $i = 1$, $s = 0$,

$$\frac{\partial a_1(0, t)}{\partial t} = \left[-\frac{\partial L(s)}{\partial s} \right]_{s=0} = \lambda_0, \quad \text{say,}$$

we have

$$(10) \qquad a_1(0, t) = \lambda_0 t.$$

Hence further, for $i = 1$ but general s,

$$\frac{\partial a_1(s, t)}{\partial t} = \frac{-\partial L(s)}{\partial s} + \varepsilon[\lambda_0 t L(s) - a_1(s, t)]$$

whence

$$(11) \qquad a_1(s, t) = \left[-\frac{\partial L(s)}{\partial s} \right]\left[\frac{1 - e^{-\varepsilon t}}{\varepsilon} \right] + \varepsilon\lambda_0 L(s)\left[\frac{t}{\varepsilon} - \frac{1}{\varepsilon^2} + \frac{e^{-\varepsilon t}}{\varepsilon^2} \right].$$

For $a_2(0, t)$ we deduce

$$\frac{\partial a_2(0, t)}{\partial t} = 2[(\lambda_0^2 + v) - \lambda_0^2]\left[\frac{1 - e^{-\varepsilon t}}{\varepsilon} \right] + 2\lambda_0^2 t,$$

$$(12) \qquad a_2(0, t) = \frac{2v}{\varepsilon}\left[t + \frac{e^{-\varepsilon t} - 1}{\varepsilon} \right] + \lambda_0^2 t^2$$

where v is the variance of λ from $G(\lambda)$. Hence for the variance κ_2 corresponding to $M_t(\theta)$ we have as t increases

$$(13) \qquad \kappa_2 \sim (\lambda_0 + 2v/\varepsilon)t - 2v/\varepsilon^2,$$

where the last term is dependent on initial conditions (for example, it is $-4v/\varepsilon^2 + (\Lambda - \lambda_0)^2/\varepsilon^2$ for $\lambda = \Lambda$ initially). It will be negligible compared with the first term for $t \gg 1/\varepsilon$.

Notice that when $2v/\varepsilon$ is σ^2, say, but v/ε^2 is negligible, the process approximates to a 'white-noise' Gauss–Poisson process.

For the calculations leading to $a_3(0, t)$ we restrict attention for simplicity to the asymptotic case when all the leading terms in the cumulants are proportional to t. We then have

$$\varepsilon a_2(s, t) + \frac{\partial a_2(s, t)}{\partial t} \sim \frac{2}{\varepsilon}\left[\frac{\partial^2 L}{\partial s^2} + \lambda_0 \frac{\partial L}{\partial s}\right] - 2\lambda_0 t \frac{\partial L}{\partial s} + \varepsilon t L(s)\left[\frac{2v}{\varepsilon} + \lambda_0^2 t\right]$$

whence

$$
\begin{aligned}
a_2(s, t) &\sim \frac{2}{\varepsilon^2}\left[\frac{\partial^2 L}{\partial s^2} + \lambda_0 \frac{\partial L}{\partial s}\right] + 2\left[vL(s) - \lambda_0 \frac{\partial L}{\partial s}\right]\left[\frac{t}{\varepsilon} - \frac{1}{\varepsilon^2}\right] \\
&\quad + \lambda_0^2 L(s)\left[t^2 - \frac{2t}{\varepsilon} + \frac{2}{\varepsilon^2}\right]
\end{aligned}
$$
(14)

and

$$
\begin{aligned}
\frac{\partial a_3(0, t)}{\partial t} &= -3\left[\frac{\partial a_2(s, t)}{\partial s}\right]_{s=0} \sim \frac{6}{\varepsilon^2}[2\lambda v + \mu_3] + 6[2\lambda_0 v + \lambda_0^3]\left[\frac{t}{\varepsilon} - \frac{1}{\varepsilon^2}\right] \\
&\quad + 3\lambda_0^3\left[t^2 - \frac{2t}{\varepsilon} + \frac{2}{\varepsilon^2}\right]
\end{aligned}
$$

where μ_3 is the third cumulant for $G(\lambda)$, whence

$$a_3(0, t) \sim \frac{6\mu_3 t}{\varepsilon^2} + \frac{6\lambda_0 v t^2}{\varepsilon} + \lambda_0^3 t^3$$
(15)

and

$$\kappa_3 = a_3(0) + 3a_2(0) + a_1(0) - 3\kappa_2\kappa_1 - \kappa_1^3 \sim \frac{6\mu_3 t}{\varepsilon^2} + \frac{6vt}{\varepsilon} + \lambda_0 t.$$
(16)

This asymptotic form of the results (not valid for small t) may be summarised as

$$A_t(0) \sim \exp\{ut(\lambda_0 + uv/\varepsilon + u^2\mu_3/\varepsilon^2 \cdots)\}.$$
(17)

Notice with (17) that its form has been expressed in terms of the first three cumulants of $G(\lambda)$, viz. λ_0, v and μ_3, and thus is easily checked from the earlier results for the simple switching model (e.g. Equation (36) of Bartlett (1982)); this simple model was symmetric between λ_1 and λ_2, but it is left as an exercise for the reader to extend it to an asymmetric version with unequal probabilities for λ_1 and λ_2, for which $\mu_3 \neq 0$).

Note also that, like the Gauss–Poisson process (see, for example, Formula

(46) of Bartlett (1982)), or the process of Barndorff–Nielsen and Yeo (1969), the switching process is not a process leading exactly to the negative binomial distribution, even if we choose the gamma distribution for $G(\lambda)$. Thus for

$$(18) \qquad dG(\lambda) = \beta^{\alpha-1}\lambda^{\alpha-1}e^{-\beta\alpha}d\lambda/\Gamma(\alpha),$$

for which $\lambda_0 = \alpha/\beta$, $v = \alpha/\beta^2$, $\mu_3 = 2\alpha/\beta^3$, Equation (16) gives

$$(19) \qquad \kappa_3 = \lambda_0 t(1 + 6/(\beta\varepsilon) + 12/(\beta\varepsilon)^2),$$

compared with the value from (18), this being, after choice of the parameters to agree on mean and variance,

$$(20) \qquad \kappa_3 = \lambda_0 t(1 + 6/(\beta\varepsilon) + 8/(\beta\varepsilon)^2).$$

However, if a distribution closer to the negative binomial is required, this may be achieved by choice of an appropriate $G(\lambda)$. For example, if

$$G(\lambda) = \tfrac{1}{2}[G_1(\lambda) + G_2(\lambda)].$$

where $G_1(\lambda)$ and $G_2(\lambda)$ are gamma distributions with the same mean, but with $\alpha_1 = c_1\alpha$, $\alpha_2 = c_2\alpha$, where

$$c_1 = 1 + 1/\sqrt{3} = 1.5773, \qquad c_2 = 1 - 1/\sqrt{3} = 0.4227$$

then we obtain the first *three* cumulants as the negative binomial.

It is also worth pointing out that the simple switching model in the asymmetric version mentioned above may also be used to generate a distribution with the same first three cumulants as the negative binomial. For if $\lambda_0 = p\lambda_1 + q\lambda_2$, where $q = 1 - p$, then $v = pq(\lambda_1 - \lambda_2)^2$ and $\mu_3 = pq(p - q)(\lambda_1 - \lambda_2)^3$. The additional relation to be satisfied by appropriate choice of λ_1, λ_2, and p is found to be

$$pq(\lambda_1 - \lambda_2) = \tfrac{3}{4}(p - q)\lambda_0,$$

or, for given λ_0 and v,

$$\sqrt{v} = \tfrac{3}{4}\lambda_0\left(\sqrt{\frac{p}{q}} - \sqrt{\frac{q}{p}}\right),$$

whence

$$(21) \qquad p/q = 1 + \tfrac{8}{9}v/\lambda_0^2 + \frac{4}{3}\left[v/\lambda_0^2\left(1 + \frac{4v}{9}\Big/\lambda_0^2\right)\right]^{\frac{1}{2}}.$$

For use later we also note that under the conditions for which the Gauss–Poisson 'white-noise' process is beginning to be a reasonable approximation, we may as a convenient and rather better approximation make use of the negative binomial model

$$(22) \qquad \Pi_t(z) = (1 - \lambda_0 u/\kappa)^{-\kappa t},$$

where κ is large.

For other purposes involving short time intervals, it might be convenient to use the more general Gauss–Poisson process with correlation term $e^{-\varepsilon t}$ (cf. Bartlett (1982), Equation (48), where there is an obvious misprint in the second term on the right, as the whole expression must be of $O(t^2)$ as t becomes small).

3. The combined doubly stochastic and mixed Poisson process

To contrast for a moment the properties of a Cox process and a mixed Poisson process, in the latter results are pooled from a mixture of individual Poisson processes, giving for a large enough set the formula

$$M_t(\theta) = E_\lambda\{e^{\lambda tu}\},$$

whence, as $u \equiv e^\theta - 1$,

$$(23) \qquad \kappa_1 = \lambda_0 t, \qquad \kappa_2 = \lambda_0 t + wt^2, \qquad \kappa_\beta = \lambda_0 t + 3wt^2 + \nu_3 t^3,$$

where $\lambda_0 = E\{\lambda\}$, and w and ν_3 are the second and third cumulants respectively. The resulting distribution is the negative binomial when the distribution of λ is the gamma distribution; but what is more important in discriminating such a process from the doubly stochastic process of Section 2 is the behaviour of the cumulants with time. Thus a convenient way of checking the behaviour of the variance in (23) is to consider the expression $(\kappa_2 - \kappa_1)/t$, which in (23) varies as wt, in contrast with Formula (17), which gives $(\kappa_2 - \kappa_1)/t$ constant (more accurately, from (13), plus a term decreasing as $1/t$).

An example will be discussed below, but to anticipate the final suggested model, this might require both types of variability to be included. In contrast with (23), the results of Section 2 correspond to a distribution of the form (17).

We now consider a combined doubly stochastic and mixed process with intensity rate

$$(24) \qquad\qquad\qquad\qquad \lambda = \lambda_1\lambda_2,$$

(the multiplicative assumption being more plausible than an additive one). It is assumed that λ_1 and λ_2 are independent factors, λ_1 being associated with variability among individual processes (e.g. variation among individuals) and λ_2 with common variation over time. It is convenient to choose $E\{\lambda_2\} = 1$, so that

$$E\{\lambda_1\lambda_2\} = E\{\lambda_1\} = \lambda_0.$$

The second and third cumulants for λ are *for given* λ_1 denoted by $v\lambda_1^2$ and $\mu_3\lambda_1^3$. Now for the combined model we have from (17) $M_t(\theta) = E_\eta\{\exp[\eta tu + u^2t\eta^2 v/\varepsilon + u^3 t\eta^3\mu_3/\varepsilon^2 + \cdots]\}$ where $\eta \equiv \lambda_1$ as a more convenient notation. Hence, if $C(\theta) \equiv E\{e^{\eta\theta}\}$,

$$M_t(\theta) = E_\eta\{[1 + u^2 t\eta^2 v/\varepsilon + u^3 \mu_3/\varepsilon^2 \cdots]e^{\eta t u}\}$$
$$= \{1 + t(\partial/\partial t)^2 v/\varepsilon + t(\partial/\partial t)^3 \mu_3/\varepsilon^2 + \cdots\}C(tu),$$

or from (23)

$$(25) \qquad \begin{aligned} M_t(\theta) &= \{1 + t(\partial/\partial t)^2 v/\varepsilon + t(\partial/\partial t)^3 \mu_3/\varepsilon^2 \cdots\} \\ &\quad \times \exp\{\lambda_0 u t + \tfrac{1}{2}wu^2 t^2 + \tfrac{1}{6}\nu_3 u^3 t^3 \cdots\} \end{aligned}$$

whence, as $u = \theta + \tfrac{1}{2}\theta^2 + \tfrac{1}{6}\theta^3 + \cdots$,

$$(26) \qquad \begin{aligned} \log M_t(\theta) &= K_0(\theta) + \tfrac{1}{2}\theta^2[\lambda_0^2 + w][2vt/\varepsilon] \\ &\quad + \tfrac{1}{6}\theta^3\{[(\lambda_0 + w) + (2\lambda_0 w + \nu_3)t][6vt/\varepsilon] \\ &\quad + [3\lambda_0 w + \nu_3 + \lambda_0^3][6\mu_3 t/\varepsilon^2]\} + \cdots \end{aligned}$$

where $K_0(\theta)$ is the cumulant function for the mixed Poisson process alone. Hence finally for the combined model

$$(27) \qquad \left\{ \begin{aligned} &\kappa_1 = \lambda_0 t, \quad \kappa_2 = (\lambda_0 t + wt^2) + 2t(\lambda_0^2 + w)v/\varepsilon, \\ &\kappa_3 = (\lambda_0 t + 3wt^2 + \nu_3 t^3) + 6t[(\lambda_0^2 + w) + (2\lambda_0 w + \nu_3)t]v/\varepsilon \\ &\quad + 6t(\lambda_0^3 + 3\lambda_0 w + \nu_3)\mu_3/\varepsilon^2. \end{aligned} \right.$$

Notice that now

$$(28) \qquad (\kappa_2 - \kappa_1)/t = 2(\lambda_0^2 + w)v/\varepsilon + wt,$$

an equation linear in t but no longer passing through the origin.

A further relation of possible interest, if we neglect the term in λ_3 which involves μ_3/ε^2, is the equation

$$(29) \qquad \frac{\kappa_3 - 3\kappa_2 + 2\kappa_1}{t^2} \sim \nu_3 t + \frac{6v}{\varepsilon}(2\lambda_0 w + \nu_3).$$

For a gamma distribution for λ_1, note that $(2\lambda_0 w + \nu_3)/(\lambda_0^2 + w) = 2w/\lambda_0$, and ν_3/w is $2w/\lambda_0$, so that if Equation (28) is $a + bt$, then (29) is $(2w/\lambda_0)(bt + 3a)$.

4. Example

As an example, some data of J. Ferreira Jr. on automobile accidents of 7842 Californian drivers over six years, quoted and analysed by Seal (1980), are discussed below. The fit by Seal of a negative binomial distribution to the six-year totals suggests that variability among drivers is the major source of heterogeneity, but it seemed a not unreasonable hypothesis that some of the variability might arise from common variability over time due to variation in

TABLE 1

Interval t	Mean κ_1	Variance[†] κ_2	Third cumulant[††] κ_3	$(\kappa_2-\kappa_1)/t$	$(\kappa_3-3\kappa_2+2\kappa_1)/t^2$
1 year	0.08240	0.08884	0.10228	0.00644	0.00056
2 years	0.16480	0.19103	0.25147	0.01311	0.00200
3 years	0.24719	0.30101	0.43099	0.01794	0.00248
6 years	0.49439	0.68107	1.2600	0.03111	0.00571

[†] Divisor $n-1=7841$

[††] Divisor $(n-1)(n-2)/n$

environmental conditions, and the dependence of the variance with time was examined by taking the separate figures for 1 year, 2 years, and 3 years (making use of all the data for each of these intervals). The results are shown in Table 1 and Figure 1; they confirm that the behaviour of the variance is largely attributable to the variation among drivers, but there is some evidence that the straight line through the points in the graph cuts the axis above the origin, giving an estimate of the variation of the common factor λ_2 over time.

The optimum fit of the straight line would be rather complex, in view of the *same* data being in use for each interval; the procedure actually adopted was to

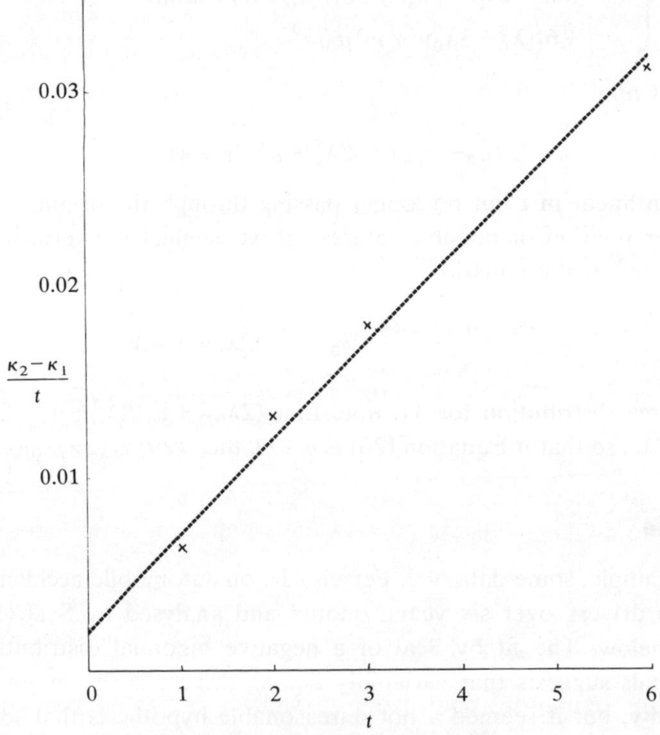

Figure 1. $(\kappa_2-\kappa_1)/t$ plotted against t (years)

estimate the slope from the first and the last points, and the mean from all the intervals. This gave the equation

$$(30) \qquad (\kappa_2 - \kappa_1)/t = 0.00493t + 0.00238.$$

This corresponds to an estimate 0.00493 for w, and, if we assume a combined model (27), a value of v/ε of $0.00238/(2 \times 0.01172) = 0.102$.

Notice that in (30) any correction term to κ_2 due to the effect of initial conditions was neglected. There is no indication, however, that the fit is not adequate, this perhaps not being surprising when any effect due to variation in λ_2 appears comparatively small, and ε is likely to be effectively large in the time-unit of one year. The adequacy of this last assumption was, however, checked by plotting the means of the *first* 1-year, 2-year, etc., intervals, and obtaining a close fit to a straight line *through the origin* (cf. Formula (10) with its modification $a_1(0) = \lambda_0 t + (\Lambda - \lambda_0)/\varepsilon$ when $\lambda = \Lambda$ initially).

It seemed of interest to examine also the third moments, which are also given in Table 1.

The values in the last column of Table 1 are obviously subject to rather large sampling fluctuations, as may be illustrated by noting that the third cumulant is appreciably affected by one anomalous observation for a driver with 11 accidents in the 6 years, the next largest value being 6 accidents (for 5 drivers). Without this observation the entry in the last column for $t = 6$ is more than halved to 0.00277! Thus, although a theoretical equation (29) was calculated relying solely on the variance estimates, together with the assumption that the third moment v_3 is that for a gamma distribution, viz.

$$(31) \qquad (\kappa_3 - 3\kappa_2 + 2\kappa_1)/t^2 = 0.000590t + 0.000854,$$

a very close fit to this line could hardly be expected (see Figure 2). The neglect of μ_3/ε^2 in (31) seemed justified in view of the comments above that

(i) variation among drivers preponderates over common variation over time,

(ii) the correlation factor $e^{-\varepsilon t}$ for λ_2 over time is small for $t \gg 1$, and the effect of initial conditions may be neglected. In this situation, the doubly stochastic model made use of is not critically dependent on the 'switching' aspects of the model, but, as already intimated in Section 2, is effectively equivalent to a Gauss–Poisson model with 'white-noise' variability σ^2 per unit time equal to $2v/\varepsilon$.

5. Further analysis and comments

As no exact demonstration of the inadequacy of the mixed Poisson model has been given so far, merely some evidence in support of the more compli-

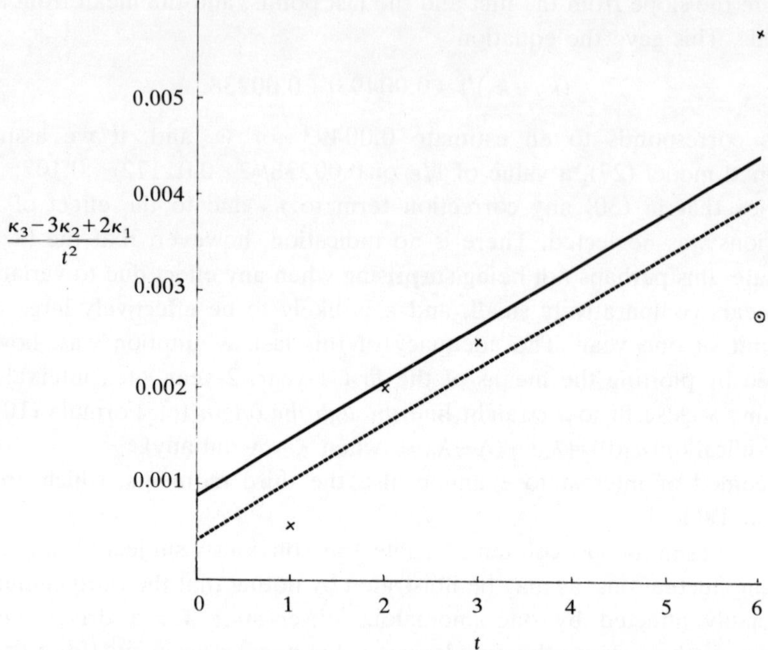

Figure 2. $(\kappa_2 - 3\kappa_2 + 2\kappa_1)/t^2$ plotted against t. The circled point for $t = 6$ excludes the anomalous driver with 11 accidents. The continuous and dotted lines refer respectively to models where the additional variance attributable to a fluctuating environment is proportional to the square of, and alternatively directly to, the intensity rate for any driver

cated model, a precise test is derived below, together with further remarks on what can legitimately be inferred about the correct model from the data under analysis.

First of all we noted that for the mixed Poisson hypothesis the results for each driver separately should constitute a Poisson process, with an intensity peculiar to that driver. As the total number n accidents for that driver is a sufficient statistic for λ, the conditional distribution for the six yearly values $x_1, x_2, \cdots, x_6$, given n, is a multinomial distribution *independent of* λ, and hence results for the same n may be combined. From the multinomial distribution the mean per year is $n/6$ and the variance $5n/36$. Hence the ratio variance/mean is also independent of n, and equal to $\frac{5}{6} = 0.833$. Moreover, the variance of an observed variance for any driver over the six years is found to be for any driver (after some algebra involving multinomial cumulants)

$$5n(n-1)/648,$$

or equivalently the variance σ_n^2 of the variance/mean ratio

$$(32) \qquad [5n(n-1)/648]/(n/6)^2 = 5(n-1)/(18n).$$

TABLE 2

n	N_n	Variance/mean	Standard error
2	595	0.8517	0.0153
3	167	0.8913	0.0333
4	54	0.9167	0.0621
5	14	0.7095	0.1260
6	5	0.8667	0.2152
11	1	0.9820	0.5025
	836	0.8617	0.0136

This result was checked by noting that the variance of the variance has the general form $An^2 + Bn$. It is 0 for $n = 1$, and easily ascertained directly to be $5/324$ for $n = 2$, for which the only types of pattern are 000002 and 000011. For N_n drivers with the same n, the standard error of the mean ratio is hence $\sigma_n / \sqrt{N}$, and for the combined mean ratio from $n = 2, 3, \cdots, 11$ (unweighted over drivers) a theoretical standard error

$$(33) \qquad \sqrt{(N_2 \sigma_2^2 + N_3 \sigma_3^2, \cdots)} / (N_2 + N_3 + \cdots).$$

The values obtained from the data are shown in Table 2, and, in spite of the absence of any relevant information from the drivers with $n = 0$ and 1, the overall mean ratio is significantly above the theoretical value 0.8333 (significance level for one tail is $P = 0.0184$). It should be noted that the excess cannot be accounted for by a simple heterogeneity in λ for each driver; for a variable λ between years, the above conditional theory is modified to a multinomial distribution with unequal probabilities $p_i = \lambda_i / \Sigma_i \lambda_i$, and this gives the same average mean $n/6$, and a variance over the six years of $\Sigma_i np_i(1 - p_i)$, which is always *less* than $5n/36$. A *larger* variance can be explained by an intrinsic larger variance for each year, this being consistent with a Cox process per driver of the type examined above, though the nature of the data does not permit discrimination between our model and other sources of heterogeneity within a year, such as independent (not common) heterogeneity for the various drivers. In order to discriminate between such different models, the data would have to be for more numerous and smaller time units for each driver; and if such data were available, the effort of correlation and initial conditions could no longer be neglected.

For the data available, with the conditions (i) a relatively small temporal variation in λ 'within' a driver, (ii) temporary variation fairly rapid compared with the minimum observed interval of one year, the switching model is, as already remarked, similar to a 'white-noise' Gauss–Poisson model, and to a somewhat better degree of approximation, to a negative binomial model for any driver. If we assumed therefore the approximate model (22), it may be shown (see Appendix) that the variance/mean ratio for given n (and $t = 1$ year) is modified approximately to $\frac{5}{6}(1 + (n-1)/\kappa)$. A more sensitive test of the excess variance/mean ratio might therefore be provided by taking note of the factor $n - 1$; a recalculation of the excess divided by $n - 1$ gave a mean 0.01785 with standard error 0.00907. This happens to give a significance level with the slightly *higher P* value of 0.0245, but has been noted as it would provide a rigorous estimate of κ if required.

As a variant of the model (22), a rather more convenient approximate model for some purposes would be

$$(34) \qquad \Pi_t(z) = [1 - \sigma(z - 1)]^{-\lambda_1 t/\sigma},$$

say, where σ is small. This model is different from (22) in the functional relation with the mean intensity rate; for the model (34) the variance for given λ_1 is proportional to λ_1, not λ_1^2, and thus involves an alternative assumption to (24). It has one slight advantage in leading to a closed expression for the p.g.f. assuming a gamma distribution for λ_1, viz.

$$(35) \qquad \log \Pi_t(z) = \frac{-\lambda_0}{\sigma_1} \log \left\{ 1 + \frac{\sigma_1 t}{\sigma} \log [1 - \sigma(z - 1)] \right\},$$

where

$$E\{\exp(\lambda_1 \theta)\} = (1 - \sigma_1 \theta)^{-\lambda_0/\sigma_1},$$

so that $\mathrm{Var}(\lambda_1) = \lambda_0(1 + \sigma_1)$. For σ^2 negligible (consistently with the equivalence of the process per driver to a 'white-noise' Gauss–Poisson process) this yields

$$K_t(\theta) = -(\lambda_0 t/\sigma) \log[1 - \sigma(e^\theta - 1)] + \tfrac{1}{2}(\lambda_0 \sigma_1 t^2/\sigma^2) \log^2[1 - \sigma(e^\theta - 1)]$$

$$- \tfrac{1}{3}(\lambda_0 \sigma_1^2 t^3/\sigma^3) \log^3[1 - \sigma(e^\theta - 1)] + \cdots$$

whence, neglecting σ^2,

$$(36) \qquad \begin{cases} \kappa_1 = \lambda_0 t, \quad \kappa_2 = \lambda_0 t(1 + \sigma) + \lambda_0 \sigma_1^2 t \\ \kappa_3 = \lambda_0 t(1 + 3\sigma) + 3\lambda_0 t^2 \sigma_1(1 + \sigma) + 2\lambda_0 \sigma_1^2 t^3, \end{cases}$$

or $(\kappa_2 - \kappa_2)/t = \lambda_0(\sigma + \sigma_1 t)$, $(\kappa_3 - 3\kappa_2 + 2\kappa_1)/t^2 = \lambda_0 \sigma_1(3\sigma + 2\sigma_1 t)$.

In terms of this model, the estimate of σ_1 from the $(\kappa_2 - \kappa_1)/t$ equation is 0.05983, and of σ 0.02888. However, it will be found that the predicted equation $(\kappa_3 - 3\kappa_2 + 2\kappa_1)/t^2$ provides little discriminatory evidence for one model as against the other, being the same as (31) apart from a halving of the constant term. (For the additive and rather unrealistic assumption $\lambda = \lambda_1 + \lambda_2$, the constant term would under the same approximation disappear in the equation for $(\kappa_3 - 3\kappa_2 + 2\kappa_1)/t^2$. As far as the evidence goes, it seems slightly to favour the intermediate model (shown by a dotted line in Figure 2).)

6. Concluding remark

Although the data analysed do not allow discrimination between heterogeneity over time due to a common fluctuating environment or one peculiar to each driver, it might be remarked that public overall statistics on automobile accidents, with peaks at times of bad weather conditions or busy holiday periods, would provide independent evidence on the existence of a common fluctuating environmental factor.

Appendix.　Variance/mean ratio for given n for model (22)

For (21) with $t = 1$ we have, if $\sigma = \lambda_0/\kappa$,

$$\text{(A1)} \qquad p(x_i) = \frac{(\kappa + x_i - 1)!}{x_i!\,(\kappa - 1)!}\,\sigma^{x_i}(1 + \sigma)^{-(\kappa + x_i)}$$

and

$$\text{(A2)} \qquad p(x_1, x_2, \cdots, x_6 \mid n) = \prod_{i=1}^{6} \frac{(\kappa + x_i - 1)!}{x_i!\,(\kappa - 1)!} \bigg/ \frac{(6\kappa + n - 1)!}{n!\,(6\kappa - 1)!}\,.$$

In particular,

$$\text{(A3)} \qquad \begin{cases} p(x_1 \mid n) = \dfrac{(\kappa + x_1 - 1)}{x_1!\,(\kappa - 1)!}\,\dfrac{(5\kappa + n - x_1 - 1)!}{(n - x_1)!\,(5\kappa - 1)!} \bigg/ \dfrac{(6\kappa + n - 1)!}{n!\,(6\kappa - 1)!}\,, \\[2ex] E\{x_1 \mid n\} = kn/(6\kappa) = n/6, \\[2ex] E\{x_1(x_1 - 1) \mid n\} = \dfrac{\kappa(\kappa + 1)n(n - 1)}{6\kappa(6\kappa + 1)} = \dfrac{n(\kappa + 1)(n - 1)}{6(6\kappa + 1)} \end{cases}$$

whence

$$\sigma^2(x_1 \mid n) = \frac{5n(n + 6\kappa)}{36(6\kappa + 1)} \sim \frac{5n}{36}\left(1 + \frac{n - 1}{6\kappa}\right) \quad \text{for large } \kappa,$$

or equivalently

$$\text{(A4)} \qquad \sigma^2(x_1 \mid n)/E\{x_1 \mid n\} \sim \frac{5}{6}\left(1 + \frac{n - 1}{6\kappa}\right).$$

References

BARNDORFF-NIELSEN, O. AND YEO, G. F. (1969) Negative binomial processes. *J. Appl. Prob.* **6**, 633–647.

BARTLETT, M. S. (1982) Some stochastic models in biology. *Utilitas Math.* **21**A, 291–309.

BARTLETT, M. S. (1983) On doubly stochastic processes. In *Probability, Statistics and Analysis*, ed. J. F. C. Kingman and G. E. H. Reuter, Cambridge University Press, 24–45.

SEAL, H. L. (1980) Mixed Poisson processes and risk theory (Notes of a lecture course, University of Lausanne).

A Note on the Two-Sex Population Process

J. GANI

PYKE TIN

Abstract

This paper considers the two-sex birth–death model $\{X(t), Y(t); t \geq 0\}$; an explicit solution is obtained for its probability generating function. It is shown that moments of the process can be found directly from the Kolmogorov forward equations for the probabilities. An integral equation approach is also used to throw light on the structure of the process.

BIRTH–DEATH PROCESS; RICCATI EQUATION; INTEGRAL EQUATION APPROACH

1. Introduction

Following the early work of Goodman (1953) and Joshi (1954), several authors have considered the two-sex birth–death model. Recently Tapaswi and Roychoudhury (1983) in an initial paper re-examined this process, deriving some of its moments. In a later paper (1985), at approximately the same time as ourselves (who had read their first report), they obtained the expression for the joint probability generating function (p.g.f.) of the process, and studied its marginal distributions in detail. In this note, we shall attempt to concentrate rather more on the structural aspects of the problem. It should be noted that the two-sex model is related to the birth–death model in which the parity of individuals is recorded. This was originally studied by Gani and Saunders (1976) and recently extended by Srinivasan and Ranganathan (1983); they solved the partial differential equation for the p.g.f. in the simplest case, and pointed out the problem's relationship to the two-sex model.

Let us consider a birth–death process of females $X(t) = 0, 1, 2, \cdots$ and males $Y(t) = 0, 1, 2, \cdots, t \geq 0$ for which the joint probability is denoted by

$$P_{xy}(t) = P\{X(t) = x, Y(t) = y \mid X(0) = 1, Y(0) = 0\}.$$

The infinitesimal transition probabilities are

$$P\{X(t+\delta t)=x+1,\ Y(t+\delta t)=y \mid X(t)=x,\ Y(t)=y\}=\lambda px\delta t+o(\delta t)$$

$$P\{X(t+\delta t)=x,\ Y(t+\delta t)=y+1 \mid X(t)=x,\ Y(t)=y\}=\lambda qx\delta t+o(\delta t)$$

$$P\{X(t+\delta t)=x-1,\ Y(t+\delta t)=y \mid X(t)=x,\ Y(t)=y\}=\mu x\delta t+o(\delta t)$$

$$P\{X(t+\delta t)=x,\ Y(t+\delta t)=y-1 \mid X(t)=x,\ Y(t)=y\}=\mu'y\delta t+o(\delta t).$$

The probability that there is no change is given by

$$P\{X(t+\delta t)=x,\ Y(t+\delta t)=y \mid X(t)=x,\ Y(t)=y\}$$
$$=1-(\lambda+\mu)x\delta t-\mu'y\delta t-o(\delta t)$$

where $p+q=1(p,q>0)$, and $\lambda,\mu,\mu'>0$. Thus the females give birth to both females and males who die out with different rates, while the males do not reproduce.

The forward Kolmogorov equations for this bivariate Markov process $\{X(t),Y(t);t\geq 0\}$ are derived by Tapaswi and Roychoudhury (1983) as

$$(1.1)\quad \frac{d}{dt}P_{xy}(t)=\lambda p(x-1)P_{x-1y}(t)+\lambda qxP_{xy-1}(t)-((\lambda+\mu)x+\mu'y)P_{xy}$$
$$+\mu(x+1)P_{x+1y}(t)+\mu'(y+1)P_{xy+1}(t)\qquad x,y=0,1,2,\cdots,$$

with $P_{xy}(t)=0$ when either of x or $y<0$, and $P_{10}(0)=1$. By appropriate summations, they obtain the means of $X(t)$ and $Y(t)$ from the equations (1.1), as

$$\frac{d}{dt}E(X(t))=(\lambda p-\mu)E(X(t)),$$
$$(1.2)$$
$$\frac{d}{dt}E(Y(t))=\lambda qE(X(t))-\mu'E(Y(t)),$$

whence, for the particular case where $X(0)=1$, $Y(0)=0$,

$$E(X(t))=\exp\left((\lambda p-\mu)t\right),$$
$$(1.3)$$
$$E(Y(t))=\frac{\lambda q}{\lambda p-\mu+\mu'}\left(\exp\left((\lambda p-\mu)t\right)-\exp\left(-\mu't\right)\right).$$

They show that the marginal distribution of the females $P_x(t)$ follows that for a standard birth–death process, with the well-known p.g.f.'s for the distinct cases where $\lambda p\neq\mu$ and $\lambda p=\mu$.

While several approaches to the problem are possible, two particular ones prove most rewarding: the p.g.f. approach, and the integral equation approach. We consider each in turn.

2. The p.g.f. approach

If we define the p.g.f. of the process by

$$(2.1) \qquad \phi(u, v, t) = \sum_{x,y=0}^{\infty} P_{xy}(t) u^x v^y$$

we can readily show from (1.1) that

$$(2.2) \qquad \frac{\partial \phi}{\partial t} = [\lambda p u^2 + \lambda q u v - (\lambda + \mu)u + \mu] \frac{\partial \phi}{\partial u} + u'(1-v) \frac{\partial \phi}{\partial v},$$

where $\phi(u, v, 0) = u$. The auxiliary equations of this partial differential equation are

$$(2.3) \qquad \frac{dt}{-1} = \frac{du}{\lambda p u^2 + \lambda p u v - (\lambda + \mu)u + \mu} = \frac{dv}{\mu'(1-v)} = \frac{d\phi}{0},$$

from which we can obtain

$$(2.4) \qquad (1-v) \exp(-\mu' t) = c_1$$

where c_1 is an arbitrary constant. What is now required is the solution of

$$(2.5) \qquad \frac{du}{dv} = \frac{\lambda p u^2 + \lambda q u v - (\lambda + \mu)u + \mu}{\mu'(1-v)}.$$

This is a Riccati equation for which it can be shown that the transformations

$$u = \frac{\mu'(v-1)}{\lambda p z} \frac{dz}{dv}$$

and $v = 1 - \theta$ reduce (2.5) to the linear second-order equation

$$(2.6) \qquad \theta^2 R(\theta) \frac{d^2 z}{d\theta^2} + \theta P(\theta) \frac{dz}{d\theta} + Q(\theta) z = 0.$$

Here

$$R(\theta) = 1, \qquad P(\theta) = \frac{\mu' - \mu - \lambda p - \lambda q \theta}{\mu'} \quad \text{and} \quad Q(\theta) = \frac{\lambda \mu p}{\mu'^2}.$$

It is seen that the Equation (2.6) has a regular singular point at $\theta = 0$, and $R(\theta), P(\theta), Q(\theta)$ are regular on any interval $-h < \theta < h$, $h > 0$. Moreover, $R(\theta)$ is never 0. Thus using the standard Frobenius power series method we can develop in the neighbourhood $0 < |\theta| < h$ a solution of the form

$$(2.7) \qquad z(\theta) = \sum_{n=0}^{\infty} A_n \theta^{n+s}, \qquad A_0 \neq 0.$$

Substitution of (2.7) into (2.6) results in the indicial equation

$$(2.8) \qquad s^2 - \left(\frac{\mu + \lambda p}{\mu'}\right)s + \frac{\lambda p\mu}{\mu'^2} = 0$$

and the difference equation for the coefficients

$$(2.9) \qquad \left[(s+n)^2 - \left(\frac{\mu + \lambda p}{\mu'}\right)(s+n) + \frac{\lambda p\mu}{\mu'^2}\right]A_n = \frac{(n+s-1)\lambda q}{\mu'}A_{n-1}$$
$$n = 1, 2, \cdots.$$

The roots of the indicial equation are readily seen to be $s_1 = \lambda p/\mu'$ and $s_2 = \mu/\mu'$, which will each be considered in turn. Let us first assume that $\lambda p \neq \mu$, then for $s_1 = \lambda p/\mu'$, (2.9) reduces to

$$A_n = \left(\frac{\lambda q}{\mu'}\right)\left[\frac{n\mu' + \lambda p - \mu'}{n(n\mu' + \lambda p - \mu)}\right]A_{n-1},$$

so that

$$(2.10) \qquad A_n = A_0\left(\frac{\lambda q}{\mu'}\right)^n \frac{1}{n!}\prod_{k=0}^{n-1}\left[\frac{(n-k)\mu' + \lambda p - \mu'}{(n-k)\mu' + \lambda p - \mu}\right] \qquad n = 1, 2, \cdots.$$

Then a solution of (2.6) in the neighbourhood $0 < |\theta| < h$, $h > 0$ is given by

$$(2.11) \qquad A_0 z_1(\theta) = A_0\theta^{\lambda p/\mu'}\sum_{n=0}^{\infty}\alpha_n\theta^n,$$

where, clearly $\alpha_0 = 1$, and $\alpha_n = A_n/A_0$ for $n = 1, 2, \cdots$.

Similarly, by considering the case $s_2 = \mu/\mu'$, we obtain a second solution of (2.6) in $0 < |\theta| < h$, $h > 0$ as

$$(2.12) \qquad A_0' z_2(\theta) = A_0'\theta^{\mu/\mu'}\sum_{n=0}^{\infty}\beta_n\theta^n,$$

where A_0' is now the first coefficient of the power series, which may be different from A_0, $\beta_0 = 1$ and

$$(2.13) \qquad \beta_n = \left(\frac{\lambda q}{\mu'}\right)^n \frac{1}{n!}\prod_{k=0}^{n-1}\left[\frac{(n-k)\mu' + \mu - \mu'}{(n-k)\mu' + \mu - \lambda p}\right] \qquad n = 1, 2, \cdots.$$

Adding (2.11) and (2.12), we have that for any θ in $0 < |\theta| < h$, $h > 0$ and arbitrary constants A_0, A_0',

$$(2.14) \qquad z(\theta) = A_0 z_1(\theta) + A_0' z_2(\theta)$$

is the general solution of (2.6). By substituting this into the transformations for

u and v previously used, we obtain the general solution to (2.5) as

$$(2.15) \qquad u = \frac{\mu'(v-1)}{\lambda p}\left[\frac{c_0 z_1'(1-v)+z_2'(1-v)}{c_0 z_1(1-v)+z_2(1-v)}\right],$$

where $c_0 = A_0/A_0'$ is an arbitrary constant and $z_i'(\theta) = dz_i(\theta)/d\theta$ for $i = 1, 2$.

The general solution of (2.2) is obtained by setting $\phi(u, v, t) = f(c_0, c_1)$ for an arbitrary function $f(\cdot, \cdot)$. Using standard arguments and writing

$$(2.16) \qquad (w_1 - \lambda p w_2)z_1(\theta) + \mu'\theta w_2 z_1'(\theta) = \theta^{\lambda p/\mu'} H_1(w_1, \mu'w_2, \theta)$$

and

$$(2.17) \qquad (w_1 - \lambda p w_2)z_2(\theta) + \mu'\theta w_2 z_2'(\theta) = \theta^{\mu/\mu'} H_2(w_1, \mu'w_2, \theta),$$

we finally obtain, for the general case when $X(0) = a$ and $Y(0) = b$, the p.g.f.

$$(2.18) \qquad \phi_{ab}(u, v, t) = [1 - \exp(-\mu't) + v \exp(-\mu't)]^b [\phi_{10}(u, v, t)]^a.$$

Here $\phi_{10}(u, v, t)$ is the p.g.f. for the process starting with $X(0) = 1$ and $Y(0) = 0$ given by

$$(2.19) \qquad \phi_{10}(u, v, t) = \frac{1}{\lambda p}$$

$$\times \left[\frac{\begin{array}{l} H_1(\lambda pu - \lambda p, \mu', (1-v))H_2(\mu, \mu', (1-v)\exp(-\mu't)) \\ \quad - H_1(\lambda p, \mu', (1-v)\exp(-\mu't))H_2(\lambda pu - \mu, \mu', (1-v))\exp(-(\lambda p - \mu)t) \end{array}}{\begin{array}{l} H_1(\lambda pu - \lambda p, \mu', (1-v))H_2(1, 0, (1-v)\exp(-\mu't)) \\ \quad - H_1(1, 0, (1-v)\exp(-\mu't))H_2(\lambda pu - \mu, \mu', (1-v))\exp(-(\lambda p - \mu)t) \end{array}}\right].$$

Note that the initial b males are subject to a pure death process; if $a = 1$, $b = 0$, the solution to (2.2) is simply $\phi(u, v, t) = \phi_{10}(u, v, t)$, the function indicated above. By considering the first two components of (2.3), Tapaswi and Roychoudhury (1985) obtain from the Riccati equation, after suitable transformations, a congruent hypergeometric differential equation and derive a solution equivalent to (2.19) in terms of Kummer's functions $M(a, b, z)$ and $U(a, b, z)$ (see Abramowitz and Stegun (1972), p. 504).

Let us now consider the case $\lambda p = \mu$. Here since $s_1 = s_2 = \lambda p/\mu'$ we find only one solution of the form (2.7). But it is known that

$$(2.20) \qquad A_0'z(\theta) = A_0'z_1(\theta)\log(\theta) + \sum_{n=0}^{\infty}\left(\frac{\partial A_n(s)}{\partial s}\right)_{s=\lambda p/\mu'} \theta^{n+(\lambda p/\mu')}$$

is also a solution of (2.6), where $A_n(s)$ satisfies the recurrence equation (2.9).

Thus, by methods similar to those used earlier, we can show that in the case

$$\lambda p = \mu,$$

$$\phi_{10}(u, v, t) = \frac{1}{\lambda p}$$
$$\times \left[\frac{B_1(u, v, t) - tH_1(\lambda pu - \lambda p, \mu', (1-v))H_1(\lambda p, \mu', (1-v)\exp(-\mu't))}{B_2(u, v, t) - tH_1(\lambda pu - \lambda p, \mu', (1-v))H_1(1, 0, (1-v)\exp(-\mu't))} \right]$$

(2.21)

where the functions $B_i(u, v, t)$, $i = 1, 2$, are given by

$$B_1(u, v, t) = H_1(\lambda pu - \lambda p, \mu', (1-v))H_1(1, 0, (1-v)\exp(-\mu't))$$
$$+ H_1(\lambda p, \mu', (1-v)\exp(-\mu't))H_1(1, 0, (1-v))$$
$$+ \sum_{n=0}^{\infty} \frac{1}{\mu'}\left(\frac{\partial A_n(s)}{\partial s}\right)_{s=\lambda p/\mu'} [(\lambda p + n\mu_1')H_1(\lambda pu - \lambda p, \mu', (1-v)\exp(-\mu't))$$
$$- (\lambda pu - \lambda p + n\mu')H_1(\lambda p, \mu', (1-v)\exp(-\mu't))],$$

$$B_2(u, v, t) = H_1(1, 0, (1-v)\exp(-\mu't))H_1(1, 0, (1-v))$$
$$+ \sum_{n=0}^{\infty} \frac{1}{\mu'}\left(\frac{\partial A_n(s)}{\partial s}\right)_{s=\lambda p/\mu'} [H_1(\lambda Pu - \lambda p, \mu', (1-v)\exp(-\mu't))$$
$$- (\lambda pu - \lambda p + n\mu')H_1(1, 0, (1-v)\exp(-\mu't))].$$

Setting $v = 1$ in (2.18) and (2.21) we obtain directly the well-known p.g.f.'s for the linear birth–death process $X(t)$ with the birth rate λp and death rate μ, in both the cases $\lambda p \neq \mu$ and $\lambda p = \mu$.

Similarly we can obtain an expression for the p.g.f. of the process $Y(t)$ by putting $u = 1$ in (2.18) and (2.21). This leads to an expression which Tapaswi and Roychoudhury (1985) have discussed in detail.

Equations (2.19) and (2.21) enable us to investigate the probability of ultimate extinction for the general process $\{X(t), Y(t)\}$. Letting $t \to \infty$ in (2.19) and (2.21), it is seen that the probability of extinction is 1 if $\lambda p \leq \mu$ and $\mu/\lambda p$ if $\lambda p > \mu$, exactly as one would expect from what is basically a birth–death-type model.

3. Means, variances and covariance

Tapaswi and Roychoudhury (1983) consider the means of the process starting with a males and b females at the initial time $t = 0$. They derive, from (1.4) the results (see also Equation (1.3))

$$E(X(t)) = a \exp((\lambda p - \mu)t)$$

(3.1)

$$E(Y(t)) = b \exp(-\mu't) + \frac{\lambda qa}{\lambda p - \mu + \mu'}(\exp((\lambda p - \mu)t) - \exp(-\mu't)).$$

These can equally well be obtained from (2.2) subject to the initial conditions $X(0) = a$, $Y(0) = b$, as can also the variances

(3.2)
$$\sigma_x^2(t) = \text{Var}\,(X(t)) = a\,\frac{\lambda p + \mu}{\lambda p - \mu}\,\exp\,((\lambda p - \mu)t)[\exp\,((\lambda p - \mu)t) - 1]$$

$$\sigma_y^2(t) = \text{Var}\,(Y(t)) = b\,\exp\,(-\mu't) + (b + \alpha + \beta)\,\exp\,(-\mu't)$$

$$+ (\lambda + \alpha\,\exp\,((\lambda p - \mu)t))\,\exp\,((\lambda p - \mu)t)$$

where

$$\alpha = a\,\frac{\lambda p + \mu}{\lambda p - \mu} \cdot \frac{2\lambda p - 2\mu + \mu'}{\mu'}\left(\frac{\lambda q}{\lambda p - \mu + \mu'}\right)^2$$

$$\beta = a\left[\frac{\lambda q}{\lambda p - \mu + \mu'} - 2\,\frac{\lambda p + \mu}{\lambda p - \mu} \cdot \frac{\lambda^2 q^2}{\mu'(\lambda p - \mu + 2\mu')}\right].$$

We present briefly here alternative derivations of $E(Y(t)) = M_y(t)$,

$$\sigma_y^2(t) = E(Y(t)Y(t-1)) + E(Y(t)) - E(Y(t))^2 = L_y(t) + M_y(t) - M_y(t)^2$$

and $E(X(t)Y(t)) = M_{xy}(t)$. Observing that for $u = 1$, $v = 1$, $M_y(t) = \partial\phi/\partial v$, $M_{xy}(t) = \partial^2\phi/\partial u\partial v$ and $L_y(t) = \partial^2\phi/\partial v^2$, we see from Equation (2.2) that

(3.3)
$$\frac{\partial M_y(t)}{\partial t} = \lambda q E(X(t)) - \mu' M_y(t)$$

(3.4)
$$\frac{\partial M_{xy}(t)}{\partial t} = (\lambda p - \mu - \mu')M_{xy}(t) + \lambda q E(X(t)^2)$$

(3.5)
$$\frac{\partial L_y(t)}{\partial t} = 2\lambda q M_{xy}(t) - 2\mu' L_y(t).$$

By solving these equations and using the initial conditions $X(0) = a$, $Y(0) = b$, we readily find $E(Y(t))$ as in (3.1) and $\text{Var}\,(Y(t))$ as in (3.2).

Similarly solving Equation (3.4) and using (3.1), we obtain

(3.6)
$$\text{Cov}\,(X(t)Y(t)) = M_{xy}(t) - M_x(t)M_y(t)$$

$$= a\lambda q\,\exp\,((\lambda p - \mu)t)$$

$$\times\left[\frac{\lambda p + \mu}{\lambda p - \mu}\left\{\frac{\exp\,((\lambda p - \mu)t)}{\lambda p - \mu + \mu'} - \frac{1}{\mu'}\right\} + \frac{(\lambda p + \mu)\,\exp\,(-\mu't)}{\mu'(\lambda p - \mu + \mu')}\right].$$

This method is closely related to the method of Goodman (1953) and also to its extension by Srinivasan and Ranganathan (1983) to the parity-dependent birth–death model.

4. The integral equation approach

If we consider the two-sex birth–death process, we recognize that the important factor is the number of females $X(t)$ at any time t. If this number is fixed for the period $[0, t]$ at $X(\tau) = x$, $0 \leq \tau \leq t$, then we can readily show that the p.g.f. of the process, conditional on $X(\tau) = x$ and $Y(0) = y$, is given by

$$\phi(u, v, t \mid X(\tau) = x, Y(0) = y) = u^x(1 - \exp(-\mu't) + v\exp(-\mu't))^y$$

$$(4.1) \qquad \times \exp\left(-\frac{\lambda q}{\mu'}x(1 - \exp(-\mu't))(1 - v)\right).$$

We note that the initial males $Y(0) = y$ follow a pure death process, while new males are produced as a non-homogeneous Poisson process.

This enables us to derive an integral equation for the p.g.f. $\phi_{10}(u, v, t)$ of (2.19) or (2.21) by arguments similar to those used in branching process theory. Suppose we begin with $X(0) = 1$ and $Y(0) = 0$, then starting with 1 female at $t = 0$ either (a) there will be no change in the number of females in $[0, t]$ or (b) in some interval $(\tau, \tau + \delta\tau)$ the female dies out, or (c) a second female is born in this interval. In the first case, we shall have from (4.1) that the p.g.f. for (a) will be

$$(4.2) \qquad u\exp(-(\lambda p + \mu)t) \cdot \exp\left(-\frac{\lambda q}{\mu'}(1 - \exp(-\mu't))(1 - v)\right).$$

In case (b), the p.g.f. will be

$$(4.3) \qquad \int_0^t \exp(-(\lambda p + \mu)\tau)\mu \exp\left(-\frac{\lambda q}{\mu'}(1 - \exp(-\mu'\tau))\right.$$

$$\left. \times (1 - [1 - \exp(-\mu'(t - \tau)) + v\exp(-\mu'(t - \tau))])\right)d\tau$$

where no event occurs up to τ, the single female dies out in $(\tau, \tau + \delta\tau)$ and the males produced up to time τ with p.g.f.

$$\exp\left(-\frac{\lambda q}{\mu'}(1 - \exp(-\mu'\tau))(1 - v)\right)$$

each begin to die out with p.g.f. $(1 - \exp(-\mu'(t - \tau)) + v\exp(-\mu'(t - \tau)))$ in the remaining interval (τ, t). We see that (4.3) can be simplified to

$$\int_0^t \exp(-(\lambda p + \mu)(t - \tau))\mu \exp\left(-\frac{\lambda q}{\mu'}(\exp(-\mu'\tau) - \exp(-\mu't))(1 - v)\right)d\tau.$$

In the final case (c), the p.g.f. will be

$$\int_0^t \exp(-(\lambda p + \mu)\tau)\lambda p$$

$$(4.4) \qquad \times \exp\left(-\frac{\lambda q}{\mu'}(\exp(-\mu'(t - \tau)) - \exp(-\mu't))(1 - v)\right)\phi_{10}^2(u, v, t - \tau)d\tau$$

$$= \int_0^t \exp(-(\lambda p + \mu)(t - \tau))\lambda p$$

$$\times \exp\left(-\frac{\lambda q}{\mu'}(\exp(-\mu'\tau) - \exp(-\mu't))(1 - v)\right)\phi_{10}^2(u, v, \tau)d\tau,$$

where the same argument applies for the males produced up to time τ, and the new female born in $(\tau, \tau + \delta\tau)$ together with the first female, will give rise to the joint p.g.f. $\phi_{10}^2(u, v, t-\tau)$.

If we add the three components (4.2), (4.3) and (4.4) we finally obtain that

$$\phi_{10}(u, v, t) = u \exp\left(-(\lambda p + \mu)t + \frac{\lambda q}{\mu'}(1 - \exp(-\mu' t))(v - 1)\right)$$

$$(4.5) \qquad + \int_0^t \exp\left(-(\lambda p + \mu)(t - \tau) + \frac{\lambda q}{\mu'}(\exp(-\mu'\tau) - \exp(-\mu' t))(v - 1)\right)$$

$$\times [\mu + \lambda p \phi_{10}^2(u, v, \tau)]d\tau.$$

This equation, as in the simpler birth–death process, is equivalent to the Kolmogorov equation (2.2) for the p.g.f. It can be used to obtain equations for the moments of the process by expansion of the various functions. To illustrate this, let us concentrate on the males $Y(t)$, whose marginal p.g.f. $\phi_{10}(1, v, t) = \psi(v, t)$ is given by

$$\psi(v, t) = A(t) \exp((B(0) - B(t))(v - 1))$$

$$(4.6) \qquad + \int_0^t A(t - \tau) \exp((B(\tau) - B(t))(v - 1))[\mu + \lambda p \psi^2(v, \tau)]d\tau,$$

where

$$(4.7) \qquad A(t) = \exp(-(\lambda p + \mu)t), \qquad B(t) = \frac{\lambda q}{\mu'}\exp(-\mu' t).$$

Expanding the functions about $v = 1$, equating coefficients of $(v - 1)^n$ in (4.6), and using Leibniz's formula for derivatives of $\psi^2(1, \tau)$, we find that

$$(4.8) \qquad \frac{\partial^n \psi(1, t)}{\partial v^n} = A(t)[B(0) - B(t)]^n + \int_0^t A(t - \tau)\bigg[\mu(B(\tau) - B(t))^n$$

$$+ \lambda p \sum_{j=0}^n \binom{n}{j}(B(\tau) - B(t))^{n-j} \sum_{r=0}^j \binom{j}{r}\frac{\partial^r \psi(1, t)}{\partial v^r} \cdot \frac{\partial^{j-r}\psi(1, t)}{\partial v^{j-r}}\bigg]d\tau,$$

where the highest derivative of ψ on the right-hand side of (4.8) is also $\partial^n\psi(1, t)/\partial v^n$. Hence we can write (4.8) in the simplified form

$$(4.9) \qquad \frac{\partial^n \psi(1, t)}{\partial v^n} = F_n(t) + 2\lambda p \int_0^t A(t - \tau)\frac{\partial^n \psi(1, \tau)}{\partial v^n}\,d\tau,$$

where $F_n(t)$ is a known function of t containing terms which involve derivatives of $\psi(v, t)$ with highest order $n - 1$. We can therefore, using Laplace transforms or otherwise, obtain higher moments of the distribution progressively, starting from $\partial\psi(1, t)/\partial v$ and moving from $\partial^{n-1}\psi(1, t)/\partial v^{n-1}$ to $\partial\psi^n(1, t)/\partial v^n$.

5. Concluding remarks

The two-sex model, despite its apparent ease of formulation, has a fairly complex structure. Its p.g.f. as given in (2.19) and (2.21) involves some functions which are not easy to manipulate. Its moments can be derived, in principle, but again with heavy algebra. Nevertheless, it leads to further problems, of interest not only in population processes but also in queueing situations. We hope to explore some of these, and in particular the cases with immigration and with several categories of individuals, rather than just males and females.

Acknowledgements

We wish to thank Drs Tapaswi and Roychoudhury for communicating their papers to us, and for their helpful cooperation in our joint endeavours. Thanks are also due to the Office of Naval Research and the Fulbright Foundation for supporting this research.

References

ABRAMOWITZ, M. AND STEGUN, I. A. (1972) *Handbook of Mathematical Functions*. Dover Publications, New York.

GANI, J. AND SAUNDERS, I. W. (1976) On the parity of individuals in a branching process. *J. Appl. Prob.* **13**, 219–230.

GOODMAN, L. A. (1953) Population growth of the sexes. *Biometrics* **9**, 212–225.

JOSHI, D. D. (1954) Les processus stochastiques en démographie. *Publ. Inst. Statist. Univ. Paris* **3**, 153–177.

SRINIVASAN, S. K. AND RANGANATHAN, C. R. (1983) Parity dependent population growth models. *J. Math. Phys. Sci.* **17**, 279–292.

TAPASWI, P. K. AND ROYCHOUDHURY, R. K. (1983) A new approach to the distribution problems arising in the studies of population growth of sexes. Technical Report, 151 Calcutta.

TAPASWI, P. K. AND ROYCHOUDHURY, R. K. (1985) A solution to the distribution problems arising in the studies of a two-sex population process. *Stoch. Proc. Appl.* **19**, 359–370.

On the Use of Time Series Representations of Population Models

C. C. HEYDE

Abstract

Many population models which are far from stationarity can nevertheless be written in autoregressive format, perhaps with random coefficient. It is the thesis of this paper that procedures developed for stationary time series models are a useful guide to inferential results for population processes and may indeed be directly applicable. The illustrations concentrate on estimation of the matrix of mean vital rates in an age-structured population.

AUTOREGRESSION; BRANCHING PROCESS; POPULATION MODEL; SEQUENTIAL CONTROL OR DESIGN MODEL; VITAL RATES

1. Introduction and discussion

Population models rarely deal with stationary or asymptotically stationary regimes, which are the principal domain of time series methodology. Nevertheless, it can be advantageous to describe population models in time series format for the purpose of analysis, as estimation procedures developed for time series models are often quite robust and still applicable.

In this paper we shall be concerned with age-structured populations whose vital rates may vary stochastically in time. Let Y_t be the (column) vector consisting of the numbers of individuals in each of m age classes at time t, while $\mathscr{F}_t$ denotes a past history σ-field which includes (and may equal) that generated by $\{Y_k, k \leq t\}$. We shall consider population systems which evolve according to equations of the kind

$$(1) \qquad E(Y_t \mid \mathscr{F}_{t-1}) = MY_{t-1} + \lambda \quad \text{a.s.}$$

where M is an $m \times m$ matrix of (vital rate) parameters and λ is a non-negative constant m-vector (possibly 0) and we shall focus on the estimation of M. It should be noted that, assuming finite variances of $\{Y_t\}$ as we shall do throughout the paper, the best predictor (in the least squares sense) of Y_{t+1} based on $\mathscr{F}_t$ is $MY_t + \lambda$ and if $\lambda = 0$ the best predictor of Y_{t+k} based on $\mathscr{F}_t$ is $M^k Y_t$.

From (1) we may write

$$(2) \qquad\qquad Y_t = M_t Y_{t-1} + \lambda_t + \varepsilon_t$$

where

$$E(M_t \mid \mathscr{F}_{t-1}) = M \quad \text{a.s.,} \qquad E(\lambda_t \mid \mathscr{F}_{t-1}) = \lambda \quad \text{a.s.,} \qquad E(\varepsilon_t \mid \mathscr{F}_{t-1}) = 0 \quad \text{a.s.}$$

and if $\lambda_t = 0$ a.s. this is in standard form as a (random coefficient) first-order autoregression. These have been extensively studied in the stationary case by Nicholls and Quinn [12]. The fundamental difference here is that ε_t will in general depend on the population vector Y_{t-1} and the process $\{Y_t\}$ may be far removed from stationarity.

In the model (2), $\{M_t\}$ represents a sequence of (possibly random) matrices of vital rates, $\{\lambda_t\}$ a sequence of recruitment (immigration) vectors, and $\{\varepsilon_t\}$ is a sequence of stochastic disturbance vectors. Such a model could originate, for example, from a multitype (Bienaymé–) Galton–Watson branching process with immigration in random environments. Here the different types correspond to different age classes and M_t denotes the matrix of offspring means of single individuals of the different ages in generation t.

Many particular cases of the model (2) have been extensively studied although not from the point of view of this paper. Special mention may be made of the (ordinary) multitype Galton–Watson process both with and without immigration (e.g. Mode [11], Athreya and Ney [4], Harris [7]) and the case where $\{M_t\}$ is a stationary ergodic sequence of random matrices (e.g. Heyde and Cohen [9], Heyde [8]). Some restrictions are, not surprisingly, necessary in order that M may be estimated.

The following results and illustrations are just intended to be indicative. The object is to emphasize the significance of describing population models such as branching ones in time series form. We shall explore the applications of the theorem given below which is an extension of the principal result of Anderson and Taylor [2]. Other procedures which are familiar in time series analysis could also have been treated although with rather less transparent results. Examples are maximum likelihood estimation based on the assumption of multivariate normality for the distributions of M_t, λ_t and ε_t and tests for the randomness of the coefficients M_t, λ_t (provided sufficiently concrete assumptions are made about the distribution of ε_t).

Theorem. Let $y_t = B' x_t + u_t$, $t = 1, 2, \cdots$ where B is a $p \times m$ matrix of parameters, x_t is a p-component vector and y_t and u_t are m-component stochastic vectors, the disturbances u_t being unobservable. We suppose that $E(u_t \mid \mathscr{F}_{t-1}) = 0$ a.s. where $\mathscr{F}_t$ is the σ-field generated by $u_1, u_2, \cdots, u_t$, $x_1, x_2, \cdots, x_{t+1}$ and $\mathscr{F}_0$ is generated by x_1. Write $A_t = \sum_{s=1}^{t} x_s x_s'$ and suppose that A_t is positive definite with probability 1 for $t > T_0$. Then, for $T > T_0$,

$$\hat{B}_T = A_T^{-1} \sum_{t=1}^{T} x_t y_t' \xrightarrow{\text{a.s.}} B$$

as $T \to \infty$ if any of the following sets of conditions (C1), (C2) or (C3) are satisfied:

- (C1) (a) $\lim_{T \to \infty} A_T^{-1} = 0$ a.s.
  - (b) The ratio of the largest to the smallest eigenvalues of A_T is bounded uniformly in T with probability 1.
  - (c) $E(u_t u_t' \mid \mathscr{F}_{t-1}) = O(1\,1')$ a.s.
    as $t \to \infty$, 1 denoting the m-component vector $(1, \cdots, 1)'$.
- (C2) (a) x_t is non-negative, $t = 1, 2, \cdots$.
  - (b) For some $\rho > 1$, $\rho^{-t} x_t$ converges almost surely to a positive random vector.
  - (c) $E(u_t u_t' \mid \mathscr{F}_{t-1}) = O(\rho^{2t\alpha} 1\,1')$ a.s.
    as $t \to \infty$ for some $\alpha < 1$.
- (C3) (a) $\lim_{T \to \infty} T^{-1} A_T = D$ a.s. where D is positive definite.
  - (b) $E x_t x_t' \leq C$, a matrix of finite constants.
  - (c) $E(u_t u_t' \mid \mathscr{F}_{t-1}) = O(|x_t|^{2\alpha} 1')$ a.s.
    for some $\alpha < 1$, $|x_t|^{2\alpha}$ denoting the vector $(|x_{1t}|^{2\alpha}, \cdots, |x_{mt}|^{2\alpha})'$.

It should be noted that $\hat{B}_T$ is the least squares estimate of B based on T observations and that the use of this particular functional form is ubiquitous in time series analysis.

The first part of the theorem ((C1)) is a general-purpose one and a minor extension of Theorem 1 of [2] in which their assumption $E \operatorname{tr} A_q^{-1}$ for some $q > p$ is avoided. The second and third parts of the theorem ((C2), (C3)) deal specifically with situations of geometric growth or approximate stationarity which may arise for population processes, the former being common and the latter usually only occurring in the case where recruitment is allowed. These parts avoid the complications inherent in the provision of a matrix version of the Kronecker lemma (for which see Anderson and Taylor [2], Anderson and Moore [1]).

2. Proof of theorem

Part (1). The proof of Theorem 1 of [2] can be followed except that the argument on boundedness of the variance of the martingale z_T (for which $E \operatorname{tr} A_q^{-1} < \infty$ is required) is replaced by a convergence condition involving conditional variances via, say, Theorem 2.15 of Hall and Heyde [6].

Part (2). Again we need to show that

$$(3) \qquad \hat{B}_T - B = A_T^{-1} \sum_{t=1}^{T} x_t u_t' \xrightarrow{\text{a.s.}} 0$$

as $T \to \infty$ and since $\rho^{-t} x_t$ converges almost surely to a positive random vector as $t \to \infty$ it suffices to establish that

$$(4) \qquad \rho^{-2T} \sum_{t=1}^{T} \rho^t u_t \xrightarrow{\text{a.s.}} 0$$

as $T \to \infty$.

Let u_{kt} be the kth element in the vector u_t, $1 \leq k \leq m$. Then, again applying Theorem 2.15 of [6], we have $\sum_{t=1}^{\infty} \rho^{-t} u_{kt}$ converges almost surely since $\sum_{t=1}^{\infty} \rho^{-2t} E(u_{kt}^2 \mid \mathscr{F}_{t-1})$ converges in view of

$$E(u_{kt}^2 \mid \mathscr{F}_{t-1}) = O(\rho^{2\alpha t}) \quad \text{a.s.}$$

and the Kronecker lemma gives

$$\rho^{-2T} \sum_{t=1}^{T} \rho^t u_{kt} \xrightarrow{\text{a.s.}} 0$$

as $T \to \infty$. This holds for each k, $1 \leq k \leq m$, and (4) follows.

Part (3). In view of the condition 3(a), $\hat{B}_T \xrightarrow{\text{a.s.}} B$ if

$$T^{-1} \sum_{t=1}^{T} x_t u_t' \xrightarrow{\text{a.s.}} 0$$

as $T \to \infty$ which in turn holds if

$$(5) \qquad T^{-1} \sum_{t=1}^{T} x_{it} u_{jt} \xrightarrow{\text{a.s.}} 0$$

for each $1 \leq i, j \leq m$, where x_{it} and u_{jt} are, respectively, the ith and jth elements in the vectors x_t and u_t. However, yet again applying Theorem 2.15 of [6] and the Kronecker lemma, we have that (5) holds if

$$\sum_{t=1}^{\infty} t^{-2} x_{it}^2 E(u_{jt}^2 \mid \mathscr{F}_{t-1}) < \infty \quad \text{a.s.}$$

and thus, in view of 3(c), if

$$(6) \qquad \sum_{t=1}^{\infty} t^{-2} x_{it}^2 |x_{jt}|^{2\alpha} < \infty \quad \text{a.s.}$$

But, using 3(b) we have

$$(7) \qquad \sum_{t=1}^{\infty} t^{-2/(1+\alpha)} x_{it}^2 < \infty \quad \text{a.s.}$$

for any $1 \leq i \leq m$ and thus in particular

$$(8) \qquad t^{-1/(1+\alpha)} |x_{jt}| \xrightarrow{\text{a.s.}} 0$$

as $t \to \infty$. Combining (7) and (8) gives (6), which completes the proof.

3. Applications

The applications of the theorem given below are all based on the model (2) and we note at the outset that $\sum_{t=1}^{T} Y_{t-1} Y_{t-1}'$ will be positive definite with probability 1 for T sufficiently large. This follows since otherwise there would exist a fixed non-zero m-component vector α such that $\alpha' Y_{t-1} = 0$ for all $t = 1, 2, \cdots$ which is precluded for stochastic $\{Y_i\}$ which is genuinely m-dimensional.

1. *Supercritical multitype Galton–Watson processes and their extensions.* The multitype Galton–Watson process can be described in the form (2) with $M_t \equiv M$ and $\lambda_t \equiv 0$ for all t. Here $M = (m_{ij})$ is the matrix of first moments $m_{ij} = E(Y_1^{(i)} \mid Y_0 = e_j)$, e_j being the vector whose jth component is 1 and whose other components are 0 while $Y_1^{(i)}$ is the number of individuals of type i in generation 1.

Suppose that the process is positive regular (meaning that all the elements of M^N are strictly positive for some $N > 0$). Then, if $\rho > 1$ is the maximal eigenvalue of M, and u and v are the right and left eigenvectors of M corresponding to ρ and normalized so that $u'v = 1$ and $u'1 = 1$ it is well known (e.g. Harris [7], Theorem 9.2, p. 44) that in the case of finite variance

$$(9) \qquad \rho^{-t} Y_t \xrightarrow{\text{a.s.}} uW$$

as $t \to \infty$ where W is a non-negative and non-degenerate random variable. Conditional on the non-extinction set $\{W > 0\}$ this deals with Condition (C2)(b) of the theorem and for Condition (C2)(c) it is noted that

$$E(\varepsilon_t \varepsilon_t' \mid \mathscr{F}_{t-1}) = E(Y_t Y_t' \mid \mathscr{F}_{t-1}) - M Y_{t-1} Y_{t-1}' M'$$

$$(10) \qquad = \sum_{i=1}^{m} V_i Y_{t-1}^{(i)}$$

using, for example, the considerations which lead up to Equation (4.2), p. 37 of

Harris [7]. Here V_i denotes the matrix $\mathrm{Var}\,(Y_1 \mid Y_0 = e_i)$ and $Y_{t-1}^{(i)}$ is the number of individuals of type i in generation $t-1$.

From (9) and (10) we see that (C2)(c) of the theorem holds with $\alpha = \tfrac{1}{2}$. Condition (C2)(a) is trivially satisfied and the theorem gives that, conditional on non-extinction,

$$(11) \qquad \left(\sum_{t=1}^{T} Y_{t-1} Y'_{t-1} \right)^{-1} \sum_{t=1}^{T} Y_{t-1} Y'_t$$

is a strongly consistent estimator of M'.

A different strongly consistent estimator of M, based on a knowledge of all parent–offspring combinations, has been discussed by Asmussen and Keiding [3] (Theorem 5.1). The virtue of (11) is that only the Y_i, $0 \le i \le T$, are required.

The estimator (11) continues to be strongly consistent for M' under various extensions of the Galton–Watson model to incorporate other types of stochastic disturbances. The principal constraint is supplied by Condition (C2)(c) which, as noted above, is satisfied for $\alpha = \tfrac{1}{2}$ in the Galton–Watson case (while only $\alpha < 1$ is required).

A direct proof of (C2)(b) is not possible from Equation (2) without further information but the following argument indicates that, for positive regular M with maximal eigenvalue $\rho > 1$, little additional specification is required. Let v be the (positive) left eigenvector of M associated with ρ. Then, from (1) with $\lambda = 0$,

$$E(v' Y_t \mid \mathscr{F}_{t-1}) = v' M Y_{t-1} = \rho v' Y_{t-1} \quad \text{a.s.}$$

so that $\{\rho^{-t} v' Y_t, \mathscr{F}_t, t \ge 0\}$ is a positive martingale and the martingale convergence theorem gives almost sure convergence of $\rho^{-t} v' Y_t$ as $t \to \infty$. Positivity of this limit with non-zero probability and of the individual components in the vector $\rho^{-t} Y_t$ as $t \to \infty$ are, however, not assured in general. Uniform integrability arguments provide the standard route for checking that the limits are not almost surely 0. For example, if the vector Y_0 is non-random and not identically 0,

$$E(\rho^{-t} v' Y_t) = \rho^{-t} v' M^t Y_0 = v' Y_0 > 0$$

and if $\rho^{-2t} E(v' Y_t)^2 \le C < \infty$ for some constant C then $\{\rho^{-t} v' Y_t, t \ge 0\}$ is uniformly integrable and

$$E\left(\lim_{t \to \infty} \rho^{-t} v' T_t \right) = v' Y_0 > 0$$

(e.g. Theorem 5.4, p. 32 of Billingsley [5]) which ensures that $\lim_{t \to \infty} \rho^{-t} v' Y_t$ is not almost surely 0.

In the case where $M_t \not\equiv M$ the requirement of (C2) unfortunately forces

$M_t \xrightarrow{\text{a.s.}} M$ as $t \to \infty$. Suppose $\rho^{-t} Y_t \xrightarrow{\text{a.s.}} W > 0$. Then (2) with $\lambda_t \equiv 0$ gives

$$\text{(12)} \qquad \rho^{-t} Y_t = \rho^{-1} M_t \rho^{-(t-1)} Y_{t-1} + \rho^{-t} \varepsilon_t$$

while $\rho^{-t} \varepsilon_t \xrightarrow{\text{a.s.}} 0$ in view of (C2)(c) and taking the limit as $t \to \infty$ in (12) leads to $M_t \xrightarrow{\text{a.s.}} M$ and $W = \rho^{-1} M W$. This last condition ensures that ρ is an eigenvalue of M and that W is of the form cu where c is a positive constant and u is a right eigenvector of M corresponding to the eigenvalue ρ.

2. *Multitype Galton–Watson processes with immigration and their extensions.* The multitype Galton–Watson process with immigration can be described in the form (2) with $M_t \equiv M$ and λ_t as independent and identically distributed random vectors. In order to use the theorem we rewrite the model in the form

$$\text{(13)} \qquad y_t = M x_t + u_t$$

where

$$y_t = Y_t - \lambda, \qquad x_t = Y_{t-1}, \qquad u_t = \varepsilon_t + \lambda_t - \lambda.$$

First we note that variances are straightforward to calculate in (13). Details are given in Quine [13] from whose results we deduce that

$$\text{(14)} \qquad E(u_t u_t' \mid \mathscr{F}_{t-1}) = O(Y_{t-1} 1') \quad \text{a.s.}$$

Now suppose that the Galton–Watson process in question is positive regular and let ρ be the maximal eigenvalue of M. If $\rho > 1$ then it follows from e.g. Theorem 2.1 and 2.2 of Hoppe [10] that $\rho^{-t} Y_t$ converges almost surely to a random vector which is non-zero with positive probability. This result, together with (14) ensures that, conditional on non-extinction,

$$\text{(15)} \qquad \left(\sum_{t=1}^{T} Y_{t-1} Y_{t-1}' \right)^{-1} \sum_{t=1}^{T} Y_{t-1} (Y_t - \lambda)' \xrightarrow{\text{a.s.}} M'$$

using (C2). Furthermore, it is readily seen that λ can be replaced in (15) by $T^{-1} \sum_{t=1}^{T} \lambda_t$ without disturbing the asymptotic behaviour of the estimator.

On the other hand, if $\rho < 1$ then $\{Y_t\}$ is asymptotically stationary and it is not difficult to establish that as $T \to \infty$,

$$\text{(16)} \qquad T^{-1} \sum_{t=1}^{T} Y_{t-1} Y_{t-1}' \xrightarrow{\text{a.s.}} D + \mu\mu'$$

for a certain positive definite matrix $D + \mu\mu'$ with

$$\mu = \lim_{t \to \infty} E Y_t = M\mu + \lambda,$$

while

$$(17) \qquad T^{-1} \sum_{t=1}^{T} Y_{t-1} Y_t' \xrightarrow{\text{a.s.}} DM + \mu\mu'$$

(Quine and Durham [14]). Equations (16) and (17) combine to again yield (15) and of course the theorem of this paper could also have been invoked using (C3).

A different strategy can, however, be used to estimate M, when $\rho < 1$, without knowledge of the specific immigration components. In this case $\{Y_t\}$ is asymptotically stationary and $\hat{Y} = T^{-1} \sum_{t=1}^{T} Y_t$ is a strongly consistent estimator of the stationary mean. Here it is advantageous to write the model in the form

$$\mathcal{Y}_t = M\mathcal{Y}_{t-1} + u_t$$

where

$$\mathcal{Y}_t = Y_t - EY_t, \qquad u_t = \varepsilon_t + \lambda_t - \lambda.$$

It is easily checked with the aid of (16) and (17) that

$$\left(\sum_{t=1}^{T} (Y_{t-1} - \hat{y})(Y_{t-1} - \hat{y})' \right)^{-1} \sum_{t=1}^{T} Y_{t-1}(Y_t - \hat{y})'$$

is a strongly consistent estimator of M (Quine and Durham [14]).

As for the ordinary Galton–Watson model, the estimator (15) continues to hold under various extensions of the model with immigration to incorporate other types of stochastic disturbances. When $\rho > 1$ the previous considerations still apply; in this case $\{\rho^{-t} v' Y_t, \mathcal{F}_t, t \geq 0\}$ is a positive submartingale. On the other hand, when $\rho < 1$, the estimator will hold for a variety of cases in which the departure from asymptotic stationarity is not drastic.

References

[1] ANDERSON, B. D. O. AND MOORE, J. B. (1976) A matrix Kronecker lemma. *Linear Algebra Appl.* **15**, 227–234.

[2] ANDERSON, T. W. AND TAYLOR, J. B. (1979) Strong consistency of least squares estimates in dynamic models. *Ann. Statist.* **7**, 484–489.

[3] ASMUSSEN, S. AND KEIDING, N. (1978) Maringale central limit theorems and asymptotic estimation theory for multitype branching processes. *Adv. Appl. Prob.* **10**, 109–129.

[4] ATHREYA, K. B. AND NEY, P. E. (1972) *Branching Processes.* Springer-Verlag, New York.

[5] BILLINGSLEY, P. (1968) *Convergence of Probability Measures.* Wiley, New York.

[6] HALL, P. AND HEYDE, C. C. (1980) *Martingale Limit Theory and its Application.* Academic Press, New York.

[7] HARRIS, T. E. (1963) *The Theory of Branching Processes.* Springer-Verlag, Berlin.

[8] HEYDE, C. C. (1984) On inference for demographic projection of small populations. *Proceedings of the Neyman–Kiefer Conference, Berkeley, CA, 1983*, ed. R. Olshen. Wadsworth and IMS, to appear.

[9] HEYDE, C. C. AND COHEN, J. E. (1985) Confidence intervals for demographic projections based on products of random matrices. *Theoret. Popn Biol.* **27**.

[10] HOPPE, F. M. (1976) Supercritical multitype branching processes. *Ann. Prob.* **4**, 393–401.

[11] MODE, C. J. (1971) *Multitype Branching Processes.* Elsevier, New York.

[12] NICHOLLS, D. F. AND QUINN, B. G. (1982) *Random Coefficient Autoregressive Models: An Introduction.* Lecture Notes in Statistics **11**, Springer-Verlag, Berlin.

[13] QUINE, M. P. (1970) The multi-type Galton–Watson process with immigration. *J. Appl. Prob.* **7**, 411–422.

[14] QUINE, M. P. AND DURHAM, P. (1977) Estimation for multitype branching processes. *J. Appl. Prob.* **14**, 829–835.

ANOVA Models with Random Effects: An Approach via Symmetry

T. P. SPEED

Abstract

The standard ANOVA models with random effects for multi-indexed arrays of random variables with an arbitrary nesting structure on the indices are considered from the viewpoint of symmetry. It is found that the covariance matrix of such an array has sufficient symmetry to permit viewing the usual components of variance as a generalised spectrum and the linear models of random effects as a generalised spectral decomposition.

COMPONENT OF VARIANCE; GROUP INVARIANCE; EXCHANGEABILITY; GENERALISED SPECTRUM ANALYSIS

1. Introduction

In his 1965 monograph *Group Representations and Applied Probability* Ted Hannan included a section (§5.2) entitled '*Analysis of variance*' in which he presented results of Professor Allan James which derived certain analyses of variance from symmetry considerations. The following section, entitled '*Stationary second-order processes*', contained material which Hannan described as 'mathematically equivalent to that discussed in the last application'. The setting was a stochastic process $x(t)$ defined on an index set T acted upon by a transitive permutation group G and satisfying the condition (of zero mean and) $\mathbb{E}\{x(s)x(t)\} = \mathbb{E}\{x(gs)x(gt)\}$ where $s, t \in T$ and $g \in G$.

Not long before this John Tukey (1961) had published his '*Discussion emphasising the connection between analysis of variance and spectrum analysis*' and the conclusion one came to after reading both these fascinating works was that, in some sense, ANOVA and the spectral analysis of weakly stationary time series were the same thing. At the time this topic was not discussed in any great generality nor, one might add, with any attempt at completeness, but the connection was nonetheless clear and it has proved useful to many workers in both areas since then.

As someone who has long been intrigued by this connection, I find it a very great pleasure to be permitted to include the following contribution to its clarification in this volume honouring Ted Hannan.

2. Outline of results and illustrative examples

ANOVA concerns multiply indexed arrays of random variables with an arbitrary system of *nesting* on the indices where we mean, loosely speaking, that an index i_q *is nested within* an index i_p (equivalently, i_p *nests* i_q) if i_q is only meaningful within the levels of i_p. With the introduction of appropriate permutation groups below, this notion becomes more precise. If we denote this relationship of indices by $q \geqq p$ (equivalently, $p \leqq q$), then any system of nesting can be described by a partially-ordered set $(P, \leqq)$ over the set P of index labels. Our interest is in arrays $y = (y_i : i \in \mathscr{I})$ of random variables where the multiple index $i = (i_p : p \in P)$ belongs to $\mathscr{I} = \prod \{\mathscr{I}_p : p \in P\}$, a product of countably infinite index sets labelled by the elements p of a nesting partially ordered set $(P, \leqq)$, and in this section we illustrate our approach and results by discussing the two cases depicted in Figure 1. In discussing these simple cases we shall revert to the more common usage y_{ij} rather than use $y_{i_1 i_2}$, and take $\mathscr{I} = \mathbb{N} \times \mathbb{N}$, where $\mathbb{N} = \{1, 2, 3, \cdots\}$.

The group G of permutations of $\mathscr{I}$ which plays a central role in our discussion is what we term the *restricted generalised wreath product* $G = \prod^* (G_p : p \in P)$ of the groups G_p which, for each $p \in P$, consist of the permutations of $\mathscr{I}_p$ of finite support, i.e. the *finitary* permutations of $\mathscr{I}_p$. For details of the (unrestricted) generalised wreath product construction we refer to Bailey et al. (1983) and references therein; to obtain *restricted* generalised wreath products we consider only finitary permutations in the component groups and, wherever a set of maps from a set into a group is defined by the construction, we consider only those maps which send all but finitely many points of their domain to the identity. With this restriction we find that most of the results of Bailey et al. (1983) carry over and we shall refer to the ones we need in due course. Letting S_∞ denote the group of all finitary permutations of the countably infinite set $\mathbb{N} = \{1, 2, 3, \cdots\}$ we note that for the simple crossing, $G = S_\infty \times S_\infty$ (direct or Cartesian product of groups), whilst the simple nesting has

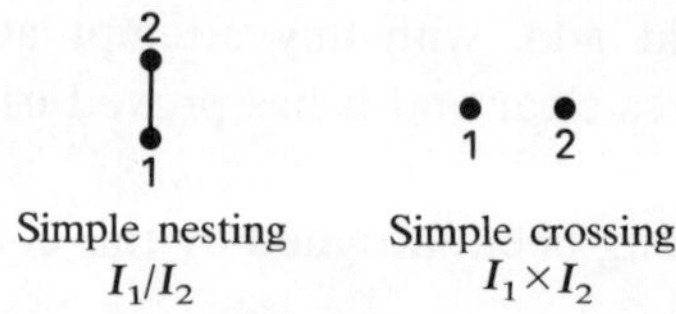

Figure 1

$G = S_\infty \text{wr}^* S_\infty$, the * restricting the wreath product group $S_\infty \text{wr} S_\infty$ to permutations which leave all but finitely many first indices fixed, and which, independently within finitely many first indices, permute only finitely many second indices.

When we have an array $y = (y_i : i \in \mathcal{I})$ of random variables and a group G acting on the index set $\mathcal{I}$, we can consider the property of *weak G-stationarity*, this being the set of constraints

$$(2.1a) \qquad\qquad \mu(i) = \mu(i^g),$$

$$(2.1b) \qquad\qquad \Gamma(i, j) = \Gamma(i^g, j^g),$$

where $\mu(i) = \mathbb{E}\{y_i\}$, $\Gamma(i, j) = \text{Cov}(y_i, y_j)$, $(i, j) \in \mathcal{I} \times \mathcal{I}$ and $g \in G$. In a case like the present one, where G acts *transitively* on $\mathcal{I}$, (2.1a) implies that μ is constant, and we will often assume in such cases that this constant value is 0. However an important question which arises immediately is the following.

(a) *Describe the class* γ *of all matrices* Γ *satisfying* (2.1b).

In the case of ordinary (weakly) stationary processes this question is answered by Bochner's theorem and in Theorem A below we give a complete answer to it in our context as well. We find that the most general Γ satisfying (2.1b) has the form

$$\Gamma(i, j) = \sum_{b \in L(P)} \varphi_b \zeta(b, a(i, j))$$

where $a(i, j)$ denotes the orbit of G on $\mathcal{I} \times \mathcal{I}$ containing (i, j), $\{\varphi_b\}$ is a uniquely defined set of non-negative parameters, $L(P)$ is a set whose structure will be clarified shortly, and ζ a function on $L(P) \times L(P)$.

The next question which arises naturally when continuing the analogy with weakly stationary processes is the following.

(b) *Is there an orthogonal decomposition of* y_i *($i \in \mathcal{I}$) corresponding to the decomposition of the (permutation) representation of G in the Hilbert space spanned by* $\{y_i : i \in \mathcal{I}\}$?

In Theorem B below we answer this question by proving that under (2.1a), (2.1b) we have the decomposition

$$y_i = \mu + \sum_{b \in L(P)} \varepsilon_{i(b)}$$

where

$$\mathbb{E}\{\varepsilon_{i(b)} \varepsilon_{j(c)}\} = \begin{cases} \varphi_b & \text{if} \quad b = c \text{ and } i_p = j_p, \quad p \in b; \\ 0 & \text{otherwise.} \end{cases}$$

We turn now to an illustration of our methods and results in the case of $\mathbb{N}/\mathbb{N}$ and $\mathbb{N} \times \mathbb{N}$. Taking the simple nesting first, it is easy to see that the orbits of $G = S_\infty \text{wr}^* S_\infty$ on $\mathbb{N}/\mathbb{N}$ have characteristic functions (matrices) as follows:

$$A_{\{1,2\}}((i, j), (i', j')) = \begin{cases} 1 & \text{if} \quad i = i', \quad j = j' \\ 0 & \text{otherwise;} \end{cases}$$

$$A_{\{1\}}((i, j), (i', j')) = \begin{cases} 1 & \text{if} \quad i = i', \quad j \neq j' \\ 0 & \text{otherwise;} \end{cases}$$

$$A_\phi((i, j), (i', j')) = \begin{cases} 1 & \text{if} \quad i \neq i' \\ 0 & \text{otherwise;} \end{cases}$$

i.e. $A_{\{1,2\}} = I \otimes I$, $A_{\{1\}} = I \otimes (J - I)$ and $A_\phi = (J - I) \otimes J$ where $I = I_\infty$ and $J = J_\infty$ are the identity and all-1 matrix, respectively, over $\mathbb{N}$. A general G-invariant covariance matrix Γ over $\mathbb{N}/\mathbb{N}$ thus takes the form

$$(2.2) \qquad \Gamma = aI \otimes I + bI \otimes (J - I) + c(J - I) \otimes J$$

where a, b and c are such that Γ is positive definite. We will see shortly that the condition for this to hold is $a \geq b \geq c \geq 0$, with at least one of the inequalities being strict, i.e. $a > 0$ if $b = c = 0$, or $b > 0$ if $c = 0$. Notice that a simple way of obtaining a positive definite matrix Γ of the form (2.2) is the following: for uncorrelated arrays ε, (ε_i) and (ε_{ij}) of uncorrelated effects with variances σ_ϕ^2, $\sigma_{\{1\}}^2$ and $\sigma_{\{1,2\}}^2$, respectively, we write

$$(2.3) \qquad y_{ij} = \mu + \varepsilon + \varepsilon_i + \varepsilon_{ij} \quad (i, j) \in \mathbb{N}/\mathbb{N}.$$

For in this case it is clear that the dispersion matrix $\Gamma = \mathbb{D}y$ of $y = (y_{ij})$ has the form (2.2) with $a = \sigma_\phi^2 + \sigma_{\{1\}}^2 + \sigma_{\{1,2\}}^2$, $b = \sigma_\phi^2 + \sigma_{\{1\}}^2$ and $c = \sigma_\phi^2$.

Let us prove that the construction (2.3) just given is the most general way of obtaining such Γ, thereby answering Questions (a) and (b) for the simple nesting structure $\mathbb{N}/\mathbb{N}$. Our method of proof uses the *truncation* $\Gamma^{(m,n)}$ of Γ to $i \leq m$ and $j \leq n$, for in this form the spectral decomposition of $\Gamma^{(m,n)}$ is evident. Indeed if we write

$$(2.4) \qquad \Gamma^{(m,n)} = aI_m \otimes I_n + bI_m \otimes (J_n - I_n) + c(J_m - I_m) \otimes J_n$$

then this expression is easily rearranged into the form $\xi_\phi S_\phi + \xi_{\{1\}} S_{\{1\}} + \xi_{\{1,2\}} S_{\{1,2\}}$ where

$$\xi_\phi = a + (n - 1)b + n(m - 1)c$$

$$\xi_{\{1\}} = a + (n - 1)b - c$$

$$\xi_{\{1,2\}} = a - b$$

are the non-negative eigenvalues of $\Gamma^{(m,n)}$; and

$$S_\phi = \frac{1}{m} J_m \otimes \frac{1}{n} J_n$$

$$S_{\{1\}} = \left(I_m - \frac{1}{m} J_m\right) \otimes \frac{1}{n} J_n$$

$$S_{\{1,2\}} = I_m \otimes \left(I_n - \frac{1}{n} J_n\right)$$

are the corresponding projectors of multiplicities

$$d_\phi = \operatorname{tr} S_\phi = 1$$

$$d_{\{1\}} = \operatorname{tr} S_{\{1\}} = m - 1$$

and

$$d_{\{1,2\}} = \operatorname{tr} S_{\{1,2\}} = m(n-1).$$

Notice that we have temporarily omitted the dependence of the ξ's, S's and d's upon (m, n). The next step is to rewrite $\Gamma^{(m,n)}$ as

(2.5)
$$\Gamma^{(m,n)} = \left[\frac{a-b}{mn} + \frac{b-c}{m} + c\right] J_m \otimes J_n + \left[\frac{a-b}{n} + b - c\right]$$
$$\times \left(I_m - \frac{1}{m} J_m\right) \otimes J_m + [a - b] I_m \otimes \left(I_n - \frac{1}{n} J_n\right)$$

and let $m, n \to \infty$. We obtain the following rearrangement of (2.2):

(2.6)
$$\Gamma = cJ \otimes J + (b - c)I \otimes J + (a - b)I \otimes I$$

and the knowledge that we must have $c \geq 0$, $b \geq c$ and $a \geq b$ with at least one inequality strict. Choosing $\sigma_\phi^2 = c$, $\sigma_{\{1\}}^2 = b - c$ and $\sigma_{\{1,2\}}^2 = a - b$ proves the assertion we made above, but we can go further. Write $V_1(\mathbb{N}/\mathbb{N})$ for the class of all positive definite matrices of the form (2.2) which have $a = 1$, a compact convex set in the pointwise topology. From Figure 2 it is clear that the *extreme points* of this convex set are $b = c = 0$, $b = c = 1$ and $c = 0$, $b = 1$, i.e. the extremal elements of $V_1(\mathbb{N}/\mathbb{N})$ are $I \otimes I$, $I \otimes J$ and $J \otimes J$, and so (2.6) is also a (unique) extreme point representation. These assertions are easily proved without having to appeal to diagrams.

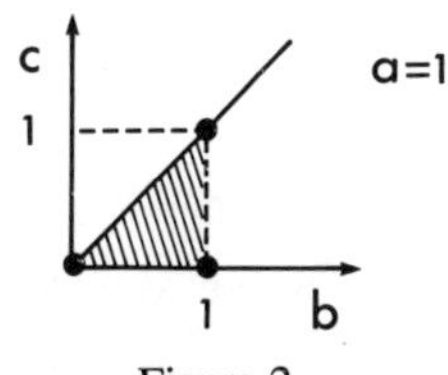

Figure 2

Now let us suppose that we have an array (y_{ij}) of random variables indexed by $\mathbb{N}/\mathbb{N}$ with zero mean and covariance matrix of the form (2.2). Our final introductory result concerning the simple nesting shows that such an array y must arise in the manner described in (2.3) above. More precisely, we can construct ε, (ε_i) and (ε_{ij}) from the array (y_{ij}) so that (2.3) holds. If we introduce the notation

$$y_{i\cdot}^{(m,n)} = \frac{1}{m} \sum_{j=1}^{n} y_{ij}$$

$$y_{\cdot\cdot}^{(m,n)} = \frac{1}{mn} \sum_{i=1}^{m} \sum_{j=1}^{n} y_{ij}$$

then the following are not too difficult to prove for $1 \le i \le m \le m'$, $1 \le j \le n \le n'$:

(2.7a) $\qquad \|y_{\cdot\cdot}^{(m,n)} - y_{\cdot\cdot}^{(m',n')}\|^2 = (a-b)\left[\dfrac{1}{mn} - \dfrac{1}{m'n'}\right] + (b-c)\left[\dfrac{1}{m} - \dfrac{1}{m'}\right]$

(2.7b) $\quad \|(y_{i\cdot}^{(m,n)} - y_{\cdot\cdot}^{(m,n)}) - (y_{i\cdot}^{(m',n')} - y_{\cdot\cdot}^{(m',n')})\|^2$

$$= (a-b)\left[\left(\frac{1}{n} - \frac{1}{n'}\right)\left(1 - \frac{1}{m}\right) + \frac{1}{n'}\left(\frac{1}{m} - \frac{1}{m'}\right)\right] + (b-c)\left[\frac{1}{m} - \frac{1}{m'}\right]$$

(2.7c) $\qquad \|(y_{ij} - y_{i\cdot}^{(m,n)}) - (y_{ij} - y_{i\cdot}^{(m',n')})\|^2 = (a-b)\left[\dfrac{1}{n} - \dfrac{1}{n'}\right]$.

It can be seen that the left-hand sides of these equations are all of the form $\|S_a^{(m,n)} y_{ij} - S_a^{(m',n')} y_{ij}\|^2$, $a = \phi$, $\{1\}$ or $\{1, 2\}$, where the S_a are the projectors introduced above.

It follows from (2.7a) and (2.7c) that $\{y_{\cdot\cdot}^{(m,n)}\}$ and $\{y_{i\cdot}^{(m,n)}\}$ both converge in mean square and so we can write

$$\varepsilon = \text{m.s.} \lim_{m,n} y_{\cdot\cdot}^{(m,n)}$$

$$\varepsilon_i = \text{m.s.} \lim_{m,n} (y_{i\cdot}^{(m,n)} - y_{\cdot\cdot}^{(m,n)})$$

and

$$\varepsilon_{ij} = \text{m.s.} \lim_{m,n} (y_{ij} - y_{i\cdot}^{(m,n)}).$$

With these definitions we see that the spectral decomposition

(2.8) $\qquad y_{ij} = y_{\cdot\cdot}^{(m,n)} + (y_{i\cdot}^{(m,n)} - y_{\cdot\cdot}^{(m,n)}) + (y_{ij} - y_{i\cdot}^{(m,n)}) \qquad i \le m,\ j \le n,$

of y_{ij} relative to $\Gamma^{(m,n)}$ converges in mean square as $m, n \to \infty$ to

(2.9) $\qquad\qquad\qquad y_{ij} = \varepsilon + \varepsilon_i + \varepsilon_{ij}$

where $\varepsilon \perp \varepsilon_i$, $\varepsilon \perp \varepsilon_{ij}$ and $\varepsilon_i \perp \varepsilon_{ij}$ follow from the orthogonality of the terms in (2.8). The further orthogonalities $\varepsilon_i \perp \varepsilon_{i'}$ and $\varepsilon_{ij} \perp \varepsilon_{i'j'}$ when $i \neq i'$ and $(i, j) \neq (i', j')$ respectively are readily proved and it follows from the same calculations that $\|\varepsilon\|^2 = c$, $\|\varepsilon_i\|^2 = b - c$ and $\|\varepsilon_{ij}\|^2 = a - b$. Thus we have answered Question (b) for the case of a simple nesting.

With this background it is easy to summarise the corresponding results for the simple crossing: $G = S_\infty \times S_\infty$ acting on $\mathbb{N} \times \mathbb{N}$. The most general G-invariant covariance matrix Γ over $\mathbb{N} \times \mathbb{N}$ takes the form

$$(2.10) \qquad \Gamma = aI \otimes I + bI \otimes (J - I) + c(J - I) \otimes I + d(J - I) \otimes (J - I)$$

where a, b, c and d are such that Γ is positive definite, and here the conditions are $d \geq 0$, $b \geq d$, $c \geq d$ and $a + d \geq b + c$, again with at least one strict inequality. As with the simple nesting, one can generate such covariance matrices via arrays

$$(2.11) \qquad y_{ij} = \mu + \varepsilon + \varepsilon_i + \varepsilon_j + \varepsilon_{ij}$$

where the ε, (ε_i), (ε_j) and (ε_{ij}) are orthogonal sets of orthogonal effects having variances σ_ϕ^2, $\sigma_{\{1\}}^2$, $\sigma_{\{2\}}^2$ and $\sigma_{\{1,2\}}^2$ respectively, and it is also possible to prove a theorem answering Question (b) showing that *all* arrays (y_{ij}) of random variables having a covariance matrix of the form (2.10) must be representable in the form (2.11). The details of the proof in this case are very similar to those already sketched for $\mathbb{N}/\mathbb{N}$, the main difference being the presence of four terms instead of three.

3. Preliminaries for the general theory

Our first concern is in the *orbits* of the restricted generalised wreath product group G of finitary symmetric groups on the set of all *pairs* $(i, j) \in \mathcal{I} \times \mathcal{I}$ where $\mathcal{I} = \prod \{\mathcal{I}_p : p \in P\}$ is the basic index set on which G acts. It follows by arguments identical to those in §5 of Bailey et al. (1983) that these are labelled by the elements of the *distributive lattice* $L(P)$ of all *hereditary subsets* of P, i.e. subsets $a \subseteq P$ with the property $p \in a$ and $q \leq p$ implies $q \in a$. For such a subset $a \subseteq P$ we find that $i = (i_p : p \in P)$ and $j = (j_p : p \in P)$ satisfy $(i, j) \in \mathcal{O}_a$, the orbit labelled by a, if $i_p = j_p$ for all $p \in a$, and there is *no* $b \in L(P)$ strictly including a for which $i_p = j_p$ for all $p \in b$.

A more useful description of these orbits now follows. In order to frame it we use the notation $i(a) = (i_p : p \in a)$, $a \subseteq P$, and the *Möbius function* μ of the lattice $L(P)$ (see Aigner (1979) for a general discussion of this useful notion). For an element $b \in L(P)$ we write

$$(3.1) \qquad R_b(i, j) = \begin{cases} 1 & \text{if} \quad i(b) = j(b) \\ 0 & \text{otherwise,} \end{cases}$$

the *relationship matrix* (cf. James (1957)) corresponding to indices in b, and then we may define the closely related *association matrices*

$$(3.2) \qquad A_a = \sum_b \mu(a, b) R_b.$$

It is easy to see that $A_a(i, j) = 1$ if and only if $(i, j) \in \mathcal{O}_a$, $a \in L(P)$, and that these matrices generalise those introduced in the previous section to arbitrary nesting of indices. Because of the fact that μ is inverse to the *zeta function:* $\zeta(b, a) = 1$ if $b \subseteq a$, $\zeta(b, a) = 0$ otherwise, we have

$$(3.3) \qquad R_b = \sum_a \zeta(b, a) A_a = \sum_{a \subseteq b} A_a.$$

There is in fact a simple formula for μ (whose values are always 0, 1 or -1) but we will not need it in what follows, referring the interested reader to Aigner (1979) for more details. Similarly we do not need the explicit matrix form of A_a, $a \in L(P)$ but it is nonetheless of interest to give it:

$$A_a = \bigotimes_{p \in a} I \otimes \bigotimes_{q \in \min(P \setminus a)} (J - I) \otimes \bigotimes_{r \in \mathrm{rest}(a)} J$$

where I and J are the identity and all-1 matrices, respectively, on $\mathcal{I}_p = \mathbb{N}$, $p \in P$, $\min(P \setminus a)$ denotes the *minimal* elements of the sub-poset $(P \setminus a, \leq)$ and $\mathrm{rest}(a) = P \setminus \{a \cup \min(P \setminus a)\}$.

In much of what follows we will be *truncating* our matrices at $n_p < \infty$, $p \in P$, and we shall describe this as truncating at $n = (n_p : p \in P)$, writing the corresponding matrices as $R_b^{(n)}$, $A_a^{(n)}$, etc. The index set in this case is $\mathcal{I}^{(n)} = \prod_{p \in P} \{1, 2, \cdots, n_p\}$, and by $n \uparrow \infty$ we mean $n_p \to \infty$, $p \in P$, in which case $\mathcal{I}^{(n)} \uparrow \mathcal{I}$.

4. Theorem A and its proof

Our aim in this section is to answer Question (a) by characterising all G-invariant positive definite covariance matrices over $\mathcal{I} = \prod \{\mathcal{I}_p : p \in P\}$. By making use of the description of the orbits of G on $\mathcal{I} \times \mathcal{I}$ given in Section 2 above, we see that this is the same as analysing the structure of the class $\mathcal{V}$ of all Γ of the form

$$(4.1) \qquad \Gamma = \sum_a \gamma_a A_a$$

where $\{\gamma_a : a \in L(P)\}$ are such that Γ is positive definite. Let $\mathcal{V}_1$ denote the subclass of $\mathcal{V}$ consisting of elements with diagonal element $\gamma_P = 1$.

Theorem A. (i) The class $\mathcal{V}_1$ is a finite-dimensional convex set with extreme points

$$\{R_b : b \in L(P)\}.$$

(ii) For every $\Gamma \in \mathcal{V}$, $\varphi_b = \sum_a \mu(a, b)\gamma_a \geqq 0$, $b \in L(P)$, and we have the unique representation

$$\Gamma = \sum_b \varphi_b R_b.$$ (4.2)

Proof. From (4.1) and (3.2) we deduce that

$$\Gamma = \sum_a \gamma_a \left\{ \sum_b \mu(a, b) R_b \right\} = \sum_b \left\{ \sum_a \mu(a, b)\gamma_a \right\} R_b$$

and so (ii) will follow from (i) as soon as we prove that $\varphi_b \geqq 0$, $b \in L(P)$. We do this by truncating at n, normalising, and then passing to a suitable limit.

The following spectral decomposition is proved in Speed and Bailey (1982):

$$\Gamma^{(n)} = \sum_a \gamma_a A_a^{(n)} = \sum_b \xi_b^{(n)} S_b^{(n)}$$ (4.3)

where $\Gamma^{(n)}$ and $A_a^{(n)}$ are the truncations of Γ and A_a at n,

$$\xi_b^{(n)} = \sum_a p_{ba}^{(n)} \gamma_a$$ (4.4)

$$p_{ba}^{(n)} = \sum_c \zeta(a \vee b, c)\mu(a, c)\bar{n}_c, \qquad \bar{n}_c = \prod_{p \not\in c} n_p,$$ (4.5)

$$S_b^{(n)} = \frac{1}{n} \sum_a q_{ab}^{(n)} A_a^{(n)}, \qquad n = \prod_{p \in P} n_p = |\mathcal{I}^{(n)}|,$$ (4.6)

and

$$q_{ab}^{(n)} = \sum_c \zeta(c, a \wedge b)\mu(c, b)n_c, \qquad n_c = \prod_{p \in c} n_p.$$ (4.7)

Now let us write $d_b^{(n)} = q_{Pb}^{(n)}$. It is easy to check from (4.5) and (4.7) that as $n \uparrow \infty$

$$n^{-1} d_b^{(n)} p_{ba}^{(n)} \to \mu(a, b)$$

and

$$[d_b^{(n)}]^{-1} q_{ab}^{(n)} \to \zeta(b, a).$$

Thus, if we define $\varphi_b^{(n)} = n^{-1} d_b^{(n)} \xi_b^{(n)}$, Equation (4.4) implies that

$$0 \leqq \varphi_b^{(n)} \to \sum_a \mu(a, b)\gamma_a = \varphi_b,$$ (4.8)

which proves that $\varphi_b \geqq 0$. Similarly we can use (4.3) and (4.6) to see that if

$A_a(i, j) = 1$, we have

$$\xi_b^{(n)} S_b^{(n)}(i, j) = n^{-1} d_b^{(n)} \xi_b^{(n)} [d_b^{(n)}]^{-1} q_{ab}^{(n)}$$
$$\rightarrow \varphi_b \zeta(b, a) = \varphi_b R_b(i, j).$$

This proves (ii) apart from the uniqueness assertion.

We turn now to the proof of (i), beginning with the observation that the $\{R_b\}$ are non-negative definite on $\mathscr{I}$. This follows from the fact that

$$(4.9) \qquad P_b^{(n)} = \frac{1}{\bar{n}_b} R_b^{(n)}$$

is the projector onto $\sum_{a \subseteq b} \mathscr{R}(S_a^{(n)})$, where $\mathscr{R}(\cdot)$ denotes the range of the operator in parentheses, see Speed and Bailey (1982); a direct proof is also possible. Next let us suppose that for some $c \in L(P)$ we have

$$R_c = \alpha U + \beta V$$

where $\alpha > 0$, $\beta > 0$, and $U, V \in \mathscr{V}_1$. Then $U = \sum_a \gamma_a^U A_a$ and $V = \sum_a \gamma_a^V A_a$ whence $\varphi_b^U = \sum_a \mu(a, b) \gamma_a^U \geqq 0$, $b \in L(P)$ and similarly for φ_b^V and φ_b^R, where $\varphi_b^R = \alpha \varphi_b^U + \beta \varphi_b^V$, $b \in L(P)$. But $\varphi_b^R = 1$ if $b = c$ and $= 0$ otherwise, and all φ's are non-negative, and so we must have $\varphi_b^U = \varphi_b^V = 0$ for all $b \neq c$. But this means that $\varphi_c^U = \varphi_c^V = 1$ since $1 = \gamma_P^U = \sum_b \varphi_b^U$ and similarly for φ_c^V, whence $U = V = R_c$ proving (i), (ii) and the theorem.

A by-product of the proof is the relation

$$(4.10) \qquad \xi_b^{(n)} = \sum_{c \supseteq b} \bar{n}_c \varphi_c$$

which will be useful later. It follows from (4.4) and the definition of the φ's.

5. Theorem B and its proof

Let us suppose that we have an array $y = (y_i : i \in \mathscr{I})$ of (real-valued) random variables with zero mean and covariance matrix $\Gamma = \mathbb{D}y$ belonging to the class $\mathscr{V}$ described in the previous section. In this section we show that there *is* an orthogonal decomposition of such random variables y_i ($i \in \mathscr{I}$) corresponding to the decomposition of the permutation representation of G in the Hilbert space spanned by $\{y_i : i \in \mathscr{I}\}$. Thus our concept of weak G-stationarity has an associated spectral decomposition, although there appears to be no general theory such as that for locally compact abelian groups acting on themselves, or for weakly symmetric spaces, Hannan (1965), which includes our class of groups and their associated permutation representations. As our proof is essentially *ad hoc*, we describe the steps in some detail before formulating and proving Theorem B.

Denote by $\mathcal{H}$ the Hilbert space spanned by $\{y_i : i \in \mathcal{I}\}$ and equipped with the inner product extending $\langle y_i, y_j \rangle = \Gamma(i, j)$, $(i, j) \in \mathcal{I} \times \mathcal{I}$. Our immediate aim is to define certain (closed linear) subspaces $\{\mathcal{S}_b : b \in L(P)\}$ of $\mathcal{H}$ which will turn out to be (i) G-invariant, (ii) G-irreducible, (iii) pairwise orthogonal and in addition they will satisfy (iv) $\sum_b \mathcal{S}_b = \mathcal{H}$. These subspaces $\mathcal{S}_b$ will be defined as the closed linear span of certain elements $S_b y_i$, $i \in \mathcal{I}$, of $\mathcal{H}$ which we shall construct as mean-square limits, and in the process we will find that the map extending $y_i \to S_b y_i$ is the projector of $\mathcal{H}$ onto $\mathcal{S}_b$. More precisely, let $n = (n_p : p \in P)$ be a truncation index and for $i \in \mathcal{I}^{(n)}$, $b \in L(P)$ write

$$(5.1) \qquad S_b^{(n)} y_i = \sum_{j \in \mathcal{I}^{(n)}} S_b^{(n)}(i, j) y_j.$$

The following lemma has a straightforward proof which we omit.

Lemma. For truncations at m and $n \geq m$ and $i, j \in \mathcal{I}^{(m)}$ with $A_a(i, j) = 1$, we have

$$(5.2) \qquad \langle S_b^{(m)} y_i, S_b^{(n)} y_j \rangle = \frac{1}{m} q_{ab}^{(m)} \xi_b^{(n)} \frac{\bar{m}_b}{\bar{n}_b}.$$

Corollary. If b^* denotes the set of maximal elements of $b \in L(P)$, we have:

$$(5.3) \qquad \begin{aligned} \|S_b^{(m)} y_i - S_b^{(n)} y_i\|^2 = \sum_{c \geq b} \Bigg\{ &\prod_{b^*} (1 - m_p^{-1}) \left[\frac{1}{m_{c \backslash b}} - \frac{1}{n_{c \backslash b}} \right] \\ &+ \frac{1}{n_{c \backslash b}} \left[\prod_{b^*} (1 - n_p^{-1}) - \prod_{b^*} (1 - m_p^{-1}) \right] \Bigg\} \varphi_c. \end{aligned}$$

Proof. Using the lemma we can expand the left-hand side as

$$m^{-1} d_b^{(m)} \xi_b^{(m)} + n^{-1} d_b^{(n)} \xi_b^{(n)} - 2 m^{-1} d_b^{(m)} \xi_b^{(n)} (\bar{n}_b)^{-1} \bar{m}_b$$

$$= m^{-1} d_b^{(m)} [\xi_b^{(m)} - (\bar{n}_b)^{-1} \xi_b^{(n)} \bar{m}_b] + n^{-1} \xi_b^{(n)} [d_b^{(n)} - m_b^{-1} d_b^{(m)} n_b]$$

and this simplifies to the right-hand side when we use (4.10) above and the result $d_b^{(n)} = n_b \prod_{b^*} (1 - n_p^{-1})$.

It is clear from (5.3) that $\{S_b^{(n)} y_i\}$ is a Cauchy sequence as $n \uparrow \infty$ and we write $S_b y_i$ for its mean-square limit. Further, we define $\mathcal{S}_b$ to be the closed linear span of $\{S_b y_i : i \in \mathcal{I}\}$, and extend the definition of S_b to the whole of $\mathcal{H}$ by linearity and closure. This is straightforward since $\|S_b y_i\|^2 \leq \|y_i\|^2$, $i \in \mathcal{I}$. Indeed we find from (5.2) and (4.8) that

$$\|S_b y_i\|^2 = \lim_n \|S_b^{(n)} y_i\|^2 = \lim_n n^{-1} q_{Pb}^{(n)} \xi_b^{(n)} = \varphi_b \leq \gamma_P = \|y_i\|^2.$$

A similar calculation proves that S_b is the orthogonal projection of $\mathcal{H}$ onto $\mathcal{S}_b$, for clearly $S_b y_i \in \mathscr{S}_b$, $i \in \mathscr{I}$, and if $j \in \mathscr{I}$,

$$\langle y_i - S_b y_i, S_b y_j \rangle = \lim_n \left[\langle y_i, S_b^{(n)} y_j \rangle - \langle S_b^{(n)} y_i, S_b^{(n)} y_j \rangle \right] = 0.$$

With these preliminaries we can now state and prove the following result.

Theorem B. The subspaces $\{\mathscr{S}_b\}$ are invariant and irreducible under the action of G, and

$$(5.4) \qquad \mathscr{H} = \oplus \{ \mathscr{S}_b : b \in L(P) \}.$$

Further, in the orthogonal decomposition

$$(5.5) \qquad y_i = \sum_b S_b y_i$$

we have $S_b y_i = S_c y_j$ if $b = c$ and $i(b) = j(c)$, and $S_b y_i \perp S_c y_j$ otherwise.
Finally, $\| S_b y_i \|^2 = \varphi_b$, $b \in L(P)$.

Proof. We begin with a proof that the subspaces $\mathscr{S}_b$ are G-invariant, i.e. invariant under the linear transformations (which we also denote by g) extending $g y_i = y_{i^g}$, $i \in \mathscr{I}$, $g \in G$. Note that for any element $g \in G$ there is an $m = m(g)$ large enough to ensure that $i^g = i$ for all $i \in \mathscr{I} \setminus \mathscr{I}^{(m)}$. Then for any $i \in \mathscr{I}$ we can choose n such that $n \geq m(g)$ and $i \in \mathscr{I}^{(n)}$, in which case

$$g S_b^{(n)} y_i = g \sum_{j \in \mathscr{I}^{(n)}} S_b^{(n)}(i, j) y_j$$

$$= \sum_{j \in \mathscr{I}^{(n)}} S_b^{(n)}(i, j) y_{j^g}$$

$$= \sum_{j \in \mathscr{I}^{(n)}} S_b^{(n)}(i^g, j^g) y_{j^g} = S_b^{(n)} y_{i^g}$$

and then $g S_b = S_b g$ follows in the limit.

Before we complete the proof of irreducibility of the $\{\mathscr{S}_b\}$, let us prove the rest of Theorem B. In an obvious notation we have

$$I^{(n)} = \sum_b S_b^{(n)} \quad \text{on } \mathscr{H}^{(n)}$$

from which we readily conclude that $I = \sum_b S_b$ on $\mathscr{H}$, and it is clear that $\langle S_b y_i, S_c y_j \rangle = \lim_n \langle S_b^{(n)} y_i, S_c^{(n)} y_j \rangle = 0$ if $b \neq c$. If $b = c$, then the lemma together with results of the previous section show that the limit is $\varphi_b R_b(i, j)$. In particular, $\mathscr{S}_b$ is spanned by the distinct elements of $\{ S_b y_i : i \in \mathscr{I} \}$ which are labelled by $\mathscr{I}(a) = \prod \{ \mathscr{I}_p : p \in a \}$, and these elements are all orthogonal and have equal squared length φ_b.

Now suppose that $\mathscr{K}$ is a G-invariant subspace of $\mathscr{S}_b$ and put $z_i = K y_i$ where K is the orthogonal projection onto $\mathscr{K}$. Then $z = (z_i)$ is a weakly G-stationary array of random variables as well, for

$$\begin{aligned}
\langle z_i, z_j \rangle &= \langle gz_i, gz_j \rangle && \text{since } \langle \cdot, \cdot \rangle \text{ is } G\text{-invariant} \\
&= \langle gKy_i, gKy_j \rangle && \text{by definition of } z \\
&= \langle Kgy_i, Kgy_j \rangle && \text{since } gK = Kg \\
&= \langle Ky_{i^g}, Ky_{j^g} \rangle \\
&= \langle z_{i^g}, z_{j^g} \rangle.
\end{aligned}$$

It follows that all the constructions we have carried out for $y = (y_i)$ have analogues for $z = (z_i)$ and, in particular, we can obtain analogues $\{\tilde{\mathcal{S}}_b\}$ of the subspaces $\{\mathcal{S}_b\}$ constructed above. However a simple calculation shows that $\tilde{S}_b K = K S_b$ for

$$\begin{aligned}
\tilde{S}_b z_i &= \lim_n \sum_{j \in \mathcal{G}^{(n)}} S_b^{(n)}(i, j) z_j \\
&= \lim_n \sum_{j \in \mathcal{G}^{(n)}} S_b^{(n)}(i, j) K y_j && \text{by definition of } z_j \\
&= \lim_n K \sum_{j \in \mathcal{G}^{(n)}} S_b^{(n)}(i, j) y_j && \text{since } K \text{ is linear} \\
&= K \lim_n \sum_{j \in \mathcal{G}^{(n)}} S_b^{(n)}(i, j) y_j && \text{since } K \text{ is a contraction} \\
&= K S_b y_i.
\end{aligned}$$

We now prove that for any $j \in \mathcal{G}$ and $b \in L(P)$ we have $S_b z_j = S_b y_j$ or 0, i.e. that $S_b K = S_b$ or 0. Write z_j in terms of the orthogonal bases of $\{\mathcal{S}_b : b \in L(P)\}$ as follows:

$$(5.6) \qquad z_j = \sum_b \sum_{i(b)} \lambda_{i(b)} \varepsilon_{i(b)}$$

where $\varepsilon_{i(b)} = S_b y_i$ and $\lambda_{i(b)} = \langle z_j, \varepsilon_{i(b)} \rangle / |\varepsilon_{i(b)}|^2$. The application of K to both sides leaves the left-hand side unchanged whilst it must, by $\tilde{S}_b K = K S_b$, convert the right-hand side into the spectral decomposition of z_j:

$$(5.7) \qquad z_j = \sum_c \tilde{\varepsilon}_{j(b)}$$

where $\tilde{\varepsilon}_{j(b)} = \tilde{S}_b z_j$. It follows by comparing (5.6) and (5.7) and using the properties of the decomposition (5.7) already established that $\lambda_{i(b)} = 0$ unless $i(b) = j(b)$ in which case $\lambda_{i(b)} = 1$. In other words,

$$\langle S_b K y_j, S_c y_i \rangle = \begin{cases} \varphi_b & \text{if} \quad b = c \text{ and } i(b) = j(b) \\ 0 & \text{otherwise,} \end{cases}$$

which proves that $S_b K = S_b$ unless $K S_b = 0$, in which case we will have deduced nothing. This proves the irreducibility of the $\{\mathcal{S}_b\}$ and completes the proof of the theorem.

6. Closing remarks

It is easy to see that the description of all G-invariant positive definite matrices on $\mathcal{I}$ given by Theorem A carries over to not-necessarily-positive-definite matrices, although naturally the non-negativity of the $\{\varphi_b\}$ no longer holds. With minor modifications to the proofs we are then able to give multivariate analogues of Theorems A and B. (The mathematical reason for this is the fact that all the irreducible representations included in our permutation representation of G are of real type.)

A variant on the constructions we have described would be the ANOVA underlying a diallel cross, cf. James (1982). In this case we have S_∞ or $S_\infty \times S_2$ acting on $\mathbb{N} \times \mathbb{N}$ or $\mathbb{N} \times \mathbb{N} \backslash \Delta$ where Δ is the diagonal, and arguments just like those presented above will lead to random effects models for the various diallel crosses. Similarly triallel and double-cross models may be derived.

We note in closing that although it appears that we have done nothing more than rederive the standard ANOVA models from considerations of symmetry, the combinatorial tools we have used to do so have many more uses. For example, it is straightforward to give expressions for the best linear predictors of the random effects based upon suitable finite subarrays; expected mean-square identities are easily obtained; recursive algorithms for partitioning a data array into its spectral components (strata) can be derived, and so on, but we must leave the details of these results to another time.

References

AIGNER, M. (1979) *Combinatorial Theory*. Springer-Verlag, New York.

BAILEY, R. A., PRAEGER, C. E., ROWLEY, C. A. AND SPEED, T. P. (1983) Generalised wreath products of permutation groups. *Proc. London Math. Soc.* (3) **47**, 69–82.

HANNAN, E. J. (1965) Group representations and applied probability. *J. Appl. Prob.* **2**, 1–68.

JAMES, A. T. (1957) The relationship algebra of an experimental design. *Ann. Math. Statist.* **28**, 993–1002.

JAMES, A. T. (1982) Analysis of variance determined by symmetry and combinatorial properties of zonal polynomials. In *Statistics and Probability: Essays in Honor of C. R. Rao*, ed. G. Kallianpur, P. R. Krishnaiah and J. K. Ghosh, North-Holland, Amsterdam, 329–341.

SPEED, T. P. AND BAILEY, R. A. (1982) On a class of association schemes derived from lattices of equivalence relations. In *Algebraic Structures and Applications*, ed. P. Schultz, C. E. Praeger and R. P. Sullivan, Marcel Dekker, New York, 45–54.

TUKEY, J. W. (1961) Discussion, emphasising the connection between analysis of variance and spectrum analysis. *Technometrics* **3**, 1–29.

Random Cyclic Transformations
of Points

J. G. VEITCH
G. S. WATSON*

Abstract

We consider the action of independent and identically distributed $n \times n$ circulants $S_1, S_2, \cdots$ on $V = [v_1, \cdots, v_n]$ whose columns are the positions of n points in $\mathbb{R}^d$. The positions of the n points after m transformations are the columns of $W^{(m)} = VS_1 \cdots S_m$. We describe, in several ways, the shape of the configuration of the points $W^{(m)}$ as $m \to \infty$. When $n = 3, 4$ and $d = 2$, a special discussion in terms of Moebius transformations is given.

CIRCULANT; SHAPE; STOCHASTIC GEOMETRY

1. Introduction

Berlekamp et al. (1965) began an interesting non-statistical literature by generalizing the following problem (proposed by Rosenman (1932)). Let P_i be a closed polygon with n vertices and P_{i+1}, the polygon obtained by joining, in the same order, the midpoints of the sides of P_i. What are the properties of the sequence? They suggested first that the vertices of P_{i+1} should be formed cyclically, using the same convex combination. Thus if $v_{j,i}$ are vectors to the vertices of P_i, the vectors $v_{j,i+1}$ are given by

$$v_{j,i+1} = s_0 v_{j,i} + s_1 v_{j+1,i} + \cdots + s_{n-1} v_{j+n-1,i}$$

where the subscripts are interpreted modulo n and s_i are non-negative and $\sum_0^{n-1} s_k = 1$. They very briefly discussed generalizations to polyhedra.

Schoenberg (1950) had earlier been stimulated by Rosenman's problem and examined the planar problem with complex $s_0, \cdots, s_{n-1}$ by using the finite Fourier transform. Davis (1979), whose book on circulants stimulated our

---

* The senior author takes great pleasure in dedicating this paper to his old friend, Ted Hannan. May these circulants remind him of the early days of time series and the implied finite Fourier transform of more recent days!

interest, discussed the planar case by the equivalent method of circulants. Both authors exploit the notion of representing the vertices of the polygons by points $z_1, \cdots, z_n$ in the complex plane.

Watson (1983a) introduced a stochastic element when $n = 3$ by supposing that $z_{j,i+1} = s_i z_{j,i} + (1 - s_i) z_{j+1,i}$ where $j + 1$ is as usual interpreted modulo $n = 3$ and the sequence of random variables $\{s_i\}$ are independent and identically distributed (i.i.d.) on $(0, 1)$. The extension to s any real random variable is trivial. The metrical properties of such random sequences of triangles (e.g. area and orientation) turn out to be of less interest than the shape of the triangles. Since the inscribed triangle at each step clearly tends to the centroid of the initial triangle, it is convenient to rescale so as to have a non-vanishing asymptotic triangle. Following Kendall (1984) this was measured by a complex number $b, |b| \leq 1, 0 \leq \arg(b) \leq \pi/3$. It was shown that b follows a random walk on a circle about the origin in the unit disc and so usually becomes uniformly distributed as the number of transformations goes to infinity.

In this paper, we consider first the deterministic iteration applied to n points in $\mathbb{R}^d$ and then use random i.i.d. transformations. Two limit theorems are given in Section 2 which describe the asymptotic configuration of n points in two distinct ways.

In the special case of $d = 2$, when one is interested only in shapes, one is tempted to renormalize each time so that two of the points are placed on 0 and 1 in the complex plane. For three points, (triangles), this leads to an elegant application of Moebius transformations and the change of shape at each step may be seen as a rotation of the Riemann sphere. This representation partially extends to quadrilaterals and produces elegant pictures in both cases but it breaks down for more than four points. This aspect of the problem is discussed in Section 3.

2. Basic results

Suppose n points in $\mathbb{R}^d$ are denoted by the column vectors $v_1, \cdots, v_n$. This will be called a configuration. To get a figure they must be joined in some way. If $d = 2$ and the straight line joins are from v_1 to v_2, v_2 to $v_3, \cdots, v_n$ to v_1, it is a polygon, an n-gon. If $d = 3$, the v_i might be vertices of a certain polyhedra. For the moment we shall just consider configurations. Suppose that a new configuration $w_1, \cdots, w_n$ is generated by the cyclic rules

$$(2.1) \qquad w_j = s_0 v_j + s_1 v_{j+1} + \cdots + s_{n-1} v_{j+n-1}, \qquad j = 1, \cdots, n$$

where the subscripts are interpreted modulo n. Write the configurations as $d \times n$ matrices $W = [w_1, \cdots, w_n]$ and $V = [v_1, \cdots, v_n]$ and the transformation (2.1) as a circulant matrix $S = \operatorname{circ}(s_0, s_{n-1}, s_{n-2}, \cdots, s_1)$ in the notation of

Davis (1979). Then (2.1) may be written compactly as

$$(2.2) \qquad\qquad W = VS.$$

There are several ways to study the effect on configuration V of many successive repetitions of this transformation with possibly different circulants.
The eigenvalues and eigenvectors of S, $S\boldsymbol{u}_k = \lambda_k \boldsymbol{u}_k$ are given (if $\omega = \exp(2\pi i/n)$) by

$$(2.3) \qquad \begin{aligned} \boldsymbol{u}_k^{t} &= [1, \omega^k, \omega^{2k}, \cdots, \omega^{(n-1)k}] \\ \lambda_k &= s_0 + s_{n-1}\omega^k + \cdots + s_1 \omega^{(n-1)k} \end{aligned}$$

for $k = 1, \cdots, n$. If we write $\Omega = [\omega^{jk}]$ where j and k run from 0 to $n-1$ and observe that if $*$ denotes conjugate transpose, $\Omega^*\Omega = nI_n$, then

$$(2.4) \qquad S = \frac{1}{n} \sum_{k=1}^{n} \lambda_k \boldsymbol{u}_k \boldsymbol{u}_k^* = \frac{1}{n} \Omega D(\lambda) \Omega^*$$

where $D(\lambda)$ is a diagonal matrix with entries $\lambda_1, \cdots, \lambda_n$. Thus the product of m circulants S_j is simply (2.4) with λ_k replaced by the product of the kth eigenvalues from each circulant, $\prod_{j=1}^{m} \lambda_k^j$.

Any configuration V, a $d \times n$ real matrix, can be written as $V = C\Omega^*$, with $C = n^{-1}V\Omega$, a $d \times n$ complex matrix. The columns of Ω are $\boldsymbol{u}_1, \cdots, \boldsymbol{u}_n$ so if the columns of C are denoted by $\boldsymbol{c}_k$, $V = \sum_{k=1}^{n} \boldsymbol{c}_k \boldsymbol{u}_k^*$. We shall demand that the centroid of the points $v_1, \cdots, v_n$ is always at the origin in $\mathbb{R}^d$, i.e. $\sum_1^n v_k = 0$. Since $\boldsymbol{u}_n^* = [1, 1, \cdots, 1]$, this implies that $\boldsymbol{c}_n = 0$. Hence all configurations of interest to us can be written as

$$(2.5) \qquad V = \sum_1^{n-1} \boldsymbol{c}_k \boldsymbol{u}_k^*.$$

Further, since $\boldsymbol{u}_{n-k} = \bar{\boldsymbol{u}}_k$ and V is a real matrix, $V = \sum_1^{n-1} \bar{\boldsymbol{c}}_{n-k} \boldsymbol{u}_k^*$, so that $c_{n-k} = \bar{c}_k$. Thus we can write V as the sum of real $d \times n$ matrices F_k,

$$V = \sum_1^{[n/2]} F_k, \qquad \left[\frac{n}{2}\right] = \text{integer part of } \frac{n}{2}$$

$$(2.6) \qquad F_k = \boldsymbol{c}_k \boldsymbol{u}_k^* + \bar{\boldsymbol{c}}_k \bar{\boldsymbol{u}}_k^*$$

$$F_{[n/2]} = \boldsymbol{c}_{[n/2]} \boldsymbol{u}_{[n/2]}^* \quad (n \text{ even}).$$

Writing $\boldsymbol{c}_k = \boldsymbol{a}_k + i\boldsymbol{b}_k$, $\boldsymbol{a}_k, \boldsymbol{b}_k \in \mathbb{R}^d$, the configuration F_k has n points on the curve

$$(2.7) \qquad f_k(t) = 2(\boldsymbol{a}_k \cos kt + \boldsymbol{b}_k \sin kt)$$

for $t = 2\pi(j/n)$, $j = 1, \cdots, n$. This curve is an ellipse in the two-plane spanned by $\boldsymbol{a}_k$ and $\boldsymbol{b}_k$; but if $\boldsymbol{a}_k$ and $\boldsymbol{b}_k$ are linearly dependent the curve is degenerate. We may think of the figure obtained by joining these points cyclically as an affine transformation in the two-space of a regular n-gon. Thus V may, in general, be thought of as made up of $[n/2]$ convex polygons in various two spaces of $\mathbb{R}^d$.

The effect of repeated transformations (2.2), say of $S_1, \cdots, S_m$, is to give the configuration

$$(2.8) \qquad W^{(m)} = VS_1 \cdots S_m = CD(\lambda^{(m)})\Omega^*$$

where $\lambda_k^{(m)} = \lambda_k^1 \cdots \lambda_k^m$. Thus

$$(2.9) \qquad W^{(m)} = \sum_{k=1}^{[n/2]} F_k^{(m)}$$

where

$$(2.10) \qquad F_k^{(m)} = \lambda_k^{(m)} c_k u_k^* + \bar{\lambda}_k^{(m)} \bar{c}_k \bar{u}_k^*.$$

Writing

$$\lambda_k^{(m)} = r_k^{(m)} \exp{(i\varphi_k^{(m)})}, \quad \varphi_k^{(m)} = \arg{(\lambda_k^1 \lambda_k^2 \cdots \lambda_k^m)} = \sum_{j=1}^m \arg{(\lambda_k^j)}, \quad c_k = a_k + ib_k,$$

the points in the configuration $F_k^{(m)}$ lie on the curve

$$(2.11) \qquad f_k^{(m)}(t) = 2r_k^{(m)}(\boldsymbol{a}_k \cos{(kt)} + \boldsymbol{b}_k \sin{(kt)})$$

but rotated by $\varphi_k^{(m)}$, and for $t = 2\pi(j/n)$, $j = 1, \cdots, n$. The ellipse lies in the subspace spanned by $\boldsymbol{a}_k$ and $\boldsymbol{b}_k$—in fact this is the same ellipse as (2.7), but with semi-axes multiplied by $r_k^{(m)}$ and rotated by $\varphi_k^{(m)}$. Alternatively, we may regard the ellipse as fixed and the points as moving, i.e. being at the points $(2\pi(jk/n) + (\varphi_k^{(m)}/k))$.

If the points in the configuration $F_k^{(m)}$ are joined cyclically in the case $d = 2$, the n-gons formed are either simple polygons (when $k = 1$) or stars. If $k = [n/2]$, the figure is degenerate. It is of interest to represent geometrically these n-gons. Given k, we can represent the entire family of possible n-gons inscribed in ellipse (2.7) by parametrizing the angle of rotation φ, i.e. $P_k(\varphi)$ is the cyclic join of the configuration

$$(2.12) \qquad F_k(\varphi) = 2(\boldsymbol{a}_k \cos{(kt + \varphi)} + \boldsymbol{b}_k \sin{(kt + \varphi)}).$$

Let $\boldsymbol{a}_k = [a_1, a_2]^t$ and $\boldsymbol{b}_k = [b_1, b_2]^t$. The ellipse and the inscribed figure are the affine transformation of a circle of unit radius with vertices on the circle at equi-angular intervals $2\pi j(k/n)$. The affine transformation is obviously

$$(2.13) \qquad \begin{bmatrix} u \\ v \end{bmatrix} = 2 \begin{bmatrix} a_1 & b_1 \\ a_2 & b_2 \end{bmatrix} \begin{bmatrix} x \\ y \end{bmatrix}.$$

Notice that the inscribed figure prior to the affine transform is a regular polygon or regular star with n vertices. The joins are tangent to an inner circle

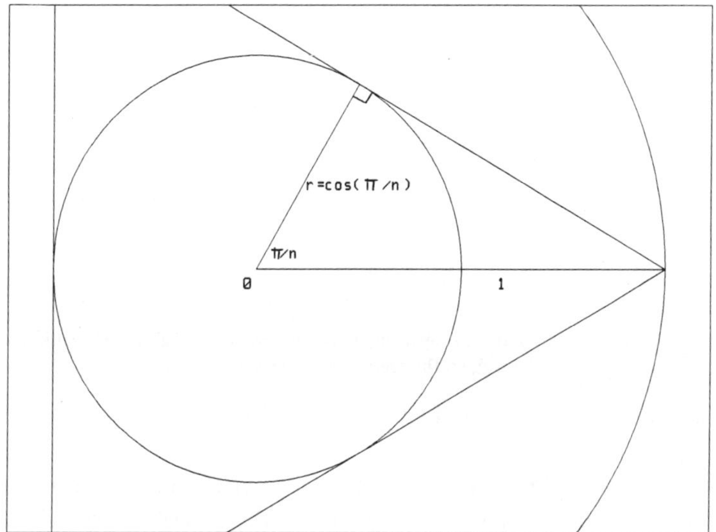

Figure 1. The inner circle to the regular polygon for $n = 3$, prior to the affine transformation (2.13)

with radius given by modulus of $\cos(\pi(k/n))$, as in Figure 1. Since an affine map preserves the ratios of the lengths of any two parallel vectors, this inner circle will be mapped by the transform to an ellipse inside ellipse (2.7), where the ratio of the axes is fixed at $\cos(\pi(k/n))$. Clearly we can rotate ellipse (2.7) so that one semi-axis is given by the vector $|\boldsymbol{a}_k|(1, 0, 0, \cdots, 0)^t$ and the other is given by $|\boldsymbol{b}_k|(0, 1, 0, \cdots, 0)^t$. Effectively, $d = 2$, so ignoring the other coordinates, the outer ellipse is given by

$$(2.14) \qquad f_1(t) = 2(|\boldsymbol{a}_k|\cos kt, |\boldsymbol{b}_k|\sin kt)^t$$

and the inner ellipse by

$$(2.15) \qquad f_2(t) = 2\left(\cos\left(\pi\frac{k}{n}\right)\right)(|\boldsymbol{a}_k|\cos kt, |\boldsymbol{b}_k|\sin kt)^t.$$

A geometric picture of the family $P_k(\varphi)$ (defined by cyclically joining configuration $F_k(\varphi)$ in (2.12)) is simply that of all n-gons with vertices on the outer ellipse and joins tangent to the inner ellipse.

This construction combined with (2.11) and rescaling by $1/r_k^{(m)}$ can be used to generate any sequence of such n-gons. An example is illustrated in Figure 2 ($n = 3$, $k = 1$). In this figure $\boldsymbol{a}_1 = [2, 0]^t$ and $\boldsymbol{b}_1 = [0, 1]^t$. Two triangles are given; the second is a transform of the first where $\arg(\lambda_1)$ is about $\pi/6$. A similar picture for quadrilaterals ($n = 4$, $k = 1$) is given in Figure 3. These static pictures cannot convey the excitement of watching the evolution of the figures on the screen of a graphics terminal. An alternative geometric representation

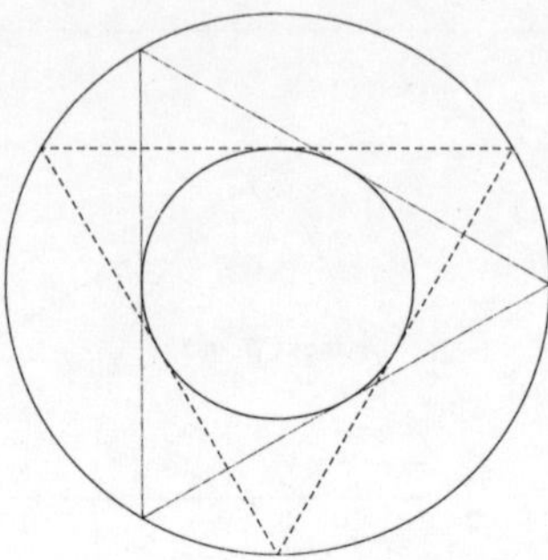

Figure 2. The figure F_k (when $n = 3$) showing two successive triangles. (The initial triangle is equilateral, so the transformation is invariant.)

for the case $d = 2$ is given in Davis (1979), by using complex numbers to represent the vertices.

It is obvious that, if $|\boldsymbol{a}_k| = |\boldsymbol{b}_k|$, the ellipse is a circle and the inscribed figure merely rotates around. For example, if $n = 3$, the inscribed triangle is always equilateral. Notice such regular figures are the only invariant figures in the sense that their shape remains constant under any transformation S_j. This suggests it would also be interesting to follow the 'shape' of the configuration under repeated transformations of the type (2.1). This is taken up again below.

In the deterministic literature, which is surveyed above, circulants S_j are fixed, say S, and in the typical case there is a single k for which $|\lambda_l| < |\lambda_k| = |\lambda_{n-k}|$. Then the kth configuration in (2.9) dominates for m large.

It is also possible to follow some metrical properties of the configurations. If for example, $n = d + 1$, the volume of the polyhedron is $(d+1)^{-1} \left| \det \begin{bmatrix} v \\ u_n^t \end{bmatrix} \right|$. Since $u_n^t S = u_n^t$, the volume of $W = VS$ is $|\det S|$ times the volume of V. Thus one must study $\prod_1^m |\det S_j|$. It seems, from Watson (1983a), that there is little interest in these properties.

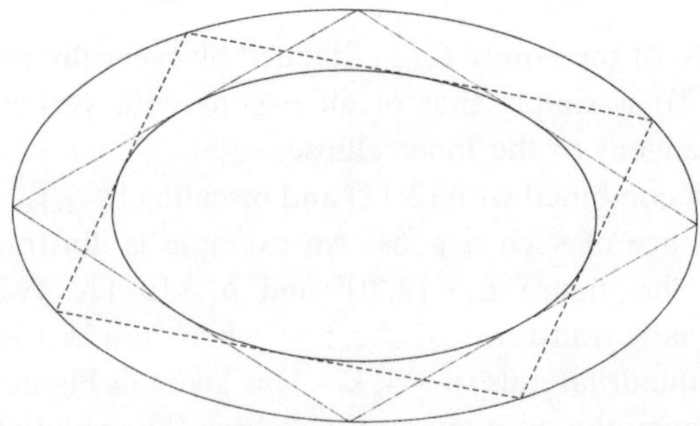

Figure 3. The figure F_k (when $n = 4$) showing two successive parallelograms.

To take up the discussion on shape again, Kendall (1984) has defined this concept so that the shape of n points in $\mathbb{R}^d$ is a point on a certain manifold. It is the motion of this point that we want to define. Clearly the shape of V is the same as the shape of $tHVP^pQ^q$ where t is any real number, H is a rotation on $\mathbb{R}^d$, P a cyclic permutation matrix with first row $(0, 1, 0, \cdots, 0)$ and $p = 0, 1, \cdots, n-1$, Q a matrix reversing the order of a vector and $q = 0, 1$. The $d \times n$ matrix V has a singular-value decomposition which, for $d < n$, may be written

$$V = \sum_1^d \alpha_j e_j r_j^t$$

where $e_1, \cdots, e_d$ are the orthonormal in $\mathbb{R}^d$ and $r_1, \cdots, r_d$ are orthonormal in $\mathbb{R}^n$ and $\alpha_j \geqq 0$. The vectors r_j must be orthogonal to u_n since $V u_n = 0$. The matrix $R = [r_1, \cdots, r_d]$ must therefore lie on a Stiefel manifold usually denoted by $V_{d,n-1}$. Because V is equivalent to tHV, only the ratios of the α_j are relevant to shape and the e_j are irrelevant. The equivalence relation identifying V with VP^pQ^q divides the Stiefel manifold into $2n$ equivalence classes. For a fuller discussion, and measures on the shape manifold, the reader must turn to Kendall's paper. In particular, his use of the Gaussian to get these measures might be expected in view of the intimate connections of Stiefel manifolds with the Gaussian—see e.g. James (1954), Muirhead (1982), Watson (1983b). Watson (1983a) gives a simple discussion of shape for $n = 3$, $d = 2$.

In this paper we need little more because we only handle the case where, as $m \to \infty$, the kth term in (2.9) dominates, i.e. the modulus of $\lambda_k^{(m)}$ (which equals the modulus of $\lambda_{n-k}^{(m)}$) will dominate the other eigenvalues. Thus d is effectively 2 and we need only discuss the shape of F_k and $F_k^{(m)}$. We may apply an orthogonal transformation in $\mathbb{R}^d$ to F_k so that the last $d-2$ coordinates in c_k are 0. Subtracting i times the second row from the first, F_k is now equivalent to a row vector of n complex numbers $\zeta^t = [\bar{\zeta}_1, \cdots, \bar{\zeta}_n]$. Operating similarly on $c_k u_k^* + \bar{c}_k \bar{u}_k^*$, the right-hand side of (2.6), we get the form $\bar{f}_k u_k^* + f_k \bar{u}_k^*$, say, where f_k is a complex number. Conjugating and transposing,

$$(2.16) \qquad \zeta_k = f_k u_k + \bar{f}_k \bar{u}_k.$$

Doing the same to $F_k^{(m)}$, we find

$$(2.17) \qquad \zeta_k^{(m)} = f_k \bar{\lambda}_k^{(m)} u_k + \bar{f}_k \lambda_k^{(m)} \bar{u}_k.$$

With this formulation, the shape of ζ is unchanged by multiplication with a complex number, conjugation or by any cyclic permutation, which corresponds to replacing u_k by $\omega^{kp} u_k$, $p = 0, \cdots, n-1$, where ω is $\exp(2\pi/N)$. Notice that conjugation is included as we allowed all orthogonal transformations in getting this form; this includes reflection. So our definition of shape gives as equivalent all similar figures and their reflections. Since $f_k \omega^{kp}$ and its conjugate both lead to equivalent shapes, we may take $|f_k| \leqq 1$ and assume it parametrizes shape. From (2.17) we see that the shape of $\zeta_k^{(m)}/|\lambda_k^{(m)}|$ moves around a circle of fixed radius within the unit disc, which is divided into $2N(k, n)$ sectors, where $N(k, n)$ is the order of the group generated by ω^k. If n and k are coprime the disc will be divided into $2n$ sectors corresponding to n cyclic permutations by one reflection, as $f_k,\ \omega^{kp} f_k,\ (p = 0, \cdots, n-1)$, and $\bar{f}_k$ are all equivalent shapes. If k and n have greatest common divisor, $l \geqq 1$, the figure formed has n/l vertices, traversed l times each. If we are interested only in the figure formed by cyclically joining points of the configuration, we can reduce the problem to that of an (n/l)-gon.

The above discussion of the division of the unit disc into sectors remains unchanged even if we do not normalize to take $f_k \leqq 1$. Then (2.17) are points on an ellipse with the ratio of its semiaxes identical to the ellipse given by (2.7), and this ellipse is similarly divided into sectors. So if we are only interested in shape, we can geometrically specify any member of the family $P_k(\varphi)$ by taking a point with argument between 0 and $\pi/N(k, n)$ on the outer ellipse (2.14) and drawing in the n-gon with joins tangent to the inner ellipse (2.15).

We turn now to the case where $S_1, S_2, \cdots$ are independent and identically distributed copies of S. The vector $\lambda \in \mathbf{C}^n$ of eigenvalues $\lambda_k(S)$, $k = 1, \cdots, n$ is defined by

$$(2.18) \qquad\qquad\qquad\qquad \lambda = \Omega s,$$

where $s \in \mathbb{R}^n$ is a random vector with elements $s_0, \cdots, s_{n-1}$. It would be interesting to characterize the class of joint distributions of $s_0, \cdots, s_{n-1}$ for which

$$(2.19) \qquad\qquad \prod_{j=1}^{m} \left| \frac{\lambda_l^j}{\lambda_k^j} \right| \xrightarrow{\ \mathbf{P}\ } 0, \quad \text{for all} \quad l \neq k, n-k$$

or equivalently by the law of large numbers,

$$(2.20) \qquad\qquad E \log|\lambda_l^j| - E \log|\lambda_k^j| < 0, \quad \text{for all} \quad l \neq k, n-k.$$

If the distribution of S is such that

$$(2.21) \qquad \text{dist of arg } \lambda_k(S) \text{ is not concentrated on a regular polygon}$$

it will also be true (see Feller (1971), p. 274) that $\sum_{j=1}^{m} \arg \lambda_k(S_j) \xrightarrow{\ \mathcal{D}\ } \Phi$, where $e^{i\Phi}$ is uniformly distributed on the unit circle. We will be interested in distributions for which both (2.20) and (2.21) are true. It is not empty as may be seen by taking $s_0 = t,\ s_1 = 1 - t,\ s_j = 0,\ j \geqq 2$ and t distributed on $[0, 1)$. Then we can state the following result.

Theorem 1. *If the transformation* (2.1) *is repeated with i.i.d. copies of S satisfying* (2.20), (2.21), *then asymptotically the points in the configuration*

$W^{(m)}/\prod_{j=1}^{m} |\lambda_k(S_j)|$ *lie on the ellipse*

$$(2.22) \qquad\qquad f_k^{\infty}(t) = 2(\boldsymbol{a}_k \cos(kt) + \boldsymbol{b}_k \sin(kt))$$

rotated by a Φ*, a random angle uniformly distributed on* $(0, 2\pi)$*, and where* $t = 2\pi(j/n)$, $j = 1, \cdots, n$.

Proof. Put W in form (2.5), $W = \sum_{l=1}^{n-1} c_l \boldsymbol{u}_l^*$. Then the normalized mth iterate

$$W^{(m)}\Big/\prod_{j=1}^{m} |\lambda_k(S_j)| = c_k \prod \frac{\lambda_k(S_j)}{|\lambda_k(S_j)|} \boldsymbol{u}^* + \bar{c}_k \prod \frac{\bar{\lambda}_k(S_j)}{|\lambda_k(S_j)|} \bar{\boldsymbol{u}}^* + o_p(1)$$

by (2.20). Hence assumption (2.20) reduces us to the case $c_l = 0$ for l not equal to k or $n-k$. Looking back at the derivation of (2.11) and noting (2.21) implies Φ is uniform, the theorem follows immediately.

In other words the theorem says that, asymptotically, points are spaced equi-angularly on an ellipse rotated by a random angle. The alternative way of viewing the sequence is in terms of the motion of the shape parameter.

Theorem 2. Under the conditions of Theorem 1, the shape parameter executes a random walk on a circle within the disc $|f_k| \leq 1$*. The asymptotic distribution of the shape parameter is uniform on this circle.*

Proof. The argument given in the proof of Theorem 1 means we can reduce to the case where c_l is 0, $l \neq k$. Then (2.17) implies that the shape parameter of the mth iterate $W^{(m)}$ is given by $f_k, \exp({}^i\varphi_k^{(m)})$, where $\varphi_k^{(m)}$ is given by $\sum_{j=1}^{m} \arg \lambda_k(S_j)$. So (2.21) and the assumed independence of the S_j imply the theorem.

3. Nesting triangles and quadrilaterals by Moebius transformations

In this section we consider a different geometric interpretation for some special cases of transformation (2.1) for triangles and quadrilaterals in the plane. First, we look at triangles, so $d = 2$, $n = 3$ and assume circulants have top row $(s, 1-s, 0)$ s real. A circulant of this form induces a transformation (2.1) called *nesting*, since the new figure is nested inside the old. If we fix the first two vertices of the triangle to be 0 and 1 by translating and rescaling by a complex number, then we get all triangular shapes by allowing the third vertex to be an arbitrary complex number z. This is of course not a 1–1 map.

If we nest a triangle inside a given triangle $(0, 1, z)$, and rescale and translate the first two vertices of the nested triangle to 0 and 1, the third vertex is given by the linear fractional map or Moebius transformation

$$(3.1) \qquad\qquad f_s(z) = \frac{sz - (1-s)}{(1-s)z - (1-2s)}.$$

Map (3.1) has fixed points given by solving $z = f_s(z)$, namely

$$(3.2) \qquad v_1 = \tfrac{1}{2} + \frac{\sqrt{3}}{2} i \quad \text{and} \quad v_2 = \tfrac{1}{2} - \frac{\sqrt{3}}{2} i$$

independent of s.

Put a sphere (denoted by S_0^2) on the complex plane with south pole at the point $\tfrac{1}{2}$ and with diameter $\sqrt{3}/2$. Then points p on S_0^2, can be represented by (x_1, x_2, x_3), where p is directly above the complex number $x_1 + ix_2$. Let ψ be the stereographic projection got by mapping points p in S to points $\psi(p)$ in $\mathbf{C}$. ψ is graphically represented by a ray from the north pole $(\tfrac{1}{2}, 0, \sqrt{3}/2)$ through p to the complex plane, striking it at $\psi(p)$. This map identifies the north pole with the point at ∞ in the usual compactification of $\mathbf{C}$.

Any linear fractional map f induces a map of $S \to S$ by $\Psi = \psi^{-1} \circ f \circ \psi$. Then (3.2) implies any map induced on S by a linear fractional of the form (3.1) must leave fixed two points on the ends of a certain diameter D through the equator of S, namely the points $\psi^{-1}(v_1)$ and $\psi^{-1}(v_2)$. We also know that a map of the form (3.1) maps $\tfrac{1}{2}$ to $(3s - 2)/(3s - 1)$ (i.e. it stays on the real axis). ψ^{-1} of this point is on a great circle lying on a plane which bisects D, so we can find a rotation $\varphi(s)$ about D (leaving the two endpoints of D fixed) which maps the south pole $(\psi^{-1}(\tfrac{1}{2}))$ to $\psi^{-1}((3s - 2)/(3s - 1))$. Now it is well known (e.g. see Rudin (1970), Chapter 14, pp. 268–271) that for any two ordered triples in $\mathbf{C}$ there is only one linear fractional map mapping the (ordered) elements of the first triple to their corresponding member in the second triple. Hence there is a unique linear fractional map from $(v_1, v_2, \tfrac{1}{2})$ to $(v_1, v_2, (3s - 2)/(3s - 1))$. It is also well known (e.g. see Rudin (1970), Chapter 14, pp. 268–271) that any linear fractional on $\mathbf{C}$ is a superposition of translations, rotations, dilations, and inversions of $\mathbf{C}$, and these correspond to translations rotations, dilations, and inversions of the standard sphere S^2. Now S_0^2 is a translation and dilation of S^2, and if we rotate S_0^2 about D by $\varphi(s)$, this corresponds to a unique linear fractional map which coincides with f_s at three points, and hence everywhere.

So Ψ_s the map induced on S_0^2 by f_s is uniquely specified by a rotation about D, where the angle of rotation is given by

$$\varphi(s) = \arctan \left(\frac{\dfrac{3s - 2}{3s - 1} - \dfrac{1}{2}}{\dfrac{1}{2} - \dfrac{\sqrt{3}}{4}} \right)$$

$$= \arctan \left(\frac{4}{2 - \sqrt{3}} \frac{(6s - 5)}{(6s - 2)} \right).$$

Given any triangle $(0, 1, z)$, a sequence of shapes obtained by map (3.1) (i.e. nesting) for values (s_i) can be thought of as points lying on a circle in S_0^2 obtained by rotating $\psi^{-1}(z)$ about the diameter D. This circle will usually be mapped to a circle in $\mathbf{C}$ by ψ. Figure 4 graphically illustrates this.

The circle in S_0^2 is defined by its 'radius', defined for example by

$$(3.3) \qquad |\psi^{-1}(z) - \psi^{-1}(v_1)|$$

where we regard points on S_0^2 as points in $\mathbb{R}^3$ and use Euclidean distance. This is fixed by the initial triangle. We give a lemma.

Lemma. If s_i are i.i.d. and the distribution of the angles $w_i = \varphi(s_i)$ is not concentrated on a finite number of rational fractions of 2π, then the asymptotic

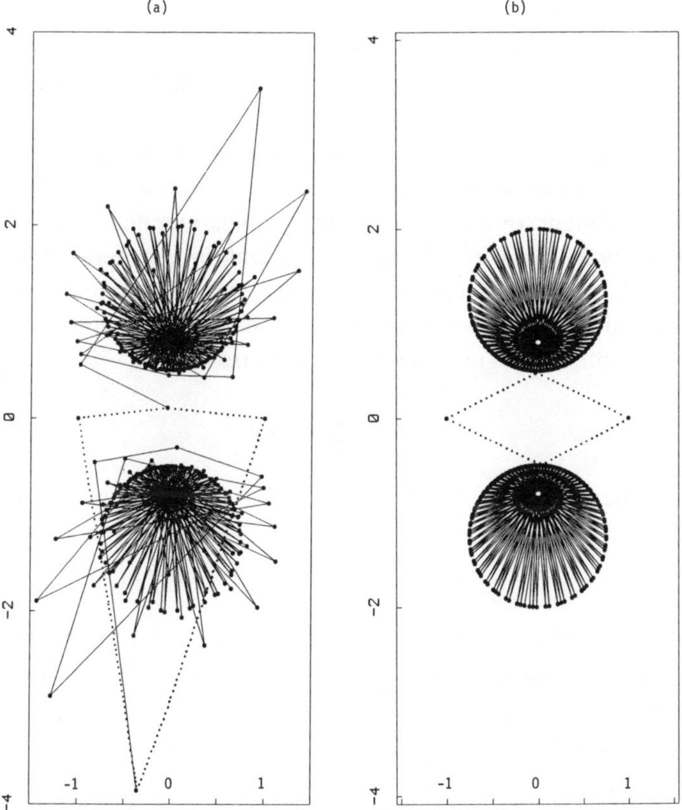

Figure 4. Successive iterates of the upper and lower vertices of a quadrilateral, rescaled at each step so two vertices remain at $(-1, 0)$ and $(1, 0)$ respectively. The solid lines joint successive iterates: dotted lines show the initial quadrilaterals. In (a) an irregular initial quadrilateral is used; in (b) a parallelogram is used. In (a) one can clearly see the convergence to the limit cycles of parallelograms. In (b) the figure is regular, so the upper and lower vertices immediately move on circles

distribution of the sum

$$(3.4) \qquad \varphi_n = \sum_1^n w_i$$

tends to a uniform distribution on $(0, 2\pi]$.

For a proof see Feller (1971), p. 274.

The above discussion and the lemma imply the following result.

Theorem 3.1. Let (s_i) *be an i.i.d. real sequence, satisfying the conditions of the lemma, then given an initial triangle* $(0, 1, z_0)$ *the sequence of shapes* $(0, 1, z_i)$ *generated by nesting using* s_i *in formulation* (3.1) *tends in distribution to the uniform measure on the set*

$$\{w : |w - \psi^{-1}(f_0)| = |\psi^{-1}(z_0) - \psi^{-1}(v_1)|\}$$

using metric (3.3) *on* S_0^2.

The proof is obvious in light of the lemma and the previous discussion identifying nesting with a rotation of S_0^2.

There is a partial extension to quadrilaterals. First, we return to the discussion in Section 2. Suppose that d, the dimension, is fixed at 2. Then we can represent any configuration by a vector of complex numbers, where each point (x, y) is given by $x + iy$. Then if a new configuration is generated by (2.1), the matrix W in (2.2) is simply represented by a vector of complex numbers. Then the discussion given at the beginning of Section 2 gives an analogous equation to (2.5) for complex configurations W with centroid 0,

$$(3.5) \qquad W = \sum_1^{n-1} c_k \mathbf{u}_k^*$$

where c_k are the arbitrary complex numbers and the $\mathbf{u}_k$ are given by (2.3). This is fully expressed as

$$(3.6) \qquad W = c_1(1, i, -1, -i)^* + c_3(1, -i, -1, i)^* + c_2(1, -1, 1, -1)^*.$$

If we take the cyclic join of W to be a quadrilateral, and if $c_2 = 0$, then it is easy to see that the quadrilateral is a parallelogram. Conversely, any parallelogram with centroid 0 is of form (3.6) with $c_2 = 0$.

From the discussion in Section 2, if (2.20) holds for $k = 1$ the asymptotic behavior of the shape of configuration W under circulant transformations S_j will depend only on c_1, c_3, since the coefficient of $\mathbf{u}_2$ tends to 0 in relation to the other two coefficients, (2.20) certainly holds whenever a new vertex is given by a linear combination of two adjacent vertices (or circulants S_j have top row $(s_j, 1 - s_j, 0, 0)$, s_j a real less than 1). So if we assume (2.20) we need only concern ourselves with the asymptotic behavior of parallelograms.

Given any initial parallelogram, rescale, rotate and translate it so that two opposing vertices are on -1 and 1. Apply S_j to the figure and if the new parallelogram is again rescaled and rotated so that the vertices on -1 and 1 remain stationary, then the vertex, say w, above the real axis is mapped to:

$$(3.7) \qquad f^{(1)}_{s_j}(w) = \frac{(1 - s_j) + s_j w}{s_j - (1 - s_j)w}$$

and the vertex $-w$ below the real axis is mapped to $-f^{(1)}_{s_j}(w)$. Hence the behavior of the vertex below the real axis is always a $180°$ rotation of the vertex above the real axis. It is clear that fixed points of this map are i and $-i$. Also, (3.7) is a Moebius transformation, so the discussion about triangles is directly applicable to these parallelograms. Notice that a vertex in the upper half-plane remains there under (3.7); so the motion of a vertex in the upper half-plane under a circulant transform can be represented as a rotation of one hemisphere of a Riemann sphere by an angle (the hemisphere is just those points mapped by a stereographic projection onto the upper half-plane). Similarly, the motion of the vertex in the lower half-plane is clearly identified with a rotation of the other hemisphere by the same angle, but in the reverse direction.

Figure 4 shows the evolution of an irregular and a regular quadrilateral where $s_j = 0.05$ (we use an expression similar to (3.7), but which correctly handles the fact that, in the irregular case, the vertices above and below the real axis are not negatives of each other). In particular, we see the convergence, in the irregular case, to limit circles. The limit circles are reached immediately in the parallelogram case.

The above discussion implies that a version of Theorem 3.1 holds, but for a different Riemann sphere (one with base on 0 of height 2), and where one identifies parallelograms by restricting attention to only the hemisphere mapping to the upper half-plane. Unfortunately, the preceding discussion suggests that the geometrical interpretation given to circulant transforms for triangles and parallelograms will not easily generalize to more points or higher dimensions.

Acknowledgements

This work was partially supported by a contract and two grants to Princeton University; one from the Department of Energy DE-AC02-81ER10841 (J.V.), and two grants from the National Science Foundation MSC821381 (G.S.W.) and MSC8304732. The first and last grants also provided the computing equipment used.

References

BERLEKAMP, E. R., GILBERT, E. N. AND SINDERS, F. W. (1965) A polygon problem. *Amer. Math. Monthly* **72**, 233–241.

DAVIS, P. J. (1979) *Circulant Matrices*. Wiley, New York.

FELLER, W. (1971) *An Introduction to Probability Theory and its Applications*, Vol. 2, 2nd edn. Wiley, New York.

JAMES, A. T. (1954) Normal multivariate analysis and the orthogonal group. *Ann. Math. Statist.* **25**, 40–75.

KENDALL, D. G. (1984) Shape manifolds, Procrustean metrics and complex projective spaces. *Bull. London Math. Soc.* **16**, 81–121.

MUIRHEAD, R. (1982) *Aspects of Multivariate Statistical Theory*. Wiley, New York.

ROSENMAN, M. (1932) Problem 3547. *Amer. Math. Monthly* **39**, 239.

RUDIN, W. (1970) *Real and Complex Analysis*. McGraw-Hill, New York.

SCHOENBERG, I. J. (1950) The finite Fourier series and elementary geometry. *Amer. Math. Monthly* **57**, 390–404.

WATSON, G. S. (1983a) Random triangles. In *Proc. 2nd Internat. Workshop on Stereology and Stochastic Geometry, ed. E. B. Jensen and J. G. Gundersen*, Memoirs No. 6, Institute of Mathematics, University of Aarhus, 183–199.

WATSON, G. S. (1983b) Limit theorems on high dimensional spheres and Stiefel manifolds. In *Studies in Econometrics, Time Series and Multivariate Analysis*, ed. S. Karlin, T. Amemiya, and L. Goodman, Academic Press, New York, 559–570.

The Risk-Sensitive Certainty Equivalence Principle

PETER WHITTLE

Abstract

A risk-sensitive certainty equivalence principle is deduced, expressed in Theorem 1, for a model with linear dynamics and observation rules, Gaussian noise and an exponential-quadratic criterion of the form (2). The senses in which one is now to understand certainty equivalence and the separation principle are discussed.

RISK-SENSITIVITY; OPTIMAL CONTROL

1. The classical certainty-equivalence principle

The classical certainty-equivalence principle, due to Theil (1957), holds a central place in the theory of dynamic optimization, and simplifies many analyses. The principle is associated with linear/quadratic/Gaussian (LQG) structure; and it is not at all dependent on assumptions of state structure.

It may be stated in the following form, in which form it is proved in Whittle (1982). One makes the following assumptions:

(a) The actions u_t to be taken over a finite horizon $(0 \leqq t < h)$ take values in finite-dimensional vector spaces.

(b) The action u_t can be a function only of observables W_t at t: previous actions $U_{t-1} = \{u_0, u_1, \cdots, u_{t-1}\}$ and observation history $Y_t = \{y_0, y_1, \cdots, y_t\}$, $(0 \leqq t < h)$.

(c) The cost function C is a quadratic function $Q(U_{h-1}, \xi)$ of the control sequence $U_{h-1} = \{u_0, u_1, \cdots, u_{h-1}\}$ and an exogenous noise vector ξ, positive definite in U_{h-1} for all ξ.

(d) ξ is normally distributed with known parameters, independent of policy.

(e) Each y_t is a policy-independent linear function of ξ.

The classic certainty-equivalence principle. Suppose that one wishes to choose policy π so as to minimise $E_\pi(C)$. Then, under Assumptions (a)–(e) above the optimal value of u_t is determined by minimising $Q(U_{h-1}, \xi^{(t)})$ with respect to

$u_t, u_{t+1}, \cdots, u_{h-1},$ *where*

$$(1) \qquad\qquad \xi^{(t)} = E(\xi \mid Y_t).$$

That is, one substitutes for ξ one's best current estimate of it, $\xi^{(t)}$, and then proceeds just as one would in the full-information case: to optimise freely with respect to all decisions as yet unmade. The determination of the current control action, u_t, is indeed optimal. The preliminary determination of future actions u_τ $(\tau > t)$ is valuable, in that it makes explicit the idea of a provisional forward plan. Even though this will be revised as later observations become available, the idea of a continually-revised forward plan corresponds very well to human intuition and practice. The proof of the principle also yields in passing that the value of $Q(U_{h-1}, \xi^{(t)})$ minimised with respect to $u_t, u_{t+1}, \cdots, u_{h-1}$ is just the W_t-dependent part of minimal expected cost conditional on W_t.

Note that the certainty equivalence principle has two features:

(i) *Conversion to free form.* A minimisation with respect to functions $u_\tau(W_\tau)$ $(\tau \geq t)$ is replaced by a free minimisation with respect to constants u_τ.

(ii) *Separation.* Optimisation of estimation and control are separated, in that the estimate (1) is derived without reference to the determination of the u_t, and the u_t are determined as they would be in the full-information case, with the simple substitution of $\xi^{(t)}$ for ξ. The estimate (1) itself has optimal properties, as the minimum variance unbiased estimate. However, it also turns out to be the estimate required for optimal control.

The vector ξ is considered to summarise all exogenous random effects such as process noise, observation noise or the randomness in a reference course which one is trying to make the system follow.

2. The risk-sensitive certainty-equivalence principle

Suppose the criterion that $E_\pi(C)$ be minimised is modified to the criterion that

$$(2) \qquad\qquad \gamma_\pi(\theta) = -\frac{2}{\theta} \log \left[E \exp \left(-\frac{\theta C}{2} \right) \right]$$

be minimised, with respect to π. Here θ is a scalar parameter, the *risk-sensitivity* parameter. The case $\theta = 0$ corresponds to the *risk-neutral* case

$$\gamma_\pi(\theta) = E_\pi(C).$$

In the case $\theta > 0$ the optimiser is *risk-seeking*; he is more concerned to reduce the frequent occurrence of moderate values of C than the occasional occurrence of large values. In the case $\theta < 0$ the reverse is the case, and the optimiser is *risk-averse*.

It is remarkable that the features of the well-known risk-neutral case have a version for the risk-sensitive case, although these are transformed sufficiently that they are not immediately evident. This analysis was developed for the state-structured case by Whittle (1981), (1982), (1983). However, state structure has little relevance, and we now give the general version.

Suppose that the exogenous noise vector ξ is normally distributed with zero mean and covariance matrix V. Define

$$D = \xi' V^{-1} \xi \tag{3}$$

recognisable as occurring in the exponent of the ξ-density, and define the *total stress*,

$$S = C + \frac{1}{\theta} D. \tag{4}$$

This is a spontaneously occurring combination when one evaluates the expectation in (2). C is the component of stress due to cost (i.e. to departures of u from 0) and D the component due to implausibility, (i.e. to departures of ξ from 0).

We shall use the term *extremisation* to denote an operation which is minimisation when $\theta \geqq 0$ and maximisation when $\theta < 0$.

We shall also introduce a hypothetical observation y_h which completes one's knowledge of ξ. Since this is observed only after all actions have been taken it does not affect the specification at all, but is only a formal convenience.

Theorem 1. The risk-sensitive certainty equivalence principle. Suppose that one wishes to choose policy π to minimise criterion (2). Then under Assumptions (a)–(e) above the optimal value of u_t is determined by simultaneously minimising S with respect to $u_t, u_{t+1}, \cdots, u_{h-1}$ and extremising it with respect to $y_{t+1}, y_{t+2}, \cdots, y_h$. In words: one obtains an optimal current decision by minimising stress with respect to all decisions currently unmade and extremising it with respect to all quantities currently unobservable.

This principle really does reduce to the risk-neutral version as $\theta \downarrow 0$. The stress-extremising value of unobservables tends to that minimising D for a given value of Y_t, this leads exactly to the estimate $\xi^{(t)}$ of ξ.

One can ask in what sense this is a certainty equivalence principle. It certainly has the property (i) above, of conversion to free form. That is, minimisation with respect to policies π, i.e. with respect to functions u_t of W_t $(t = 0, 1, \cdots, h - 1)$, has been replaced by a free minimisation/extremisation, of stress with respect to relevant decisions/unobservables.

It does not have property (ii), of separation. Determination of optimal control u_t and of an effective current estimate of ξ are intertwined in the minimisation/extremisation of stress.

This fact is inevitable. If 'separation' is a meaningful concept at all, it must surface in another and less evident form. This it does, as indicated in Whittle (1981). If one has state structure, then one can decompose stress into *past stress* and *future stress* at time t, and carry out the minimisation/extremisation of past and future stress separately for a prescribed value of x_t. These two calculations can both be carried out recursively in very much the classical risk-neutral pattern. The calculation of past stress condenses observation history to a form sufficient for state-estimation; the calculation of future stress determines what the optimal control $u_t(x_t)$ would be if state x_t were observable. *Separation* indeed manifests itself in that these two calculations are thus decoupled. One couples them by then choosing x_t so as to extremise total stress. If the resultant estimate is $\hat{x}_t$, then Theorem 1 implies that the optimal value of u_t is $u_t(\hat{x}_t)$. This is very obviously a certainty-equivalence statement.

One can make a further observation independent of state structure.

Theorem 2. The optimiser is respectively optimistic and pessimistic in the risk-seeking $(\theta > 0)$ and risk-averse $(\theta < 0)$ cases in that he behaves as if unobservables would take values to his advantage and disadvantage respectively. There is a negative value $\bar{\theta}$ of θ such that

$$(5) \qquad \inf_{\pi} \gamma_{\pi}(\theta) \begin{cases} < +\infty & \theta > \bar{\theta} \\ = +\infty & \theta < \bar{\theta}. \end{cases}$$

This value marks a degree of pessimism at which the optimiser becomes convinced of inability to control.

3. Proofs

It is convenient to assume V non-singular and the observations y_t linearly independent. This can always be achieved by jettisoning of redundant observations. The densities $f(Y_t)$ relative to Lebesgue measure then exist. Define the quantities

$$(6) \qquad G(W_t) = \operatorname*{ext}_{\pi} [-f(Y_t) E_{\pi}(e^{-\theta C/2} \mid W_t)]$$

where 'ext' indicates the operation of extremising: with respect to π in this case.

The dynamic programming equation now becomes

$$(7) \qquad G(W_t) = \operatorname*{ext}_{u_t} \int G(W_{t+1}) dy_{t+1} \qquad (0 \leq t < h)$$

with terminal condition

$$(8) \qquad G(W_h) = -\exp\left[-\frac{\theta}{2}(d_h + S)\right]$$

where d_t indicates a term dependent only upon t. The extremising value of u_t in (7) is the optimal value.

We shall now verify by induction that expression (8) generalises to

$$(9) \qquad G(W_t) = -\exp\left[-\frac{\theta}{2}(d_t + \phi(W_t))\right].$$

Here $\phi(W_t)$ is a t-dependent quadratic function of U_{t-1} and Y_t obeying the recursion

$$(10) \qquad \phi(W_t) = \min_{u_t} \operatorname{ext}_{y_{t+1}} \phi(W_{t+1})$$

with terminal condition

$$(11) \qquad \phi(W_h) = S,$$

the minimising value of u_t in (10) being the optimal value. This then implies that

$$(12) \qquad \phi(W_t) = \min_{u_t, u_{t+1}, \cdots} \operatorname{ext}_{y_{t+1}, y_{t+2}, \cdots} S$$

with the minimising value of u_t being the optimal choice, and so proves Theorem 1.

Evaluation (11) begins the induction; assume then (9) to be true at $t+1$ and substitute this expression for $G(W_{t+1})$ in the integral of (7). It follows then from the evaluation of normal integrals that, if $\theta\phi(W_{t+1})$ is strictly convex in y_{t+1}, then

$$(13) \qquad \int \exp\left[-\frac{\theta}{2}\phi(W_{t+1})\right] dy_{t+1} = \exp\left\{\max_{y_{t+1}}\left[-\frac{\theta}{2}\phi(W_{t+1})\right] + \cdots\right\}.$$

Here $+\cdots$ indicates a purely t-dependent term, in fact equal to $\frac{1}{2}r\log 2\pi + \frac{1}{2}\log|R|$, where R is the matrix of the quadratic form in y_{t+1} in the exponent of the integrand and r is its dimension. Relations (7) and (13) between them imply validity of (9), (10) and (12) in this case, and so the validity of Theorem 1.

The first statement of Theorem 2 follows from the fact that S is minimised or maximised with respect to unobservables according as θ is positive or negative. To prove the existence of a lower critical value $\bar{\theta}$, note that if the quadratic form $\theta\phi(W_{t+1})$ is not strictly convex in y_{t+1} then (13) will still hold with both sides having the evaluation $+\infty$. By the definition of $G(W_t)$ this then implies that

$$E_\pi(e^{-\theta C/2}) \propto \int e^{-\theta S/2} d\xi = +\infty$$

for all π. The same result will hold for smaller values of θ, since the expectation is decreasing in θ.

The existence of a threshold value $\bar\theta$ of θ such that $\gamma_\pi(\theta) = +\infty$ for all π and $\theta \leq \bar\theta$ is thus established. The critical value must be negative, since certainly $\gamma_\pi(\theta) < \infty$ for $\theta > 0$.

The first significant progress with the risk-sensitive case was made by Jacobson (1973), (1977). His analysis remained incomplete, however, largely because he had no analogue of the certainty-equivalence principle. The analysis of this paper finds application in Whittle and Kuhn (1986).

References

JACOBSON, D. H. (1973) Optimal stochastic linear systems with exponential performance criteria and their relation to deterministic differential games. *IEEE Trans. Automatic Control* **AC-18**, 124–131.

JACOBSON, D. H. (1977) *Extensions of Linear-Quadratic Control, Optimization and Matrix Theory.* Academic Press, New York.

THEIL, H. (1957) A note on certainty equivalence in dynamic planning. *Econometrica* **25**, 346–349.

WHITTLE, P. (1981) Risk-sensitive linear/quadratic Gaussian control. *Adv. Appl. Prob.* **13**, 764–777.

WHITTLE, P. (1982) *Optimization over Time*, Vol. 1. Wiley Interscience, New York.

WHITTLE, P. (1983) *Prediction and Regulation by Linear Least Square Methods*, 2nd edn. University of Minnesota Press, Minneapolis.

WHITTLE, P. AND KUHN, J. (1986) A Hamiltonian formulation of risk-sensitive linear/quadratic/Gaussian control. *Int. J. Control.* To appear.

PART 7

ALGORITHMS AND COMPUTATIONS

On the Equivalence of Two Stochastic Approaches to Spline Smoothing

CRAIG F. ANSLEY
ROBERT KOHN

Abstract

Wahba (1978) and Weinert et al. (1980), using different models, show that an optimal smoothing spline can be thought of as the conditional expectation of a stochastic process observed with noise. This observation leads to efficient computational algorithms. By going back to the Hilbert space formulation of the spline minimization problem, we provide a framework for linking the two different stochastic models. The last part of the paper reviews some new efficient algorithms for spline smoothing.

SIGNAL EXTRACTION; DIFFUSE PRIOR; STOCHASTIC DIFFERENTIAL EQUATION; NON-LINEAR REGRESSION; KALMAN FILTER

1. Introduction

It is shown by Kimeldorf and Wahba (1970a,b), (1971), Wahba (1978) and Weinert et al. (1980) that optimal smoothing splines .can be obtained by smoothing an appropriate stochastic process observed with noise. By taking advantage of the structure of the stochastic process to reexpress it in state space form, efficient algorithms for spline smoothing can be obtained as shown by Weinert et al. (1980), Wecker and Ansley (1983) and Kohn and Ansley (1984). Because of frequent citation we refer to Weinert et al. (1980) as WBS and Kohn and Ansley (1984) as KA.

The key to obtaining the correspondence between smoothing splines and stochastic processes lies in expressing the spline smoothing problem as a minimization problem in a reproducing kernel Hilbert space, and then using the reproducing kernel as the covariance matrix of the stochastic process. Details are given in Sections 3 and 4.

Wahba (1978) and WBS choose different stochastic processes. Wahba's (1978) stochastic process is the solution to a stochastic differential equation with a diffuse prior distribution on the initial conditions. See Section 7 for further details. WBS use a stochastic model in which the signal is correlated with the noise. Although both stochastic processes satisfy the same differential equation, it is somewhat surprising that such differing treatments of initial conditions give rise to the same smoothing spline.

It is the purpose of this paper to provide a better understanding of the link between the WBS and Wahba stochastic processes. We show that one of the terms in the Hilbert space inner product in WBS can be multiplied by an arbitrary positive factor k, and that the resulting generalized WBS stochastic model gives the same smoothing spline for all values of k. The value of k corresponds to a scale factor for the prior variance of the initial conditions, and can be chosen arbitrarily large. The Wahba stochastic model is then a modification of the generalized WBS model in which there is no signal–noise correlation, but which gives the optimal spline when $k \to \infty$. The above results are obtained in Sections 3 to 7 of the paper.

In Section 8 we give a brief survey of efficient algorithms for spline smoothing. In Section 9 we give a Bayesian interpretation of smoothing splines, and in Section 10 we discuss some general results of Kohn and Ansley (1983) showing that in general when we smooth a stochastic process observed with noise we obtain a spline function.

2. Some definitions and assumptions

Assumption 2.1. Let L be the nth-order linear differential operator

$$(2.1) \qquad L = d^n/dt^n - a_{n-1}(t)d^{n-1}/dt^{n-1} - \cdots - a_0(t)$$

where the coefficient $a_j(t)$ $(j = 0, \cdots, n-1)$ has j continuous derivatives on $[0, T]$.

Definition 2.1. Let H be the space of real functions that are n times differentiable on $[0, T]$ with nth derivative that is square integrable on $[0, T]$. Let H_0 be the null space of L with respect to H; i.e., $H_0 = \{g \in H : Lg = 0\}$.

Assumption 2.2.
(i) For $i = 1, \cdots, N$, λ_i is an extended Hermite–Birkhoff linear functional on H of the form

$$(2.2) \qquad \lambda_i(g) = \sum_{j=0}^{n-1} h_{ij} g^{(j)}(t_i)$$

with $g \in H$, and $0 \leq t_1 \leq t_2 \leq \cdots \leq t_n \leq T$. Here and below $g^{(j)}(t)$ denotes $d^j g(t)/dt^j$.

(ii) Amongst the linear functionals $\lambda_1, \cdots, \lambda_N$, there are n that are linearly independent over H_0.

Assumption 2.3.

(i) $r(t)$ is a positive and continuous function of t for $t \in [0, T]$.

(ii) Q is a diagonal matrix having $\rho_1, \cdots, \rho_N$ on the diagonal, with $\rho_i > 0$, $i = 1, \cdots, N$.

The class of functionals described by (2.2) is used in WBS and also in Weinert and Sidhu (1979). In the state space framework of Section 5, the elements of the state vector at t correspond to the derivatives $f^{(j)}(t)$ $(j = 0, \cdots, n-1)$, so that $\lambda_i(f)$ is just a linear combination of the elements of the state vector at $t = t_i$, and fits neatly into the observation equation of the state space model.

We now define an optimal Lgr smoothing spline.

Definition 2.2. Suppose we are given the observations $y(1), \cdots, y(N)$ at the points t_i, $i = 1, \cdots, N$, with $0 \leq t_1 \leq t_2 \leq \cdots \leq t_N \leq T$, and also the weight function $r(t)$. The optimal Lgr smoothing spline, call if $\hat{f}$, of $y(1), \cdots, y(N)$ with respect to L, Q, r and the functionals $\lambda_1, \cdots, \lambda_N$, is that $f \in H$ which minimizes

$$(2.3) \qquad \sum_{i=1}^{N} \rho_i^{-1} [y(i) - \lambda_i(f)]^2 + \int_0^T \frac{1}{r(t)} [Lf(t)]^2 dt.$$

The existence of a unique minimizing solution $\hat{f}$ is guaranteed by Assumption 2.2(ii); see Anselone and Laurent (1968) or WBS. Equation (2.3) consists of the usual sum-of-squares loss function together with

$$(2.4) \qquad \int_0^T \frac{1}{r(t)} [Lf(t)]^2 dt$$

which is called a smoothness penalty function and is a weighted integral of $(Lg)^2$. If $r(t)$ is a constant equal to r, then $\hat{f}$ is called an Lg smoothing spline. (See for example WBS and Wahba (1978).) We allow for non-uniform weighting so that there can be differing weightings on smoothness in disjoint subintervals of $[0, T]$.

To motivate the above definitions and assumptions, and help the reader obtain a better grasp of them, we look at polynomial splines and in particular at the cubic spline, perhaps the most popular spline used in applications.

Example 2.1. Suppose that $\lambda_i(f) = f(t_i)$ for $i = 1, \cdots, N$, $L = d^n/dt^n$, $\rho_i = \sigma^2$ (constant) for $i = 1, \cdots, N$, and $r(t) = \sigma^2 r$ (a constant). Then (2.3) becomes

$$(2.5) \qquad \sigma^{-2} \left\{ \sum_{i=1}^{N} [y(i) - f(t_i)]^2 + \frac{1}{r} \int_0^T [f^{(n)}(t)]^2 dt \right\}.$$

In (2.5) r controls the weights given to the sum of squares component versus the smoothness penalty. $\int_0^T [f^{(n)}(t)]^2 dt$ provides a measure of the smoothness of

$f^{(n)}(t)$ and hence $f(t)$. Provided at least n of the t_i are distinct, the minimization problem (2.5) has a unique solution $\hat{f}$ which is a natural polynomial spline of degree $2n-1$; see Schoenberg (1964) and Wahba (1978). For $n=2$, $\int_0^T [f^{(2)}(t)]^2 dt$ is a measure of the curvature of $f(t)$ in $[0, T]$ and $\hat{f}$ is then a cubic spline.

The motivation for minimizing (2.5) with $n=2$ is that we have an unknown function $f(t)$ observed with noise at the points $t = t_i$. Thus

$$(2.6) \qquad\qquad y(i) = f(t_i) + e(i), \qquad i = 1, \cdots, N$$

where $\{e(i)\}$ is an i.i.d. $(0, \sigma^2)$ sequence of random variables.

We would like to estimate $f(t)$ from our observations $y(1), \cdots, y(N)$ so that $f(t)$ is as smooth as possible, smoothness being measured by curvature.

The most common source of dependence of the functionals $\lambda_1, \cdots, \lambda_N$ occurs when we have several observations at the same values of t. For example, if $t_i = t_{i+1}$, then $\lambda_i(f) = f(t_i) = f(t_{i+1}) = \lambda_{i+1}(f)$ so that $\lambda_i = \lambda_{i+1}$. If n values of the t_i are distinct, then n of the functionals $\lambda_i(f) = f(t_i)$ are linearly independent, giving a unique solution to (2.5).

3. Spline smoothing as a minimum norm problem

The results in this section follow closely the corresponding results in WBS but there are two generalizations. The first is that we introduce a class of inner products, and hence reproducing kernels, indexed by a parameter k, any one of which will give us an optimal Lgr smoothing spline. This will give us the crucial link between the Wahba and WBS stochastic models. The next generalization is that we allow $r = r(t)$ to be a function of t. Because of Assumption 2.2(ii) there is no loss of generality in assuming for now that $\lambda_1, \cdots, \lambda_n$ are linearly independent over H_0. If they are not then we can simply reorder $\lambda_1, \cdots, \lambda_n$ in such a way that the first n functionals are now linearly independent. WBS assume $\lambda_1, \cdots, \lambda_N$ are linearly independent over H_0, but in later papers, Sidhu and Weinert (1979), (1984) note that this assumption is unnecessary, although they omit the algorithmic details concerned with reordering the λ_i.

For $\theta_1, \cdots, \theta_n$ real numbers, and $\theta = (\theta_1, \cdots, \theta_n)'$ we define the augmented space H^+ by

$$H^+ = \{(f, \theta) : f \in H \text{ and } \theta \in R^n\}.$$

We now define a class of inner products on H^+ indexed by the parameter $k > 0$,

$$
\begin{aligned}
\langle (f, \theta), (g, \phi) \rangle = &\int_0^T \frac{1}{r(t)} [Lf(t)][Lg(t)] dt \\
&+ \frac{1}{k} \sum_{i=1}^n (\theta_i + \lambda_i f)(\phi_i + \lambda_i g) + \sum_{i=1}^N \rho_i^{-1} \theta_i \phi_i.
\end{aligned}
$$
(3.1)

Denote the corresponding norm on H^+ by $\| \cdot \|$.

As in WBS we can check that (3.1) is a proper inner product and H^+ is a Hilbert space. In WBS $k = 1$ and $r(t) = r$ a constant.

Now define the functionals λ_i^+, $i = 1, \cdots, N$, on H^+ as

$$\lambda_i^+(f, \theta) = \lambda_i(f) + \theta_i.$$

Then it can be shown as in WBS that the λ_i^+ are bounded linear functionals so that there exist $h_i^+ \in H^+$, $i = 1, \cdots, N$ satisfying

$$\lambda_i^+(f, \theta) = \langle (f, \theta), h_i^+ \rangle.$$

The following theorem provides a general solution to the spline smoothing problem.

Theorem 3.1. Minimizing (2.3) is equivalent to minimizing $\|(f, \theta)\|$ subject to $\langle (f, \theta), h_i^+ \rangle = y_i$, $i = 1, \cdots, N$. The minimizing value of (f, θ) is

$$(\hat{f}, \hat{\theta}) = \sum_{i=1}^{N} d_i h_i^+$$

where $(d_1, \cdots, d_N)' = S^{-1}y$. $y = (y(1), \cdots, y(N))'$ and S is the $N \times N$ matrix with (i, j)th element $\langle h_i^+, h_j^+ \rangle$. Furthermore, $(\hat{f}, \hat{\theta})$ is independent of the parameter k.

Proof. Except for the independence of $\hat{f}$ and $\hat{\theta}$ from k, the proof is the same as in WBS. Now note that $\lambda_i(f) + \theta_i = y_i$ for $i = 1, \cdots, n$, and y_i is fixed. The independence from k follows immediately because the middle term in (3.1) is independent of f.

Let $h_1^+ = (h_i(t), \omega_{i1}, \cdots, \omega_{iN})$. Then

$$\hat{f}(t) = \sum_{i=1}^{N} h_i(t) d_i = h(t)' S^{-1} y$$

so that $\hat{f}(t)$ is a linear function of y. Furthermore, $\hat{\theta}_i = y_i - \lambda_i(\hat{f})$ is the ith residual.

Now let $z_1(t), \cdots, z_n(t)$ be a basis of H_0 such that $\lambda_i\{z_j(t)\} = \delta_{ij}$, $i, j = 1, \cdots, n$; $\delta_{ij} = 1$ if $i = j$ and 0 otherwise.

Let $G(t, s)$ be the Green's function for L so that $LG(t, s) = \delta(t - s)$ and $\lambda_i\{G(t, \cdot)\} = 0$ for $i = 1, \cdots, n$; $\delta(t - s)$ is the Dirac delta function. See WBS for details on $G(t, s)$.

Following WBS we now show that H^+ is a reproducing kernel Hilbert space and define the reproducing kernel.

Let

$$R_0(t, s) = \sum_{i=1}^{n} (k + \rho_i) z_i(s) z_i(t)$$

$$R_1(t, s) = \int_0^T r(u) G(t, u) G(s, u) \, du$$

$$R(t, s) = R_0(t, s) + R_1(t, s)$$

$$R_t^+(s) = [R(t, s), -Q(^{z(t)}_0)], \qquad t,s \in [0, T]$$

$$R_j^+(s) = [-i_j' Q(^{z(s)}_0), Q i_j], \qquad j = 1, \cdots, N; s \in [0, T].$$

$z(s) = (z_1(s), \cdots, z_n(s))'$ and i_j is the N-dimensional vector having 1 in the jth position and zeros elsewhere.

Then exactly as in WBS we have the following result.

Theorem 3.2.

$$\langle R_t^+, (f, \theta) \rangle = f(t), \qquad t \in [0, T]$$

$$\langle R_j^+, (f, \theta) \rangle = \theta_j, \qquad j = 1, \cdots, N$$

and

$$h_i^+ = \lambda_i R_t^+(\cdot) + R_i^+, \qquad i = 1, \cdots, N.$$

Therefore, $R_t^+, R_j^+, j = 1, \cdots, N$ is a reproducing kernel Hilbert space for H^+. Theorems 3.1 and 3.2 show that there is a class of inner products and corresponding reproducing kernels indexed by the parameter k, any one of which gives the solution to the Lgr spline smoothing problem.

4. A generalized WBS stochastic model

For the rest of this paper, we will take all random variables to have zero mean and be jointly Gaussian. Because R^+ is a reproducing kernel, we can find a random function $f(t)$ on $[0, T]$ and random variables $e(i)$, $i = 1, \cdots, N$, with the following covariances:

$$(4.1) \qquad E[f(t)f(s)] = \langle R_t^+, R_s^+ \rangle, \qquad t,s \in [0, T]$$

$$(4.2) \qquad E[f(t)e(j)] = \langle R_t^+, R_j^+ \rangle, \qquad t \in [0, T], \qquad j = 1, \cdots, N$$

and

$$(4.3) \qquad E[e(i)e(j)] = \langle R_i^+, R_j^+ \rangle.$$

It follows from Section 3 that

$$E[f(t)f(s)] = R(s, t)$$

(4.4)
$$= \sum_{i=1}^{n} (k + \rho_i) z_i(t) z_i(s) + \int_0^T r(\zeta) G(t, \zeta) G(s, \zeta) d\zeta$$

(4.5)
$$E[f(t)e(j)] = -\rho_j z_j(t), \qquad j = 1, \cdots, n$$
$$= 0 \qquad\qquad j > n$$

(4.6)
$$E[e(i)e(j)] = -\rho_i \delta_{ij}, \qquad i, j = 1, \cdots, N.$$

We see from (4.4) that the covariance function of $f(t)$ depends on k. Let

(4.7)
$$y(i) = \lambda_i(f) + e(i), \qquad i = 1, \cdots, N$$

which is a signal plus noise model. As in WBS we can use the correspondence established by (4.1)–(4.3) to show that the optimizing spline $\hat{f}$ of Theorem 3.1 equals $E[f(t) \,|\, y(1), \cdots, y(N)]$. This result follows easily if we note that

$$\begin{aligned}
\text{Cov}\,[f(t), y(i)] &= E[f(t)\lambda_i(f)] + E[f(t)e(i)] \\
&= \langle R_t^+, \lambda_{is} R_s^+ \rangle + \langle R_t^+, R_i^+ \rangle_{H^+} \\
&= \langle R_t^+, h_i^+ \rangle_{H^+}
\end{aligned}$$

and

$$\begin{aligned}
\text{Cov}\,[y(i), y(j)] &= E[\{\lambda_i(f) + e(i)\}\{\lambda_j(f) + e(j)\}] \\
&= \langle (\lambda_{it} R_t^+ + R_i^+), (\lambda_{jt} R_t^+ + R_j^+) \rangle \\
&= \langle h_i^+, h_j^+ \rangle.
\end{aligned}$$

Therefore, the conditional expectation of $f(t)$ does not depend on k although $f(t)$ does.

In preparation for the next section, we note that from (4.4),

$$\text{Cov}\,\{Lf(t), Lf(s)\} = r(t)\delta(t - s)$$

so that

(4.8)
$$Lf(t) = \sqrt{r(t)} \, dW(t)/dt$$

where $W(t)$ is the zero-mean Wiener process with $W(0) = 0$ and $\text{Var}\,[W(1)] = 1$.

For the remainder of this paper, we make the following assumption.

Assumption 4.1.
 (i) $\rho_i = \sigma^2$ for $i = 1, \cdots, N$, so that $Q = \sigma^2 I$, and
 (ii) $r(t) = r\sigma^2$ for all t, where r is constant.
 In (4.7) we now have $\text{Var}\, e(i) = \sigma^2$ for all t, while (4.8) becomes

(4.9)
$$Lf(t) = \sigma\sqrt{r} \, dw(t)/dt.$$

Assumption 4.1 simplifies the notation considerably, but it is clear that all results in the following sections extend immediately to the more general case.

5. A state space representation for the WBS stochastic model

We now express the stochastic model of Section 4 in state space form. This is the key to efficiently computing $\hat{f}(t)$.

From (4.8) and a standard representation of linear differential equations (see for example Coddington and Levinson (1955), p. 82, WBS or KA), it follows that we can write

$$(5.1) \qquad dx(t)/dt = C(t)x(t) + bdW(t)/dt$$

$$(5.2) \qquad f(t) = l'x(t)$$

where $x(t)$ is the n-dimensional state vector $x(t) = [f(t), f^{(1)}(t), \cdots, f^{(n-1)}(t)]'$, $b = \sigma\sqrt{r}(0, 0, \cdots, 1)'$ and $l = (1, 0, 0, \cdots, 0)'$.

The state transition matrix $C(t)$ is given by

$$C(t) = \left[\begin{array}{c|c} 0 & I_{n-1} \\ \hline a_{n-1}(t) & a_{n-2}(t) \cdots a_0(t) \end{array}\right].$$

Now let $\chi(t)$ be the $n \times n$ matrix solution of the differential equation

$$d\chi(t)/dt = C(t)\chi(t), \qquad t \in [0, T]$$

with initial condition $\chi(0) = I_n$. Then $\chi(t)$ exists and is unique. Let $\phi(t, s) = \chi(t)\chi(s)^{-1}$. Then, as in KA

$$(5.3) \qquad x(t) = \phi(t, s)x(s) + w(t, s)$$

where

$$(5.4) \qquad w(t, s) = \int_s^t \phi(t, v)bdW(v)$$

for any $s, t \in [0, T]$ with $s \leq t$.

From (2.2) $\lambda_i(f) = H(i)x(t_i)$ for all i, where $H(i) = (h_{i,0}, \cdots, h_{i,n-1})$. Therefore, from (4.7) and (5.3) we rewrite our stochastic model in state space form as

$$(5.5) \qquad y(i) = H(i)x(t_i) + e(i), \qquad i = 1, \cdots, N$$

$$(5.6) \qquad x(t_{i+1}) = \phi(t_{i+1}, t_i)x(t_i) + w(t_{i+1}, t_i).$$

For later use we now adopt the following notation. For $i = 1, \cdots, N$, let

$$x(t \mid i) = E[x(t) \mid y(1), \cdots, y(i)]$$

and

$$(5.7) \qquad S_x(t \mid i) = E\{[x(t) - x(t \mid i)][x(t) - x(t \mid i)]'\}/\sigma^2.$$

Therefore, $x(t\,|\,i)$ is the best predictor of $x(t)$ conditional on $y(1),\cdots,y(i)$ and $S_x(t\,|\,i)$ is the prediction error variance. We define $f(t\,|\,i)$ and $S_f(t\,|\,i)$ similarly.

Example 5.1. To illustrate the above general discussion, let L and the λ_i be the same as in Example 2.1. It is straightforward to check that

$$\chi(t) = \begin{bmatrix} 1 & t & t^2/2 & \cdots & t^{n-1}/(n-1)! \\ & 1 & t & & t^{n-2}/(n-2)! \\ & & 1 & & \vdots \\ & & & & t \\ & & & \ddots & \vdots \\ 0 & & & & 1 \end{bmatrix}$$

and that $\phi(t, s) = \chi(t - s)$.

6. Further analysis of the WBS stochastic model

We now demonstrate directly why $f(t\,|\,N)$ (which is also the optimal smoothing spline) is independent of k. From Section 5 we have for all i

$$(6.1) \qquad \begin{aligned} \lambda_i(f) &= H(i)x(t_i) \\ &= H(i)\phi(t_i, 0)x(0) + H(i)w(t_i, 0). \end{aligned}$$

Let $\eta = [\lambda_1(f), \cdots, \lambda_n(f)]'$, $y^{(n)} = [y(1), \cdots, y(n)]'$ and $e^{(n)} = [e(1), \cdots, e(n)]'$. Then,

$$(6.2) \qquad \eta = Mx(0) + \xi$$

where the matrix M is $n \times n$ with ith row $H(i)\phi(t_i, 0)$ and ξ is $n \times 1$ with ith element $H(i)w(t_i, 0)$. From (2.6)

$$(6.3) \qquad y^{(n)} = \eta + e^{(n)}$$

$$(6.4) \qquad = Mx(0) + v^{(n)}$$

where $v^{(n)} = e^{(n)} + \xi$.

Now define the $N \times 1$ vector $z = [y(n+1), \cdots, y(N), x(t)']'$ for some fixed t. Then we can write

$$(6.5) \qquad z = Tx(0) + \zeta$$

where T is $N \times n$ with ith row $H(i+n)\phi(t_{i+n}, 0)$ for $i = 1, \cdots, N-n$ and the last n rows of T are $\phi(t, 0)$. ζ is $N \times 1$ with ith element $H(i+n)\phi(t_{i+n}, 0) \times w(t_{i+n}, 0) + e(i)$ for $i = 1, \cdots, N-n$ and the last n elements of ζ are $w(t, 0)$. From (6.2) and (6.5) we obtain

$$(6.6) \qquad z = B\eta + \zeta - B\xi$$

where $B = TM^{-1}$ is $N \times n$.

We can now check, using (4.4)–(4.6), that the following hold:

$$\mathrm{Var}\,(\eta) = (1+k)\sigma^2 I_n, \qquad \mathrm{Cov}\,(\eta, e^{(n)}) = -\sigma^2 I_n,$$

$$\mathrm{Var}\,(e^{(n)}) = \sigma^2 I_n, \qquad \mathrm{Cov}\,(\eta, \zeta) = \mathrm{Cov}\,(\eta, \zeta) = 0.$$

Therefore from (6.2) and (6.6)

$$\mathrm{Var}\,[y^{(n)}] = \sigma^2 k I_n \quad \text{and} \quad \mathrm{Cov}\,(z, y^{(n)}) = \sigma^2 k B,$$

so that

$$(6.7) \qquad\qquad\qquad E[z \mid y^{(n)}] = By^{(n)}$$

and

$$(6.8) \qquad\qquad\qquad z - E[z \mid y^{(n)}] = \zeta - Bv^{(n)}$$

with

$$(6.9) \qquad\qquad\qquad \mathrm{Var}\,[z - E(z \mid y^{(n)})] = \mathrm{Var}\,(\zeta - Bv^{(n)}).$$

The right sides of (6.7) and (6.9) do not depend on k so that the distribution of z conditional on $y^{(n)}$ does not depend on k. This implies that the distribution of $x(t)$ conditional on $y(1), \cdots, y(N)$ does not depend on k; i.e., $x(t \mid N)$ and $S_x(t \mid N)$ do not depend on k. It follows that $f(t \mid N) = l'x(t \mid N)$ and $S_f(t \mid N) = l'S_x(t \mid N)l$ do not depend on k.

Now in (6.3) suppose more generally that $\mathrm{Var}\,(\eta) = kQ_1 + Q_0$ and $\mathrm{Cov}\,(\eta, e^{(n)}) = Q_2$. The matrices Q_0, Q_1, and Q_2 are all $n \times n$ and do not depend on k, with Q_1 positive definite. Suppose further that η is uncorrelated with $W(t)$ and the $e(i)$, $i > n$. Then η is uncorrelated with ξ and ζ, and

$$\mathrm{Cov}\,(z, \eta) = B(kQ_1 + Q_0) + BQ_2$$

and

$$\mathrm{Var}\,(y^{(n)}) = kQ_1 + Q_0 + Q_2 + Q_2' + \sigma^2 I_n.$$

It is straightforward to check that

$$E[z \mid y^{(n)}] = [B + O(1/k)]y^{(n)}$$

and

$$z - E[z \mid y^{(n)}] = \zeta - Bv^{(n)} + O(1/k)y^{(n)}$$

so that

$$\mathrm{Var}\,[z - E[z \mid y^{(n)}]] = \mathrm{Var}\,[\zeta - Bv^{(n)}] + O(1/k).$$

Therefore in the limit as $k \to \infty$, $E[z \mid y^{(n)}]$ and $\mathrm{Var}\,\{z - E[z \mid y^{(n)}]\}$ are the same as for the WBS stochastic model. Further, $\lim_{k \to \infty} f(t \mid N)$ and $\lim_{k \to \infty} S_f(t \mid N)$ are independent of Q_0, Q_1 and Q_2 and equal to their values under the WBS stochastic model. The choice of $Q_2 = 0$ indicating zero correlation between signal and noise is of particular importance.

We see that what is obtained by the special stochastic structure of WBS can be obtained by the more general model if we let $k \to \infty$, i.e., if we let η and hence $x(0)$ become diffuse. In the next section we use the observations above to explain the equivalence of the WBS and Wahba stochastic models.

7. The Wahba stochastic model

Wahba (1978), (1983a) assumes the following stochastic structure:

$$Lf = \sigma \sqrt{r} \, dW(t)/dt$$

$$y(i) = \lambda_i(f) + e(i)$$

with $x(0) \sim N(0, kI_n)$ and independent of $W(t)$ and the $e(i)$. The $e(i)$ are i.i.d. $(0, \sigma^2)$. By taking $Q_0 = \text{Var}(\zeta)$, $Q_1 = MM'$ and $Q_2 = 0$ we deduce from Section 6 that $x(t \mid N)$ and $S_x(t \mid N)$ for the WBS stochastic model are equal to $\lim_{k \to \infty} x(t \mid N)$ and $\lim_{k \to \infty} S_x(t \mid N)$ respectively for the Wahba model.

The WBS and Wahba models are essentially the same except that the special correlation between signal and noise for WBS ensures that $f(t \mid N)$ is the optimal smoothing spline for all k. In the Wahba model this is achieved by letting $k \to \infty$, thus making $x(0)$ diffuse.

It is perhaps a little surprising that $S_x(t \mid N)$ for WBS is equal to the limiting value of $S_x(t \mid N)$ for the Wahba model. We make use of this in Section 9.

8. Computation of smoothing splines

An important contribution of WBS was to take advantage of the stochastic structure (4.8) to rewrite it in the space form (5.1) and (5.2) and consequently (5.6). This allows the use of the Kalman filter and the fixed interval smoothing algorithms to obtain $x(t \mid N)$ and $S_x(t \mid N)$ and hence $f(t \mid N)$ and $S_f(t \mid N)$ in $O(N)$ operations. Because $\lambda_i(f)$ is correlated with $e(i)$, $i = 1, \cdots, N$, in the WBS model, the Kalman filter is started at $t = t_n$ and the fixed interval smoothing algorithm is used to get $x(t \mid N)$ for $t \geq t_n$. An auxiliary subroutine is needed to obtain $x(t \mid N)$ for $t = 0$ to t_n. If the first n linear functionals $\lambda_1, \cdots, \lambda_n$ are not linearly independent over H_0, then we proceed as follows. Let m be the smallest integer such that $(\lambda_1, \cdots, \lambda_m)$ contains n functionals that are linearly independent over H_0. Then the Kalman filter and fixed interval smoothing algorithms are applied for $t \geq t_m$ to obtain $x(t \mid N)$ and $S_x(t \mid N)$. Additional subroutines are required for $t \leq t_m$, details being given in Ansley and Kohn (1984).

Consider now the Wahba stochastic model with $x(0) \sim N(0, kI_n)$. For a fixed k we can obtain $x(t \mid N)$ and $S_x(t \mid N)$ for $t_1 \leqq t \leqq t_N$ by applying the Kalman filter and fixed interval smoothing algorithms because $x(t)$ and $e(i)$ are uncorrelated for all t and i. However, the optimal spline is given by $\lim_{k \to \infty} x(t \mid N)$, whereas the outlined procedure gives us a numerical value for $x(t \mid N)$ only for a given k so the limiting operation cannot be carried out. The simple device of choosing k large is not recommended because it may result in numerical instability.

KA solved this problem by showing that for all t and j

$$(8.1) \qquad x(t \mid j) = x^{(0)}(t \mid j) + O(1/k)$$

$$(8.2) \qquad S_x(t \mid j) = kS_x^{(1)}(t \mid j) + S_x^{(0)}(t \mid j) + O(1/k)$$

with $x^{(0)}(t \mid j)$, $S_x^{(1)}(t \mid j)$ and $S_x^{(0)}(t \mid j)$ independent of k. Furthermore, with m defined as above, $S_x^{(1)}(t \mid j) = 0$ for $j \geqq m$ and all t. KA derives a modified Kalman filter algorithm giving recursions for $x^{(0)}(t_i \mid i)$, $S_x^{(1)}(t_i \mid i)$ and $S_x^{(0)}(t_i \mid i)$ for $i = 1, \cdots, N$. For $i \geqq m$, $S_x^{(1)}(t_i \mid i) = 0$ and the modified Kalman filter reduces to the ordinary Kalman filter. Similarly, a modified fixed interval smoothing algorithm is derived that allows us to compute $x^{(0)}(t \mid N)$ and $S_x^{(0)}(t \mid N)$.

Now from (8.1) and (8.2)

$$\lim_{k \to \infty} x(t \mid j) = x^{(0)}(t \mid j)$$

and for $j \geqq m$

$$\lim_{k \to \infty} S_x(t \mid j) = S_x^{(0)}(t \mid j)$$

and so $l'x^{(0)}(t \mid N)$ is the optimal smoothing spline. It is the canonical representation (8.1) and (8.2) that allows us first to filter and smooth and only afterwards take the limit.

We mention at this stage that the results in KA are an application of more general results obtained in Ansley and Kohn (1983a,b) for state space models with diffuse initial conditions.

A third algorithm based essentially on the Wahba stochastic model is given by Wecker and Ansley (1983). (We refer to this paper as WA.) WA take $x(0)$ to be an unknown constant vector to be estimated by maximum likelihood, the likelihood being based on the stochastic model.

In comparing the above three algorithms we first note that all three are $O(N)$, N being the sample size. WBS requires auxiliary algorithms at the beginning of the series especially if the first n linear functionals $\lambda_1, \cdots, \lambda_n$ are not linearly independent over H_0. The KA and WA algorithms in contrast can handle any pattern of dependence among the linear functionals $\lambda_1, \cdots, \lambda_N$. Because $x(0)$ is treated as a constant by WA, their algorithm requires that $x(0)$ be estimated by generalized least squares and that the variability due to estimation of $x(0)$ be incorporated in the computation of $S_x(t \mid N)$.

So far we have assumed that the smoothing parameter r is known. Usually it is not and has to be estimated from the data. A maximum likelihood approach treating r as a parameter of the stochastic model is suggested in WA and KA, with the likelihood being computed using the Kalman filter. For details see WA and KA. Even when the smoothing parameter r is estimated by maximum likelihood the above algorithms are still $O(N)$.

An alternative way of estimating r which can be carried out using the above algorithms is generalized cross-validation (GCV) due to Craven and Wahba (1979). Preliminary theoretical and empirical results by Wahba (1983b) suggest that at least in large samples GCV gives more satisfactory results than maximum likelihood. It is unclear at this stage, however, whether for moderate sample sizes GCV performs any better than maximum likelihood. It is important to note that the above algorithms become $O(N^2)$ if GCV is used, so that spline smoothing with GCV can be significantly more costly than with maximum likelihood estimation of r. We note that an approximate GCV method requiring only $O(N)$ operations has recently been proposed by Silverman (1984).

To complete this section we recall from Section 5 that $x(t) = [f(t), f^{(1)}(t), \cdots, f^{(n-1)}(t)]'$, so that $x_j(t \mid N)$ gives an estimate of $f^{(j-1)}(t)$.

9. A Bayesian interpretation of spline smoothing

Wahba (1978), (1983a) suggests that we regard the stochastic model on $f(t)$ described in Section 7 as a prior distribution on the unknown function $f(t)$. Thus our prior tells us that $f(t)$ has $(n-1)$ continuous derivatives and $[f(0), \cdots, f^{(1)}(0), \cdots, f^{(n-1)}(0)]' \sim N(0, kI_n)$. If we let $k \rightarrow \infty$, then we are placing a diffuse prior at $t = 0$ on $f(t)$ and its first $(n-1)$ derivatives. Because the variance of $dW(t)/dt$ is infinite we are also placing a diffuse prior on the nth derivative of $f(t)$.

The Wahba prior on $f(t)$ just described is not implausible. A diffuse prior on $f(t)$ and its first $(n-1)$ derivatives at $t = 0$ indicates a lack of knowledge about the starting values for $f(t)$. Similarly, a diffuse prior on the nth derivative of $f(t)$ indicates a lack of knowledge about the nth derivative.

With the above prior, the posterior distribution of $x(t)$ is $N[x(t \mid N), S_x(t \mid N)]$, and in particular the posterior distribution of $f(t)$ is $N[f(t \mid N), S_f(t \mid N)]$. This immediately suggests how to obtain Bayesian confidence intervals for $f(t)$ and its first $(n-1)$ derivatives.

In the WBS stochastic model $f(t)$ is correlated with the noise, so that the prior distribution on the starting conditions in the WBS stochastic model is not, in our opinion, as plausible as Wahba's. On the other hand, the WBS prior is proper whereas Wahba's is diffuse.

Because it is often mathematically more convenient to work with a proper prior rather than a diffuse one, the WBS and Wahba priors present important complementary tools for spline smoothing.

10. Smoothing a stochastic process

It is at first a little surprising that conditional expectations from the Wahba and WBS stochastic models should give us a function that is as smooth as a spline. Kohn and Ansley (1983) show that this smoothness is a general property of stochastic processes, the key requirement being that $W(t)$ has uncorrelated increments; normality is not required. Thus $f(t \mid N)$ is a spline, its smoothness determined by n, the order of L, and the functionals $\lambda_1, \cdots, \lambda_N$. In addition, they show that $S_f(t \mid N)$ is smooth, having approximately twice as many continuous derivatives as $f(t \mid N)$.

References

ANSELONE, P. M. AND LAURENT, P. J. (1968) A general method for the construction of interpolating or smoothing spline functions. *Numer. Math.* **12**, 66–82.

ANSLEY, C. F. AND KOHN, R. (1983a) State space models with diffuse initial conditions I: Filtering and likelihood. Technical Report #13, Statistics Research Center, Graduate School of Business, University of Chicago.

ANSLEY, C. F. AND KOHN, R. (1983b) State space models with diffuse initial conditions II: Smoothing and continuous time processes. Technical Report #15, Statistics Research Center, Graduate School of Business, University of Chicago.

ANSLEY, C. F. AND KOHN, R. (1984) The equivalence of two stochastic approaches to spline smoothing. Technical Report #19, Statistics Research Center, Graduate School of Business, University of Chicago.

CODDINGTON, E. A. AND LEVINSON, N. (1955) *Theory of Ordinary Differential Equations.* McGraw-Hill, New York.

CRAVEN, P. AND WAHBA, G. (1979) Smoothing noisy data with spline functions: Estimating the correct degree of smoothing by the method of generalized cross-validation. *Numer. Math.* **31**, 317–403.

KIMELDORF, G. AND WAHBA, G. (1970a) A correspondence between Bayesian estimation on stochastic processes and smoothing by splines. *Ann. Math. Statist.* **41**, 495–502.

KIMELDORF, G. AND WAHBA, G. (1970b) Spline functions and stochastic processes. *Sankhyā* A **32**, 173–180.

KIMELDORF, G. AND WAHBA, G. (1971) Some results on Tchebycheffian spline functions. *J. Multivariate Anal. Appl.* **33**, 82–95.

KOHN, R. AND ANSLEY, C. F. (1983) On the smoothness properties of the best linear unbiassed estimate of a stochastic process observed with noise. *Ann. Statist.* **11**, 1011–1017.

KOHN, R. AND ANSLEY, C. F. (1984) A new algorithm for spline smoothing and interpolation based on smoothing a stochastic process. *SIAM J. Scient. Statist. Comput.* To appear.

SCHOENBERG, I. J. (1964) Spline functions and the problem of graduation. *Proc. Nat. Acad. Sci. USA* **52**, 947–950.

SIDHU, G. S. AND WEINERT, H. L. (1979) Vector-valued Lg-splines I: Interpolating splines. *J. Math. Anal. Appl.* **70**, 505–529.

SIDHU, G. S. AND WEINERT, H. L. (1984) Vector-valued Lg-splines II: Smoothing splines. *J. Math. Anal. Appl.* **101**, 380–396.

SILVERMAN, B. W. (1984) A fast and efficient cross-validation method for smoothing parameter choice in spline regression. *J. Amer. Statist. Assoc.* **79**, 584–589.

WAHBA, G. (1978) Improper priors, spline smoothing and the problem of guarding against model errors in regression. *J. R. Statist. Soc.* B **40**, 364–372.

WAHBA, G. (1983a) Bayesian confidence intervals for the cross validated smoothing spline. *J. R. Statist. Soc.* B **45**, 133–150.

WAHBA, G. (1983b) A comparison of GCV and GML for choosing the smoothing parameter in the general spline smoothing problem. Technical Report #712, Department of Statistics, University of Wisconsin, Madison.

WECKER, W. E. AND ANSLEY, C. F. (1983) The signal extraction approach to nonlinear regression and spline smoothing. *J. Amer. Statist. Assoc.* **78**, 81–89.

WEINERT, H. L., BYRD, R. H. AND SIDHU, G. S. (1980) A stochastic framework for recursive computation of spline functions: Part II, Smoothing splines. *J. Optimization Theory Appl.* **30**, 255–268.

WEINERT, H. L. AND SIDHU, G. S. (1979) A stochastic framework for recursive computation of spline functions: Part I, Interpolating splines. *IEEE Trans. Information Theory* **IT-24**, 45–50.

Data Structures—The Problem of Time Series Computing

JOHN HENSTRIDGE

Abstract

The scope of statistical packages is largely defined by the data structures supported. It is claimed that none of the standard structures used in major packages is wholly suitable for time series data and this is the major reason why time series has not benefited from modern developments in statistical computing.

STATISTICAL COMPUTING; PACKAGE DESIGN

Most major statistical packages include some procedures for time series analysis. However, none of these procedures is widely regarded as being particularly good and in general the time series sections have all the appearance of being 'add ons' to an existing package which is in other respects reasonably consistent. It is profitable to investigate this and to see why no major time series package has emerged.

Data structures in existing packages

Payne and Nelder (1976) proposed that one criterion which could be used to assess packages is the range of data structures which they support. This criterion can be part of a measure of the versatility of a package—the other major factor influencing versatility is the input language structure (see for example Becker and Chambers (1978)).

As Payne and Nelder pointed out, many major packages have as their principal data structure the data matrix. The columns of the matrix are referred to as the variables and the rows as the observations or cases. This concept was widely used in early packages, such as SPSS (Nie et al. (1975)) and SAS (Goodnight (1982)), for two related reasons. Firstly, the structure matched the matrix formulation of regression, the most common statistical computation. Secondly and more importantly, these packages were designed to process large data sets and they did this by storing the bulk of the data as a disk file and sequentially accessing the rows of the data matrix one at a time.

The alternative structure for a package involves defining a set of statistical data types together with a more complicated method of organising their storage and access. An example is GENSTAT (Alvey et al. (1977)) with its 'VARIATE' data type which is effectively a vector. A data matrix is then conceptually formed by having a list of vectors all of the same length. Unfortunately many packages which have such a versatile data structure have been implemented by storing all the data in the computer's main memory, thus restricting them to relatively small data sets. Alternatives to this are usually heavily machine-dependent and operating-system-dependent.

A simple statistical data structure can be defined in terms of the type of values taken by its elements and the set of indices of the elements (the domain). In the case of the data matrix the values are row vectors and the domain is the set $(1, 2, \cdots, n)$.

The data matrix and similar variable structures assume that the elements of each data structure have the same domain, $(1, 2, \cdots, n)$. Furthermore it is usually assumed that the ordering of this set is not important—at the very least the indices are exchangeable. Some packages which handle special data have developed other data types—for example RGSP (Yates (1976)) has unsurpassed facilities for manipulating tables. However, the simple data matrix or vector data type appears superficially to be appropriate for time series, and hence there have been many attempts to fit time series procedures into programs which support only such a structure.

Time series structure

This standard data structure ignores several important features of time series, namely:

1. The order is important—the domain is an ordered set. In most cases the only meaningful permutation which can be applied is a translation corresponding to a time lead or lag.
2. The domain need not start at time 1. In fact in one analysis there may be series with a variety of starting times and finishing times.
3. It is possible that certain binary operations can be carried out with series with a different indexing set. For example if one series is defined for times 1 to 15 and a second one is defined for times 5 to 20, then it is meaningful to form their sum, a series defined for times from 5 to 15. In addition, operations such as applying moving-average filters give series which are shorter than the original.

4. Arithmetic operations often use elements of series corresponding to different times. Since such operations usually involve leads and lags, this property is closely related to the first point above. All filtering operations are like this as are procedures for performing lagged regressions and estimating correlation functions.

5. In some applications the time interval between elements of a series may differ both within and between series. For example, econometric data is commonly a mixture of monthly, quarterly and annual series.

6. Fourier methods and frequency domain techniques lead to new data structures indexed by frequency rather than by time.

Most existing packages attempt to overcome these problems by fitting time series data into a vector or data matrix structure matrix and using some of the following techniques:

1. Either vectors of differing lengths are supported or an attribute giving the number of valid values is attached to each vector. In many cases it is still assumed that all series start at time 1.

2. Standard length vectors are used with the ends (corresponding to times before and after domain of the series) padded out with missing values. This is an attractive option in some ways, since many major packages already have procedures for treating missing values. The technique can readily manage binary operations with series of differing index set since for each time if the element of either series is missing, the result is set to missing.

3. The ordered nature of the index set of a series is usually ignored except for some display purposes—a 'time series plot' is sometimes provided which provides a plot of the values against the index. Special facilities may exist for applying leads or lags to series. Sometimes, where disk file storage, is used only lags are provided for. Applying a lead to one series is performed by lagging all the others!

4. Differing time intervals are simply not allowed for.

5. Frequency domain techniques are severely limited and are usually restricted to deriving spectra for immediate display.

One package, TSA (Henstridge (1980), (1982)), was developed to address some of these problems. A special SERIES data type is used with attributes of starting time and length (or end time). The facilities for manipulation recognise this and dynamically determine the corresponding attributes of the result. In addition, all the procedures readily permit the introduction of leads and lags. However, no attempt is made to handle series with differing time intervals between observations beyond providing several commands for interpolation so that all series can be reduced to the same time base.

TSA highlights another problem with time series, since at a formal level it only supports series and not ordinary vectors. It is common in time series data analysis to form sorted residuals after fitting a model. While these are derived from a series the time ordering has clearly been lost. However, in TSA it is still possible to perform a time series operation such as estimating an autocorrelation function of such data. A language which properly supports time series and tries to prevent the user from performing illogical operations would require both series and vector data types and a well-defined interface between them.

Other data structures for time series

For general time series applications, more than just series need to be supported by a package. At the moment there is no general consensus on which structures are needed and how they are best implemented. TSA has Fourier transforms and ARIMA filters as standard data types, together with facilities for manipulating them. Few other packages have anything.

The Fourier transform of a series is usually a derived data structure. It is formed from data initially in the package. Hence the package is in a position to impose restrictions on such a structure. The transformation is usually implemented by means of the fast Fourier transform algorithm, resulting in the transform being defined at equally spaced frequencies and its length being a highly composite number. A difficulty for most packages is that the transform has complex values. Few areas of statistics apart from times series require any complex arithmetic.

In one sense a filter is not data but it is useful to have a facility to store and manipulate them since they are often combined together and manipulated in various ways. They also provide a natural way of storing time series ARIMA models.

Future developments

The future of time series computing is not clear since, as pointed out above, most statistical computing is associated with major packages which have not been properly designed for time series. At the moment the choice for the user is between using a standard package or writing procedures in a general high-level language such as FORTRAN. The current hardware trend towards computers with large central memory or virtual memory should allow packages to break away from the constraint of disk file storage and thus have more flexible data structures. However, this will benefit time series computing only if standards are developed by users of time series for what data structures are required.

References

ALVEY, N. G. ET AL. (1977) *GENSTAT, a General Statistical Program.* Numerical Algorithms Group, Oxford.

BECKER, R. A. AND CHAMBERS, J. M. (1978) Design and implementation of the *S* system for interactive data analysis. *Proc. IEEE Compsac* **78**, 626–629.

GOODNIGHT, J. H. (1982) *SAS User's Guide.* SAS Institute, Cary, North Carolina.

HENSTRIDGE, J. D. (1980) TSA, a package for time series analysis and manipulation. In *COMPSTAT-80*, ed. M. M. Barritt and D. Wishart, Physica Verlag, Wien.

HENSTRIDGE, J. D. (1982) *TSA, an Interactive Package for Time Series Analysis.* Numerical Algorithms Group, Oxford.

NIE, N. H. (1975) *SPSS, Statistical Package for the Social Sciences*, 2nd edn. McGraw-Hill, New York.

PAYNE, R. W. AND NELDER, J. A. (1976) Data structures in statistical computing. *Proc. 9th Biometric Conf.* **2**, 191–208.

YATES, F. (1976) *The Rothamsted General Survey Program.* Program Library Unit, University of Edinburgh.

Orthant Probabilities and Gaussian Markov Processes

P. A. P. MORAN

Abstract

A method is given for calculating the probability that n terms of a Gaussian first-order autoregressive series lie between any specified limits where n may be as large as 20 or more.

AUTOREGRESSIVE PROCESSES

1. Introduction

Suppose that $\{X_i\}$, $i = 1, \cdots, k$, are a sequence of jointly normally distributed random variables with zero means, unit variances and correlations given by

$$(1) \qquad \rho_{ij} = \rho^{|i-j|}.$$

Suppose also that

$$-\infty \leqq a_i < b_i \leqq \infty, \quad \text{for} \quad i = 1, \cdots, k.$$

The purpose of the present paper is to point out that it is not difficult to calculate probabilities of the form

$$(2) \qquad P_k = \Pr\{a_i \leqq X_i \leqq b_i, i = 1, \cdots, k\}$$

numerically, or to calculate similar probabilities conditional on the initial value of X_1. Such a sequence can be regarded as a section of a Gaussian Markov process in discrete time. The method will be illustrated in the orthant case where $a_i = 0$, $b_i = \infty$, for all i, but this is quite inessential and no further difficulties arise in the more general case.

2. The calculation of orthant probabilities

An extensive literature exists on the evaluation of orthant probabilities for the general multivariate normal population (see Johnson and Kotz (1972) for a

---

413

survey). For $k = 2, 3$ the results are known explicitly. For $k = 4$ an effective method was devised by Plackett (1954) and developed by Gehrlein (1979). The case $k = 5$ can be expressed as the sum of five integrals with $k = 4$. The case where $k = 6$ could also be done by Plackett's method but would require considerably more computation.

For a few special types of matrix (P_{ij}) effective methods also exist for general k. The present paper gives a further such case with the added advantage of working for general values of a_i and b_i.

3. The present method

Write $p_i(x_i)$ for the conditional probability density

$$p_i(x_i \mid a_1 \leq x_1 \leq b_1, \cdots, a_{i-1} \leq x_{i-1} \leq b_{i-1}).$$

Then

$$(3) \qquad p_{i+1}(x) = \frac{1}{\sqrt{(2\pi(1-\rho^2))}} \int_{a_i}^{b_i} \exp -\{(x-\rho y)^2/(2(1-\rho^2))\} p_i(y) dy.$$

Thus the distributions $p_i(x_i)$ can be calculated numerically in succession, and a final integration gives P_k as defined by (2).

As an experiment P_k was calculated for $k = 2, \cdots, 20$, where all the $a_i = 0$ and the $b_i = \infty$, and $\rho = 0.1(0.1)0.9$, and 0.95. Since $p_1(6) = 0.000\,000\,006$ the integration was taken over the range $(0, 6)$ only. On second thoughts it might have been very slightly more accurate to extend this slightly.

Gregory's integration formula was used up to and including the eighth forward difference, the coefficients of the differences being used with their expansions to calculate weights for the first nine ordinates. The interval of integration was taken as 0.1, at first, thus requiring the calculation of 61 ordinates in each integration. No weighting was used at the upper end of the interval.

In general it is much better to use corrections based on central rather than forward differences. However, in the present case the cost in time of evaluating $p_i(x)$ for further values of $x < 0$ outweighs the advantages since the weights are easily calculated in a preliminary calculation and stored in the computer.

Gaussian integration could have been used. However, inspection of (3) shows that the first function of y in the integrand is changing quite rapidly at the main values of x and y when ρ is large, e.g. larger than 0.7. Combined with the fact that the lengths of the central intervals in Gaussian integration tend to be relatively larger, this is likely to reduce the accuracy when ρ is larger than 0.7, say.

If other values of $(a_i\ b_i)$ had been used the procedure would have been equally simple, except that if any of the b_i had been less than 6, weighting corrections would also have been needed at the upper limit of the integrals.

Notice also that the present method could be used with any correlation matrix of the form $\rho_{ij} = \alpha_i \alpha_{i+1} \cdots \alpha_{j-1}$ where $i < j$ and the α's are any numbers less than unity in absolute value.

4. The results

The calculations were made on a desk computer which is alleged by the manufacturers to work to 15 decimal places, and to have a multiplication speed of 2.85 milliseconds. The calculation of each step took approximately 11 minutes, and at each step the value of P_k was obtained by the same method of numerical integration. The values of P_2, P_3 and P_4 were also independently found; P_2 and P_3 from the explicit formulae, and P_4 from a modification of the Plackett–Gehrlein method.

Finally the calculations were repeated for $\rho = 0.6, 0.9$ using an interval of 0.05 and 121 ordinates. For $\rho = 0.6$ the differences between the results for $P_2, \cdots, P_{20}$ did not exceed 10^{-9}, and for $\rho = 0.9$ did not exceed 10^{-7}. Thus this method seems quite accurate for $0 \leq \rho \leq 0.9$.

When ρ approaches ± 1 the accuracy decreases because the factor $1 - \rho^2$ in the integrand in (3) tends to 0 so that the first factor in the integral varies rapidly between one integration point and the next.

For each value of k the ratio P_k/P_{k-1} was calculated. This converges to a value between 0 and 1, and does so quite rapidly. Thus for $\rho \leq 0.4$, P_k/P_{k-1} is constant to six decimal places for $k \geq 10$, and for $\rho = 0.7$ it was constant to five decimal places for $k \geq 17$. This behaviour can be looked at in another way.

The integral transform (3) is closely analogous to the transformation of a finite vector, x, by a matrix A say, with non-negative elements (the numerical approximation we have used is not quite this because some of the weights in the Gregory integration formula are negative). If the vector x has non-negative elements the well-known Perron–Frobenius theory shows that the root of A with largest absolute value is, under weak conditions, simple and real, and that the elements of

$$A^n x$$

for large n are asymptotically proportional to those of a vector, x_0 say, with non-negative components and such that

$$A x_0 = \lambda x_0,$$

where λ is the largest real root of A. Thus the repetition of the integral transformation (3), with $a_i = 0$, $b_i = \infty$, can be expected to produce a function

$p_n(x)$ which is asymptotically proportional to

$$\lambda^n f(x)$$

where $0 < \lambda < 1$, and $f(x)$ is a solution of the integral equation

$$(4) \qquad f(x) = \lambda^{-1} \int_0^\infty (2\pi(1-\rho^2))^{-\frac{1}{2}} \exp -\{(x-\rho y)^2/(2(1-\rho^2))\} f(y)\, dy.$$

λ thus corresponds to the limiting value of P_n/P_{n-1}, and $f(x)$ to the limit of the normalised function

$$p_n(x) \Big/ \int_0^\infty p_n(u)\, du.$$

A theorem of this type is given by Harris (1963) but his conditions are not quite general enough to apply to the present case and the results of later much more elaborate theorems are required.

5. Continuous-time process

Much work has been done on approximate solutions of problems of the above type for Gaussian processes, $X(t)$, where t is continuous and the serial correlation function, $\rho(t)$, satisfies restricted conditions. If

$$(5) \qquad\qquad\qquad\qquad \rho(t) = \exp -\alpha t$$

the process is, with probability 1, continuous but without a derivative. Probabilities of the form

$$(6) \qquad\qquad\qquad\qquad P\{a \leqq x(t) \leqq b;\ 0 \leqq t \leqq T\}$$

could be estimated approximately by the procedure outlined above using a discrete approximation. This provides an upper bound to (6) but if we attempt to improve the approximation by making the size of the discrete intervals smaller the correlation between neighbouring values of $X(t)$ approaches unity and the numerical procedure becomes inaccurate.

Most of the work on calculating probabilities of the type (6) has been done for Gaussian processes in which

$$(7) \qquad\qquad\qquad\qquad \rho(t) = 1 - at^2 + o(t^2)$$

(see, for example, Longuet-Higgins (1963) and Slepian (1963)). Such processes have first-order derivatives and so much of the interest has been concentrated on the distribution of the intervals between zeros, since such a distribution then has a meaning. These processes are not Markovian and the method of the present paper cannot be used to obtain an approximate answer.

References

GEHRLEIN, W. V. (1979) A representation for quadrivariate normal positive orthant probabilities. *Commun. Statist. B Simulation and computation* **B8**, 349–358.

HARRIS, T. E. (1963) *The Theory of Branching Processes.* Springer-Verlag, Berlin.

JOHNSON, N. L. AND KOTZ, S. (1972) *Distributions in Statistics: Continuous Multivariate Distributions.* Wiley, New York.

LONGUET-HIGGINS, M. S. (1963) Bounding approximations to the distribution of intervals between zeros of a stationary Gaussian process. *Proc. Symp. Time Series Analysis*, ed. M. Rosenblatt. Wiley, New York.

PLACKETT, R. L. (1954) A reduction formula for normal multivariate integrals. *Biometrika* **41**, 351–360.

SLEPIAN, D. (1963) On the zeros of Gaussian noise. *Proc. Symp. Time Series Analysis*, ed. M. Rosenblatt. Wiley, New York.

An Algorithm for Exponential Fitting Revisited

M. R. OSBORNE

G. K. SMYTH

Abstract

An algorithm for exponential fitting is presented which exploits the separable regression structure and a reparametrization. The algorithm has proved very satisfactory, and theoretical reasons for this are developed.

PRONY'S METHOD; SEPARABLE REGRESSION; NON-LINEAR LEAST SQUARES; GAUSS–NEWTON ALGORITHM; REPARAMETRIZATION; NON-LINEAR EIGENPROBLEM

1. Introduction

The problem of fitting a sum of exponentials to observed data by least squares is well known to cause numerical difficulties. For example, Wilson (1983) cites a number of problems encountered in applying standard non-linear least squares procedures to data which will be summarized below. The problems included premature termination of the search for a minimum, inability to handle ill-conditioning, unreliable covariance matrices for the estimates, and failure to give adequate error indications. The data, summarized by Figure 1.1, consist of 33 equispaced observations on a simple chemical reaction of the form $A \rightarrow B \rightarrow C$, and are modelled by

$$\mathscr{E}(y_i) = \alpha_1 + \alpha_2 \exp\left(-\beta_2 t_i\right) + \alpha_3 \exp\left(-\beta_3 t_i\right).$$

This presents a dilemma because the model appears to describe a well-determined physical process sampled in an interval in which all terms in the model are well represented. The chemist has some justification for considering the numerical difficulties to be shortcomings of available techniques.

In fact good non-linear least squares algorithms do find acceptable numerical values on this problem (this means that values consistent with those obtained by other methods are obtained, not that tight confidence intervals are suggested), but often take a surprisingly large number of iterations from quite

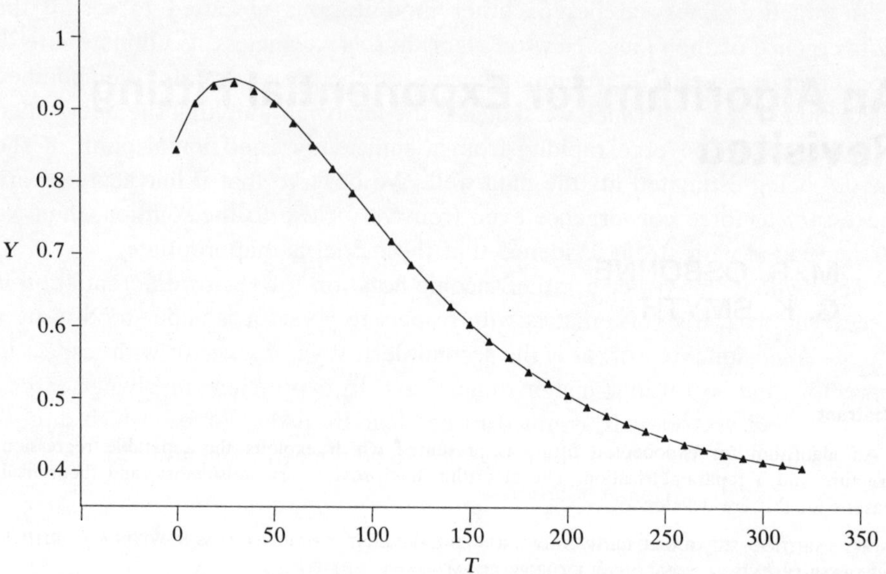

Figure 1.1. 33 equispaced observations on a chemical reaction, and the curve obtained from fitting a sum of exponentials by least squares. The curve fits the data well—the appearance that it lies slightly above the data is an artifact of the plotting process

close initial guesses. For example the method of Osborne (1976) starting from $\alpha_1 = 0.5$, $\alpha_2 = -1.0$, $\alpha_3 = 1.5$, $\beta_2 = 0.01$, $\beta_3 = 0.02$, takes twenty iterations to give parameter estimates $\hat{\alpha}_1 = 0.3754$, $\hat{\alpha}_2 = -1.4647$, $\hat{\alpha}_3 = 1.9358$, $\hat{\beta}_2 = 0.01287$, and $\hat{\beta}_3 = 0.02212$.

Here we re-examine an algorithm proposed by Osborne (1975) which converges to estimates of similar accuracy in three iterations using basically the same—at most first derivative—information. This algorithm is closely related to Prony's (1795) method for separating the linear and non-linear parameters. The condition for a stationary value of the transformed problem leads to a non-linear eigenvalue problem which is solved by simple iteration. The success of the method is attributed to the resulting reparametrization rather than to the separation process itself. Returning from the new to the old parameters does require solving a polynomial equation and is therefore another possible source of ill-conditioning; this problem is pursued by Smyth (1985).

Our aim in this paper is to contribute towards an understanding of the performance of the new algorithm by relating it to the Gauss–Newton method for non-linear least squares. We review the convergence theory for the Gauss–Newton algorithm in the next section and then develop the new algorithm in the context of separable regression.

In practice, line searches or other modifications are used to secure the convergence of the Gauss–Newton algorithm. Nevertheless its ultimate rate of convergence close to the solution depends on the properties of the unmodified algorithm. The result which we sketch is the important one that the unmodified algorithm will converge rapidly, from a sufficiently good initial point, if the model being estimated fits the data well. We observe that if line searches are necessary to force convergence even from very close to the solution, then we are provided with strong evidence that the model is inappropriate.

Throughout this paper a rather laconic notation for vector differentiation is used. The derivative of a matrix with respect to a vector is understood to be a three-dimensional tensor, as is the second derivative of a vector with respect to a vector, and so on into higher dimensions. In expressions involving tensors, matrices and vectors it is assumed to be clear from the context which dimensions relate to which. As an example, let A be a matrix function of a vector x, and y any vector. The expression

$$y^{\mathrm{T}}\nabla_x^2 A y$$

is to be interpreted as the matrix whose ijth element is

$$y^{\mathrm{T}}\frac{\partial^2 A}{\partial x_i \partial x_j} y$$

and the many like examples are treated similarly.

2. Least squares estimates and the Gauss–Newton algorithm

Suppose the observed vector $y^{\mathrm{T}} = (y_1, \cdots, y_n)$ to have mean

$$(2.1) \qquad \mu = \mathscr{E}(y) = A(\beta)\alpha$$

and covariance matrix

$$\mathscr{V}(y) = \sigma^2 I$$

where $A_{ij} = \exp(-\beta_j t_i)$, $i = 1, \cdots, n$, $j = 1, \cdots, p$ and $t_i = t_1 + (i-1)h$.

It is well known that under general conditions the least squares estimates $\begin{pmatrix}\hat{\alpha}\\\hat{\beta}\end{pmatrix}$ which minimize

$$\phi(\alpha, \beta) = (y - \mu)^{\mathrm{T}}(y - \mu)$$

are consistent and asymptotically normal. Furthermore the Gauss–Newton iteration for calculating them is asymptotically stable. Jennrich (1969) gives sufficient conditions; these are basically that all inner products of the form $(1/n)f^{\mathrm{T}}g$ have finite limits, where f and g represent μ or first or second derivatives of μ with respect to the components of α and β, and in particular that

$$(2.2) \qquad H = \lim_{n \to \infty} \frac{1}{n} \nabla_{(\alpha,\beta)} \mu^{\mathrm{T}} \nabla_{(\alpha,\beta)} \mu$$

exists and is positive definite.

In exponential fitting we have

$$(2.3) \qquad \nabla_{(\alpha,\beta)} \mu^{\mathrm{T}} \nabla_{(\alpha,\beta)} \mu = \begin{bmatrix} A^{\mathrm{T}} A & A^{\mathrm{T}} \partial_\beta \mu \\ \partial_\beta \mu^{\mathrm{T}} A & \partial_\beta \mu^{\mathrm{T}} \partial_\beta \mu \end{bmatrix}$$

and the components of $A^{\mathrm{T}} A$, $A^{\mathrm{T}} \partial_\beta \mu$ and $\partial_\beta \mu^{\mathrm{T}} \partial_\beta \mu$ are

$$\sum_{s=1}^{n} \exp\left(-(\beta_i + \beta_j) t_s\right),$$

$$\sum_{s=1}^{n} t_s \exp\left(-(\beta_i + \beta_j) t_s\right),$$

and

$$\sum_{s=1}^{n} t_s^2 \exp\left(-(\beta_i + \beta_j) t_s\right) \text{ respectively.}$$

For condition on H to be satisfied we require that the β_i be distinct and that t_1 and h be chosen appropriately. In fact, if t_1 is considered fixed then h must approach 0 in such a way that $t_n - t_1 = (n-1)h$ is also constant. This may be interpreted as saying that sampling of a transient system must be done while something is happening.

An important part of the proof that the least squares estimates are asymptotically normally distributed consists of writing (let θ represent α and β)

$$(2.4) \qquad \frac{1}{2n} \nabla_\theta^2 \phi = \frac{1}{n} \nabla_\theta \mu^{\mathrm{T}} \nabla_\theta \mu - \frac{1}{n} \nabla_\theta^2 \mu^{\mathrm{T}} (y - \mu)$$

and observing that the first term converges to H while the second has almost sure limit 0. The positive definite matrix H is then the asymptotic covariance matrix of $n^{\frac{1}{2}} \hat{\theta}$.

The terms on the right-hand side of (2.4) are also important in showing the asymptotic stability of the Gauss–Newton iteration. That iteration may be written as

$$\theta^{(i+1)} = F(\theta^{(i)})$$

where

$$F(\theta^{(i)}) = \theta^{(i)} + (\nabla_\theta \mu^{\mathrm{T}} \nabla_\theta \mu)^{-1} \nabla_\theta \mu^{\mathrm{T}} (y - \mu)_{\theta = \theta^{(i)}}.$$

The condition for a stationary point $\hat{\boldsymbol{\theta}}$ to be a point of attraction is that

$$\|\nabla_\theta F(\hat{\boldsymbol{\theta}})\| < 1$$

which written explicitly is

$$\|(\nabla_\theta\boldsymbol{\mu}^\mathrm{T}\nabla_\theta\boldsymbol{\mu})^{-1}\nabla_\theta^2\boldsymbol{\mu}^\mathrm{T}(\mathbf{y}-\boldsymbol{\mu})\|_{\theta=\hat{\theta}} < 1.$$

This condition is shown to be asymptotically satisfied by observing as above that $(1/n)\nabla_\theta^2\boldsymbol{\mu}^\mathrm{T}(\mathbf{y}-\boldsymbol{\mu})$ tends to 0 almost surely while $(1/n)\nabla_\theta\boldsymbol{\mu}^\mathrm{T}\nabla_\theta\boldsymbol{\mu}$ has limit H.

3. Exponential fitting as a separable regression

The model (2.1) is separable in the sense that for any fixed $\boldsymbol{\beta}$ it is easy to minimize $\phi(\boldsymbol{\alpha}, \boldsymbol{\beta})$ separately with respect to the linear parameters to give $\boldsymbol{\alpha} = \hat{\boldsymbol{\alpha}}(\boldsymbol{\beta})$. Substituting back into ϕ gives a reduced function objective $\psi(\boldsymbol{\beta})$ depending only upon the non-linear parameters. In general the evidence in favour of solving the reduced problem seems inconclusive: the advantage of reduced dimension is often offset by increased difficulty in manipulating the reduced objective function. (See Golub and Pereyra (1973), who have made a range of numerical studies.) In the case of exponential fitting, however, it can be useful especially after reparametrization.

For fixed $\boldsymbol{\beta}$, $\phi(\boldsymbol{\alpha}, \boldsymbol{\beta})$ is minimized by

$$\hat{\boldsymbol{\alpha}}(\boldsymbol{\beta}) = (A^\mathrm{T}A)^{-1}A^\mathrm{T}\mathbf{y}.$$

Substitution into ϕ gives the reduced objective function

$$(3.1) \qquad \psi(\boldsymbol{\beta}) = \phi(\hat{\boldsymbol{\alpha}}(\boldsymbol{\beta}), \boldsymbol{\beta}) = \mathbf{y}^\mathrm{T}(I - A(A^\mathrm{T}A)^{-1}A^\mathrm{T})\mathbf{y}.$$

As general results on reduced objective functions we have that

$$\nabla_\beta\psi(\hat{\boldsymbol{\beta}}) = \partial_\beta\phi|_{\substack{\alpha=\hat{\alpha} \\ \beta=\hat{\beta}}}$$

and

$$\nabla_\beta^2\psi(\hat{\boldsymbol{\beta}}) = [\partial_\beta^2\phi - \partial_\beta\partial_\alpha\phi(\partial_\alpha^2\phi)^{-1}\partial_\alpha\partial_\beta\phi]_{\substack{\alpha=\hat{\alpha} \\ \beta=\hat{\beta}}}.$$

The last result can be obtained either from direct calculations as in Richards (1961), or by observing that since $\nabla_\beta^2\psi(\hat{\beta})$ estimates the inverse covariance of $\hat{\boldsymbol{\beta}}$ its inverse must be the appropriate component of $[\nabla_{(\alpha,\beta)}^2\phi(\hat{\boldsymbol{\alpha}}, \hat{\boldsymbol{\beta}})]^{-1}$. In any case we conclude from (2.3), (2.4) and the consistency of $(\hat{\alpha}, \hat{\beta})$ that

$$(3.2) \qquad \frac{1}{n}\{\nabla_\beta^2\psi(\hat{\boldsymbol{\beta}}) - 2(\partial_\beta\boldsymbol{\mu}^\mathrm{T}\partial_\beta\boldsymbol{\mu} - \partial_\beta\boldsymbol{\mu}^\mathrm{T}A(A^\mathrm{T}A)^{-1}A^\mathrm{T}\partial_\beta\boldsymbol{\mu})\} \to 0 \quad \text{a.s.}$$

To set up the reparametrization, we note that $I - A(A^\mathrm{T}A)^{-1}A^\mathrm{T}$ is the projection onto the $(n-p)$-dimensional space orthogonal to the columns of A. It could equally well be written then as $X(X^TX)^{-1}X^T$ where X is any $n \times (n-p)$ matrix of rank $(n-p)$ such that $X^TA = 0$.

Appropriate forms for X are readily derived. We follow a suggestion of N. E. Day to define $X_{ii}, \cdots, X_{(i+p)i}$ as the coefficients of $\xi_1, \cdots, \xi_{p+1}$ in the expansion

$$
\begin{vmatrix}
\xi_1 & A_{i1} & A_{ip} \\
\cdot & \cdot & \cdot \\
\cdot & \cdot \cdots & \cdot \\
\cdot & \cdot & \cdot \\
\xi_{p+1} & A_{(i+p)1} & A_{(i+p)p}
\end{vmatrix}
= \sum_{j=1}^{p+1} X_{(i+j-1)i}\xi_j
$$

and to set

$$
X =
\begin{bmatrix}
X_{11} & & & & \\
\cdot & X_{22} & & & \\
\cdot & \cdot & & & \\
X_{(p+1)1} & \cdot & \cdot & & \\
 & X_{(p+2)2} & \cdot & \cdot & \\
 & & X_{(n-p)(n-p)} & & \\
 & & \cdot & & \\
 & & \cdot & & \\
 & & X_{n(n-p)} &
\end{bmatrix}.
$$

This has a particularly simple form for the exponential fitting problem. Consider the case $p = 2$. Then

$$
\begin{vmatrix}
\xi_1 & \exp(-\beta_1 t_i) & \exp(-\beta_2 t_i) \\
\xi_2 & \exp(-\beta_1(t_i+h)) & \exp(-\beta_2(t_i+h)) \\
\xi_3 & \exp(-\beta_1(t_i+2h)) & \exp(-\beta_2(t_i+2h))
\end{vmatrix}
$$

$$
= \exp(-(\beta_1+\beta_2)t_i)
\begin{vmatrix}
\xi_1 & 1 & 1 \\
\xi_2 & \exp(-\beta_1 h) & \exp(-\beta_2 h) \\
\xi_3 & \exp(-2\beta_1 h) & \exp(-2\beta_2 h)
\end{vmatrix}
$$

$$
= \exp(-(\beta_1+\beta_2)t_i)(\exp(-\beta_2 h) - \exp(-\beta_1 h))
$$
$$
\times (\xi_1 \exp(-(\beta_1+\beta_2)h) - \xi_2(\exp(-\beta_1 h)+\exp(-\beta_2 h))+\xi_3).
$$

As common factors are not relevant we can take

$$
X_{ii} = \exp(-(\beta_1+\beta_2)h)
$$

$$
X_{(i+1)i} = -(\exp(-\beta_1 h)+\exp(-\beta_2 h))
$$

$$
X_{(i+2)i} = 1.
$$

These are the elementary symmetric functions of $\exp(-\beta_1 h)$ and $\exp(-\beta_2 h)$, and hence are also the coefficients of the polynomial having these quantities as roots.

The general case is similar. We set

$$X = \begin{bmatrix} c_1 \\ \vdots & & \ddots \\ c_{p+1} & & & c_1 \\ & & & \vdots \\ & & & c_{p+1} \end{bmatrix}$$

where the polynomial

$$c_{p+1}z^p + c_p z^{p-1} + \cdots + c_1$$

has roots

$$z_i = \exp(-\beta_i h) \qquad i = 1, \cdots, p.$$

It generalizes a method of Prony to observe that ψ can be parametrized in terms of $c^{\mathrm{T}} = (c_1, \cdots, c_{p+1})$ instead of β. The apparent extra degree of freedom is removed by noting that $X(X^{\mathrm{T}}X)^{-1}X^{\mathrm{T}}$ does not depend on the magnitude of c so that we can adjoin the condition $c^{\mathrm{T}}c = 1$ without loss of generality. In subsequent manipulations the value of the new parametrization will be that X is linear in c. In terms of the new parameters the problem becomes

$$(3.3) \qquad \begin{aligned} &\text{minimize} \quad \psi(c) = y^{\mathrm{T}}X(X^{\mathrm{T}}X)^{-1}X^{\mathrm{T}}y \\ &\text{subject to} \quad c^{\mathrm{T}}c = 1. \end{aligned}$$

The method of Prony would be (approximately) to minimize the simple sum of squares of the modified residuals $X^{\mathrm{T}}y$, but in this form even consistency has not been proved, and simulations show poor recovery of actual parameter values.

4. An algorithm

Necessary conditions for a solution of (3.3) are

$$(4.1) \qquad \nabla_c \psi(c) - 2\lambda c = 0$$

where λ is a Lagrange multiplier. Direct differentiation yields

$$(4.2) \quad \nabla_c \psi = 2y^{\mathrm{T}}\nabla_c X(X^{\mathrm{T}}X)^{-1}X^{\mathrm{T}}y - 2y^{\mathrm{T}}X(X^{\mathrm{T}}X)^{-1}\nabla_c X^{\mathrm{T}}X(X^{\mathrm{T}}X)^{-1}X^{\mathrm{T}}y$$

which we can write in the form $\nabla_c \psi(c) = 2B(c)c$ with

$$(4.3) \quad B(c) = y^{\mathrm{T}}\nabla_c X(X^{\mathrm{T}}X)^{-1}\nabla_c X^{\mathrm{T}}y - y^{\mathrm{T}}X(X^{\mathrm{T}}X)^{-1}\nabla_c X^{\mathrm{T}}\nabla_c X(X^{\mathrm{T}}X)^{-1}X^{\mathrm{T}}y$$

by noting that since X is linear in c,

$$X = \sum_{j=1}^{p+1} c_j \frac{\partial X}{\partial c_j}.$$

The necessary conditions then reduce to

(4.4) $[B(c) - \lambda I]c = 0$

and $B(c)$ may be written more concisely as

$$B(c) = Y^{\mathrm{T}} M^{-1} Y - V^{\mathrm{T}} V$$

where $M = X^{\mathrm{T}} X$, and Y and V are the $(n-p) \times (p+1)$ and $n \times (p+1)$ matrices given by

$$Y = \begin{bmatrix} y_1 & & y_{p+1} \\ \vdots & \cdots & \vdots \\ y_{n-p} & & y_n \end{bmatrix}$$

and

$$V = \begin{bmatrix} v_1 & & \\ \vdots & & \\ v_{n-p} & \ddots & v_1 \\ & \ddots & \vdots \\ & & v_{n-p} \end{bmatrix}$$

where $v = M^{-1} Y c$.

Remark (i). It is easy to show that $\lambda = 0$. In fact

$$c^{\mathrm{T}} B(c) c = 0$$

identically in c.

Remark (ii). The set of possible c span the set of all coefficients of non-trivial polynomials of degree $\leqq p$. However, the image of the set of all possible β is the set of all polynomials of strict order p and distinct positive roots. Thus (3.3) entertains a strictly larger set of comparisons and so may give a lower minimum than (3.1). However, it follows from remark (i) and the discussion in Section 3 of Osborne (1975) that if β minimizes (3.1) then the corresponding c satisfies the necessary conditions (4.1) with $\lambda = 0$. It corresponds then to a stationary point of (3.3) even if not a strict minimum.

Algorithm. The suggested algorithm has the form:
(i) Set the initial guess $c^{(1)}$, $i := 1$; (One possibility is the eigenvector of $Y^{\mathrm{T}} Y$ corresponding to the eigenvalue least in magnitude.)
(ii) Find the eigenvalue least in magnitude of

$$(B(c^{(i)}) - \lambda^{(i+1)} I) c^{(i+1)} = 0$$

$$c^{(i+1)\mathrm{T}} c^{(i+1)} = 1;$$

(iii) Test convergence; (For example by testing the convergence of $\lambda^{(i+1)}$ to 0.)

(iv) $i := i+1$, go to (ii).

Example (Osborne (1975)). Table 4.1 presents results obtained by applying the algorithm to simulated data distributed independently and normally with means

$$\mu_i = 0.45 - 1.15 \exp(-0.0115t_i) + 1.85 \exp(-0.0225t_i)$$

and common variance σ^2. The initial guess was set by taking $\beta_2 = -0.01$ and $\beta_3 = -0.02$, and the stopping criterion was $|\lambda^{(i+1)}| \leq 10^{-15} \|B(c^{(1)})\|$. It will be seen the algorithm performs very satisfactorily, and that the number of iterations tends to decrease as n increases and as σ decreases.

We now add theoretical support to this numerical evidence by relating this algorithm to the Gauss–Newton process. A complete proof of the asymptotic stability of the algorithm under conditions similar to Jennrich (1969) will be published elsewhere.

We note that an iteration of the algorithm has the form

$$F(\theta^{(i+1)}, \theta^{(i)}) = 0$$

TABLE 4.1

Summary of results for the algorithm applied to simulated data (data generated using $\beta_2 = -0.0115$, $\beta_3 = -0.0225$)

n	σ	Number of iterations	$-\beta_2$	$-\beta_3$	ψ	$\hat{\sigma} = \sqrt{\psi(n-5)}$
33	0.001	3	0.011152	0.022883	0.2016E-4	8.5 E-4
	0.005	3	0.009914	0.024255	0.5052E-4	4.2 E-3
	0.01	3	0.008589	0.025726	0.2021E-2	8.5 E-2
	0.05	5	0.000401	0.033919	0.5056E-1	4.2 E-2
65	0.001	2	0.011761	0.022197	0.4580E-4	8.7 E-4
	0.005	3	0.012995	0.020792	0.1145E-2	4.4 E-3
	0.01	3	0.016245	0.017324	0.4579E-2	8.7 E-3
	0.05	3	complex pair		0.1142E 0	4.4 E-2
129	0.001	2	0.011488	0.022545	0.1077E-3	9.3 E-4
	0.005	2	0.011449	0.022721	0.2691E-2	4.57E-3
	0.01	2	0.011417	0.022923	0.1077E-1	9.3 E-3
	0.05	3	0.011909	0.023835	0.2691E 0	4.57E-2
257	0.001	2	0.011532	0.022467	0.2290E-3	9.5 E-4
	0.005	2	0.011659	0.022336	0.5729E-2	4.8 E-3
	0.01	2	0.011820	0.022169	0.2291E-1	9.5 E-3
	0.05	2	0.013411	0.020527	0.5730E 0	4.8 E-2

where $\boldsymbol{\theta}^{(i)T} = (\boldsymbol{c}^{(i)T}, \lambda^{(i)})$ and $F(\boldsymbol{u}, \boldsymbol{v})$ is a non-linear iteration function. A sufficient condition for a solution of (4.4) $\hat{\boldsymbol{\theta}}$ to be a point of attraction is

$$\|\nabla_u F(\hat{\theta})^{-1}\nabla_v F(\hat{\theta})\| < 1,$$

and a direct calculation gives this condition explicitly as

$$\left\|\begin{bmatrix} B(\hat{c}) & \hat{c} \\ \hat{c}^{T} & 0 \end{bmatrix}^{-1} \begin{bmatrix} (\nabla_c B(\hat{c}))\hat{c} & 0 \\ 0 & 0 \end{bmatrix}\right\| < 1.$$

Since $\hat{c}$ is the only eigenvector of $B(\hat{c})$ with zero eigenvalue, and since $\hat{c}^{T}(\nabla_c B(\hat{c}))\hat{c} = 0$, this condition is also equivalent to

$$(4.5) \qquad\qquad \|B^{+}(\hat{c})^{-1}(\nabla_c B(\hat{c}))\hat{c}\| < 1.$$

Now it follows from $\nabla_c \psi = 2B(c)c$ that

$$(4.6) \qquad\qquad \nabla_c^2 \psi = 2B(c) + 2(\nabla_c B(c))c.$$

It further follows from $\nabla_\beta c^{T}\nabla_c^2 \psi \nabla_\beta c = \nabla_\beta^2 \psi$, (3.2) and the consistency of $\hat{\boldsymbol{\beta}}$ that

$$(4.7) \quad \begin{aligned} \frac{1}{n}\{&\nabla_\beta c^{T}(B(c) + (\nabla_c B(c))c)\nabla_\beta c \\ &- (\partial_\beta \mu^{T}\partial_\beta \mu - \partial \mu^{T}A(A^{T}A)^{-1}A^{T}\partial_\beta \mu)\} \to 0 \quad \text{a.s.} \end{aligned}$$

We show below, by direct calculation, that

$$(4.8) \qquad\qquad \mathscr{E}(B(c)) = \mu^{T}\nabla_c X(X^{T}X)^{-1}\nabla_c X^{T}\mu$$

and

$$(4.9) \qquad\qquad \mathscr{E}(\nabla_c B(c)c) = \boldsymbol{0}.$$

But then

$$\begin{aligned} \partial_\beta \mu^{T}\partial_\beta \mu - \partial_\beta \mu^{T}A(A^{T}A)^{-1}A^{T}\partial_\beta \mu &= \partial_\beta \mu^{T}X(X^{T}X)^{-1}X^{T}\partial_\beta \mu \\ &= \nabla_\beta c^{T}\mu^{T}\nabla_c X(X^{T}X)^{-1}\nabla_c X^{T}\mu\nabla_\beta c \\ &= \nabla_\beta c^{T}\mathscr{E}(B(c))\nabla_\beta c \end{aligned}$$

where the second-last step follows from differentiating $X^{T}\mu = 0$ to get $X^{T}\partial_\beta \mu = -\nabla_\beta X^{T}\mu$. If we further invoke (2.2), that is that the data are equally spaced over a fixed interval, we have

$$(4.10) \quad \lim_{n \to \infty} \frac{1}{n}\nabla_\beta c^{T}(B(c) + \nabla_c B(c)c)\nabla_\beta c = \lim_{n \to \infty} \frac{1}{n}\nabla_\beta c^{T}\mathscr{E}(B(c))\nabla_\beta c \quad \text{a.s.}$$

which is a positive definite matrix. Although $\nabla_\beta c$ depends on the sampling interval h as well as $\boldsymbol{\beta}$ (in fact it tends to 0 if h does), it always has rank p, as is shown in the proof of Theorem 3.1 of Osborne (1975). Both $\nabla_\beta c^{\mathrm{T}}$ and $\mathscr{E}(B(c))$ have one null eigenvector, which is c.

There is then an analogy between (4.6) and (2.3), which break the Hessians up into a term of interest and a term with expectation 0. The final fact, necessary to prove (4.5) and to make the analogy closer, is that $(1/n)\nabla_\beta c^{\mathrm{T}} B(c) \nabla_\beta c$ itself converges almost surely. Although highly plausible, the proof of this involves using the structure of $X^{\mathrm{T}} X$ to bound the eigenvalues of $(X^{\mathrm{T}} X)^{-1}$, and is not pursued further here. The effect of (4.7) then is to relate $B(c)$ to the inverse asymptotic covariance matrix of $\hat\beta$: if the eigenvalues of the latter are large, as they will be in large samples because of the consistency of $\hat\beta$, so will the non-zero eigenvalues of $B(c)$.

We now prove (4.8) and (4.9). Further differentiation of (4.2) yields

$$
\begin{aligned}
\nabla_c^2 \psi = {} & 2y^{\mathrm{T}} \nabla_c X (X^{\mathrm{T}} X)^{-1} \nabla_c X^{\mathrm{T}} y \\
& - 2y^{\mathrm{T}} X (X^{\mathrm{T}} X)^{-1} \nabla_c X^{\mathrm{T}} \nabla_c X (X^{\mathrm{T}} X)^{-1} X^{\mathrm{T}} y \\
& - 4y^{\mathrm{T}} \nabla_c X (X^{\mathrm{T}} X)^{-1} \nabla_c X^{\mathrm{T}} X (X^{\mathrm{T}} X)^{-1} X^{\mathrm{T}} y \\
& - 4y^{\mathrm{T}} \nabla_c X (X^{\mathrm{T}} X)^{-1} X^{\mathrm{T}} \nabla_c X (X^{\mathrm{T}} X)^{-1} X^{\mathrm{T}} y \\
& + 4y^{\mathrm{T}} X (X^{\mathrm{T}} X)^{-1} \nabla_c X^{\mathrm{T}} X (X^{\mathrm{T}} X)^{-1} \nabla_c X^{\mathrm{T}} X (X^{\mathrm{T}} X)^{-1} X^{\mathrm{T}} y \\
& + 2y^{\mathrm{T}} X (X^{\mathrm{T}} X)^{-1} X^{\mathrm{T}} \nabla_c X (X^{\mathrm{T}} X)^{-1} \nabla_c X^{\mathrm{T}} X (X^{\mathrm{T}} X)^{-1} X^{\mathrm{T}} y \\
& + 2y^{\mathrm{T}} X (X^{\mathrm{T}} X)^{-1} \nabla_c X^{\mathrm{T}} X (X^{\mathrm{T}} X)^{-1} X^{\mathrm{T}} \nabla_c X (X^{\mathrm{T}} X)^{-1} X^{\mathrm{T}} y.
\end{aligned}
\tag{4.11}
$$

The first two terms of (4.11) can be identified as $2B(c)$ so the rest comprise $2(\nabla_c B(c))c$. Term-by-term expectations are easily obtained by recalling that

$$
X^{\mathrm{T}} y = X^{\mathrm{T}} (y - \mu)
$$

and that

$$
\mathscr{E}\{(y - \mu)^{\mathrm{T}} G (y - \mu)\} = \mathrm{tr}\,\{G \mathscr{E}(y - \mu)(y - \mu)^{\mathrm{T}}\}
$$

for any matrix G. Therefore

$$
\begin{aligned}
\mathscr{E}(\nabla_c^2 \psi) = {} & 2\sigma^2 \mathrm{tr}\,\{\nabla_c X (X^{\mathrm{T}} X)^{-1} \nabla_c X^{\mathrm{T}}\} + 2\mu^{\mathrm{T}} \nabla_c X (X^{\mathrm{T}} X)^{-1} \nabla_c X^{\mathrm{T}} \mu \\
& - 2\sigma^2 \mathrm{tr}\,\{\nabla_c X (X^{\mathrm{T}} X)^{-1} \nabla_c X^{\mathrm{T}}\} \\
& - 4\sigma^2 \mathrm{tr}\,\{\nabla_c X (X^{\mathrm{T}} X)^{-1} \nabla_c X^{\mathrm{T}} X (X^{\mathrm{T}} X)^{-1} X^{\mathrm{T}}\} \\
& - 4\sigma^2 \mathrm{tr}\,\{X (X^{\mathrm{T}} X)^{-1} \nabla_c X^{\mathrm{T}} X (X^{\mathrm{T}} X)^{-1} \nabla_c X^{\mathrm{T}}\} \\
& + 4\sigma^2 \mathrm{tr}\,\{X (X^{\mathrm{T}} X)^{-1} \nabla_c X^{\mathrm{T}} X (X^{\mathrm{T}} X)^{-1} \nabla_c X^{\mathrm{T}}\} \\
& + 2\sigma^2 \mathrm{tr}\,\{\nabla_c X (X^{\mathrm{T}} X)^{-1} \nabla_c X^{\mathrm{T}} X (X^{\mathrm{T}} X)^{-1} X^{\mathrm{T}}\} \\
& + 2\sigma^2 \mathrm{tr}\,\{\nabla_c X (X^{\mathrm{T}} X)^{-1} \nabla_c X^{\mathrm{T}} X (X^{\mathrm{T}} X)^{-1} X^{\mathrm{T}}\}.
\end{aligned}
\tag{4.12}
$$

Identifying the first three terms of (4.12) as $2\mathscr{E}(B(c))$ and the rest as $2\mathscr{E}(\nabla_c B(c)c)$ demonstrates (4.8) and (4.9).

References

GOLUB, G. H. AND PEREYRA, V. (1973) The differentiation of pseudo-inverses and nonlinear least squares problems whose variables separate. *SIAM J. Numer. Anal.* **10**, 413–432.

JENNRICH, R. I. (1969) Asymptotic properties of non-linear least squares estimators. *Ann. Math. Statist.* **40**, 633–643.

OSBORNE, M. R. (1975) Some special nonlinear least squares problems. *SIAM J. Numer. Anal.* **12**, 571–592.

OSBORNE, M. R. (1976) Nonlinear least squares—the Levenberg algorithm revisited *J. Austral. Math. Soc.* B **19**, 343–357.

RICHARDS, F. S. G. (1961) A method of maximum-likelihood estimation. *J. R. Statist. Soc.* B **23**, 469–475.

SMYTH, G. K. (1985) Separable Parameters and Coupled Iterations in Nonlinear Estimation, Ph.D. Thesis, Australian National University, Canberra.

WILSON, S. R. (1983) Benchmark data sets for the flexible evaluation of statistical software. *Computational Statistics and Data Analysis* **1**, 29–39.

List of Contributors

AN HONG-ZHI, Institute of Applied Mathematics, Chinese Academy of Science, Beijing, China.

CRAIG F. ANSLEY, Graduate School of Business, University of Chicago, 1101 East 58th Street, Chicago, IL 60637, USA.

M. S. BARTLETT (Emeritus Professor of Biomathematics, University of Oxford) c/o Department of Biomathematics, University of Oxford, 5 South Parks Road, Oxford OX1 3UB, UK.

PETER BLOOMFIELD, Department of Statistics, North Carolina State University, Raleigh, NC 27695–8203, USA.

DAVID R. BRILLINGER, Department of Statistics, University of California, Berkeley, CA 94720, USA.

MURRAY A. CAMERON, CSIRO, Division of Mathematics and Statistics, PO Box 218, Lindfield, NSW 2070, Australia.

CHEN ZHAO-GUO, Institute of Applied Mathematics, Chinese Academy of Sciences, Beijing, China.

M. DEISTLER, Institut für Ökonometrie und Operations Research, Technische Universität Wien, Argentinierstrasse 8/119, A-1040 Wien, Austria.

J. DURBIN, Department of Statistical and Mathematical Sciences, The London School of Economics and Political Science, Houghton Street, London WC2A 2AE, UK.

J GANI, Statistics Program, Department of Mathematics, University of California, Santa Barbara, CA 93106, USA.

JOHN HENSTRIDGE, Siromath Pty Ltd, 384 Rokeby Road, Subiaco, WA 6008, Australia.

S. M. HERAVI, Department of Mathematics, The University of Manchester Institute of Science and Technology, PO Box 88, Manchester M60 1QD, UK.

C. C. HEYDE, Department of Statistics, University of Melbourne, Parkville, VIC 3052, Australia.

YUZO HOSOYA, Faculty of Economics, Tohoku University, Kawauchi, Sendai 980, Japan.

RICHARD H. JONES, Department of Preventive Medicine and Biometrics, University of Colorado Health Sciences Center, 4200 East Ninth Avenue, Denver, CO 80262, USA.

KOICHI KATSURA, The Institute of Statistical Mathematics, 4-6-7 Minami-Azabu, Minato-Ku, Tokyo, Japan.

ROBERT KOHN, Graduate School of Business, University of Chicago, 1101 East 58th Street, Chicago, IL 60637, USA.

P. A. P. MORAN, The National Health & Medical Research Council Social Psychiatry Research Unit, The Australian National University, GPO Box 4, Canberra, ACT 2601, Australia.

D. F. NICHOLLS, Department of Statistics, Faculty of Economics and Commerce, The Australian National University, GPO Box 4, Canberra ACT 2601, Australia.

YOSIHIKO OGATA, The Institute of Statistical Mathematics, 4-6-7 Minami-Azabu, Minato-Ku, Tokyo, Japan.

M. R. OSBORNE, Department of Statistics, IAS, The Australian National University, GPO Box 4, Canberra ACT 2601, Australia.

TOHRU OZAKI, The Institute of Statistical Mathematics, 4-6-7 Minami-Azabu, Minato-ku, Tokyo, Japan.

EMANUEL PARZEN, Department of Statistics, Texas A & M University, College Station, TX 77843–3143, USA.

J. H. W. PENM, Department of Statistics, Faculty of Economics and Commerce, The Australian National University, GPO Box 4, Canberra ACT 2601, Australia.

M. B. PRIESTLEY, Department of Mathematics, The University of Manchester Institute of Science and Technology, PO Box 88, Manchester M60 1QD, UK.

B. G. QUINN, Department of Mathematics, University of Queensland, St. Lucia, QLD 4067, Australia.

JORMA RISSANEN, IBM, 5600 Cottle Road, San Jose, CA 95193, USA.

P. M. ROBINSON, Department of Economics, The London School of Economics and Political Science, Houghton Street, London WC2 2AE, UK.

M. ROSENBLATT, Department of Mathematics, University of California, San Diego, La Jolla, CA 92093, USA.

RITEI SHIBATA, Department of Mathematics, Keio University, 3-14-1 Hiyoshi, Kohoku, Yokohama, 223 Japan.

G. K. SMYTH, Department of Statistics, IAS, The Australian National University, GPO Box 4, Canberra ACT 2601, Australia.

VICTOR SOLO, Department of Statistics, Harvard University, Science Center, One Oxford Street, Cambridge, MA 02138, USA.

T. P. SPEED, CSIRO Division of Mathematics and Statistics, GPO Box 1965, Canberra, ACT 2601, Australia.

KATSUTO TANAKA, Faculty of Economics, Hitotsubashi University, Kunitachi 186, Tokyo, Japan.

R. D. TERRELL, Department of Statistics, Faculty of Economics and Commerce, The Australian National University, GPO Box 4, Canberra ACT 2601, Australia.

P. J. Thomson, Department of Mathematics, Victoria University of Wellington, Private Bag, Wellington, New Zealand.

Pyke Tin, Department of Mathematics, University of Rangoon, Rangoon, Burma.

J. G. Veitch, Department of Statistics, Princeton University, PO Box 37, NJ 08544, USA.

G. S. Watson, Department of Statistics, Princeton University, PO Box 37, NJ 08544, USA.

Peter Whittle, Statistical Laboratory, University of Cambridge, 16 Mill Lane, Cambridge, CB2 1SB, UK.

Index